Handbook of Climatology

Handbook of Climatology

Edited by **Vivian Moritz**

SYRAWOOD
PUBLISHING HOUSE
New York

Published by Syrawood Publishing House,
750 Third Avenue, 9th Floor,
New York, NY 10017, USA
www.syrawoodpublishinghouse.com

Handbook of Climatology
Edited by Vivian Moritz

International Standard Book Number: 978-1-68286-204-9 (Hardback)

Printed in China.

Contents

Preface

In my initial years as a student, I used to run to the library at every possible instance to grab a book and learn something new. Books were my primary source of knowledge and I would not have come such a long way without all that I learnt from them. Thus, when I was approached to edit this book; I became understandably nostalgic. It was an absolute honor to be considered worthy of guiding the current generation as well as those to come. I put all my knowledge and hard work into making this book most beneficial for its readers.

There has been rapid progress in the researches that cover different aspects of climatology. Scientists and meteorologists have increasingly focused upon analyzing the climatic variations and meteorological measurements accumulated over a long period of time to understand different weather patterns and correlations. This book analyzes various researches to assess climatic changes, factors effecting global climate and the characteristics of regional climates. It includes topics on climate modeling and simulation, global wind characteristics, monitoring climate change, etc. which are of utmost significance and bound to provide incredible insights to readers. It aims to equip students and experts with the advanced topics and upcoming concepts in this area.

I wish to thank my publisher for supporting me at every step. I would also like to thank all the authors who have contributed their researches in this book. I hope this book will be a valuable contribution to the progress of the field.

Editor

1

Atmospheric boundary layer wind profile at a flat coastal site – wind speed lidar measurements and mesoscale modeling results

R. Floors[1], E. Batchvarova[1,2], S.-E. Gryning[1], A. N. Hahmann[1], A. Peña[1], and T. Mikkelsen[1]

[1]Wind Energy Division, Risø National Laboratory for Sustainable Energy, Technical University of Denmark, P.O. Box 49, 4000 Roskilde, Denmark

[2]National Institute of Meteorology and Hydrology, Sofia 1784, Bulgaria

Abstract. Wind profiles up to 600 m height are investigated. Measurements of mean wind speed profiles were obtained from a novel wind lidar and compared to model simulations from a mesoscale model (WRF-ARW v3.1). It is found that WRF is able to predict the mean wind profile rather well and typically within 1–2 $m\,s^{-1}$ to the individual measured values. WRF underpredicts the normalized wind profile, especially for stable conditions. The effect of baroclinicity on the upper part of the wind profile is discussed.

1 Introduction

The wind profile above the surface layer and up to a kilometre within the atmospheric boundary layer is presently a challenge for wind assessment studies and air pollution applications, as mean wind speed instruments with high resolution in time and space have been rarely available until recently. Previously, surface layer scaling was often used to compare observed and theoretical wind profiles (Holtslag, 1984) and more recently a mixing length approach has been compared with lidar and wind profiler measurements above the surface layer (Gryning et al., 2007; Peña et al., 2010; Gryning et al., 2011). However, commonly available mesoscale meteorological models seem to have difficulties predicting the mean wind profile accurately due to complex phenomena taking place in nature and not accounted for in the local scaling based ABL parameterization schemes (Draxl et al., 2010). Here we report our findings from two weeks, 15 to 30 September 2010, of novel observations of wind speed up to 600 m height at Høvsøre, Denmark. The study was carried out with a newly deployed pulsed wind lidar (Leosphere WLS70) and the measurements were compared with simulations of the Weather Research and Forecasting (WRF) model (Skamarock et al., 2008).

2 Methods

2.1 Measurements

The measurements were carried out at the National Test Station of Wind Turbines, which is located about 1.8 km east of the shoreline (for a map see Gryning et al., 2007). Except for the presence of the coastline, the terrain surrounding Høvsøre is very flat and homogeneous. It mainly consists of grass, crops and a few shrubs. On site, there is one dedicated meteorological mast of 116.5 m height and two light towers of 160 m height each. For the meteorological mast the wind speed is measured with Risø cup anemometers at nominal heights of 10, 40, 60, 80, 100, 116.5 m. The wind direction is measured with wind vanes at 10, 60 and 100 m. To extend the height range in which data are available, observations from the 160 m top level at the nearest light tower are also used. Both the light tower and the meteorological mast are equipped with METEK Scientific USA-1 sonic anemometers. These are installed at heights: 10, 20, 40, 60, 80, 100 and 160 m. The sampling frequency of the sonic measurements is 20 Hz. All sonic co-variances are based on 10-min averages of linearly de-trended time series. In addition, for extended height measurements, a pulsed wind lidar (WLS70) has been operating at the site since April 2010. The wind lidar measures wind speed and direction from 100 m above the ground and every 50 m up till 1 to 2 km height. The maximum measurement height is dependent on the attainable 10-min averaged Carrier to Noise (CNR) ratio. The upper measuring height is usually determined by the cloud base, where the lidar signal (1.55 µm) largely weakens. The wind lidars Doppler shift based wind measurements are processed into 10-min average quantities. The wind lidar is equipped with a rotating silicon prism providing an optical scanning cone of 15 degrees. One 360 degree full scan (rotation) is performed about every 30 s.

2.2 WRF model description

The wind profiles were predicted using a research real-time forecast system based on the WRF ARW model version 3.1 (Skamarock et al., 2008), developed by the National Centre for Atmospheric Research (NCAR). It is a numerical weather prediction and atmospheric simulation system designed for both research and operational applications. Here, a set up of three domains covering Northern Europe is used (for a map see Draxl et al. (2010)), with a horizontal grid size of 18, 6 and 2 km. Data for initial and boundary conditions comes from analyses and forecasts of the National Center for Environmental Prediction (NCEP, USA) global GFS model. The grid point from the 2 km resolution innermost grid at Høvsøre is located 700 m East and 2000 m North of the measuring point (mast and WLS70). The model calculates the meteorological parameters at 37 vertical levels from the surface to pressure level 100 hPa. Nine of these levels are within the height range of 600 m that is analysed in this study and the first model level is at 14 m. The physical options of model setup include the Yonsei University (YSU) planetary boundary layer (PBL) scheme (Hong et al., 2006), the Noah land surface scheme (Chen and Dudhia, 2001) and the Thompson microphysics scheme (Thompson et al., 2004). A comparison with other PBL parameterizations can be found in Draxl et al. (2010). The YSU scheme is the one of the most commonly used parameterizations and we want to investigate its performance in predicting mean wind profiles for wind resource assessments. The model is run every 24 h. After the 6 h spin-up of the model, the forecast up to 30 h is used to generate a continuous hourly time series.

3 Results

3.1 Comparison between cup and wind lidar measurements

The analysis of the measurements is limited to the period, 15 to 30 September 2010, because the lidar was running without many interruptions in this period. It constitutes a first analysis of the data in order to get an impression of the challenges and possibilities of wind profile lidar measurements and WRF wind simulations of up to several hundreds of meters. Firstly we evaluate the quality of the wind lidar measurements by comparing wind speed derived from the lidar to cup anemometer measurements at the meteorological mast at 100 m height, Fig. 1. The agreement can be seen to be good

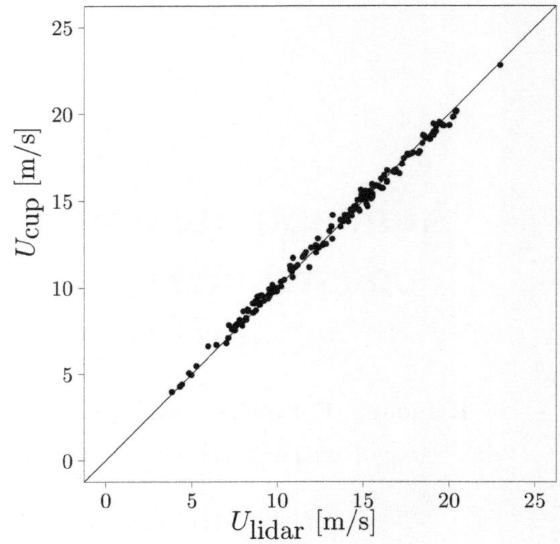

Figure 1. Comparison of wind speed measurements from the wind lidar and cup anemometer at 100 m height for 10 min means.

with a correlation coefficient $R^2 = 0.9995$ and gain of 0.9982 (through origo). It is clear from these data, that the WLS70 wind lidar is a precise remote sensing based tool for measurements of mean wind speed.

3.2 Mean wind profile

The wind speed at 100 m during the 2 week observation period is shown in Fig. 2 together with the prediction of the wind speed by the WRF model. It is a good period for model evaluation because of the considerable variation in the wind speed, from less than 2 m s^{-1} to more than 20 m s^{-1}. It can be seen that the WRF model simulations predicts the variation of the wind speed relatively well, showing no obvious systematic deviation neither for small nor high wind speeds. The comparison is further illustrated at both 100 and 600 m height in scatter plots, Figs. 3 and 4, which indicates that the wind speed is slightly underpredicted by the WRF model at both 100 m and 600 m and typically is within 1–2 m s^{-1} of the individual measured values.

The normalized mean wind speed profile for the two week period is shown on Fig. 5. Each measurement of the wind speed is normalized with the sonic-anemometer measured friction velocity at 20 m height and then all normalized wind speeds are averaged. The normalized WFR wind profile is derived in a similar way using the friction velocity calculated by the WRF model. Despite the under-prediction by the WRF model it can be seen that the general shape of the wind profile agrees rather well with the measurements. Considering the rather good wind speed prediction, Figs. 3 and 4, this suggests that the friction velocity in WRF is overpredicted. A reason for this can be that u_{*0} is a parameterized value in WRF and is an average for the grid box, whereas the measured u_{*0} is a local value.

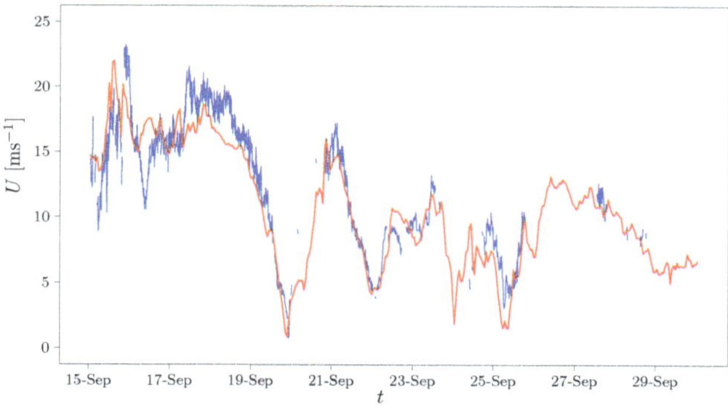

Figure 2. Time series of the wind speed at 100 m height from the WRF model simulations at the Høvsøre mast (red) and the measurements with the wind lidar (blue) for the period 15–31 September 2010. The lidar measurements are filtered with a Carrier to Noise Ratio (CNR) larger than −22 dB. The lidar worked well during the whole period, the missing measurements are caused by the filtering.

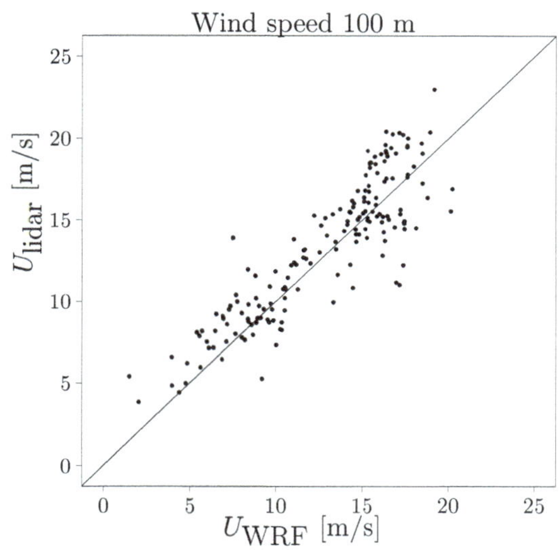

Figure 3. Comparison of the prediction of the wind speed from the simulation by the WRF model and measurements from the wind lidar at 100 m.

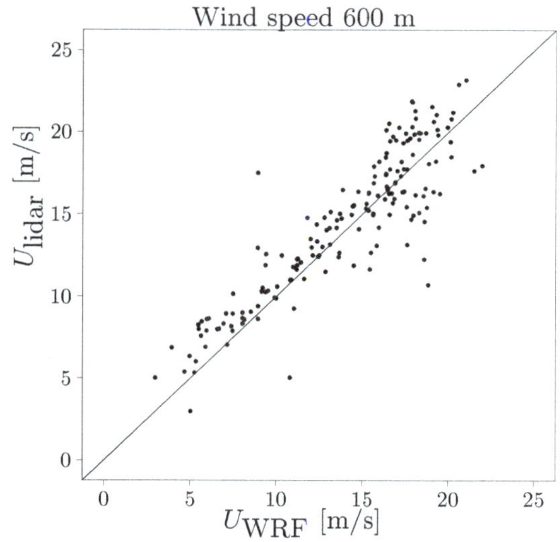

Figure 4. Comparison of the prediction of the wind speed from the simulation by the WRF model and measurements from the wind lidar at 600 m.

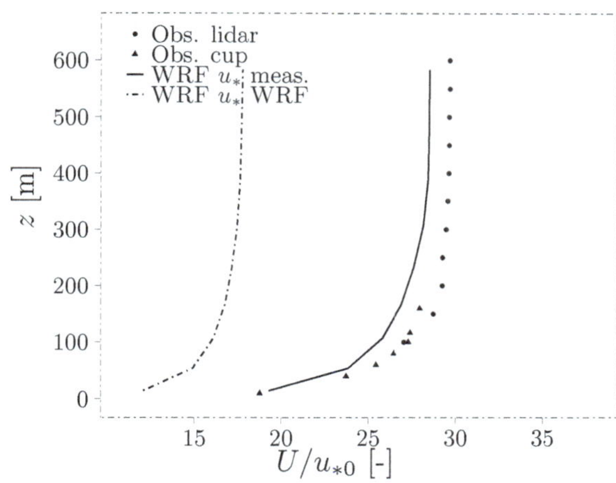

Figure 5. Normalized mean wind speed profile for the entire period, the full line shows the WRF simulations normalized with the observed u_{*0} and the dashed dotted line with the modelled u_{*0}, the triangles are cup anemometer measurements at the mast and the dots are wind lidar measurements.

3.3 Effect of stability

Whereas Fig. 5 represents a composite for all stabilities of the normalized wind speed profile, the variability of the wind profile with atmospheric stability is investigated in this section. The measured wind profiles are divided into stability classes according to the Obukhov length, L, Table 1. The Obukhov length is derived from sonic co-variance measurements of turbulence at 20 m height at the meteorological mast. For the WRF model, it is derived from the modelled heat flux, friction velocity and temperature.

Figure 6 shows the normalized wind speed profiles sorted into stability classes. The wind profiles order as expected from surface layer theory with the most unstable profile to the left, having the smallest wind speed gradient, and the stable wind profiles to the right with larger wind speed gradients. For very stable conditions the measured wind speed profile shows a characteristic low level jet at around 200 m height. It is also characteristic for several of the measured profiles that above 500 m the normalized wind speed decreases slightly with height, which suggest that the atmosphere is baroclinic and not barotropic as assumed in traditional wind profile parameterizations such as Blackadar (1962). Future attempts will be done in order to estimate the baroclinicity shear S:

$$S = \left(\frac{dU_g}{dz}\right)\frac{\kappa h}{u_{*0}}, \qquad (1)$$

Table 1. The stability classes according to Obukhov length, L.

Nr.	Stability class name	Obukhov length interval [m]	# of obs.
1	Unstable (u)	$-100 \geq L \geq -200$	10
2	Near unstable/neutral (nu)	$-200 \geq L \geq -500$	17
3	Neutral (n)	$\lvert L \rvert \geq 500$	97
4	Near stable/neutral (ns)	$200 \geq L \geq 500$	22
5	Stable (s)	$50 \geq L \geq 200$	17
6	Very stable (vs)	$10 \geq L \geq 50$	8

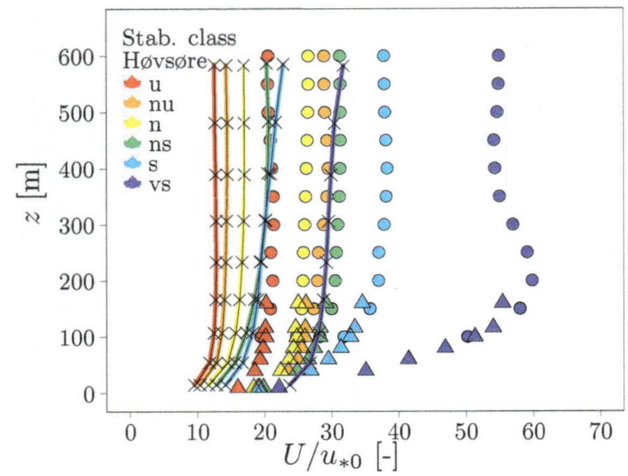

Figure 6. Profiles of normalized wind speeds as function of stability (Table 1). The full lines represent WRF simulations and the symbols measurements as in Fig. 5. The friction velocity that is used to normalize the wind speed for the measurements is taken from the meteorological mast, and for WRF from the simulation.

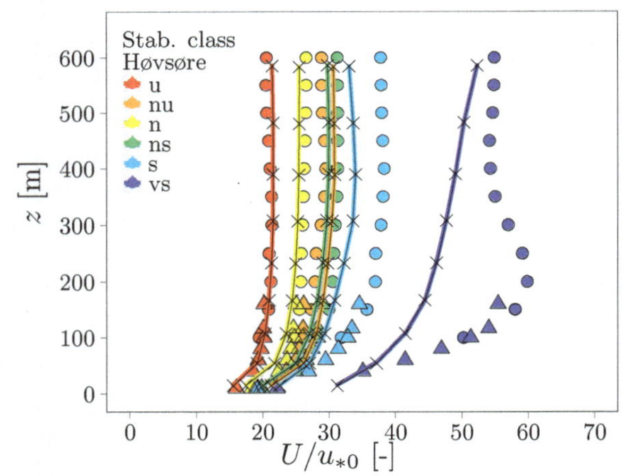

Figure 7. As Fig. 6, except that u_{*0} is taken from the measurements.

from the gradient of the wind profile in the upper part by use of slightly modified version of the model for the wind profile proposed by Gryning et al. (2007). Here U_g is the geostrophic wind speed at the top of the boundary layer, κ is the von Karman constant and h the height of the boundary layer. The effect of baroclinicity is seen in wind profiles for both unstable and stable conditions. Figure 6 also shows wind profiles from the WRF model, normalized with u_{*0} from WRF and classified according to the modelled L (Table 1). It is seen that the wind profiles vary less with atmospheric stability as compared to the measured wind profiles. This is due to the poor prediction of u_* and L in PBL parameterizations of WRF (Pagowski, 2004). Figure 7 shows wind profiles from WRF normalized with the measured u_* at Høvsøre and classified according to Table 1 using measured L. Compared to Fig. 6 the agreement between normalized modelled and measured wind profiles improves. However, the model still underpredicts the normalized wind profile for stable conditions. It is interesting to note that although the stability dependence on the wind profiles is poorly modelled in this setup of WRF, it was concluded in Sect. 3.2 that the averaged wind profile agrees rather well. This suggests that the applied configuration of the WRF model can be used to derive the climatological mean wind profile but not the variability around the mean value. Looking at the wind profile from the perspective of a Weibull distribution, the WRF model is expected to predict the scaling parameter A rather well, but not the shape parameter k for the wind profile.

4 Conclusions

From recent new observations of the mean wind profile from the ground and up to 600 m height with a pulsed wind lidar, the wind profiles were derived and compared with WRF model simulations. When averaged over all stability conditions the simulated mean wind profile agrees well with the measurements, although the normalized wind profile was underpredicted by WRF. When the normalized wind profiles were sorted in atmospheric stability classes the WRF model simulations of the normalized wind profiles showed less variability with atmospheric stability as compared to measurements. Use of u_{*0} from WRF results in a general underestimation of the normalized wind profile, while use of the measured u_{*0} leads to underprediction for stable conditions only. It was observed that the wind speed decreased with height in the upper part of the wind profile for several of the stability classes. The effect is likely caused by baroclinicity. From the perspective of representing the wind profile as a the Weibull distribution, the WRF model simulations in this configuration might be adequate to estimate the profile of the scaling parameter A but caution should be taken with respect to the shape parameter k.

Acknowledgements. Contributing to this effort are the Danish Research Agency Strategic Research Council (Sagsnr. 2104-08-0025) "Tall wind project" and the EU FP7-People-IEF VSABLA (PIEF-GA-2009-237471). It is also related to COST Action ES1002 (WIRE).

Edited by: L. Wald

References

Blackadar, A.: The vertical distribution of wind and turbulent exchange in a neutral atmosphere, J. Geophys. Res., 67, 3095–3102, 1962.

Chen, F. and Dudhia, J.: Coupling an advanced land surface-hydrology model with the Penn State-NCAR MM5 modeling system. Part I: Model implementation and sensitivity, Mon. Weather Rev., 129, 569–585, 2001.

Draxl, C., Hahmann, A., Peña, A., Nissen, J., and Giebel, G.: Validation of boundary-layer winds from WRF mesoscale forecasts with applications to wind energy forecasting, in: 19th Symposium on Boundary Layers and Turbulence, Keystone, Colorado, Paper 1B.1, 7 pp., 2010.

Gryning, S., Batchvarova, E., Brümmer, B., Jørgensen, H., and Larsen, S.: On the extension of the wind profile over homogeneous terrain beyond the surface boundary layer, Bound.-Lay. Meteorol., 124, 251–268, 2007.

Gryning, S., Batchvarova, E., Quante, M., and Matthias, V.: Evaluation of vertical profiles in mesoscale meteorological models based on observations for the COST728 study of winter 2003 PM episodes in Europe, in: Air Pollution Modeling and its Application, edited by: Steyn, D. and Trini-Casteli, S., Springer Science+Business Media B.V, Dordrecht, The Netherlands, in print, 2011.

Holtslag, A.: Estimates of diabatic wind speed profiles from near-surface weather observations, Bound.-Lay. Meteorol., 29, 225–250, 1984.

Hong, S., Noh, Y., and Dudhia, J.: A new vertical diffusion package with an explicit treatment of entrainment processes, Mon. Weather Rev., 134, 2318–2341, 2006.

Pagowski, M.: Some comments on PBL parameterizations in WRF, in: The Joint WRF/MM5 Users' Workshop, Boulder, CO, 1.13, 9 pp., 2004.

Peña, A., Gryning, S., and Hasager, C.: Comparing mixing-length models of the diabatic wind profile over homogeneous terrain, Theor. Appl. Climatol., 100, 325–335, 2010.

Skamarock, W., Klemp, J., Dudhia, J., Gill, D., Barker, D., Duda, M., Huang, X., Wang, W., and Powers, J.: A description of the Advanced Research WRF version 3, Technical Note TN-4751STR, NCAR, 2008.

Thompson, G., Rasmussen, R., and Manning, K.: Explicit Forecasts of Winter Precipitation Using an Improved Bulk Microphysics Scheme. Part I: Description and Sensitivity Analysis, Mon. Weather Rev., 132, 519–542, 2004.

2

The seasonal characteristics of the breeze circulation at a coastal Mediterranean site in South Italy

S. Federico[1,2], L. Pasqualoni[3], A. M. Sempreviva[1,4], L. De Leo[2], E. Avolio[2], C. R. Calidonna[1], and C. Bellecci[3,2]

[1]Institute of Atmospheric Sciences and Climate, Italian National Research Council (ISAC-CNR), Section of Lamezia Terme, Italy
[2]CRATI Scrl, Rende, Italy
[3]Engineering Department, University of Rome "Tor Vergata", Italy
[4]Wind Energy Department, Risoe-Danish Technical University, Roskilde, Denmark

Abstract. We present a study on the characteristics of the sea breeze flow at a coastal site located in the centre of the Mediterranean basin at the southern tip of Italy. This study is finalized to add new data on breeze circulations over a narrow peninsula and present a unique experimental coastal site at about 600 m from the coastline in a flat open area at the foot of a mountain chain located in a region of complex orography. We study the seasonal behaviour of the sea-land breeze circulation by analysing two years of hourly data of wind speed and direction, temperature, radiation and relative humidity from a surface meteorological station, eighteen-months data from a wind profiler, and two-year data from the ECMWF analysis.

Results show that breezes dominate the local circulation and play a major role for the local climate. They are modulated by the season, through the sea-land temperature difference and the large-scale flow. The large-scale forcing acts in phase with the diurnal breeze and opposes the nocturnal breeze.

In summer, the daytime difference between the land surface temperature and the SST (Sea Surface Temperature) reaches its maximum, while the nigh-time difference has its minimum. This causes a strong, frequent and intense diurnal breeze and a weak nocturnal breeze.

In winter and fall the nocturnal difference between the sea and land surface temperature reaches a maximum value, while the diurnal difference is at its minimum value. This causes a strong, frequent and intense nocturnal breeze despite of the large-scale forcing that is usually opposed to local-scale flow.

1 Introduction

Modeling the adjustment of the onshore flow inland from the coastal discontinuity in meso-scale models is still a challenge being limited by the resolution of the model grid and by the availability of comprehensive databases. A special case of flow adjustment happens during the development of the sea and land breeze system that is a meso-scale feature also involving offshore coastal areas. Sea breeze is developing on fine weather days due to the sea-land temperature difference that produces a pressure gradient between the sea and the land that forces an onshore flow. The opposite circulation, occurring at night, is called land breeze, if the temperature gradient between sea and land is reversed. The observed sea-breeze intensity that can reach 10 ms^{-1} (Pielke, 2002) varies depending on several factors such as large-scale winds, atmospheric static stability, cloud coverage, land use etc. The land breeze is a shallower and weaker phenomenon than the sea breeze because the planetary boundary layer (PBL) over land is stably stratified at night. Defant (1951) presents an excellent qualitative description of sea and land breeze development in absence of large-scale wind.

In a mountainous region, local wind patterns can develop because of the differential heating between the ground surface and the free atmosphere at the same elevation some distance away. A larger diurnal temperature variation usually occurs at the ground exposed to the sun, so that, at daytime, elevated terrain becomes a heat source, while at the night it

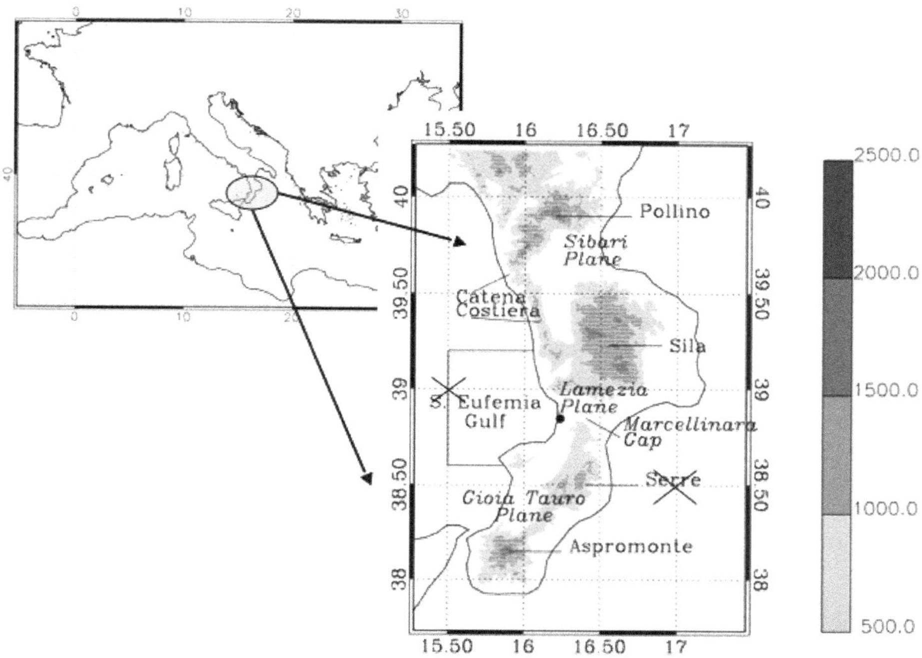

Figure 1. Calabria region in the central Mediterranean. Left: topography (m, gray shading) with topographical features cited into the text. The black dot shows the location of the experimental field. Crosses denote grid points considered in Table 2.

is an elevated heat sink. Soil temperature differences might intensify the temperature gradient between two atmospheric columns above elevated peaks and valleys. As a consequence of the temperature gradient generated, during fair weather days an upslope flow (anabatic winds) develops at daytime while, while down-slope (katabatic) winds develop during nigh time.

Because of the specificity of the breeze circulation at a particular site, several papers and books can be found in the bibliography on sea breeze and local winds, some of them reported in Simpson (1994) and Pielke (2002).

This study investigates local circulations at a site in the Central Mediterranean at the west coast of the narrow Calabria peninsula (Fig. 1), the southernmost tip of Italy. Calabria extends for 300 km between 38° and 40° latitude north and between 15°30′ and 17°15′ longitude east. The region is bounded by the Tyrrhenian Sea (west) and by the Ionian Sea (east and south). The Apennines Mountains run north-south and are characterized by five main topographical features reaching 1500–2000 m elevations: Pollino, Catena Costiera, Sila, Serre and Aspromonte. The average width of the region is about 50 km in the west-east direction.

The physical characteristics of the region, i.e. the presence of elevated mountain range along the peninsula in a symmetric position between the Tyrrhenian and the Ionian seas, provide a rather unique condition for breeze development and convergence. Sea breeze and land breeze might act in phase with upslope and down-slope winds to determine stronger and more persistent breeze system. Federico et al. (2000, 2003), using a theoretical and modelling approach, studied the relative role of the sea-land and mountain-valley contrasts for the local thermal convergence pattern over an idealized mountainous peninsula. They showed that the topography plays a fundamental role in determining the meso-scale flows. The presence of mountains highly enhances breezes and, for an idealized peninsula with the aspect (i.e. west-east width and mountains height) of Calabria, the energetic contribution of the mountain-valley contrast is larger than the sea-land contrast. Finally, the limited width of the peninsula, allows two situations (i) the upslope – down-slope breeze and the sea-land breeze merge (ii) the Tyrrhenian and the Ionian Sea breeze can converge at the centre of the peninsula creating cloud formation and very local precipitations (De Leo et al., 2008).

There is another important factor that favours the breeze development in Calabria: the Mediterranean climate. Calabria geographical position is at the centre of the Mediterranean where the atmospheric pressure field can determine calm synoptic conditions (Trigo, 1999, 2002; Bolle, 2003), especially in summer and fall (Conte et al., 1989; Colacino et al., 1990; Piervitali et al., 1999; Federico et al., 2009). These conditions favour the breeze development. The favourable conditions for the breeze development in the Mediterranean have been discussed and reported in several papers (Doran (1979) for Israel; Kallos et al. (1993) for Greece; Caccia et al. (2004) for France; and Mastrantonio et al. (2008) for Italy).

Table 1. Instruments, parameters measured and working periods.

Instrument	Data	Operational period
Surface Meteorological station	Wind speed and direction; Temperature; Global radiation; Precipitation; Relative Humidity; Pressure.	Permanently installed. From 1 January 2007 to 31 December 2008.
RADAR WP	Wind vertical profiles of the three wind components.	Permanently installed. From 1 June 2008.

While previous studies over Calabria were based on modelling the breezes, this paper focuses on experimental data collected at the Joint CRATI/ISAC-CNR site of Lamezia Terme, at the centre of the Tyrrhenian coast (Fig. 1) and at the end of the only west-east valley (the Marcellinara gap) connecting the Tyrrhenian and Ionian seas.

The paper is organized as follows: in Sect. 2, we shortly introduce the instrumental setup and the available database; in Sect. 3, we show the results and we then provide final remarks in Sect. 4.

2 Measurements

In this study, we use observations from different instruments operating at the experimental site located at 8 m a.s.l. and at 600 m from the coast http://www.crati.it/altre_attivita_crati/Lamezia_experimental_field_web.pdf. A list and the operation time for each instrument are shown in Table 1.

2.1 Surface station

A surface meteorological station operates since June 2006 measuring temperature T, wind speed U, and direction DIR at 10 m a.g.l.; precipitation Pr, relative humidity H, atmospheric pressure P, and global solar radiation Rad are collected at the height of 2 m; soil temperature has been measured at 10 cm depth (from 1 January 2008). Data are available every 15 min but, in this paper, we use hourly averages. Missing data, caused by instruments maintenance, for the two years are less than 5%. The range value of the wind speed measured by the anemometer installed at the site is 0.5 ms^{-1}–55 ms^{-1}. Velocities less than 0.5 ms^{-1} (wind calm) with $f = 12\%$ have been discarded.

2.2 Radar wind profiler

A radar wind profiler is operating at the experimental site since 1 June 2008. The radar sounds the lower troposphere, usually up to 3 km, and measures the horizontal and vertical wind components thanks to one vertical and two oblique beams slanted at off-zenith angle of 15.5°. The operating frequency is 1290 MHz (about 23 cm wavelength). Returned echoes are due to air masses refractive index fluctuation advected by the wind. The dataset consists of 30 min average of the wind components. The vertical resolution is 100 m, the minimum range gate is 150 m and the vertical range is 2–5 km depending on atmospheric conditions. Accuracy of wind measurements is <1 ms^{-1} for the horizontal wind components, 0.5 m^{-1} for the vertical component and less than 10° for the direction. Here, we use one-year data from June 2008 to May 2009.

2.3 Sea Surface Temperature (satellite observations)

The Tyrrhenian Sea Surface skin Temperature (SST), within the S. Eufemia Gulf (Fig. 1), facing the experimental site, was estimated using data disseminated by EUMETSAT Ocean & Sea Ice Satellite Application Facility (www.osi-saf.org). The original dataset consists of 12-hourly means centred on 00:00 and 12:00 h UTC with a horizontal resolution of about 10 km for 2007 and 2008. A general overview of the dataset, the algorithms and processing scheme is given by O&SI SAF (2006). SST is computed by space averaging the pixel values inside the square area in Fig. 1, representative of the portion the sea relevant for the development of the local circulation. It should be emphasized that the Rossby radius, i.e. the e-folding distance of the sea breeze from the coastline, is about 100 km for the latitude and static stability typical of Calabria peninsula (Abe and Yoshida, 1982). Because the width of Calabria is less than 100 km, the sea-land and mountain-valley breezes from both sides of the peninsula collide and an effective Rossby radius is established, which is about 50 km for Calabria (Federico et al., 2000). The sea-land and upslope-downslope breezes have and extension of one effective Rossby radius and the width of the sea in front of Lamezia Terme considered in our analysis (Fig. 1) is about 50 km.

Data from O&SI SAF are flagged according to the quality of the estimated SST and of the information on the processing conditions. The pixels where the SST is calculated are labelled on a five level scale: 5 = "excellent", 4 = "good", 3 = "acceptable", 2 = "bad", 1 = "erroneous" depending on pixel distance from clouds, SST climatology, sun glint effect etc. The "erroneous" confidence level may be attributed to a parameter missing due to a failure of the algorithm or present but with a very low level of confidence. For this study, we only used pixels with a confidence level greater than 3.

3 Results

3.1 Breeze climatology for 2007–2008

Figure 2a shows the wind rose recorded at the surface station for the period of two complete years i.e. 2007–2008. Winds blow from two main directions: W-WSW and NE-ENE. These directions account for 71% of the observed winds and their occurrence is similar, which gives to the wind rose a quite symmetrical aspect with respect to the axes intersection: 34% from NE-ENE, 37% from W-WSW. Wind speeds are low and the occurrence of $U < 6\,\mathrm{ms}^{-1}$ represents more than 90%. The threshold $6\,\mathrm{ms}^{-1}$ was chosen after a subjective inspection of measurements from different sources (including ECMWF surface and 850 hPa analyses). In particular, it was evident that surface winds larger than $6\,\mathrm{ms}^{-1}$ are attained when moderate-high surface pressure gradient are present over the area. On these days, the breeze circulation does not develop and the local circulation is dominated by the large-scale wind.

In order to study the coastal flow regime, it is necessary to analyse the data above the breeze circulation. Figure 2b shows the wind rose at 850 hPa, for the same period as in Fig. 2a (from 1 January 2007 to 31 December 2008), derived from the ECMWF (European Centre for Medium Weather range Forecast) operational analysis at 850 hPa (Uppala et al., 2005). These data, available every 6 h, are representative of the large-scale circulation above the breeze re-circulation system because about 45% of the winds come from the NW sector. Beside the expected higher wind speed at 850 hPa, the comparison between Fig. 2a and b reveals the importance of local circulations at the surface because: a) the directions in Fig. 2a are well focused (W-WSW and NE-ENE) while at 850 hPa the wind direction range is more spread; b) the NE-ENE components that might account form the channelled wind by the Marcellinara gap and the land breeze, are much more frequent at the surface (34% and 7% at the surface and at 850 hPa respectively); and c) the W-WSW directions frequency at the surface is twice that at 850 hPa (37% and 18% respectively).

To further confirm the difference between the circulation at the surface and at levels above the breeze, we show (Fig. 2c) the wind rose recorded by the wind profiler at 1.7 km a.g.l. (roughly 800 hPa) from 1 June 2008 to 31 May 2009. The comparison between Fig. 2b and c shows the similarity of the wind distribution at high level from model and data and their difference with the surface wind distribution in Fig. 2a. The discrepancy between the wind distribution in Fig. 2b and c are due to the not overlapping periods taken into consideration for the analysis of the data from the ECMWF model and the wind profiler. The analysis of data from an overlapping time window would have been advisable for conclusively supporting our discussion. However, isolating a complete year of the wind profiler data has allowed to take into consideration the wind seasonal variation, which has been

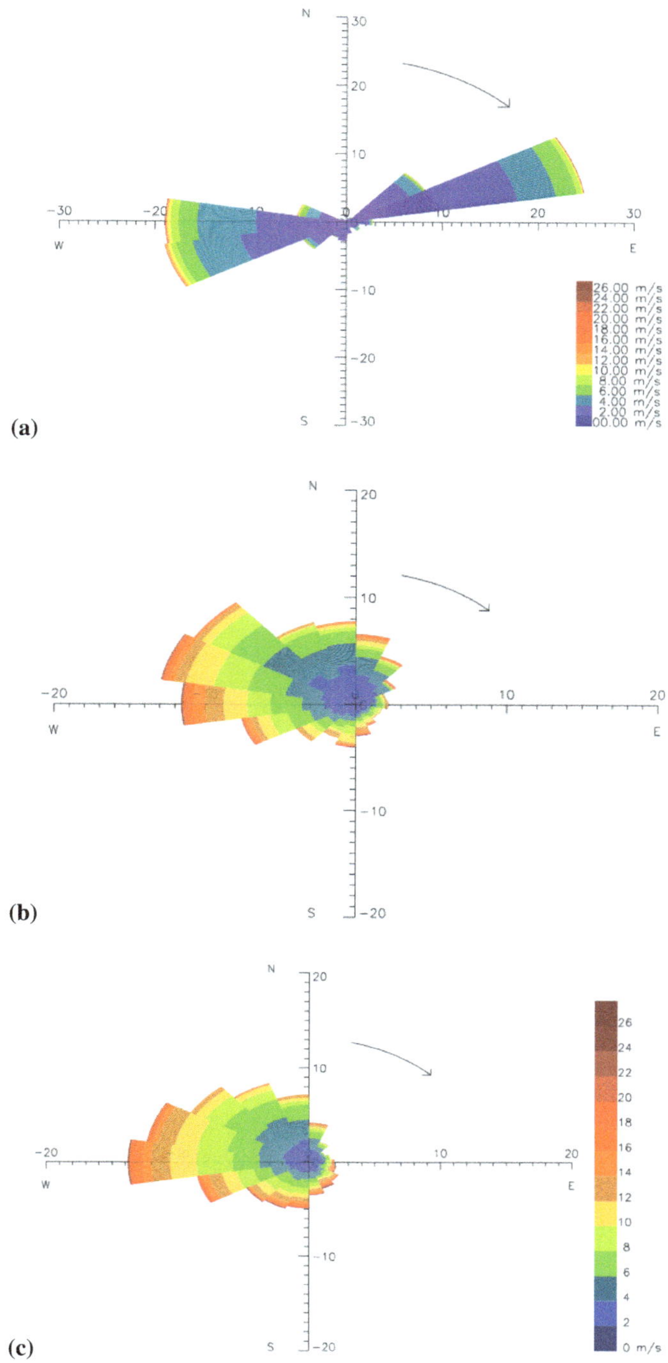

Figure 2. **(a)** Wind rose recorded at the Lamezia Terme surface station from 1 January 2007 to 31 December 2008. Color shading shows the wind intensity (ms^{-1}). **(b)** as in (a) for ECMWF analysis at 850 hPa from 1 January 2007 to 31 December 2008. **(c)** Wind rose recorded 1700 m a.g.l. by the UHF wind profiler form 1 June 2008 to 31 May 2009.

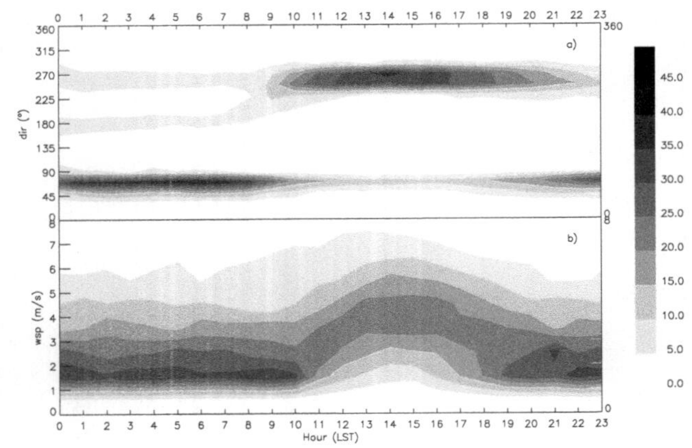

Figure 3. **(a)** Frequency of wind direction as a function of day hours recorded at the Lamezia Terme surface station for 2007–2008. Grey shading shows the frequency of occurrence (%). **(b)** as in (a) but for the wind speed. LST = UTC +1 h.

shown to have a smaller inter-annual variation from the analysis of the two years of the ECMWF data (not show).

The results shown in Fig. 2 clearly reveal that local effects are fundamental to determine the surface circulation in that Mediterranean Area. Considering the geographical position of the experimental site located at a flat coastal area close to the coastline with behind mountains, two main local effects could be important: breeze circulation and the channelling of air masses trough the Marcellinara gap (Fig. 1).

To quantify the role of the breeze circulation, we show the distributions of the wind direction and speed as a function of the hours of the day in Fig. 3a and b. The role of the breeze circulation is evident: during the day, winds blow from W-WSW while at the night they are mainly from NE-ENE. The presence of the 270 and 180 sector winds during night are likely tied to the large-scale westerly winds and to the Scirocco wind occurrence respectively. The onset of diurnal breeze circulation starts at about 10:00 and ends at 19:00 LST (LST = UTC + 1 h). These values are modulated by the season (not shown): in summer the diurnal breeze starts at 08:00 and ends at 23:00 LST while in winter, the diurnal breeze blows from 12:00 to 18:00 LST. These conditions are typical of the breeze regime (Pearson, 1973, 1975; Pielke, 1974, 2002; Simpson, 1994). The directional analysis in Fig. 3 confirms that the differences between wind rose distributions at the surface and at 850 hPa are due to the breeze circulation.

3.2 Seasonal surface wind distribution

To gain a deeper knowledge of the local circulation, we show, in Fig. 4a, b, c, and d, the wind seasonal distribution recorded at the surface. We define seasons as following, winter (December, January and February); spring (March, April and May); summer (June, July and August) and fall (September, October and November). In winter (Fig. 4a), wind comes mainly from NE-ENE (45% compared to 34% in Fig. 2a) and the wind rose is more asymmetrical compared to the average conditions. Wind speed, as expected, is larger than average and it is greater than 6 ms^{-1} in more than 13% of occasions (10 % in Fig. 2a). Of all seasons, spring has the most symmetrical distribution of cases with respect to land and sea breeze i.e. W-WSW (42%) and NE-ENE (34%). For 2007–2008, we found that spring winds have the largest frequency of high wind speed (more than 17% of speed occurrence greater than 6 ms^{-1}). This result is unexpected and it might be the result of the shorter period with respect to the analyses from Colacino (1990), Colacino et al. (1997), and Bolle (2003); in these papers it was found that the climate of the Central Mediterranean, which is typical of Calabria, is characterized by windier conditions in winter compared to other seasons.

During summer, wind blows 66% of the occurrences from the sector W-WSW; this value is about twice the average conditions (Fig. 2a). The frequency of high winds is the lowest: only 3% of occurrences are greater than 6 ms^{-1}. The fall wind distribution is similar to the distribution in winter; winds mainly blow from the NE and ENE directions with 44% of the occurrences. Similarly to average conditions (Fig. 2a) about 35% of occurrences are from W-WSW and 8% of velocities are larger than 6 ms^{-1}.

Despite the rather symmetric distribution of winds, the seasonal distributions show, on average, preferential directions at the experimental site. In particular, summer circulations are mainly from W-WSW while during fall and winter directions are mainly from the NE-ENE sector.

3.3 Pressure and temperature daily variation

To further investigate the differences of the wind speed distribution according to season, we addressed the daily variation of the pressure and temperature fields in terms of differences ΔP and ΔT of their values between two consecutive hours defined as

$$\Delta P = P_{i+1} - P_i \quad \text{and} \quad \Delta T = T_{i+1} - T_i$$

where $i = 0, 23$ indicates the hour, during summer and winter.

We refer to the results showed in Simpson (1994) where, the daily variation of the pressure pattern is estimated by the difference between the mean pressure during the day and the hourly values.

In Fig. 5, we compare the hourly pressure variation in winter and summer during the day. During summer night, ΔP is small with values around 0; after sunrise it becomes mostly negative reaching its maximum absolute value at 09:00 LST. This maximum is correlated to the highest value of ΔT (not shown) and is representative of the time when the diurnal breeze fully develops. The ΔP decreases during the day and reverts its sign reaching its positive maximum at around 20:00 LST. The maximum value of ΔP is representative of

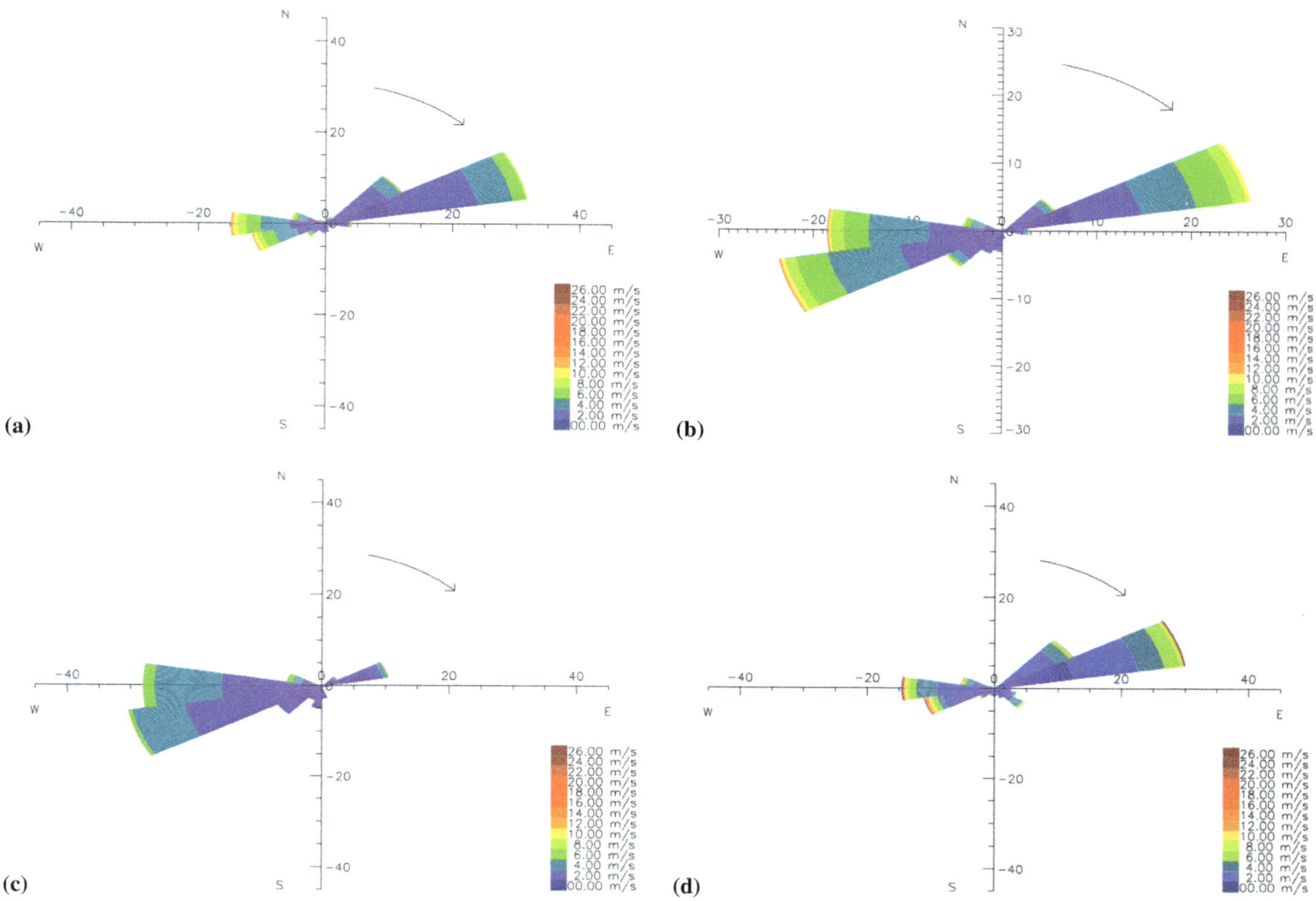

Figure 4. Wind rose recorded at the surface station for the period 2007–2008 for **(a)** winter, **(b)** spring, **(c)** summer, and **(d)** fall. As in Fig. 2a. Colour scale shows the wind intensity (ms^{-1}).

the time when the diurnal breeze forcing stops. It should be noticed that the time when the pressure forcing stops does not coincide with a reversal of the breeze circulation because of the inertia of the circulation. In particular we found that in summer the pressure forcing for the diurnal breeze stops on average at 20:00 LST while the diurnal circulation blows until 23:00 LST (see the comment of Fig. 3a and b of the previous section).

In winter the absolute value of ΔP is lower and varies less than in summer, because of the decreased solar heating. Moreover the time of the minimum (maximum) value of ΔP is delayed (anticipated), compared to summer, as a consequence of the delayed and shorter solar heating.

In Fig. 6, we compare the correlation between ΔP and ΔT in winter and summer. We note that only during summer ΔP and ΔT appear to be correlated showing that the hourly pressure variation in summer, is mainly driven by the solar radiation. In winter other factors, as the large scale forcing, likely play an important role.

3.4 The role of SST and large-scale flow on the seasonal distributions of breezes

In this section, we investigate the role of two main physical factors that could cause the differences in seasonal wind distributions: the sea-land temperature difference and the synoptic circulation in the Mediterranean area.

Figure 7 shows the monthly average of the difference ΔT between the SST (T_{sea}) in the S. Eufemia Gulf and the surface temperature T_{land} recorded at the experimental site between 11:00 and 16:00 LST (diurnal breeze) and between 00:00 and 06:00 LST (nocturnal breeze). These time windows were selected because they include the maximum speed of the diurnal and nocturnal breeze circulation, respectively, so they are suitable to characterize the thermal contrast between the sea surface and land for the two branches of breeze circulation. The ΔT of the diurnal breeze is negative, i.e. the sea is colder that the land, on average, from March to October, while it decreases from November and is positive for the whole winter. This does not exclude that, on particular days, diurnal

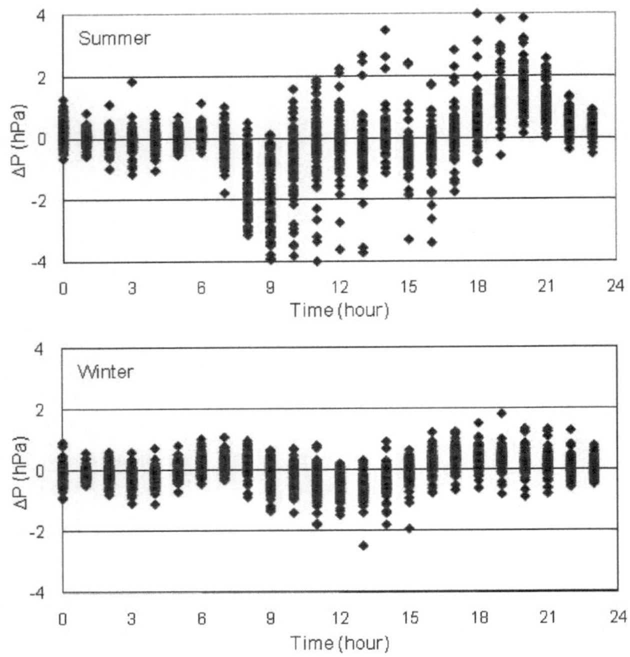

Figure 5. **(Top)** hourly ΔP during summer. **(Bottom)** hourly ΔP during winter.

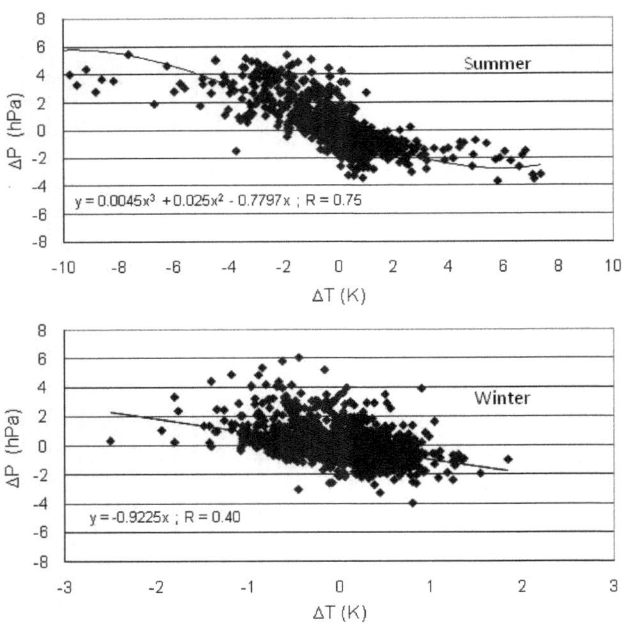

Figure 6. **(Top)** hourly ΔP versus ΔT during summer. **(Bottom)** hourly ΔP versus ΔT during winter. We note that during winter the ΔP and ΔT hourly variations appear less correlated than during summer.

breeze develops in November and during winter, and nevertheless sea breeze plays a minor role. The largest absolute difference of the diurnal sea-land temperature i.e. $\Delta T = 4\,°C$ occurs during the whole summer period.

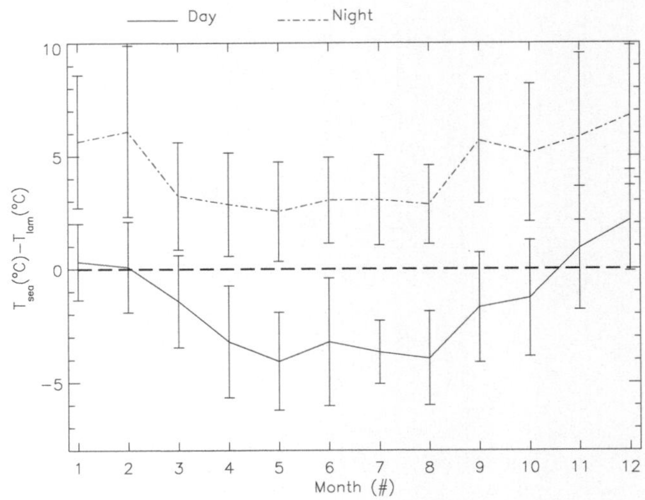

Figure 7. Monthly mean of the difference between the SST of the St. Eufemia Gulf and the air temperature measured at the station during 2007–2008. Solid curve represents the temperature difference from 11:00 to 16:00 LST; dot-dash line represents the temperature difference from 00:00 to 06:00 LST. Error bars are two times the monthly standard deviation of the temperature difference.

The monthly average of ΔT for the nocturnal breeze (00:00–06:00 LST) is always positive, i.e. the sea surface is always warmer than the land surface indicating that the nocturnal breeze can develop in all seasons. However, Fig. 7 shows that ΔT during winter and fall is twice that of spring and summer (6 °C and 3 °C respectively) and that nocturnal breeze is expected to be important in the former seasons because $\Delta T = 6\,°C$ is the maximum absolute surface temperature difference between the sea and the land.

It is not trivial to exactly quantify the impact of the large-scale flow and the air channelling through the Marcellinara gap because circulations are the result of the superposition of the wind channelling and the breeze. Nevertheless, evidence can be found by studying the 850 hPa wind distribution from the ECMWF analysis for 2007–2008. To minimize the influence of topography on wind distributions, we considered two ECMWF analysis grid points, i.e. (15.5 E, 39 N) and (17 E, 38.5 N): the former point located over the Tyrrhenian Sea in front of Lamezia Terme and the latter located over the Ionian Sea, near the east-side entrance of the Marcellinara gap (Fig. 1). The first grid point is used to evaluate the frequency of winds whose direction is between 210° and 330° (NNW/SSW), the second for winds whose direction is between 30° and 150° (NNE/SSE). It should be noticed that results are not very sensitive to the exact grid point choice. We tried several grid points over land and over the sea, bilinear interpolation to both entrances of the Marcellinara gap and inside the gap, and the same grid points at 700 hPa obtaining similar results.

Table 2. Seasonal wind occurrences f (%) for 2007–2008, from the ECMWF analysis at 850 hPa. Second and fourth columns are for grid point (15.5 E, 39 N), third and fifth columns are for grid point (17 E, 38.5 N).

Season	f (%) (15.5 E, 39 N) $210 < dir < 330$ $U > 8\,ms^{-1}$	f (%) (17 E, 38.5 N) $30 < dir < 160$ $U > 8\,ms^{-1}$	f (%) (15.5 E, 39 N) $210 < dir < 330$ $U > 8\,ms^{-1}$	f (%) (17 E, 38.5 N) $30 < dir < 160$ $U > 8\,ms^{-1}$
Winter	24	4	28	5
Spring	30	7	35	10
Summer	23	2	29	4
Fall	23	2	29	4

Table 2 shows the wind occurrence for wind speeds larger than $8\,ms^{-1}$ at 850 hPa, which should have a major effect on breeze circulation (Lyons, 1972). Using observational data, Lyons (1972) developed an index to estimate whether a lake breeze would occur. Results show that when the ratio $V^2/\Delta T$ (where V is the geostrophic wind speed (ms^{-1})) is greater than 10 the breeze will not occur. The breeze will not develop because the horizontal pressure gradient generated by the differential heating between the land and the sea is not enough to overcome the kinetic energy of the large-scale flow. The maximum ΔT for Lamezia Terme is 6 °C in fall, accordingly, $8\,ms^{-1}$ of large-scale wind should suppress the breeze development.

In Table 2, it results that occurrence of $U > 8\,ms^{-1}$ is 23–30% for winds from the NNW-SSW sector, confirming that the westerlies are more frequent than other directions (Fig. 2b). Spring has the largest occurrence (30% compared to 23–24% of other seasons). The occurrence of winds from NNW-SSW is channelled in the W-WSW directions by the Marcellinara gap.

Table 2 and Fig. 2b show that frequency of winds from the NNE-SSE sector is much lower than for the sector NNW/SSW being 2%, 2%, 4% and 7% for summer and fall, winter and spring respectively.

We also compted the wind occurrence for wind speed larger than $6\,ms^{-1}$ at 850 hPa to consider a more populated statistic. Results are similar to the $8\,ms^{-1}$ threshold: westerlies are much more frequent than easterlies, spring has the largest occurrence of high wind speeds (both easterlies and westerlies), and easterlies are infrequent in summer, fall and winter.

Fall occurrence of large-scale winds from NNE-SSE is rather low, implying that winds from NE-ENE in Lamezia Terme are the result of katabatic winds that merge with the land breeze to form the nocturnal breeze. Winter flow from NE-ENE is also due in large part to the nocturnal breeze land/mountain breezes because the synoptic wind from this direction is only 1–2% more frequent than in fall.

The symmetry of the wind distribution during spring with respect to the other seasons can be explained by two factors:

- The largest influence of the large-scale forcing (7% for $8\,ms^{-1}$, 10% for $6\,ms^{-1}$) from east, compared to the other seasons, which enhances the circulations from NE-ENE at of 4–5% and then might contrast the development of the sea breeze; and

- The progressive increasing of the mean daytime ΔT from March to June. During March, the mean sea-land temperature difference is the lowest i.e. $\Delta T = 1.5$ °C; therefore this synoptic situation would not favour the sea breeze development. On the other hand, during June, the high $\Delta T = 4$ °C would instead favour the development of the sea breeze contributing to the westerly flow occurrence.

4 Conclusions

From the combined analysis of the large-scale flow and surface in-situ and remote measurements, we can derive few general conclusions on the local circulation at Lamezia Terme:

a) The diurnal breeze circulation in spring, summer and part of fall is a superposition of the large-scale and local-scale flow.

b) The large-scale forcing usually acts in phase with the diurnal breeze and opposes the nocturnal wind development.

c) The large-scale forcing mainly drives the diurnal circulation from November and in winter.

d) The nocturnal flow is mostly due to the nocturnal breeze circulation.

While a subjective examination of each day and the account of more meteorological parameters (cloud coverage, global radiation, synoptic-scale pattern of the day etc.) should produce a finer classification of the impact of large-scale and local-scale on the daily circulation at Lamezia Terme, this analysis is out of the scope of the present paper. Nevertheless, previous results on both SST and large-scale flow clarify several aspects of wind rose of Fig. 4. In particular:

a) In summer, the daytime difference between the land surface temperature and the SST reaches its maximum (about 4 °C), while the nighttime difference reaches its minimum (absolute value 3 °C). This causes strong and frequent diurnal circulations while the nocturnal breeze is weak. The nocturnal breeze direction is opposite to the large-scale winds, which suppress its occurrence; the diurnal breeze and the large-scale winds merging causes intense diurnal flows. These factors explain the asymmetry of the wind rose in Fig. 4c.

b) In winter and fall the nocturnal difference between the sea and land surface temperature reaches a maximum (6 °C) while diurnal difference is at minimum (especially in winter). This causes strong, frequent and intense nocturnal breeze despite of the large-scale forcing that is usually opposed to local-scale flow.

c) In our definition of spring, the absolute difference between the land surface temperature and the SST during the day is varying with months: from 1.5 °C in March to 4 °C in May. This variation is likely to cause the symmetry of the distribution among westerlies and easterlies. During March the wind distribution resemble the distribution in winter and fall. During May it will have the predominant westerly shape characteristics of the summer distribution.

Work in progress addresses the turbulent characteristics of the atmospheric PBL at a coastal area in different meteorological condition. The data will be analyzed considering the temperature difference between sea and land as it is an important factor to take into account for further classifying the coastal regimes.

Acknowledgements. This work was partially funded by the project "ALPI: ALlerta Precoce Incendi boschivi" funded by the Calabria Region. We are grateful to OSI SAF and EUMETSAT for Sea Surface Temperature data. We are grateful to ECMWF and Italian Air Force for the access to MARS data (Meteorological Archive and Retrieval System).

Edited by: M. M. Miglietta

References

Abe, S. and Yoshida, T.: The effect of the width of a peninsula to the sea breeze, J. Meteorol. Soc. Jpn., 60, 1074–1084, 1982.

Bolle, H. J.: Mediterranean climate – variability and trends, Springer-Verlag, Berlin, 2003.

Caccia, J.-L., Guénard, V., Benech, B., Campistron, B., and Drobinski, P.: Vertical velocity and turbulence aspects during Mistral events as observed by UHF wind profilers, Ann. Geophys., 22, 3927–3936, 2004, http://www.ann-geophys.net/22/3927/2004/.

Conte, M., Giuffrida, A., and Tedesco, S.: The Mediterranean Oscillation, in: Impact on precipitation and hydrology in Italy Climate Water, Publications of the Academy of Finland, Helsinki, 121–127, 1989.

Colacino, M.: Mediterranean Meteorology, Proceedings NATO-ASI, edited by: Charnock, H., Reports in Meteorology and Oceanography, 40, 1–38, 1990.

Colacino, M., Conte, M., and Piervitali, E.: Elementi di climatologia della Calabria, IFA-CNR, Rome, 197 pp., 1997.

Defant, F: Local winds, Compendium of Meteorology, edited by: Maloney, T. F., American Meteorological Society, Boston, 1951.

De Leo, L., Federico, S., Sempreviva, A. M., Pasqualoni, L., Avolio, E., and Bellecci, C.: Study of the development of the sea breeze and its micro-scale structure at a coastal site using a Multi-Tone Sodar system, Earth Environmental Science, 1, 1–9, doi:10.1088/1755-1307/1/1/012054, 2008.

Doran, E.: Objective analysis of mesoscale flow fields in Israel and trajectory calculations, Israel J. Earth Sci., 28, 33–41, 1979.

Federico, S., Dalu, G. A., Bellecci, C., and Colacino, M.: Mesoscale energetics and flows induced by sea-land and mountain-valley contrasts, Ann. Geophys., 18, 235–246, 2000, http://www.ann-geophys.net/18/235/2000/.

Federico, S., Bellecci, C., and Colacino, M.: Atmospheric convergente diabatically generated in the CBL over a mountainous peninsula, Nuovo Cimento C, 24C, 223–243, 2003.

Federico, S., Avolio, E., Pasqualoni, L., De Leo, L., Sempreviva, A. M., and Bellecci, C.: Preliminary results of a 30-year daily rainfall data base in southern Italy, Atmos. Res., 94(4), 641–651, doi:10.1016/j.atmosres.2009.03.008, 2009.

Kallos, G., Kassomenos, G. P., and Pielke, R. A.: Synoptic and mesoscale weather conditions during air pollution episodes in Athens, Greece, Bound.-Lay. Meteorol., 62, 163–184, 1993.

Lyons, W. A.: The climatology and prediction of the Chicago lake breeze, J. Appl. Meteorol., 11, 1259–1270, 1972.

Mastrantonio, G., Petenko, I. , Viola, A., Argentini, S., Coniglio, L., Monti, P., and Leuzzi, G.: Influence of the synoptic circulation on the local wind field in a coastal area of the Tyrrhenian Sea, Earth Environ. Sci., 1, 1–9, doi:10.1088/1755-1315/1/1/012049, 2008.

O&SI SAF: High Latitude SST product manual, version 1.2, 10/10/2002 Met.no, Oslo (Norway), http://saf.met.no/docs/ss2_pmhlsst_v1p2.pdf (last access: 23 April 2010), 2006.

Pearson, R. A.: Properties of the sea breeze front as shown by a numerical model, J. Atmos. Sci., 30, 1050–1060, 1973.

Pearson, R. A.: On the asymmetry of the land breeze sea breeze circulation, Q. J. Roy. Meteor. Soc., 101, 529–536, 1975.

Pielke, R. A.: A three-dimensional numerical model of the sea breeze over south Florida, Mon. Weather Rev., 102, 115–139, 1974.

Pielke, R. A.: Mesoscale Meteorological Modeling, Academic Press, San Diego, 2002.

Piervitali, E., Colacino, M., and Conte, M.: Rainfall over the central-western Mediterranean Basin in the period 1951–1995. Part I: Precipitation trends, Nuovo Cimento, C, 21, 331–344, 1999.

Simpson, J. E.: Sea breeze and local winds, Cambridge University Press, Cambridge, 224 pp., 1994.

Trigo, I. F., Davies, T. D., and Grant, R. B.: Objective Climatology of Cyclones in the Mediterranean Region, J. Climate, 12, 1685–1686, 1999.

Trigo, I. F., Grant, R. B., and Davies, T. D.: Climatology of Cyclogenesis Mechanisms in the Mediterranean, Mon. Weather Rev., 130, 549–569, 2002.

Uppala, S. M., Kallberg, P. W., and Simmons, A. J.: The ERA-40 re-analysis, Q. J. Roy. Meteor. Soc., 131, 2961–3012, 2005.

Newest developments of ACMANT

P. Domonkos, R. Poza, and D. Efthymiadis

Centre for Climate Change (C3), Geography Dept., University Rovira i Virgili,
Campus Terres de l'Ebre, C/Betánia 5 Tortosa, 43500, Spain

Abstract. The seasonal cycle of radiation intensity often causes a marked seasonal cycle in the inhomogeneities (IHs) of observed temperature time series, since a substantial portion of them have direct or indirect connection to radiation changes in the micro-environment of the thermometer. Therefore the magnitudes of temperature IHs tend to be larger in summer than in winter. A new homogenisation method, the Adapted Caussinus – Mestre Algorithm for Networks of Temperature series (ACMANT) has recently been developed which treats appropriately the seasonal changes of IH-sizes in temperature time series. The performance of ACMANT was proved to be among the best methods (together with PRODIGE and MASH) in the efficiency test procedure of COST ES0601 project. A further improved version of the ACMANT is described in this paper. In the new version the ANOVA procedure is applied for correcting inhomogeneities, and with this change the iterations applied in the earlier version have become unnecessary. Some other modifications have also been made, from which the most important one is the new way for estimating the timings of IHs. With these modifications the efficiency of the ACMANT has become even higher, therefore its use is strongly recommended when networks of monthly temperature series from mid- or high geographical latitudes are subjected to homogenisation. The paper presents the main properties and the operation of the new ACMANT.

1 Introduction

The investigation of climate change and climate variability needs a large amount of observed data of high quality. In the last decades a new branch of quality control and quality improvement for observed data has been developed, it is the so-called homogenisation. The purpose of homogenisation is to filter out the effects of technical imperfectness, i.e. that of methodological or environmental changes from the results of observations (Aguilar et al., 2003; Auer et al., 2005; Gérard-Marchant et al., 2008, etc.). The most frequent form of IHs is a sudden shift (change-point) in the series of the data. The documents of technical changes and the statistical properties of datasets can be used for homogenising time series. During statistical homogenisation the spatial redundancy can be utilised for identifying and adjusting local biases. The spatial redundancy means that the same climatic sign often appears in more than one time series. Hereafter homogenisation means statistical homogenisation in this study. In spite of the spatial redundancies, observed data contains real site-specific differences, therefore the perfect homogenisation is usually impossible, but homogenisation procedures have stochastic behaviour (Sherwood, 2007; Menne and Williams, 2009; Titchner et al., 2009, etc.). The success of homogenisation depends on (i) the number and completeness of time series and their spatial correlations, (ii) the signal/noise rate which is influenced by the spatial correlations, the standard deviation of the data, and the size of IHs, (iii) the ability of homogenisation methods to identify the timings of change-points, (iv) the ability of homogenisation methods to treat with multiple structures of IHs, (v) the way of adjusting time series for detected IHs. In this study the characteristics of a new homogenisation method, the Applied Caussinus-Mestre Algorithm for homogenising Networks of Temperature series (ACMANT), are discussed. At the time of the submission of this paper, the ACMANT is still in progress. In this paper only some important parts of the method are shown in detail. The full description of the method will be published once the method is finished. However, a tested preliminary version of the method is already available for interested users. For more information one can contact the main author.

The main novelty of the ACMANT is that the seasonal changes of IHs are modelled by harmonic functions, because IHs in temperature data are often related to radiation changes, and the seasonal curve of irradiation in mid- and high latitudes can be approached well with harmonic functions. This thesis is supported also by experimental results (Domonkos and Štěpánek, 2009; Brunet et al., 2011).

2 Main properties of ACMANT

In the ACMANT the core of the detection and adjustment methods is the same as in the PRODIGE (the method of Caussinus and Mestre, 2004). The PRODIGE has a strong mathematical background, and earlier studies (Domonkos, 2006), as well as the homogenisation of COST HOME benchmark dataset (hereafter: benchmark; Venema et al., 2010) have proved that the PRODIGE is one of the most effective homogenisation methods. With the recent developments the highest efficiencies reached earlier are now significantly exceeded. The operational features and conditions for the ACMANT are as follows:

i. The use of the ACMANT is recommended especially for temperature series from mid- and high-latitudes, since its algorithm supposes quasi-harmonic annual cycle of considerable amplitude in IH-sizes

ii. A fully automated method.

iii. A relative homogenisation method, thus it can be used only for networks, and not for single time series. Reference series are always built from a minimum of two component series. Lengths of time series in a network can be different, and in that case different reference series may be used for different sections of the same candidate series.

iv. Occurrences of data-gaps are allowed up to 83% for any 30-year inside section of time series, and unlimitedly in the tails of the series (inside section means that there are no data gaps at the edges of the section).

v. The ACMANT contains separate segments for filling data-gaps and substituting outlier values. Missing data are never filled before (after) the first (last) observed value of time series.

vi. The input data-field for the ACMANT: Monthly temperature characteristics with monthly time resolution. The lengths of the original time series may be different, but the data-fields of each series are required to be converted into a common format (which format includes the same number of data for each temperature series) in a way that missing values are filled with −999.9. After preparation only 4 parameters have to be introduced before application: (a) length of time series, (b) first year of time series, (c) number of time series in the network, (d) identifier of network.

vii. The result of homogenisation is (a) timings and sizes of IHs for each series, (b) timings of outliers, (c) filled data-gaps caused by missing values or outliers inside the series, (d) homogenised time series. Sizes of IHs are characterised with two variables: (a) shift in annual means, (b) shift in the amplitude of seasonal cycle.

The operation of ACMANT is illustrated in Fig. 1.

3 Description of selected segments of ACMANT

A concise description of the method is provided here, and only segments that are primarily important in achieving higher efficiency are shown in detail.

3.1 Constructing relative time series

The spatial comparison of time series relies on the rules introduced by Peterson and Easterling (1994), with some modifications in the parameterisation.

Anomaly series ($\boldsymbol{A}_j$) are created first, by subtracting monthly means from the raw values. j is a station-identifier ($j = 1,2,\ldots J$ where J stands for the total number of time series in the network). For the candidate series $\boldsymbol{A}_j$, a relative time series ($\boldsymbol{T}_j$) is the arithmetical difference of the candidate series and the so-called reference series ($\boldsymbol{F}_j$) (Eq. 1).

$$\boldsymbol{T}_j = \boldsymbol{A}_j - \boldsymbol{F}_j \tag{1}$$

Note that in accordance with Sect. 2, (point iii.), more than one relative time series are often created for the same candidate series (this is not shown in Eq. (1) to keep the description brief). Reference series are the weighted averages of neighbouring anomaly series ($\boldsymbol{A}_i$) around the candidate series where the weights are the squared spatial correlations (r) with the candidate series (Eq. 2). Following the recommendations of Peterson and Easterling (1994), the first difference (increment) series are applied for estimating spatial correlations, since in this way the estimations are less affected by the inhomogeneities in time series.

$$\boldsymbol{F}_j = \frac{\sum_{i=1}^{J^*} r_{j,i}^2 \cdot \boldsymbol{A}_i}{\sum_{i=1}^{J^*} r_{j,i}^2} \qquad i = 1,2,\ldots J^*, J^* \le J \tag{2}$$

In the ACMANT, every $\boldsymbol{A}_i$ with $r_{j,i} \ge 0.4$ is considered for building reference, but a minimum condition for applying the homogenisation is that at least two reference composites have to exist with $r_{j,i} \ge 0.5$.

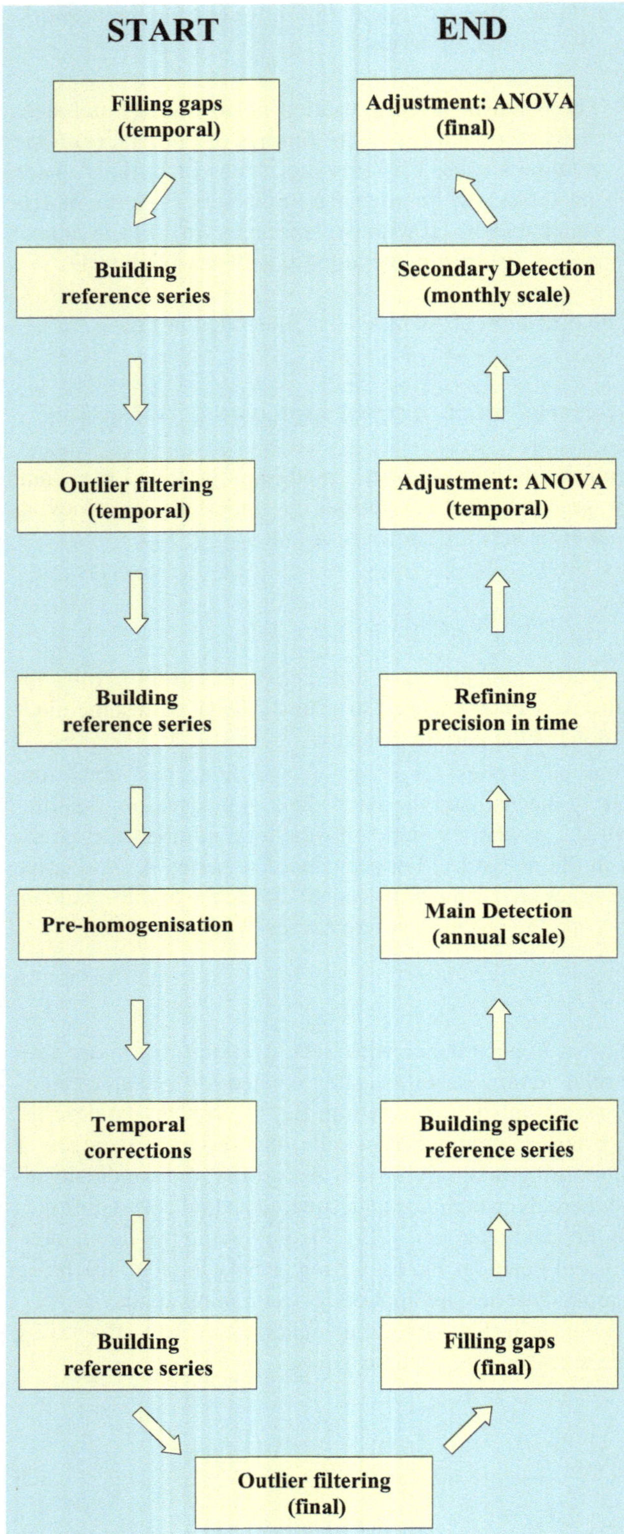

Figure 1. The operation of ACMANT.

3.2 Detecting IHs with Main Detection

Timings and sizes of IHs are searched by fitting step-functions to two annual characteristics, i.e. to annual means (***TM***) and to the range of the seasonal cycle (***TD***) in relative time series. Solutions with common timings of change-points are considered only, and the minimum sum of squared errors is searched for with the so-called dynamic programming algorithm, described by Hawkins (1972) first. Note that in the ACMANT, similarly to a large number of other methods, gradually changing biases (i.e. trend-like IHs) are represented as a series of change-points.

Let the length of time series be denoted by L, the number of change-points by K, their serial numbers by k ($k = 1,2,\ldots K$), and their timings by y_k. Note that k also shows the serial number of the section between adjacent change-points (Eqs. 3 and 4).

$$\min_{[y_1,y_2,\ldots y_K]}\left\{\sum_{k=0}^{K}\sum_{i=y_k+1}^{y_{k+1}}(tm_i-\overline{\boldsymbol{TM}}_k)^2+c_0(td_i-\overline{\boldsymbol{TD}}_k)^2\right\} \quad (3)$$

$$y_0=0, \quad y_{K+1}=L \quad (4)$$

In Eq. (3) upper stroke marks time-average, and $c_0 = 0.5$. The minimum distance between two change-points is set to be 3 years.

For using Eq. (3) K has to be set. For selecting the most appropriate K, the Caussinus – Lyazrhi criterion (Caussinus and Lyazrhi, 1997) is applied (Eqs. 5 and 6).

$$\ln\left\{1-\frac{\sum_{k=1}^{K+1}(y_k-y_{k-1})\cdot\left[(\overline{\boldsymbol{TM}}_k-\overline{\boldsymbol{TM}})^2+c_0(\overline{\boldsymbol{TD}}_k-\overline{\boldsymbol{TD}})^2\right]}{\sum_{i=1}^{L}(tm_i-\overline{\boldsymbol{TM}})^2+c_0(td_i-\overline{\boldsymbol{TD}})^2}\right\}+w \quad (5)$$

$$w=p\frac{2K}{L-1}\ln(L) \quad (6)$$

Equations (5) and (6) are calculated with each possible K, and the K providing the minimum of term (5) is retained. The Main Detection differs in three points from the classic Caussinus-Mestre detection method: (i) Step functions are fitted to two variables, (ii) the minimum distance between two change-points is 3 time-units, (iii) an extra parameter (p) is included in the penalty-term, and its value depends on J. If $J = 2$, then $p = 1.5$, if $J = 3$, then $p = 1$, and if $J > 3$, then $p = 0.75$.

The parameterisation relies on semi-empirical experiences about the changes of signal/noise rate in functions of Δy and J.

3.3 Calculation of timings of change-points with monthly preciseness

48-month wide windows are symmetrically set around the pre-estimated timings of change-points (from the Main Detection). Two-phase functions (U) are fitted to the monthly values of relative time series within each window. The functions are harmonic functions of 12-month cycle in both phases. The timing of change-point is searched in a narrower, 24-month wide window. Equation (7) shows the calculation for calendar month m.

$$u_m = \alpha + \beta \left(c_A \sin \frac{2\pi m}{12} + c_B \cos \frac{2\pi m}{12} \right) \quad (7)$$

Constants of c_A and c_B are determined in a way to have the modus of the annual cycle in the solstices. In the optimum fitting the sum of squared errors in the 48-month window is minimal. The optimum values for α, β and the timing of change-point are estimated through iterative tests.

3.4 Homogenisation-adjustment with ANOVA

In the detection process an individual IH usually causes biases in more than one time series, since during the spatial comparison each time series is used as reference composite, and homogenizers cannot pre-assume which series are homogeneous (if any). Therefore the calculation of cumulated effects of IHs needs the use of some equation system. In the ANOVA the observed values are considered to be a composition of climate-effect, station-effect and noise, and an equation-system is built and solved for the case of zero noise. Caussinus and Mestre (2004) proved that the equation system of the ANOVA provides the best estimation of IH-sizes. In that paper the full description of the ANOVA is provided.

In the ACMANT the ANOVA is applied to the two annual variables (***TM*** and ***TD***) separately, and thereafter monthly adjustments are calculated as a composition of the shift in annual mean and the relevant value of the seasonal cycle.

4 Some further details of ACMANT

The main characteristics of two further segments are presented here, these are the Pre-homogenisation and the Secondary Detection.

The aim of Pre-homogenisation is to reduce the impact of possible large IHs in the composites of reference series. The detection process is the same as in the Main Detection, but the way of the application is different: (i) Candidate series are ordered according to a pre-estimation of the severity of inhomogeneities. This ordering is based on the maximal absolute values of 5-year mean anomalies in relative time series. (ii) The Main Detection is applied starting from the least homogeneous series and proceeding always towards the more homogeneous series; (iii) Simplified adjustments (e.g. Alexandersson, 1986) are applied just after the detection of IHs in one candidate series; (iv) In building reference series for pre-homogenisation, adjusted versions of composites are used when they are available; (v) In building reference series for Pre-homogenisation, one time series is excluded from taking into account as reference-composite, namely the one for which the pre-homogenised time series will be used as reference-composite in the Main Detection. As one time series might be used as reference-composite for $J-1$ different candidate series, $J-1$ pre-homogenisations are accomplished for each time series through pre-homogenizing J sub-networks each containing $J-1$ time series. In the step of "Building specific reference series" (Fig. 1) these pre-homogenised series are used.

The aim of the Secondary Detection is to find and correct large-size but short-term biases caused by IHs. Series of accumulated anomalies are examined in time series adjusted by the ANOVA after the Main Detection. If the absolute values of accumulated anomalies exceed some arbitrarily given thresholds, IHs are searched for in monthly series, within a 60-month wide window symmetrically located around the anomaly-peak. The detection method is similar to the procedure described in Sect. 3.3, with two main differences (i) The number of change-points can be 0, 1 or 2 within a 60-month section, and the optimal choice from these three possibilities is determined by the Caussinus – Lyazrhi criterion; (ii) The length of the section between two adjacent change-points is three months minimum, and when it is shorter than 10 months, a constant function is fitted (with zero amplitude of seasonal cycle). After the Secondary Detection the application of the ANOVA is repeated both for ***TM*** and ***TD***.

5 The efficiency of ACMANT

The efficiency of homogenisation procedures can be characterised with the improvement in root mean squared errors (RMSE). We calculated the RMSE characteristics of monthly biases, annual biases and biases of linear trend-slopes for the time series of the benchmark. The efficiency is characterised by the rate of the decrease of RMSE due to the homogenisation, proportioned to the RMSE of raw time series (Eq. 8).

$$\text{Eff} = \frac{W_R - W_H}{W_R} \quad (8)$$

In Eq. (8) W_R (W_H) denotes the RMSE in raw (homogenised) time series. If a homogenisation is perfect, then the efficiency equals to 1, while in case of no change in the RMSE the efficiency is zero. Efficiencies for the ACMANT were calculated separately for the monthly, annual and trend-slope biases. All the 40 simulated temperature networks of the benchmark were used for the calculations. The results show that the efficiency of the ACMANT is 0.545 in monthly RMSE, 0.666 in annual RMSE and 0.753 in trend-slope RMSE. These values are significantly higher than those were achieved by any other method during the COST HOME experiments. The superior performance of ACMANT stems from three main properties: (i) The ACMANT adopts the best detection and adjustment segments of earlier methods. (ii) The ACMANT applies appropriate time-scales during the detection; considering that the standard deviation is smaller for annual variables than for monthly variables, the application of a relatively coarse time-resolution often yields the best results; regarding this point, a special characteristic of the ACMANT is that the seasonal changes of IH-sizes are estimated with the use of two annual characteristics only. (iii) The ACMANT has a pre-homogenisation part in which the reference composites for a particular candidate series are adjusted in a way that the spatial correlation between the candidate series and reference composites are not utilised at all.

Further verifications are still needed, because in the last phase of the development the IHs of the benchmark were known to the authors. Nevertheless, considering that the verified dataset contains 340 time series with more than one thousand IHs, the impact of previous knowledge of the time series characteristics could cause minor bias in the efficiencies calculated.

6 Conclusions

A new homogenisation method, the ACMANT has been developed. It is applicable for homogenising observational networks of monthly temperature series. While the ACMANT is based on one of the best methods that existed earlier (i.e. on the PRODIGE), it also includes a new treatment of the seasonal cycle of inhomogeneity-sizes and some other modifications. The new version of the ACMANT shows a favourably high efficiency when it is used for homogenising the COST HOME benchmark dataset. The properties described and the efficiency results obtained indicate that the ACMANT is an excellent tool for homogenising networks of temperature datasets from mid- or high latitudes.

Acknowledgements. The research was partly funded by the COST ES0601 project and by the Centre for Climate Change (C3) of University Rovira i Virgili.

Edited by: M. Brunet-India

References

Aguilar, E., Auer, I., Brunet, M., Peterson, T. C., and Wieringa, J.: WMO Guidelines on climate metadata and homogenization, WCDMP-No. 53, WMO-TD No 1186, WMO, Geneva, 2003.

Alexandersson, H.: A homogeneity test applied to precipitation data, J. Climatol., 6, 661–675, doi:10.1002/joc.3370060607, 1986.

Auer, I. and 24 coauthors: A new instrumental precipitation dataset for the greater Alpine region for the period 1800–2002, Int. J. Climatol., 25, 139–166, doi:10.1002/joc.1135, 2005.

Brunet, M., Asin, J., Sigró, J., Bañon, M., García, F., Aguilar, E., Palenzuela, J. E., Peterson, T. C., and Jones, P.: The minimization of the screen bias from ancient Western Mediterranean air temperature records: an exploratory statistical analysis. Int. J. Climatol., doi:10.1002/joc.2192, 2011.

Caussinus, H. and Lyazrhi, F.: Choosing a linear model with a random number of change-points and outliers, Ann. Inst. Statist. Math., 49/4, 761–775, doi:10.1023/A:1003230713770, 1997.

Caussinus, H. and Mestre, O.: Detection and correction of artificial shifts in climate series, J. Roy. Stat. Soc., Series C53, 405–425, doi:10.1111/j.1467-9876.2004.05155.x, 2004.

Domonkos, P.: Testing of homogenisation methods: purposes, tools and problems of implementation, Proceedings of the 5th Seminar and Quality Control in Climatological Databases, CD, 126–145, 2006.

Domonkos, P. and Štěpánek, P.: Statistical characteristics of detectable inhomogeneities in observed meteorological time series, Studia Geoph. et Geod., 53, 239–260, doi:10.007/ s11200-009-0015-9, 2009.

Gérard-Marchant, P. G. F., Stooksbury, D. E., and Seymour, L.: Methods for starting the detection of undocumented multiple changepoints, J. Climate, 21, 4887–4899, doi:10.1175/2008JCLI1956.1, 2008.

Hawkins, D. M.: On the choice of segments in piecewise approximation, J. Inst. Math. Appl., 9, 250–256, doi:10.1093/imamat/9.2.250, 1972.

Menne, M. J. and Williams Jr., C. N.: Homogenization of temperature series via pairwise comparisons, J. Climate, 22, 1700–1717, doi:10.1175/2008JCLI2263.1, 2009.

Peterson, T. C. and Easterling, D. R.: Creation of homogeneous composite climatological reference series, Int. J. Climatol., 14, 671–679, doi:10.1002/joc.3370140606, 1994.

Sherwood, S. C.: Simultaneous detection of climate change and observing biases in a network with incomplete sampling, J. Climate, 20, 4047–4062, doi:10.1175/JCLI4215.1, 2007.

Titchner, H. A., Thorne, P. W., McCarthy, M. P., Tett, S. F. B., Haimberger, L., and Parker, D. E.: Critically reassessing tropospheric temperature trends from radiosondes using realistic validation experiments, J. Climate, 22, 465–485, doi:10.1175/2008JCLI2419.1, 2009.

Venema, V., Mestre, O., and the COST HOME Team: Benchmark database, EGU General Assembly, Vienna, Austria, 3–7 May 2010, EGU CL4.6-13357, 2010.

4

Regional tendencies of extreme wind characteristics in Hungary

K. Radics[1,2], J. Bartholy[2], and Cs. N. Péliné[1,2]

[1]Geoinformation Service, Hungarian Defence Forces, Szilagyi E. fasor 7–9, 1024 Budapest, Hungary
[2]Department of Meteorology, Eötvös Loránd University, Pazmany s. 1/A, 1117 Budapest, Hungary

Abstract. Human activities have substantial effects on climate system and climate change will directly influence the global economy and society in the near future. Since change in the long-term mean climatic conditions will have significant consequences, the most important effects of climate change may come from changes in the intensity and frequency of climatic extremes. It is therefore of great interest to document and evaluate the extremes of near-surface wind that could assist in estimating the regional effects of climate change.

Recently a research started on specifying the possible changes of wind characteristics over Hungary. The study is based on 33-year-long (1975–2007) wind (speed, direction, and gust) data sets of 36 Hungarian synoptic meteorological stations. Spatial and temporal distributions of mean and extreme wind climate characteristics were estimated, wind extremes and trends were interpolated and mapped over the country. Finally, measured and reanalysis (ERA-40) wind data were compared over Hungary. This might be important in order to check the adaptability of climate simulation results in estimation of regional climate change effects besides the direct validitation of ERA-40 reanalysis data sets.

1 Introduction

Since environmental processes may be affected by global warming, changes of extreme climate events have been considered a key issue nowadays. Usually it is impossible to connect directly specific weather events to global warming. Instead, global warming is expected to cause changes in the overall distribution and intensity of extreme events, such as changes in the frequency and intensity of heavy precipitation and wind storms.

Hungary occupies the low elevated areas of the Carpathian basin. The country is predominantly flat, two-thirds of the entire territory lays below 200 m and approximately 2% is above 400 m. The latitudinal extension of Hungary (Fig. 1) – less than three degrees (45°45′–48°35′ N) – is not responsible for significant climatological differences between northern and southern regions. The country had not been the subject of extensive wind researches in the last century. However, some studies were carried out analyzing the surface and upper-air wind records spanning several decades (Tar et al., 2001). The presented study is a segment of complex wind climate research of the country, which uses statistical methods to analyze the measured wind field of the Hungarian meteorological stations.

Most scientists agree that special attention should be paid applying ERA-40 reanalysis data sets as a reference, e.g. for such model development and validation purposes where wind plays an important role (Punge and Giorgetta, 2007). Considering the main achievements of studies dealt with statistics of differences between observations and hindcasts, furthermore, regarding the fact that no corrections were applied prior to 1980 in ERA-40 data sets (Haimberger, 2005), spatially and temporally inhomogeneous measured wind speed data and reanalyzed data were compared for Hungary. In this study some preliminary results are presented demonstrating how the comparison of different datasets can improve the reliability of detected wind variability.

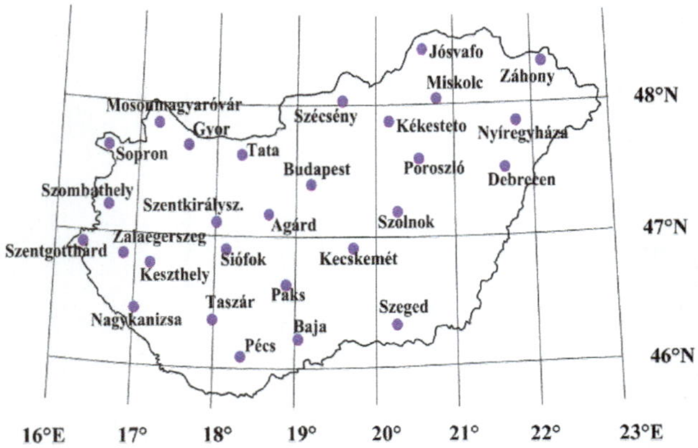

Figure 1. Geographical locations of meteorological stations involved in the analysis.

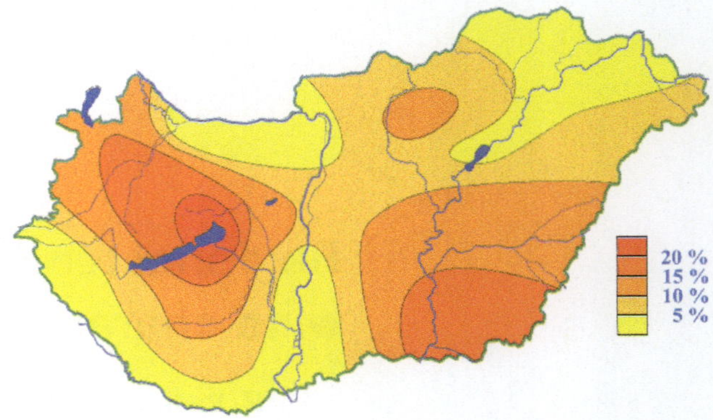

Figure 2. Relative frequencies of wind speed values greater than 5 m/s at 10 m height over Hungary.

2 Data sets

Wind data sets of 36 Hungarian synoptic meteorological stations have been analyzed, which consists of hourly wind speed, wind direction and wind gust data for the period 1975–2007.

As an achievement of technical development, automated measurement systems were installed in Hungary during the 1990s. This implies changes not only in the measuring methods and instruments, but in many cases in the location and height of observation. Furthermore, some new stations were installed, and as a consequence of urbanization, the closer and farther surroundings of measuring stations have been changed in the meantime. Due to these inhomogeneities unified analysis of the whole time series is not feasible. Therefore, (1) only 28 stations are involved in this study (Fig. 1), (2) Hungarian wind climate conditions are analyzed for two different sets of the database: the first part (1975–1994) was recorded by Fuess wind instruments; the second part (1997–2007) was measured by automated meteorological systems.

After processing the measured wind data sets (including digitisation of old instrumental records, quality control and homogenisation of measured data), time series and complex wind climate analysis were carried out.

The global multi-decadal data sets, generated by reanalysis of past observations, have become a basis of many types of atmospheric research. Reanalyses provide comprehensive information that can support a variety of applications. Thus, in the second part of the study ERA-40 reanalysis data sets from the European Centre for Medium-Range Weather Forecasts (ECMWF) were analyzed. Average wind speed (u, v) components (at 10 m height) were used for the period 1975–2002 with 6 h temporal resolution and $1^\circ \times 1^\circ$ spatial resolution over the Carpathian Basin.

3 Analyses and results

Analysing the second part (1997–2007) of measured wind data the following preliminary conclusions can be drawn (Bartholy et al., 2003; Radics and Bartholy, 2002, 2008):

- Mean wind speed is ranging between 1.47–3.84 m/s over Hungary. Therefore, the country can be considered moderately windy.
- The range of annual and daily variability of wind speed is 1–2 m/s.
- Duration of 1–3 m/s winds is the longest (1500–3000 h/year). Annual average duration of the wind speeds greater than 10 m/s is very low, less than 100 h/year generally.
- The relative frequency of the dominant wind direction is low. The frequencies of calm periods in some regions are relatively high comparing that to the regional mean values.
- Mean wind speed and wind gust values show large spatial differences over the country. The windiest region is Northwestern Hungary, while the least windy is the northeastern part of the country.
- Absolute maxima values of average wind speed and wind gust are ranging between 15–40 m/s. The largest values occurred before 2000.

The spatial distribution of wind speeds greater than 5 m/s at 10 m height is illustrated in Fig. 2. In harmony with the preliminary expectations, the maximum of relative frequency values (21.6%) occurred in the centre of Transdanubia (western part of Hungary bordered by the river Danube), while the minima (less than 4%) are found in the northeastern, southern and southwestern regions of the country.

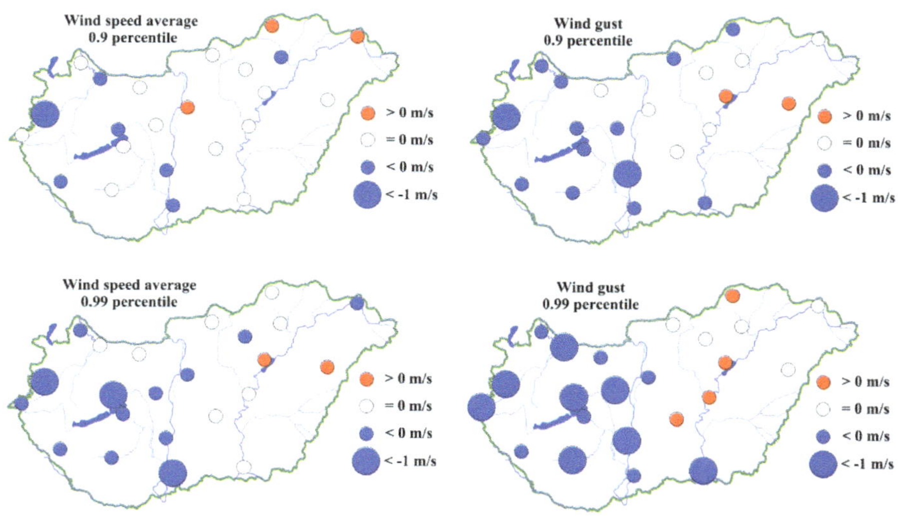

Figure 3. Spatial distribution of linear trend fitted on different percentile values (0.9 – up, 0.99 – down) calculated from three-year subseries of wind speed (left) and wind gust (right) values, 1997–2007.

Spatial and temporal distributions of extreme wind conditions were estimated by calculating three-year moving averages of different percentile values (median, 0.9, 0.95, 0.99). During the period 1975–1994 the 0.9 percentiles of wind speed did not change generally, except two stations (Keszthely and Paks) with slightly decreasing tendency. Concerning the 0.99 percentile, half of the stations shows negative tendency (mainly in Transdanubia), positive trend was found only in Zalaegerszeg. During the same time, different percentile values of wind gust decreased in the western part of the country and were constant at the eastern regions. Increasing tendencies were found only in two stations (Szeged and Nyíregyháza).

During the second period (1997–2007) the median of average wind speed was constant over the entire territory of Hungary. In cases of 0.9 and 0.99 (Fig. 3, left), decreases were found mainly in Transdanubia, and increases appeared in some stations at the eastern part of the country. Investigating temporal and spatial distributions of extreme wind gust values, the median was constant almost over the whole country. As it is demonstrated in Fig. 3 (right), the increase of the applied percentile value caused significant increase in the area and intensity of negative tendencies. In contrast, constant or slightly increasing tendencies were found over the Great Hungarian Plain.

It has already accepted that expected regional climate change can influence the Hungarian wind climate, as well, as the frequency of extreme events. Thus, the comparison of different climate scenarios with statistics of present climate parameters provides the means for estimating the potential regional effects of climate change. Therefore, the spatially and temporally inhomogeneous measured wind data and the homogeneous ERA-40 reanalysis data sets were compared (1975–2002). During the period 1975–1980 ERA-40 considerably underestimates the measured data, especially, in cases of maxima (Radics et al., 2009). Between 1988–1994 ERA-40 underestimates measured data because of the lack of night-time observations. Taking the whole period (1975–2002) into consideration, ERA-40 data significantly underestimate the maxima of measured data until 1997, and after that year the large overestimation of minima is typical. The overall degree of underestimation has decreased since 1997.

Concerning our further study of ERA-40, spatial distribution of wind speed and wind gust extremes were analyzed (1997–2002) using three-year moving averages of different percentile values. On the left side of Fig. 4, where the spatial distribution of linear trend fitted on 0.9 percentile values is shown, decreasing tendencies were found generally. Inside the country only one grid point could be characterised by a slightly increasing trend. On the other hand, increasing tendencies appeared only in some cases on the right side of Fig. 4 (0.99 percentile). Greater than 1 m/s increase were found only outside the Hungarian border.

4 Conclusions

The main goal of the presented research is to specify the variability of regional wind characteristics over Hungary. The study is based on 33-year-long (1975–2007) wind data sets of 36 Hungarian meteorological stations. Summarizing the main results, the following conclusions can be drawn.

- During the period 1975–2007 different percentile values of average wind speed and wind gust were analyzed. In cases of 0.9 and 0.99 percentiles significantly decreasing tendencies were found mainly in Transdanubia,

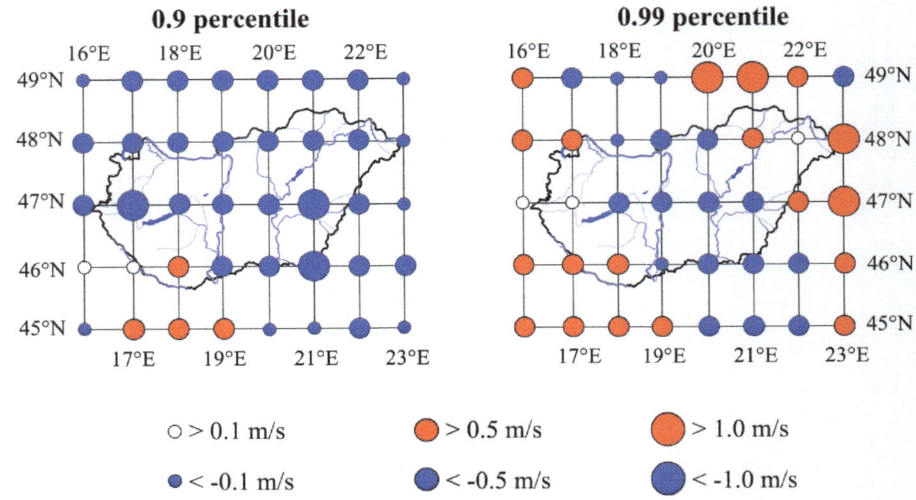

Figure 4. Spatial distribution of linear trend fitted on different percentile values (0.9 – left, 0.99 – right) calculated from three-year subseries of ERA-40 wind speed data, 1997–2002.

increasing trend appeared only in some stations at the eastern part of the country.

- The preliminary results show strong relationships between the corresponding periods of ERA-40 and station data.

- Analysing ERA-40 percentile values, generally decreasing tendencies were found. Increasing tendencies appeared only in some cases. Greater than 1 m/s increase were found only outside the country.

- In order to estimate regional climate change effects, spatially and temporally homogeneous data series of ERA-40 seems to be an appropriate tool to substitute the measured data over Hungary.

Acknowledgements. Authors are very grateful to the Geoinformation Service of Hungarian Defence Forces for the wind data bases of the Hungarian meteorological stations, to the Hungarian Meteorological Service for the ERA-40 data sets.

Research leading to this paper has been supported by the Hungarian Academy of Sciences under the program 2006/TKI/246 titled Adaptation to climate change, by the Hungarian National Science Research Foundation under grants T-049824, K-67626, K-69164 and K-78125, and by the Hungarian Ministry of Environment and Water.

Furthermore, the support of the János Bolyai Research Scholarship of the Hungarian Academy of Sciences is appreciated.

Edited by: I. Auer

References

Bartholy, J., Radics, K., and Bohoczky, F.: Present state of wind energy utilisation in Hungary: policy, wind climate, and modelling studies, Renew. Sust. Energ. Rev., 7, 175–186, 2003.

Haimberger, L.: Homogenization of the global radiosonde dataset using ERA-40 analysis feedback information, Reviews, 7, 175–186, 2005.

Punge, H. J. and Giorgetta, M. A.: Differences between the QBO in the first and in the second half of the ERA-40 reanalysis, Atmos. Chem. Phys., 7, 599–608, 2007, http://www.atmos-chem-phys.net/7/599/2007/.

Radics, K. and Bartholy, J.: Selected wind characteristics and potential use of wind energy in Hungary. Part II, Időjárás, 106, 59–74, 2002.

Radics, K. and Bartholy, J.: Estimating and modelling the wind resource of Hungary, Renew. Sust. Energ. Rev., 12, 874–882, 2008.

Radics, K., Péliné, N. Cs., and Bartholy, J.: Regional tendencies of wind climate over Hungary, New Challenges in the Field of Military Sciences, Conference Proceedings, CD ROM, 2009.

Tar, K., Makra, L., Horváth, Sz., and Kircsi, A.: Temporal change of some statistical characteristics of wind speed over the Great Hungarian Plain, Theor. Appl. Clim., 69, 69–79, 2001.

5

Vulnerability to climate change: people, place and exposure to hazard

C. W. Hutton[1]**, S. Kienberger**[2]**, F. Amoako Johnson**[3]**, A. Allan**[4]**, V. Giannini**[5]**, and R. Allen**[6]

[1]GeoData Institute, University of Southampton, UK
[2]Centre for Geoinformatics, University of Salzburg, Salzburg, Austria
[3]School of Social Sciences, University of Southampton, UK
[4]Centre for Water Law, Policy and Science, Dundee, UK
[5]Fondazione Eni Enrico Mattei, Venice, Italy
[6]ICIMOD, Katmadu, Nepal

Abstract. The Human Dimension of the Twinning European and South Asian River Basins to Enhance Capacity and Implement Adaptive Management Approaches Project (EC-Project BRAHMATWINN) is aimed at developing socio-economic tools and context for the effective inclusion of the "Human Dimension" or socio-economic vulnerability into the overall assessment of climate risk in the twinned basins of the Upper Brahmaputra River Basin (UBRB), and the Upper Danube River Basin (UDRB) . This work is conducted in the light of stakeholder/actor analysis and the prevailing legal framework.

In order to effectively achieve this end, four key research and associated activities were defined:

1. **Identifying stakeholders and actors including:** implement an approach to ensure a broad spread of appropriate stakeholder input to the assessment of vulnerability undertaken in Asia and Europe within the research activities of the project.

2. **Contextualising legal framework:** to provide an assessment of the governance framework relating to socio-environmental policy development within the study site administrative areas leading to the specific identification of related policy and legal recommendations.

3. **Spatial analysis and mapping of vulnerability:** providing a spatial assessment of the variation of vulnerability to pre-determined environmental stressors across the study areas with an additional specific focus on gender.

4. **Inclusion of findings** with the broader context of the BRAHMATWINN risk of climate change study through scenarios of hazard and vulnerability (subsequent chapters).

This study utilises stakeholder inputs to effectively identify and map relative weightings of vulnerability domains, such as health and education in the context of pre-specified hazards such as flood. The process is underpinned by an adaptation of the IPCC (2001) which characterizes *Risk* as having the components of *Hazard* (physiographic component) and *Vulnerability* (socio-economic component).

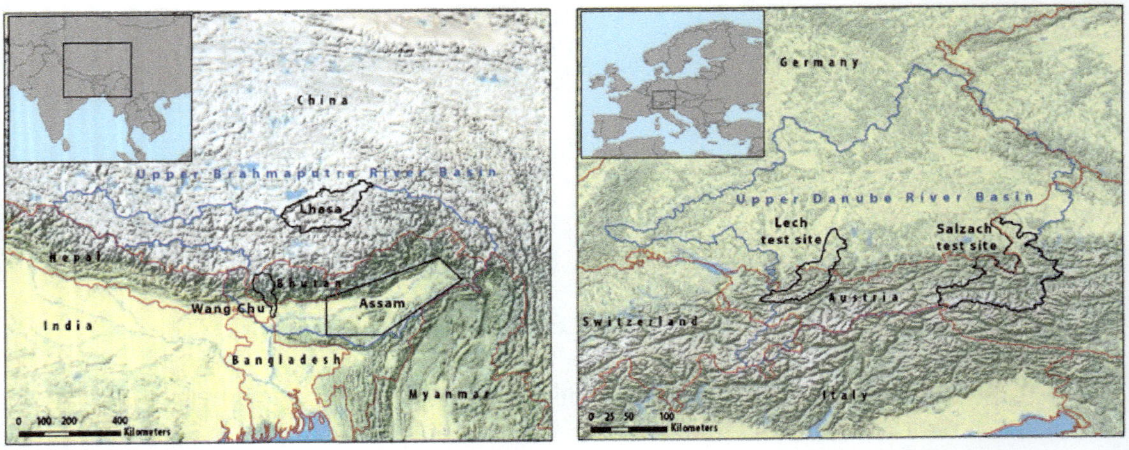

Figure 1. Outline location maps for key study sites within the BRAHMATWINN project.

1 Introduction

Climate change and its resulting effects on the environment, affects the socio-economic wellbeing of individuals, households, communities and nations. The impacts are mainly adverse, with some being irreversible, while others may be beneficial (O'Neil et al., 2001). The effects vary greatly and are felt more among the poorest of the poor and vulnerable communities due to limited resources and infrastructure available to these groups. The study presented in this chapter covers a number of case studies in the Upper Brahmaputra River Basin and the Upper Danube River Basin. These being the Assam State Brahmaputra river section (NE India), the Wang Chu River Basin in Bhutan and the Lhasa River basin in the Autonomous Region of Tibet along with the Salzach and Lech River basins of the Upper Danube in Europe. These case studies are selected on the basis that they represent basins/sub-basins under potential environmental threat due to climate change.

Vulnerability to climate change maps predominantly derived from census and land cover mapping provide a context for assessing vulnerability and the potential for adaptive capacity with regards to climate hazards in the selected twinning river basins (Fig. 1). Critical to this process has been the development of a joint (twinned) approach to the concept of vulnerability in the framework of risk and hazard as well as identifying the adaptive capacity and sensitivity components of vulnerability based upon the framework generated by the IPCC (2001) (Fig. 3). The project also includes an emphasis on gender perspective in order to developing a better understanding of the socio-economic impacts caused by climate stress on individuals, their households and communities in terms of their livelihoods, health and sanitation situations.

This chapter elucidates the methodological development of a stakeholder based flood vulnerability mapping within the study areas of the BRAHMATWINN project. As stated,

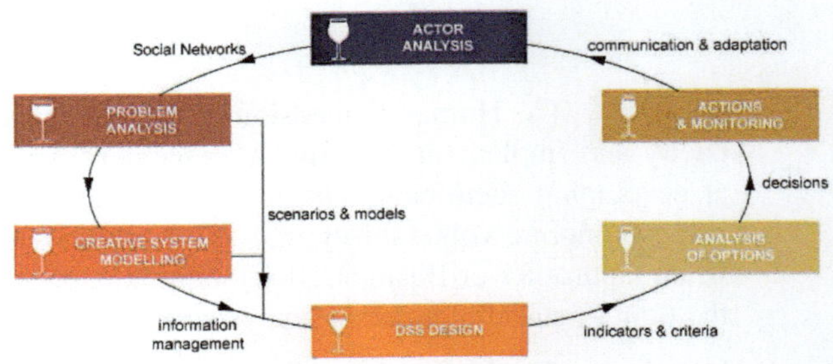

Figure 2. The NetSyMoD approach. The first 3 stages are utilsed in this approach.

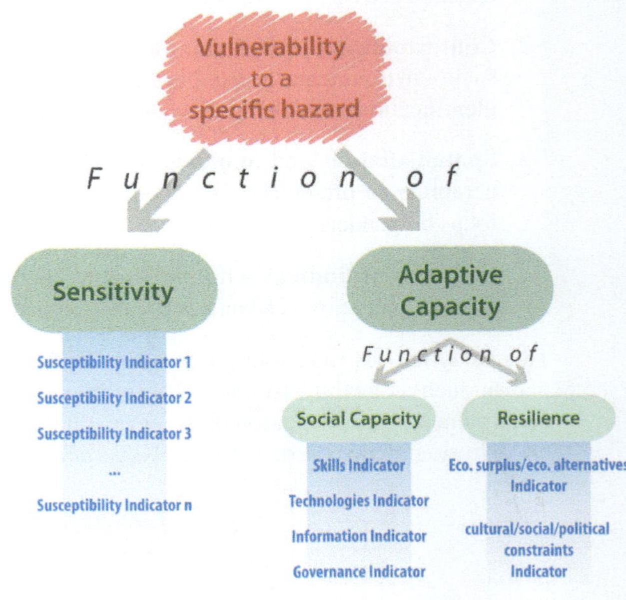

Figure 3. Conceptual model of vulnerability.

the key aim is to provide a spatial context for the assessment of socio-economic and asset based vulnerability at risk from specific hazards associated with climate change. As such a process of identifying local and regional stakeholders to underpin this process is critical to the overall sustainability of the project outputs as well as providing the context for the identification of relevant indicators and their weighting for the development of vulnerability mapping approaches.

For this purpose, the project applies the implementation of a participatory planning methodology throughout the different phases of the work. The selected participatory approach utilised within the process of eliciting stakeholder opinion is called Creative System Modelling (CSM) (Giupponi et al., 2008) which helps to identify participants and organises the structured debates in workshops, and is combined with a partial "Delphi" ranking round which allows the gathering of specific weightings for suggested domains of vulnerability. The process is supported by analysis relating to assessments of government efficacy and relevant legal structures and frameworks which providing a government context for understanding the potential impact of climate change and water resource management at a number of administrative levels in both the Asian and EU twinning basins.

Finally, all products outputs and methodological tools including data has been populated into the River Basin Information System (RBIS) which provides a common repository and organisation structure for the overall project outputs thereby enhancing the RBIS towards a Decision Information Support Tool (DIST) (BRAHMATWINN Chapter 10).

2 Role within the international BRAHMATWINN project

The role of "Human Dimension" (HD) within the BRAHMATWINN project is focused upon stakeholder, socio-economic and governance framework contribution to the effective development of Integrated Water Resources Management (IWRM) strategies. In principle the BRAHMATWINN project can be thought of as developing *Risk* scenarios for the future climate impacts in the study basins, where *Risk* is considered to have the elements of *Hazard*, which is covered in the physiographic studies and analysis of the climate change models, and *Vulnerability* (socio-economic) which are essentially the stakeholder, governance and vulnerability mapping components. As such there is a social and a physical component and the project addresses both of these to develop integrated tools to be utilised by decision makers for the purposes of policy planning, development and implementation. BRAHMATWINN Chapters 3 (physical) and 4 (social) represent this division of work which are then integrated within the broader context of the project overall in later chapters.

3 Methodology

This chapter identifies a series of quantitative and qualitative methodological approaches which when combined allow for the production of vulnerability maps in both the European and Asian context, both of which are incorporative of stakeholder opinion (to a first order) and contextualised within the legal and policy framework independently. For the sake of clarity the below methodological approach differentiates these methodologies into two key approaches (i) stakeholder processes (ii) mapping methods. The key conceptual framework utilized is demonstrated in Fig. 3 which depicts the components of vulnerability based upon the adapted principles of the IPCC (2001).

3.1 Stakeholder analysis methodologies

3.1.1 Identifying stakeholders

The project requirement was to elicit and capture the stakeholder perspective in order to develop vulnerability maps that are reflective of local expertise and understanding. In order to achieve this, a process of Actors Analysis was undertaken which is an iterative identification of actors (stakeholders, experts etc.) to be involved in the process. In addition their reciprocal relationships within the actor networks were assessed using Social Network Analysis (SNA) techniques. This process formed the formative elements of a larger project wide process of social analysis termed NetSyMoD. In sequence and in parallel, in depth problem analysis is conducted in order to acquire the basic information about the case, the possible options, etc. (Problem Analysis). Once the community of interested parties has been identified and the problem in question analysed, Creative System Modelling (CSM) techniques (e.g. cognitive mapping) are applied to produce a shared model of the system underlying the problem (Creative System Modelling). The model is a formal, albeit simplified, description of the system and its causal links, to which the problem pertains and that can be commonly understood and recognised by the actors involved. It thus brings together simulation models, perceptions and beliefs of stakeholders and policy makers, as well as decision analysis techniques and underlying models. This process was utilised by the research activities expressed in this chapter as well as forming the foundation for further more detailed analysis of social network conducted and reported in Chapters 6 and 8 respectively.

3.1.2 Government and legal framework

The methodological approach adopted relies heavily on existing projects designed to measure governance in general, such as UNDP, World Bank, Access Initiative, and Transparency International. A review of these projects has lead to an understanding of the need to separate governance into

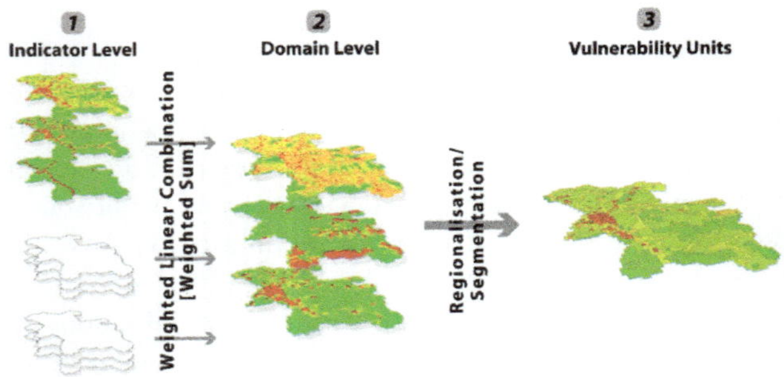

Figure 4. Workflow developed to model spatial vulnerability units in the Salzach River catchment (Kienberger et al., 2009).

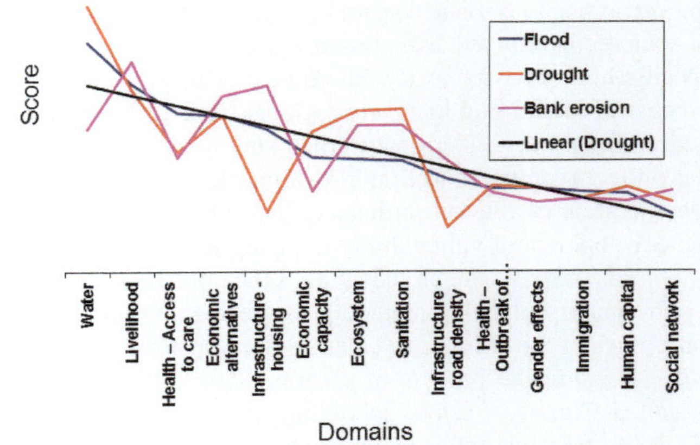

Figure 5. Domain weighting with hazard.

questions related to commitments and processes. Commitments referring to what states have adopted on paper, and processes relating to the mechanisms that are in place to ensure that the commitments are implemented within the national, regional and local context. The study has also illustrates that the best approach to gathering indicator based information is to develop a set of questions designed and structured in a way that provides key information concerning the four elements of governance, i.e, accountability, transparency, predictability and participation

3.1.3 Ranking of domains – the Delphi technique

At the stakeholder workshops, a Delphi technique was used to investigate the relative importance of each domain identified. *A domain represents a key area potentially impacting vulnerability such as health or education or different types of assets*. Each participant was asked to score the domains to sum up to a total score of 40, with the most important domain receiving the highest score and the least important receiving the lowest score. In Assam, the exercise was undertaken separately for each type of hazard (flood, drought and bank erosion) and for climate change and climate disasters for Bhutan. In the European case studies the hazard flood has been investigated solely. The sensitivity and adaptive capacity domains were ranked together to reveal their relative importance. The scores were then averaged over the number of participants and ranked in order of importance. The rankings of the domains are shown in Fig. 5. The exercise provided a preliminary classification of the different domains of sensitivity and adaptive capacity and their relative importance with regards to the different types of climate hazards. Figure 5 shows that in Assam, general access to clean water and sanitation facilities, disruption to livelihood sources, lack of access to health care and the lack of economic alternatives were ranked to contribute highly to vulnerability to climate hazards. Gender, immigration, human capital and social networks were ranked low in the final outcomes.

3.2 Vulnerability mapping method

Within a twinning framework of the BRAHMATWINN project an approach has been developed to effectively model and map vulnerability to floods in the UBRB (India, Bhutan and Tibet) and the UDRB (Salzach and Lech catchments). The full statistical method is outlined in Sharma et al. (2010). The intention of the research was to assess the socio-economic component of risk through a joint vulnerability approach. The approach reflects the wider objective and conceptualizations of the International Panel for Climate Change (IPCC) framework (2001). Within this framework vulnerability is defined as a function of adaptive capacity and sensitivity. As sensitivity and adaptive capacity are multi-dimensional constructs a wide range of possible input parameters or domains are potentially of relevance and there is a need to be pragmatic about the datasets available and the subsequent selection of indicators. To enhance the range for parameter selection, the research methodology utilises census and household survey as well as land use/land cover data to derive and assess vulnerability. For both test site areas, appropriate indicators have been chosen from these sources and represented in sub-domains and domains based upon literature and in-field interviews and substantial workshops with local experts.

3.2.1 European case study methodology: Upper Danube River Basin (UDRB) – Salzach catchment and Lech catchment

In the Salzach and Lech catchment it was aimed to model homogenous regions which share a a common property of vulnerability. The methodology developed and the results for the Salzach catchment are documented in Kienberger et al. (2009). The derivation of homogenous spatial units links to concept of Geons (Lang et al., 2008) which are defined as generic spatial objects that are homogenous in terms of a varying spatial phenomena under the influence of, and partly controlled by, policy actions. The Geon concept acts as a framework for the regionalization of continuous spatial information according to defined parameters of homogeneity

Data applied, to populate the specific indicators and domains, originates from publicly available GIS data sources and census data. The data is not only provided on the basis of different administration units, but additionally on a standardised grid (in this case 1000 m). This is a unique approach and allows visualising data not only on the arbitrary administration units but shows the characteristics in a spatial distribution. Another advantage is that in subsequent analysis steps the Modifiable Areal Unit Problem (MAUP, Unwin, 1996) is minimized. Next to the census data, data on infrastructure, the availability of early warning systems, distance to first responders and heath facilities and land use/land cover information defining different asset classes have been integrated.

In principle the methodology (Fig. 4) follows in a first step the approach of Multi Criteria Analysis (MCA), Multi Criteria Evaluation or Analytical Hierarchy Process (AHP) where a discussion on that topic and current best practices approaches is available in Carver (1991) and Malczewski (2000). This method is applied to allow the comparability of the different data layers and aggregate the data to the different sub-domain levels through weighted sum algorithms. To finally model vulnerability units the regionalisation algorithm after (Baatz and Schäpe, 2000).has been applied.

The integration of expert knowledge was maintained through the identification of indicators and sub-domains, and the weighting of the different domains which was achieved through an online Delphi exercise with stakeholders from the Salzach and Lech basins This weights have then be used within the weighted sum aggregation and within the regionalisation algorithm which also considers the integration of scores/weights.

In a final step the data has been visualised next to papers maps, also as virtual globe environments, which allow the exploration and visualisation of this hierarchically structured data.

3.2.2 Asian case study: Upper Brahmaputra River Basin (UBRB) – Assam, India (main case study), Wang Chu River Basin, Bhutan and Tibet (ART)

The data for the analysis comes from the 2001 Indian Population and Housing Census and 2001 LANDSAT data for the UBRB. For Bhutan, the 2005 Population and Housing Census and remote sensing data were used, and the analysis was conducted at the Gewog administrative level. Although the study area (the Wang Chu) covers 35 Gewogs, for the purposes comparison the analysis was conducted for all 201 Gewogs of Bhutan. The Tibetan study was limited to LANDSAT data only as there was substantial difficulty accessing any socio-economic data of relevance.

The LANDSAT data include road density, agricultural land use, distance to main settlements and health centres. The road density and agricultural land use data were extracted using a three kilometre buffer around each village/town point. Agricultural land use is expressed as the proportion of land use on the basis of commercial and non-commercial agricultural land. The distance measures (distance to main settlements and health facilities) are Euclidean distances. The data covers 14 775 towns/villages of the in UDRB.

The different domains of sensitivity and adaptive capacity were identified through literature review, field observations, and discussions with local residents, stakeholders and experts working on both the hydrological and socio-economic aspects of environmental hazards in the region. This process was used to develop an inventory of domains and indicators that can be used to profile vulnerability. Indicator selection for vulnerability profiling has been debated extensively in the literature (Jones and Andrey, 2007). The debate has focused on the justification for inclusion and exclusion of indicators. Studies such as Cutter et al. (2000); Wu et al. (2002); Odeh (2002); Brooks et al. (2005); Chakraborty et al. (2005); Jones and Andrey (2007) have used different indicators to quantify vulnerability. Some studies have used Principal Component Analysis (PCA) and correlation methods to select indicators for vulnerability profiling (Jones and Andrey, 2007; Brooks et al., 2005). The advantages and disadvantages of these methods have been discussed in the literature (Jones and Andrey, 2007). The disadvantages mentioned in the literature have focused on the reliability of the indicators to measure vulnerability in a given location.

To avoid this potential bias, a participatory process involving local residents, stakeholders and experts were employed (CSM). The criteria for selection of indicators was based on what local residents, stakeholders and experts thought appropriate, relevant and robust for quantifying a specific domain. It is worthwhile mentioning that not all indicators identified by local residents, stakeholders and experts were included in the analysis due to data limitations. The indicators selected for the analysis although may not be exhaustive for assessing environmental hazard induced vulnerability; they have been selected to ensure that it is comprehensive and relevant for the practical assessment of vulnerability to environmental hazards in the given location. The indicators although may not be replicated exactly for other regions they could be modified to suit.

4 Results

The results of the Assam basin vulnerability mapping in the Brahmaputra basin (Fig. 6) are presented as a map of community points of output vulnerability classes for flood hazard. Each point represents the centroid of a community identified on the Indian census (2001) with a population represented through the size of the circle. It was not possible to access the data regarding the actual spatial extent of the

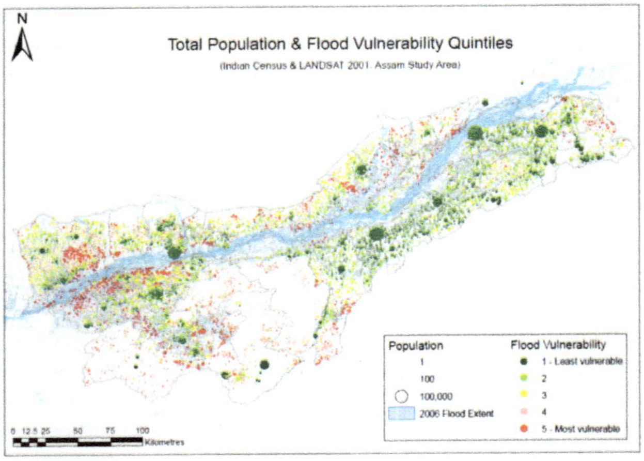

Figure 6. Vulnerability to flood based upon the geo-referenced Indian census data (2001) and a weighted indicator approach. Flood extent of a large recent flood is used as a proxy for hazard. Mapped by settlement.

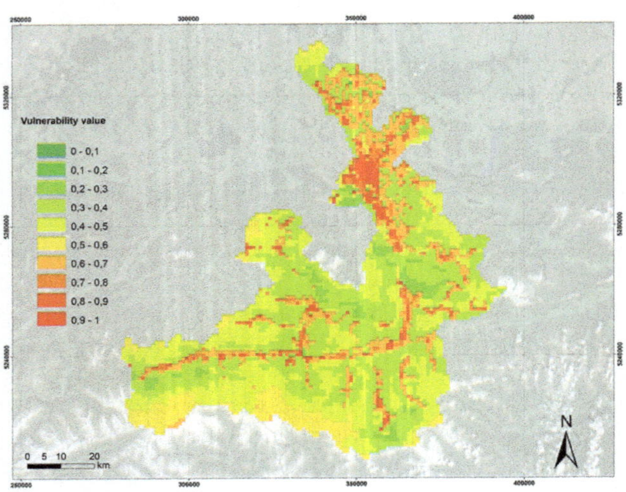

Figure 7. Bhutan case study: Climate change impacts on agricultural calendar/socioeconomic development (shift in precipitation/weather patterns) – by Gewog.

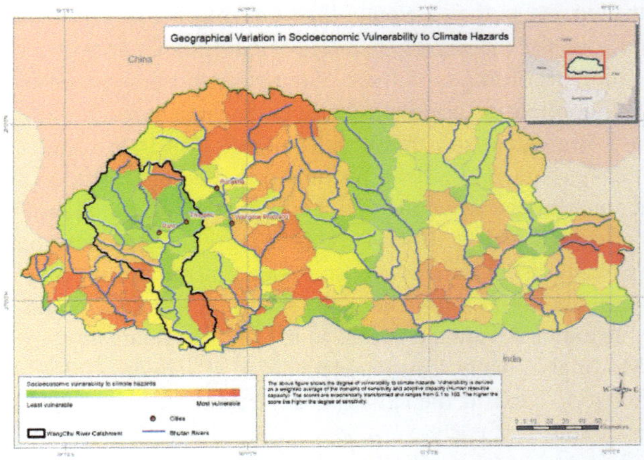

Figure 8. Tibet case study: Climate change hazard with associated impacts on agricultural calendar/socioeconomic development. Based upon LANDSAT only.

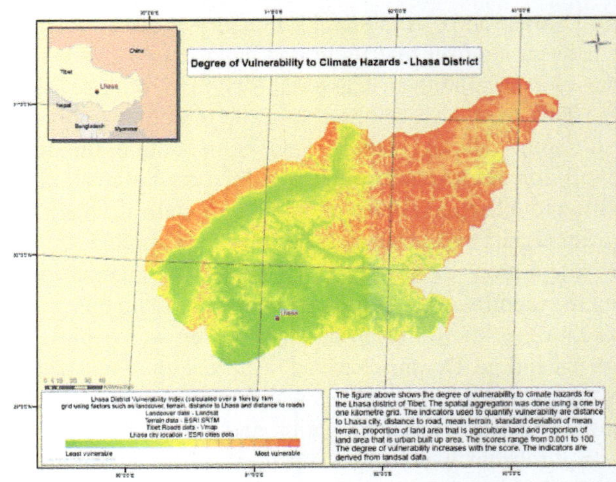

Figure 9. Salzach River case study. Socio-economic Vulnerability unit.

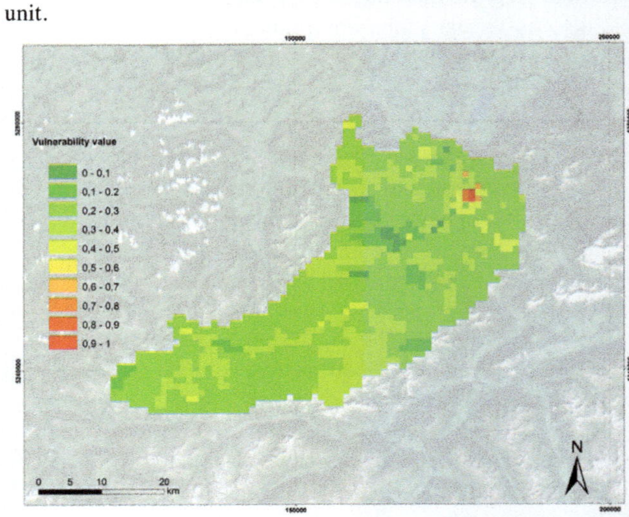

Figure 10. Lech River case study. Socio-economic Vulnerability units.

communities which would have added another valuable element. Figure 5 shows that the same process was carried out for drought and bank erosion as well with substantial weighting variations elicited form stakeholders. For Bhutan the result are constructed in a comparable fashion however the unit of analysis is the larger administrative unit out of necessity (Fig. 7). As stated earlier the output for Tibet is based only on satellite imagery and represented on a 1 × 1 km grid (Fig. 8).

In the Salzach and Lech basins (Figs. 9 and 10), due to the availability of grid-based census data, the method to delineate and identify homogenous units/areas of vulnerability has been achieved independent from administrative units (Kienberger et al., 2009). Similar to the approach applied in Assam, the vulnerability is measured on a relative scale. The

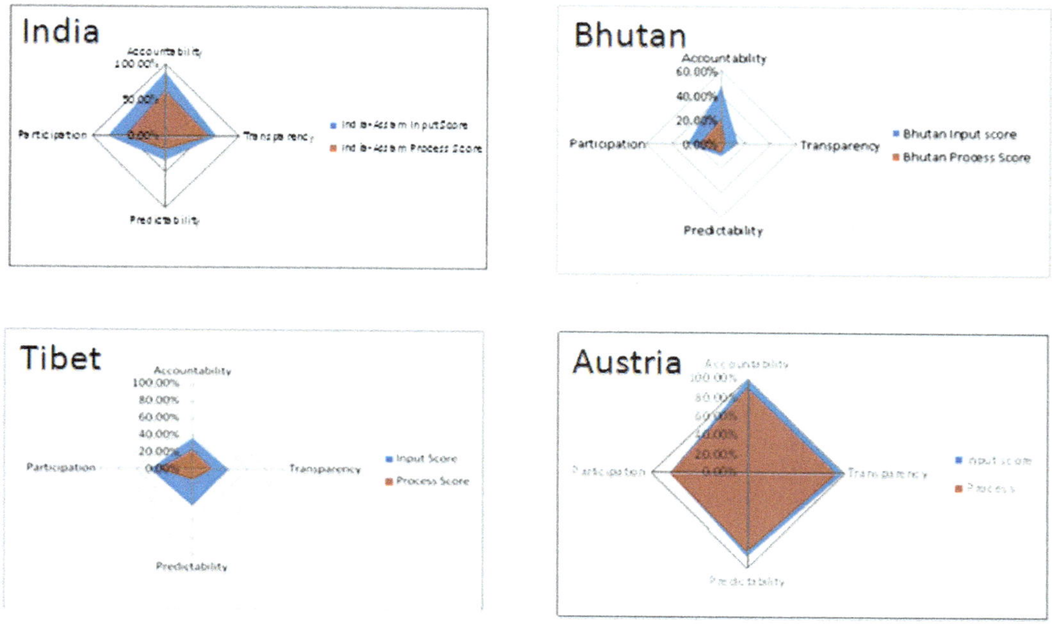

Figure 11. Attributes of the water resources government framework for the study area authorities. Input refers to the current legal framework in place and output refers to implementation.

modelling result show a high vulnerability in areas related to assets and infrastructures. Additionally, it can be observed that domains/indicator have been highly weighted through the stakeholder have a significant "imprint" on the results such as factors within the susceptibility domains "housing", "infrastructure" and "assets" and the social capacity domain "early warning" which received the highest ranks.

Both sets of maps rely upon a common conceptual framework to develop the vulnerability maps, however there is no intention to absolutely standardise approaches, indeed it is deemed of specific value to allow the selection and fulfilment of domains to be suited to the regional requirements. As such the twining referred to in the BRAHMATWINN project is limited to a conceptual level.

Figure 11 is a visualisation of the governance framework which has precedence over the study areas. The spider diagrams allow a comparison of the key attributes of good governance and provide a focus for potential enhancement of the current legal framework from the perspective of both the development of law and its application. The ultimate aim of answering the "indicator" questions is to provide a structured analysis of the extent to which existing governance arrangements within a country facilitates IWRM in the context of climate change. In separating commitments and process questions, the analysis can also assess the extent to which commitments have been realised.

5 Contribution to sustainable IWRM

Vulnerability and associated risk mapping offers decision makers a quantitative and well framed pathway into the incorporation of socio-economics to IWRM planning. To this end the contributions of Chapter 4 are seen as inputs to two key further elements of the project. The follow-up study to the work presented in Chapter 4 is presented in Chapter 8 which deals with the specific development of "what if scenarios" and Chapter 10 which identified the role of these outputs in the broad IWRM context. The main aim of this further work is to investigate the potential for vulnerability to increase or decrease depending on governments' policy directions and social values under the four scenarios related to the SRES outcomes. In Chapter 10 this outcome is considered in the context of IWRM more specifically. To do this all the individual variables that are highly correlated with the overall vulnerability score are identified and the outcomes linked to the economic and demographic development of the region as forecast for all India under the SRES scenarios. In broad conclusion the impact of GDP and population growth is seen to be highest in areas where levels of vulnerability are already high.

6 Conclusions and recommendations

A valuable outcome of the twinning approach has been the delineation of two clear paradigms in the thinking of the production and utilization of vulnerability maps. In the European context of the UDRB vulnerability is strongly characterised through assets, whereas in the Assam region of the Asian UBRB a strong focus lies on the defining characteristics of poverty at the community level. As such, in the Asian context, it is possible that a map of asset vulnerability is the inverse of a vulnerability map based upon community vulnerability. Both sit within a distinct governance and legal context which clearly delineates the complex associations between vulnerability and the strengths and weaknesses of policy and its implementation.

As indicated earlier it was not intended within the specific research context to allow a direct comparison of the vulnerabilities between the Salzach River catchment and Assam. Building on a joint conceptual framework, for each case study area specific indicators have been selected. This depends on one hand pragmatically on the availability of datasets but even more on the characteristics of vulnerability in the specific areas and the input from different stakeholder workshops, which targeted on the identification of characteristic vulnerability indicators. One element of that input that came under discussion was the role of the experts (project partners) who are not necessarily regional experts. It was felt it was important to minimise input from external sources and simply to facilitate opinion of stakeholders. To this end such domains as gender, which has a strong EU imperative were not given substantial weighting due to local expert opinion settling against them and despite partner expertise prioritising such a domain. As such the maps represent the perceptions of vulnerability of that group of individuals that were gathered together in the stakeholder meeting. This approach was a pilot and of course with more resourcing and time there would need to be a far more exhaustive exploration of the stakeholder opinion. An example of this was the strong bias towards engineers in the Assam workshop as there is a tendency for decision makers in this region to have an engineering focus. The workshop was representative of decision making in Assam but perhaps not representative of governance more broadly.

To implement successfully measures in regard to disaster risk reduction and climate change adaptation actors from national governments, provincial administrations and local authorities need to plan interventions based on location. It has been widely agreed that regional development and planning is a highly multidisciplinary such as the identification of land use changes. This integrates experts and stakeholders from different disciplines or ministries and should be best achieved through the involvement of the vulnerable population. The spatial representation of risk, visualising the extent of hazard or the distribution of vulnerability is therefore needed to successfully implement and address such strategies (Kienberger et al., 2009).

There are substantial areas of further research and effort required in this process. Firstly there needs to be work on the validation processes to test the outcomes of the maps. To some extent this has already occurred at a first order with a subsequent qualitative review by group government representatives both river basins. However a standardised sampling based approach to flood impact would be needed for full validation. Additionally, work needs to be carried out on the insertion of such approaches into the decision making processes and as well as getting a much deeper understanding of the potential to include spatial data as a proxy and/or augmentation for current commonly utilised georeferenced data sets such as the census. That said, the authors conclude that this is a practical and applicable process which provides a language of vulnerability that is relevant to the challenges of climate and socio-environmental change.

Additionally there is a substantive body of work to be undertaken to effectively include the diverse and substantial impact of governance on vulnerability and its spatial distribution. Currently the proxies for the mapping of effective governance are highly limited and as such methodologies to address this critical area are needed.

Acknowledgements. The Authors wish to thank the stakeholders of the various catchments who have contributed to this work as well as the field support of the Asian and European partners in organising the various stakeholder events which made this work possible. In addition the authors wish to acknowledge the contributions of FSU and ICIMOD more broadly in developing the foundations for this work.

References

Baatz, M. and Schäpe, A.: Multiresolution Segmentation: an optimization approach for high quality multi-scale image segmentation, in: Proceedings of the Angewandte Geographische Informationsverarbeitung XII, edited by: Strobl, J., Blaschke, T., Griesebner, G., et al., Beiträge zum AGIT Symposium Salzburg, 12–23, 2000.

Brooks, N., Adger, W. N., and Mick Kelly, P.: The determinants of vulnerability and adaptive capacity at the national level and the implications for adaptation, Global Environ. Chang., 15, 151–163, 2005.

Carver, S. J.: Integrating multi-criteria evaluation with geographical information systems, Int. J. Geogr. Inf. Syst., 5, 3, 321–339, 1991.

Chakraborty, J., Tobin, G., and Montz, B.: New process to assessing vulnerability and variability in geophysical risk and social vulnerability to natural hazards, Natural Hazards Review, 6(1), 23–33, 2005.

Cutter, S., Mitchell, J., and Scott, M.: Revealing the vulnerability of people and place: a case study of Georgetown County, South Carolina, Ann. Assoc. Am. Geogr., 90(4), 713–737, 2000.

Giupponi, C., Sgobbi, A., Mysiak, J., Camera, R., and Fassio, A.: NetSyMoD: an Integrated Approach for Water Resources Management, in: Integrated Water Management: Practical Experiences and Case studies, edited by: Meire, P. et al., Springer Netherlands, 69–93, 2008.

IPCC: Climate Change 2001: Impacts, Adaptation and Vulnerability, Contribution to the Working Group I to the third assessment report of the Intergovernmental Panel on Climate Change (IPCC), Cambridge University Press, Cambridge, 2001.

Jones, B. and Andrey, J.: Vulnerability index construction: methodological choices and their influence on identifying vulnerable neighborhoods, International Journal of Emergency Management, 4(2), 269–295, 2007.

Kienberger, S., Lang, S., and Zeil, P.: Spatial vulnerability units – expert-based spatial modelling of socio-economic vulnerability in the Salzach catchment, Austria, Nat. Hazards Earth Syst. Sci., 9, 767–778, doi:10.5194/nhess-9-767-2009, 2009.

Lang, S., Zeil, P., Kienberger, S., and Tiede, D.: Geons – policy-relevant geo-objects for monitoring high-level indicators, in: Geospatial Crossroads @ GI_Forum '08, Proceedings of the Geoinformatics Forum Salzburg, edited by: Car, A., Griesebner, G., and Strobl, J., Heidelberg, Wichmann, 180–185, 2008.

Malczewski, J.: On the Use of Weighted Linear Combination Method in GIS: Common and Best Practice Approaches, Transactions in GIS, 4(1), 5–22, 2000.

Odeh, D. J.: Natural Hazards Vulnerability Assessment for Statewide Mitigation Planning in Rhode Island, Natural Hazards Review, 3(4), 177–187, 2002.

O'Neill, B. C., MacKellar, F. L., and Lutz, W.: Population and climate change, Cambridge University Press, Cambridge, 2001.

Sharma, N., Amoako Johnson, F., Hutton, C. W., and Clark, M.: Hazard, Vulnerability and Risk on the Brahmaputra Basin: A Case Study of River Bank Erosion, in: Analysis of Climate Change Impacts on Water Resources For Developing Economies: Successes and Challenges, edited by: Adeloye, A. J. and Ojha, C. S. P., Open Journal of Hydrology, 113–114, 2010.

Unwin, D. J.: GIS, spatial analysis and spatial statistics', Prog. Hum. Geog., 20(4), 540–441, 1996.

Villagrán, J. C. V.: Vulnerability: a conceptual and methodological review, UN University Institute for Environment and Human Security, 4/2006, 64 pp., 2006.

Wu, S. Y., Yarnal, B., and Fisher, A.: Vulnerability of Coastal Communities to Sea-Level Rise: a Case Study of Cape May County, New Jersey, USA, Clim. Res., 22, 255–270, 2002.

6

"CITY 2020+": assessing climate change impacts for the city of Aachen related to demographic change and health – a progress report

C. Schneider[1], C. Balzer[3], M. Buttstädt[1], K. Eßer[5], S. Ginski[6], J. Hahmann[4], G. Ketzler[1], M. Klemme[7], A. Kröpelin[2], H. Merbitz[1], S. Michael[3], T. Sachsen[1], A. Siuda[1], M. Weishoff-Houben[3], M. F. Brunk[2], W. Dott[3], H. Hofmeister[4], C. Pfaffenbach[1], C. Roll[5], and K. Selle[6]

[1]RWTH Aachen Univ., Department of Geography, Faculty of Material Sciences and Geosciences, Germany
[2]RWTH Aachen Univ., Construction Business and Building Services, Faculty of Civil Engineering, Germany
[3]RWTH Aachen Univ., Institute of Hygiene and Environmental Medicine, Faculty of Medicine, Germany
[4]RWTH Aachen Univ., Institute for Sociology, Faculty of Social Sciences and Humanities, Germany
[5]RWTH Aachen Univ., Department of History, Faculty of Social Sciences and Humanities, Germany
[6]RWTH Aachen Univ., Institute of Planning Theory and Urban Development, Faculty of Architecture, Germany
[7]Univ. Münster, Governance Research and Applied Geography, Department of Geosciences, Germany

Abstract. The research initiative CITY 2020+ assesses the risks and opportunities for residents in urban built environments under projected demographic and climate change for the year 2020 and beyond, using the city of Aachen as a case study. CITY 2020+ develops strategies, options and tools for planning and developing sustainable future city structures. The investigation focuses on how urban environment, political structure and residential behaviour can best be adapted, with attention to the interactions among structural, political, and sociological configurations and their impacts on human health. The interdisciplinary research is organized in three clusters. Within the first cluster, strategies of older people exposed to heat stress, and their networks as well as environmental health risks according to atmospheric conditions are examined. The second cluster addresses governance questions, urban planning and building technologies as well as spatial patterns of the urban heat island. The third cluster includes studies on air quality related to particulate matter and a historical perspective of city development concerning environmental issues and climate variability. However, it turns out that research topics that require an interdisciplinary approach are best addressed not by pre-structuring the work into related sub-projects but through combining them according to shared methodological approaches. Examples illustrating this rather practical approach within ongoing research are presented in this paper.

1 Introduction

Demographers project that in Europe by 2050, nearly 30 % of the population will be over age 65 (UN World Population Prospects, 2009). An aging workforce and population combined with higher energy prices, environmental concern, and technological changes are likely to transform living and working arrangements (Grübler, 1998; Sterns, 1995).

Also by 2050, air temperatures are projected to rise on average. However, hot summer days in Central Europe are likely to exhibit an even accelerated increase (IPCC, 2007a). Due to an expected rise of mean air temperature resulting in a growing number of heat wave episodes in Europe (Meehl and Tebaldi, 2004) it is expected that the urban population will be increasingly exposed to adverse health effects (Luber and McGeehin, 2008; Koppe et al., 2004).

Weather and climate affect human health as well as well-being and productivity (Jendritzky, 2007). Heat stress, which will occur more often as a consequence of climate change has severe impacts on human health (Zebisch et al., 2005) since heat extremes can cause an increase in morbidity and mortality. This became apparent in 2003 when about 35 000 people died because of the heat wave in western Europe (Koppe et al., 2003). About 700 people died 1995 in Chicago due to a heat wave (Semanza et al., 1996). Excessive heat events do not affect people in the same way. There are specific factors that are influencing vulnerability. Healthy living adults with existing adaptation strategies are not as jeopardized as other especially vulnerable groups such as chronically ill individuals (Mücke et al., 2009). Factors can be clustered into three groups: First, cardiovascular or respiratory diseases as well as diabetes form hazardous conditions during heat waves. Second, the consumption of medicine, drugs and alcohol is positively correlated with heat-related health stress. There are third individuals that are per se more vulnerable than others and this group consists of babies, small children and older people (Jendritzky, 1998; Haines et al., 2006; Mücke et al., 2009; UBA and DWD, 2008).

In addition to these intra-individual variables of vulnerability, research shows other risk factors. A lack of information on heat stress and adaptation strategies as well as social isolation increases vulnerability and therefore results in a higher mortality risk (Pfaff et al., 2004; Ingendahl and Thieme, 2009; Mücke et al., 2009). Technologies, e.g. air conditioning, support the human body while coping with heat stress. Extended health care is important to prepare and protect individuals at risk (Zebisch et al., 2005).

The fact that especially the combination of heat and poor air quality impinges on humans' well-being accentuates the need for assessing the health risks of the residents regarding air pollutants and anomalously high summer temperatures (Roberts, 2004). In order to estimate the population exposure to air pollution and heat load, the spatial PM and temperature distributions within the city need to be investigated. Studies show that areas with enhanced PM and thermal load are not distributed uniformly in cities but appear as hot spots with varying spatial extents (Weber, 2009; Monn et al., 1997; Taha, 1997; Oke, 1987). The health effects of air pollution have been subject to intense studies in recent years. Exposure to pollutants such as airborne particulate matter has been associated with increases in mortality and hospital admissions due to respiratory and cardiovascular disease (Brunekreef and Holgate, 2002; WHO, 2006; Thornburg et al., 2009; Yang and Omaye, 2009). To date the associations often have been made using air quality standards that only encompass PM size and concentration, but the physical and chemical characteristics of the particle and/or linked components are often neglected (Hetland et al., 2004; Dagher et al., 2005).

The recent discussion on "climate-friendly urban development" (Heiland, 2008; BMVBS and BBR, 2007; Osenberg, 2007) takes up several requests from former debates: (1) is the call for sustainable urban development (Bochnig and Selle, 1993; DIFU, 2004; Hall and Pfeiffer, 2000) and circular land use management (BBR, 2006a, b; Malburg-Graf et al., 2007); (2) requests to consider demographic changes and their consequences on urban development and their implementation on site (Bürkner et al., 2007; Klemme, 2008). The local condition under which the changes are implemented and the local active protagonists can be examined by a governance perspective.

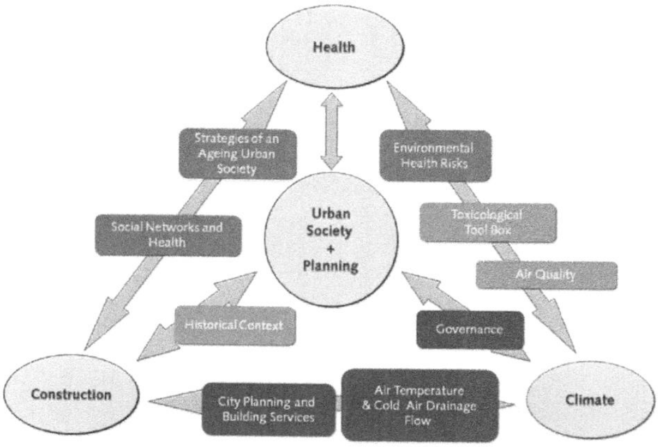

Figure 1. The structure of research project CITY 2020+. The research was initially organized into three clusters, each including three sub-projects.

Building on these studies and on previous investigations in Aachen dealing with urban climate patterns and their special characteristics (Havlik, 1981; Emonds, 1986; Ketzler, 1997; Havlik and Ketzler, 2000; Havlik, 2002), CITY 2020+ identifies the ways how microclimates in the city, health outcomes, and the urban environment are related. Within the project, the risks individuals face living and working in these conditions are assessed and new strategies are proposed based on cooperation from the fields of medicine, physical and cultural geography, sociology, history, civil engineering, planning theory and architecture for adapting the city for future needs. CITY 2020+ aims to generate strategies, to recommend options for urban planning measures and to apply tools for planning and developing sustainable future city structures. CITY 2020+ includes research projects (Fig. 1) dealing with

1. a micro-scale assessment of blockages to low-level cold-air drainage flow into the city centre by vegetation and building structures,
2. a detailed analysis of spatio-temporal air temperature patterns,
3. a micro- to meso-scale analysis of particulate matter (PM) concentrations depending on traffic, urban land use, building structure, topography, local meteorological conditions and synoptic-scale weather patterns,
4. a chemical and toxicological characterization of airborne PM and gaseous components,
5. historical viewpoints related to environment and atmospheric variability,
6. governance aspects of climate change adaptation and mitigation,

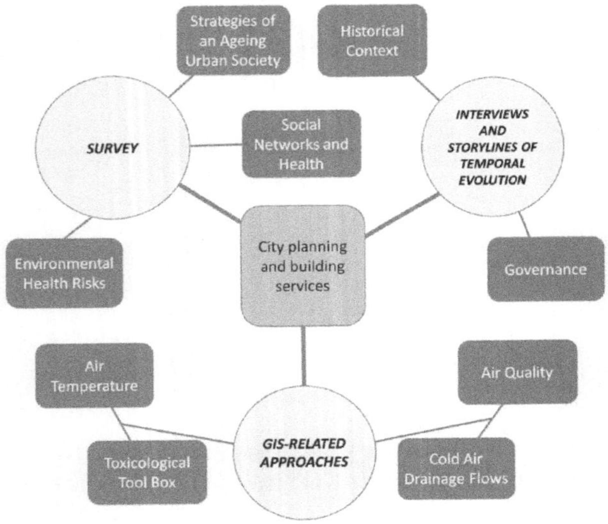

Figure 2. Project organisation related to commonly application approaches.

7. strategies of older people living and working under conditions of heat stress,
8. the importance of social networks for heat related adaptation and mitigation of individuals,
9. opportunities and risks related to building technologies for historical and modern building concepts, and
10. the impact of environmental conditions on preschool children.

Within the project CITY 2020+ it turned out that interdisciplinarity is a key issue within this highly interrelated and complex research setup. However, our experience is that interdisciplinarity does not naturally evolve within related fields but rather grows through a set of shared methods. The methods, which naturally restructure the working group of CITY 2020+, can be summarized as (1) interviews and assessments of historical lines of development, (2) surveys and (3) applications of Geographical Information Systems (GIS) (Fig. 2).

2 Interviews and storylines of temporal evolution

Aachen and its surroundings are considered an appropriate testbed because it can be expected that climate change impacts and demographic changes match average changes to be expected in Germany. In particular, Aachen shows a natural setting that is typical for German middle-sized cities situated on the fringe of a low mountain range adjacent to lowlands. Using the example of Aachen, observed climate data and climate projections allow for an analysis of the temporal evolution of air temperature in Aachen.

Especially affected by heat waves are residents in urban districts (Kuttler and Weber, 2008), which differ strongly from their surroundings due to surface sealing, building structure and anthropogenic heat release. We only considered air temperature as a simple indicator for heat stress because findings are more easily understood by the general public and other stakeholders. Moreover, the uncertainty in climate projections increases with the number of input variables (Koppe et al., 2004). A more detailed investigation on heat stress with an integrated biometeorological assessment would require the consideration of physiologically relevant parameters such as relative humidity, radiation or wind speed in addition to air temperature (see also Gabriel and Endlicher, 2011; Matzarakis and De Rocco, 2009; Havenith, 2005; Koppe et al., 2004; Mayer, 1997).

Observed air temperature is obtained from a weather station operated by the German Weather Service. Since 1930, air temperature is available on a daily time scale. Outputs of the regional climate model STARII (Orlowsky et al., 2008) serve for future estimations of air temperature for Aachen. Findings based on these data sets project a significant temperature increase of approximately 1 K for 2031–2060 compared to the late 20th century. Statistical analyses reveal that a shift in mean temperature to higher values entails pronounced changes in the frequency of extreme events (Mearns et al., 1984; Wigley, 2009). Extreme events regarding anomalously high air temperatures such as heat waves form a major source of social and economic disruptions (Kamal-Chaoui and Robert, 2009; Kunkel et al., 1999; Koppe et al., 2004; Hübler et al., 2007; IPCC, 2007b). These impacts might aggravate in the course of the current century due to a continuous or even accelerated temperature rise. For Europe different studies suggest heat waves to become more intense, longer lasting and more frequent in future (Frich et al., 2002; Meehl and Tebaldi, 2004; IPCC, 2007a). This prediction applies to Aachen only partly. The average duration of a heat wave as well as its intensity remained nearly constant since 1931 and is not expected to change significantly in future. However, expected is a future change in the frequency of heat waves. Simulations show that a doubling of heat wave situations is likely to occur for the worst case scenario A2. In this study, a heat wave is defined as a sequence of consecutive days on which the daily maximum temperature does not fall below 25 °C and exceeds 30 °C on at least three days. As the model output suggests, the temperature rise of approximately 1 K until the middle of the century produces at least 50 % more heat days ($T_{max} \geq 30$ °C) in summer. In case of the highest-warming scenario (A2) the number of hot days more than doubles whereas a slightly stronger change can be observed for tropical nights ($T_{min} \geq 20$ °C). Besides an expected increase in heat wave episodes, high nighttime temperatures caused by restrained nocturnal cooling (Black et al., 2004) contribute to a strengthening of thermal stress situations and might lead to adverse health effects or even excess human mortality as little relief is experienced at night

(Koppe et al., 2004).

Archival and picture sources allow conclusions about the historical handling of inhabitants, policymakers and urban planners with environmental conditions and climate change, and a reconstruction of the perception of air and water pollution in the 19th century. An interdisciplinary study including historical maps, pictures and a digital elevation model including three-dimensional building structure examined historical changes in the conditions of cold air drainage flows in the southern part of Aachen. A considerable decrease of positive biometeorological effects of the cold air drainage flows was found in the course of urban development between 1810 and 2010. The decline can be ascribed to increased building density and built-up area. The nocturnal cooling effect for the city centre was modeled for an equinoctial situation with clear sky conditions and high temperatures (95 %-percentile). The reduction of the cooling effect is 2.7 K and can be estimated to 3.2 K for an extreme summer situation with 30 °C at sunset (Ketzler et al., 2010).

The willingness of local actors to counteract climate change in their different fields of operation is estimated on the basis of interviews and workshops by the subproject related to Governance aspects. The perception of local actors and their attitudes towards the implementation of new strategies are analysed. In summer 2010 about 15 semi-structured expert interviews were carried out. Key-players from public administration, politicians, entrepreneurs, public utilities, housing associations and other policy and decision maker are aware of the upcoming problems of climate change. In their opinion it is very necessary to act immediately. However, their own options and resources are partially limited. For example, for many property owners it is rarely possible to invest in their buildings e.g. regarding energy-saving and refurbishment. Many owners, e.g. private owners and small companies, are not able to finance the measures. In case of obvious savings things seem to become different, especially for large scale enterprises. In the light of high energy costs, companies are willing to invest in energy saving measures. Even if the effect is the same, ultimately, the main motivation is not climate protection but cost reduction (Klemme, 2011). Regarding the strategic development of settlements and housing policy, a paradigm shift towards a more sustainable land-use management cannot be deduced. Major changes of the key strategies can hardly be noticed. The strengthening of national policy frameworks and stronger involvement of all parts of the society appears to be necessary in order to face the upcoming challenges and create more sustainable city development strategies (Klemme and Selle, 2010).

In the subproject from Civil Engineering, priority is given to the improvement of the building envelope and technical facilities. Building simulations are carried out with altered parameters in order to complement existing guidelines for energy-saving construction methods combined with considerations for enhanced thermal comfort in summer. The physical buildings' characteristics, especially the thermal transmission coefficients, were given by specifications from the German Federal Ministry Traffic, Building and Urban Development (BMVBS, 2009). These parameters are the basis for the computer simulation with a program for indoor climate and energy simulation (IDA ICE 4, www.equa.se). The assessment is combined with different settings for ventilation aiming at night-time cooling.

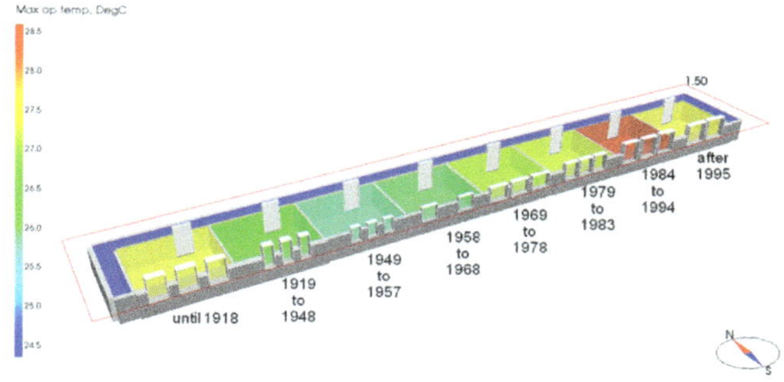

Figure 3. Maximum operative temperature depending on age and structure of buildings.

The simulations reveal that buildings built after 1958 include the most problematic structures in terms of hot season indoor climate. Older buildings contain more building mass with larger heat capacity to compensate for outdoor temperature increases. The more modern buildings incorporate much better insulation, aiming for energy savings during the winter periods. However, the trend toward much larger windows in modern buildings results in higher indoor temperatures in those buildings in summer (Fig. 3).

3 Surveys

Investigations on social networks, governance issues, building structure development and health outcomes make use of an extensive survey within the city of Aachen. To research the social situation and existing adaptation strategies concerning demographic and climate change, the fields of Cultural Geography, Sociology and Medicine conducted a comprehensive questionnaire that was sent to 8270 of Aachen's inhabitants aged 50 and older. With a response rate of 26 % ($N = 2180$ questionnaires), a comparison of Aachen's urban areas concerning for example percent of residents in older age categories, percent of chronically ill residents, the distribution of the working population and the levels social isolation in different regions is possible (Siuda et al., 2010). Other relevant topics such as the living situation, heat stress and the existence of individual adaptation strategies are also part of the survey. First results show that 70 % of the highly vulnerable group of respondents aged 80 and older has never received any information on heat stress or adaptation strategies by their general practitioner. The standard of the buildings people live in strongly influences the individual experience of heat stress, controlling for age and health status (BMVBS

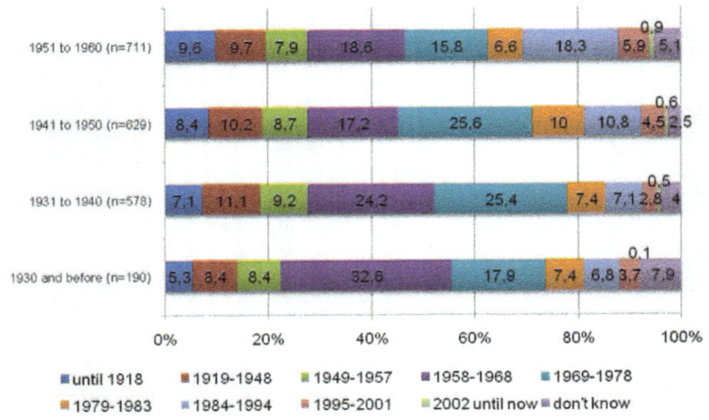

Figure 4. Year of construction of the residential buildings (coloured bars) the respondents live in, divided by age groups (y-axis representing age groups by ranges of year of birth); data from the survey on elderly people within the city of Aachen.

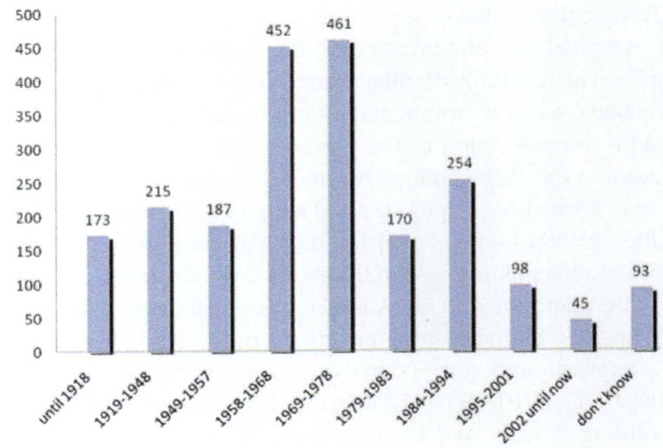

Figure 5. Numbers of buildings in different ages in Aachen.

and BBR, 2008). The younger respondents (age 50–65) are more likely to live in an old building (built 1918 or before) or a modern building (built 1984 or later). Respondents living in buildings that were built during these periods are also more satisfied with the indoor climate during summer time. The data of our survey indicate that especially the residents aged 80 and above tend to live in buildings that are not insulated properly (built between 1958–1978), which makes them even more vulnerable to heat stress (Fig. 4). In fact, most elderly people live in buildings built from 1958 to 1978 (Fig. 5). This type of building is identified through simulation (see above) to be the most problematic building type during heat waves. A straightforward possibility to reduce heat stress in summer is to exchange old windows for anti-sun glass. However, this measure may provoke problems with fungus infestation in winter in case other parts of the building are not modernized (Brunk et al., 2010a). An energy efficient refurbishment to solve problems in summer and winter is very expensive. The return on invested capital for houses built between the 1950's and 1980's will not be achieved for twenty years (Brunk et al., 2010b).

Beside the survey of the senior population the impact of climatic change on children is researched. In a cross-sectional study data on respiratory complaints, allergies and other health effects caused by air and noise pollutants are collected in the district of Aachen. Parents of about 5000 preschool children are asked to complete questionnaires during medical check-ups before school enrollment from September 2010 to June 2011. The identification of residential areas with a high proportion of vulnerable people is also relevant for the sub-projects that deal with the climatic conditions of cities. Results indicating spatial patterns can be incorporated into GIS-related approaches (see following section).

4 GIS-related approaches

As an information and evaluation tool for city planners and local authorities, maps are required which combine layers of various climatic and environmental information. Urban areas with enhanced negative health effects can therefore be detected and strategies for the abatement of these drawbacks can be developed (Ren et al., 2010).

In order to identify urban areas with a high risk exposure for various urban population groups in terms of health, measurement campaigns are carried out to capture the spatial distribution of temperature and PM concentrations in the city of Aachen (Fig. 6). Since only few permanent measurement sites exist, comprehensive information on the urban temperature and PM distribution is additionally obtained from mobile measurements. Mobile temperature measurements were initiated in March 2010 in cooperation with the municipal traffic company. The spatial and temporal pattern of the UHI and its possible enhancement by climate change is addressed employing a unique setup using GPS devices and temperature data loggers fixed to several public transport units running along different traverses across the city (Buttstädt et al., 2011). During scheduled public bus rides, temperature is recorded simultaneously throughout the day. This method provides a high spatial and temporal resolution, is cost-effective and time-saving, and can easily be transferred to other regions. Currently, temperature data from 15 measurement days are available. First analyses confirm the principally known influence of urban structures. While densely built-up areas show positive temperature anomalies with regard to temperature observations obtained from the suburban weather station Aachen-Hörn, lower temperatures occur mainly in areas covered by forest and suburban sites (Buttstädt et al., 2010). Additionally, the large data base allows detailed statistical analyses of urban structures and their temperature effect. Especially building area and building height prove to contribute to a positive

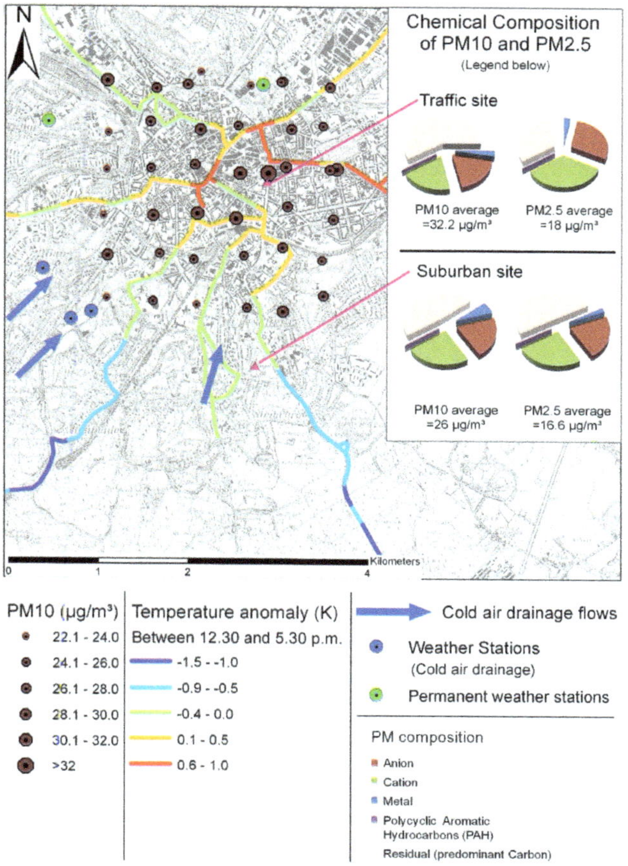

Figure 6. GIS-based investigations covering spatial PM and temperature distribution, cold air drainage flows and chemical composition of PM_{10} and $PM_{2.5}$ in the city of Aachen.

relation ($R^2 \approx 0.67$) between temperature and different spatial variables, indicated through multiple regression. The intra-urban temperature distribution and its influencing factors will be used to identify thermally affected urban districts.

A dense net of 40 urban particulate matter (PM) measurement sites is operated in order to obtain the spatial pattern of PM concentration depending on spatial variables and meteorological condition (Fig. 6). A connective study investigates the present spatial distribution of PM concentrations in Aachen. The population exposure to harmful particles is investigated by small and meso-scale measurements and geo-statistical approaches (Merbitz, 2010). Mobile measurements of PM concentrations are carried out within suburban, urban and inner city measurement campaigns since summer 2009. PM_{10} and $PM_{2.5}$ concentrations are measured at each measurement site as 1-minute averages over a total period of 10 min at a height of 1.6 m using an optical device (Environmental Dust Monitor 107, GRIMM, Germany). During a 7 month measurement campaign covering the period between September 2009 and March 2010 at each of the 40 sites between 18 and 20 measurements were carried out in order to obtain average PM_{10} and $PM_{2.5}$ distributions during the cold season. Small- and meso-scale spatio-temporal variability of PM_{10} and $PM_{2.5}$ are assessed and related to parameters such as traffic, building structure, topography, land use and meteorology. Especially for the coarse particle fraction ($PM_{10-2.5}$) a large spatial variability can be revealed, which is attributed to the influence of traffic related sources. These findings are in accordance to results of a study in Berlin showing large spatial variability for coarse particles while small particles ($PM_{2.5}$) showed more temporal than spatial variability, related to meteorological influences (Wolf-Benning et al., 2009). Results of mobile PM and temperature measurements suggest a strong dependence of temperature and PM on spatial variables, which will further be used as predictors for the identification of urban areas where both high temperatures and high levels of particulate air pollution coincide. It has been reported that PM has largest effects on mortality on hot days due to increased physiological stress and an altered physiological response to toxic agents (Roberts, 2004). In the inner city, especially the traffic density, building height and building density show significant positive correlations with PM levels and air temperature. Regarding the distribution of these parameters, areas with high risk of heat load and poor air quality are identified by means of GIS-based spatial analyses (Merbitz et al., 2011). Moreover, the urban wind field is investigated in order to identify areas with degraded ventilation conditions exacerbating the dilution of air pollutants (Merbitz et al., 2010). Distributions of PM and potential hot spots of heat stress over the total urban area are derived from the distribution of the spatial predictors applying a multiple land use regression analysis (Briggs et al., 1997, 2000; Clougherty et al., 2008; Hoek et al., 2008; Johnson et al., 2010). Models will be developed for the short-term forecast of PM levels and for the assessment of the future development of PM pollution in urban areas, considering meteorological influences (Merbitz, 2009) and climatic change.

Beside the meteorological variables, the content and composition of air pollutants continues to pose a significant threat to human health (WHO, 2006; Kinney, 2008). In this context the subproject Environmental Medicine characterizes the chemical composition and toxicological effects of gaseous and particulate airborne pollutants. This includes a full determination of inorganic and organic components, such as metals, salt, ions, total carbon (TC) and polycyclic aromatic hydrocarbons (PAH). The stationary PM10 and $PM_{2.5}$ sampling for the chemical analysis at the urban and suburban site was carried out according to DIN EN 12341 and DIN EN 14907. Analog to the geo-statistical analyses the chemical investigation reveals strong differences in the chemical composition amongst the sampling sites (Fig. 7). The main inorganic components of the airborne particles are Al, Ca, Fe, Zn, Mg, NO_3^-, SO_4^{2-}. Highest concentrations of the health relevant substance PAH were found at the traffic site. An influence of the particle size on the PAH level could not be detected. The

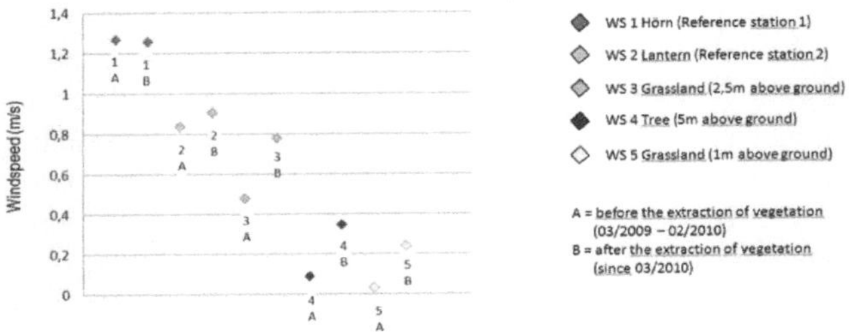

Figure 7. Average wind speed at different weather stations after the extraction of vegetation during periods with westerly winds at the reference station Aachen-Hörn (1) and a wind speed between 1 and $1.5\,m\,s^{-1}$ as a precondition for the existence of cold air drainage flow at the suburban valley stations (2–5).

combined effects of various components within PM need to be investigated in the future in more detail.

An important factor for the mitigation of PM pollution and heat is the flow of clean air, which in the city of Aachen is provided by cold air drainage, due to the valley situation. Concentrating on the diminution of thermal load situations, cold air drainage flows are identified and the amount and thickness of cold air are determined. In order to model cold air drainage flow, one sub-project focuses additionally on drawbacks that influence the flow characteristics (Sachsen et al., 2010). The analysis of the cold air drainage flow is investigated in two consecutive years with a clearing of vegetation stands in the beginning of the second year early in 2010. The investigation area in a southern suburban area of Aachen covers valleys with a very high relevance for the reduction of inner urban temperatures and air quality problems (Ketzler, 2002). Within two of these valleys, weather stations have been set up in order to obtain detailed information about the characteristics and behaviour of cold air drainage flows. The extraction of small vegetation stands at the measurement sites permit a closer look on how vegetation affects drainage flows in different heights. In order to compare the different data sets before and after the extraction a reference station (Weather station Aachen-Hörn) in Aachen is used. Measurements of wind speed and direction before and after the extraction indicate a considerable influence of porosity, height and formation on drainage flows. Figure 7 shows increased wind speeds at the weather stations in the valley within two measurement periods, one before and one after the extraction of vegetation. In many cases, areas of limited cold air influence are also affected by inferior air quality.

5 Conclusion and outlook

Cooperations between the fields of medicine, geography, sociology, history, civil engineering, and architecture provide the basis for proposing new strategies for adapting the city for future needs. The application of common methods concatenating different disciplines allow for a strong collaboration and better understanding of multidisciplinary interrelations. The results of the several fields are combined in maps to identify urban areas in which the population is extremely vulnerable regarding multiple stressors, e.g. air temperature, high level of PM and its composition, building structure and concentration of elderly persons or people that are socially isolated. While climate mitigation strategies are well established and accepted approaches, climate adaptation has not been adequately considered yet. There seem to be considerable conflicts of interests related to economic questions, the environment, and the expansion of under roof areas etc. Due to the complexity of the research topic it is conceivable to integrate other disciplines such as traffic planning, urban hydrology and economics in future research.

Acknowledgements. The Project CITY 2020+ is part of the interdisciplinary Project House HumTec (Human Technology Centre) at RWTH Aachen University funded by the Excellence Initiative of the German federal and state governments through the German Research Foundation (Deutsche Forschungsgemeinschaft, DFG).

Edited by: T. Cegnar

References

Black, E., Blackburn, M., Harrison, G., Hoskins, B., and Methven, J.: Factors contributing to the summer 2003 European heatwave, Weather, 59, 217–223, 2004.

BBR (Bundesamt für Bauwesen und Raumordnung): Perspektive Flächenkreislaufwirtschaft Kreislaufwirtschaft in der städtischen/stadtregionalen Flächennutzung – Fläche im Kreis. Ein ExWoSt-Forschungsfeld. Ergebnisse einer Akteursbefragung zur Flächenkreislaufwirtschaft in den Planspielregionen Stuttgart, Mölln, Rheinhessen-Nahe, Duisburg und Nordthüringen, Berlin, 2006a.

BBR (Bundesamt für Bauwesen und Raumordnung): Perspektive Flächenkreislaufwirtschaft. Theoretische Grundlagen und Planspielkonzeption. Band 1 der Sonderveröffentlichungsreihe zum ExWoSt-Forschungsfeld "Fläche im Kreis, Bonn, 2006b.

BMVBS (Bundesministerium für Verkehr, Bau und Stadtentwicklung): Bekanntmachung der Regeln zur Datenaufnahme und Datenverwendung im Wohngebäudebestand, Berlin, 2009.

BMVBS (Bundesministerium für Verkehr, Bau und Stadtentwicklung) and BBR (Bundesamt für Bauwesen und Raumordnung) (Eds.): Raumentwicklungsstrategien zum Klimawandel, Dokumentation der Fachtagung am 30. Oktober 2007 im Umweltforum Berlin, Bonn, Berlin, 2007.

BMVBS (Bundesministerium für Verkehr, Bau und Stadtentwicklung) and BBR (Bundesamt für Bauwesen und Raumordnung) (Eds.): Folgen des Klimawandels: Gebäude und Baupraxis in Deutschland, BBR-Online-Publication, 10, 2008.

Bochning, S. and Selle, K. (Eds.): Freiräume für die Stadt. Sozial und ökologisch orientierter Umbau von Stadt und Region, Bauverlag, Wiesbaden, 1993.

Briggs, D. J., Collins, S., Elliott, P., Fischer, P., Kingham, S., Lebret, E., Pryl, K., van Reeuwijk, H., Smallbone, K., and van der Veen, A.: Mapping urban air pollution using GIS: a regression-based approach, Int. J. Geogr. Inf. Sci., 11(7), 699–718, 1997.

Briggs, D. J., Hoogh, C. de, Gulliver, J., Wills, J., Elliott, P., Kingham, S., and Smallbone, K.: A regression-based method for mapping traffic-related air pollution application and testing in four contrasting urban environments, Sci. Total Environ., 253(1–3), 151–167, 2000.

Brunekreef, B. and Holgate, S. T.: Air pollution and health, Lancet, 360, 1233–1242, 2002.

Brunk, M., Dott, W., Balzer, C., Kröpelin, A., and Michael, S.: Entstehung von Schimmelpilzbelastung in Gebäuden in Abhängigkeit vom Lüftungsverhalten, Umweltmedizin in Forschung und Praxis, 15, 295–296, 2010a.

Brunk, M., Kröpelin, A., and Möllenhoff, N.: Energy efficient refurbishments - The necessity of political incentive systems, The REHVA European HVAC Journal, 47, 28–33, 2010b.

Bürkner, H.-J., Berger, O., Luchmann, C., and Tenz, E.: Der demographische Wandel und seine Konsequenzen für Wohnungsnachfrage, Städtebau und Flächennutzung, 2007.

Buttstädt, M., Sachsen, T., Ketzler, G. and Schneider, C.: Innerstädtische Temperaturverteilung in Aachen, Aachener Geographische Arbeiten, 47, 60–68, 2010.

Buttstädt, M., Sachsen, T., Ketzler, G., Merbitz, H., and Schneider, C.: A new approach for highly resolved air temperature measurements in urban areas, Atmos. Meas. Tech. Discuss., 4, 1001–1019, doi:10.5194/amtd-4-1001-2011, 2011.

Clougherty, J. E., Wright, R. J., Baxter L. K., and Levy, J. I.: Land use regression modeling of intra-urban residential variability in multiple traffic-related air pollutants, Environ. Health, 7–17, doi:10.1186/1476-069X-7-17, 2008.

Dagher, Z., Garçon, G., Gosset, P., Ledoux, F., Surpateanu, G., Courcot, D., Aboukais, A., Puskaric, E., and Shirali, P.: Pro-inflammatory effects of Dubkerque city air pollution particulate matter 2.5 in human epithelial lung cells (L132) in culture, J. Appl. Toxicol., 25, 166–175, 2005.

DIFU (Deutsches Institut für Urbanistik) (Eds.): Kommunen auf dem Weg zur Nachhaltigkeit, Köln, 2004.

DIN EN 12341: Air quality – Determination of the PM10 fraction of suspended particulate matter – Reference method and field test procedure to demonstrate reference equivalence of measurement methods, German version EN 12341, 1998.

DIN EN 14907: Ambient air quality – Standard gravimetric measurement method for the determination of the PM2.5 mass fraction of suspended particulate matter, German version EN 14907, 2005.

Emonds, H.: Klimatologische Beurteilungsgrundlagen zur Berücksichtigung der Luftreinhaltung bei der städtebaulichen Planung in Tallagen, Aachen, 1986.

Frich, P., Alexander, L. V., Della-Marta, P. M., Gleason, G., Haylock, M., Klein Tank, A. M. G., and Peterson, T.: Observed coherent changes in climatic extremes during the second half of the twentieth century, Clim. Res., 19, 193–212, 2002.

Gabriel, K. and Endlicher, W.: Urban and rural mortality during heat waves in Berlin and Brandenburg, Germany, Environ. Pollut., 159, 2044–2050, doi:10.1016/j.envpol.2011.01.016, 2011.

Grübler, A.: Technology and Global Change, Cambridge, 1998.

Haines, A., Kovats, R. S., Campbell-Lendrum, D., and Corvalan, C.: Climate Change and human health: impacts, vulnerability, and mitigation, Lancet, 367, 2101–2109, 2006.

Hall, P. and Pfeiffer, U.: Urban 21. Der Expertenbericht zur Zukunft der Städte, Stuttgart, München, 2000.

Havenith, G.: Temperature regulation, heat balance and climatic stress, in: Extreme weather events and public health responses, edited by: Kirch, W., Menne, B., and Bertollini, R., 69–80, 2005.

Havlik, D.: Großstädtische Wärmeinsel und Gewitterbildung – ein Beispiel anthropogener Klimamodifikation, Aachener Geographische Arbeiten, 14, 91–122, 1981.

Havlik, D.: Das Klima von Aachen, Aachener Geographische Arbeiten, 36, 1–20, 2002.

Havlik, D. and Ketzler, G.: Gesamtstädtisches Klimagutachten Aachen, Aachen, 2000.

Heiland, S.: Der Klimawandel als Herausforderung für die Landschaftsplanung. Probleme, Fragen und Lösungsansätze, Naturschutz und Landschaftsplanung, 40(2), 37–41, 2008.

Hetland, R. B., Cassee, F. R., Refsnes, M., Schwarze, P. E., Lag, M., Boere, A. J. F., and Dybing, E.: Release of inflammatory cytokines, cell toxicity and apoptosis in epithelial lung cells after exposure to ambient air particles of different size fractions, Toxicol. In Vitro, 18, 203–218, 2004.

Hoek, G., Beelen, R., Hoogh, K. de, Vienneau, D., Gulliver, J., Fischer, P., and Briggs, D.: A review of land-use regression models to assess spatial variation of outdoor air pollution, Atmos. Environ., 42(33), 7561–7578, 2008.

Hübler, M., Klepper, G., and Peterson, S.: Costs of Climate Change. The effects of rising temperatures on health and productivity in Germany, Kiel Working Paper 1321, 2007.

Ingendahl, B. and Thieme, I.: Gesundheitliche Effekte des Klimawandels – Aufklärung und Bewusstseinsbildung als wesentlicher Bestandteil von Anpassungsmaßnahmen, UMID-Themenheft Klimawandel und Gesundheit, 3, 5–7, 2009.

IPCC: Climate Change 2007: The Physical Science Basis, Contribution of Working Group I to the Fourth Assessment Report of the Intergovernmental Panel on Climate Change, edited by: Solomon, S., Qin, D., Manning, M., Chen, Z., Marquis, M., Averyt, K. B., Tignor, M., and Miller, H. L., Cambridge University Press, Cambridge, United Kingdom, New York, 2007a.

IPCC: Climate Change 2007: Impacts, Adaptation and Vulnerability, Contribution of Working Group II to the Fourth Assessment Report of the Intergovernmental Panel on Climate Change, edited by: Parry, M. L., Canziani, O. F., Palutikof, J. P., van der Linden, P. J., and Hanson, C. E., Cambridge University Press, Cambridge, 2007b.

Jendritzky, G.: Einwirkungen von Klimaänderungen auf die Gesundheit des Menschen in Mitteleuropa, in: Klimastatusbericht 1998, edited by: DWD (Deutscher Wetterdienst), 7–17, 1998.

Jendritzky, G.: Folgen des Klimawandels für die Gesundheit, in: Der Klimawandel – Einblicke, Rückblicke, Ausblicke, edited by: Endlicher, W. and Gerstengarbe, F.-W., Berlin/Potsdam, 108–119, 2007.

Johnson, M., Isakov, V., Touma, J. S., Mukerjee, S., and Ozkaynak, H.: Evaluation of land-use regression models used to predict air quality concentrations in an urban area, Atmos. Environ., 44(30),

3660–3668, 2010.

Kamal-Chaoui, L. and Robert, A. (Eds): Competitive cities and climate change, OECD Regional Development Working Papers, 2, 2009.

Ketzler, G.: Vergleichende Untersuchungen der Temperaturverhältnisse in Städten unter besonderer Berücksichtigung der Temperaturänderungsraten, Aachener Geographische Arbeiten, 31, 234 pp., 1997.

Ketzler, G.: Untersuchungen an Kaltluftströmen in kleinen stadtnahen Tälern, Aachener Geographische Arbeiten, 36, 21–64, 2002.

Ketzler, G., Eßer, K., and Sachsen, T.: Veränderung von Belüftungsfunktionen durch lokalen Kaltluftabfluss im Zuge historischer Stadtentwicklung – eine Modellstudie Aachen 1810–2010, Aachener Geographische Arbeiten, 47, 77–98, 2010.

Kinney, P. L.: Climate Change, Air Quality, and Human Health, Am. J. Prev. Med., 35(5), 459–467, 2008.

Klemme, M.: Wohnflächenentwicklung unter veränderten Rahmenbedingungen: Öffentliche Akteure zwischen Neubau und Bestand, in: Bestand? Perspektiven für das Wohnen in der Stadt, edited by: Schmitt, G. and Selle, K., Dortmund, 2008.

Klemme, M.: Akteure der lokalen Klimapolitik im Spannungsfeld von Persitenzen und Eigenlogiken, in: Anpassung an den Klimawandel – regional umsetzen!, edited by: Frommer, B., Buchholz, F., and Böhm, H. R., oekom-Verlag, München, 121–138, 2011.

Klemme, M. and Selle, K.: Gutes Klima? Klimawandel und lokale Klimapolitik aus der Governance-Perspektive. Ein Dialog in Aachen, PT_Materialien 25, 2010.

Koppe, C., Jendritzky, G., and Pfaff, G.: Die Auswirkungen der Hitzewelle 2003 auf die Gesundheit, in: Klimastatusbericht 2003, edited by: DWD (Deutscher Wetterdienst), 152–162, 2003.

Koppe, C., Kovats, S., Jendritzky, G. and Menne, B.: Heat waves: Risks and responses, Copenhagen, 2004.

Kunkel, K. E., Pielke, R. A., and Changnon, S. A.: Temporal fluctuations in weather and climate extremes that cause economic and human health impacts: a review, B. Am. Meteorol. Soc., 80(6), 1077–1098, 1999.

Kuttler, W. and Weber, S.: Urban climate and global climate change – a case study of the 'Ruhr area', Germany, Berichte des Meteorologischen Instituts der Universität Freiburg, 18, 313–319, 2008.

Luber, G. and McGeehin, M.: Climate change and extreme heat events, Am. J. Prev. Med., 35(5), 429–435, 2008.

Malburg-Graf, B., Renn, O., and Ulmer, F.: Strategien und Instrumente zur Reduzierung des Flächenverbrauchs – Ergebnisse des Ziel-30-ha-Projektes des Rates für Nachhaltige Entwicklung, in: Flächenmanagement als Instrument der integrativen Planung für ländliche Räume und der kommunalen Innenentwicklung – Beiträge des 2. Hochschultages 2007 "Strukturentwicklung ländlicher Raum in Baden-Württemberg", edited by: Malburg-Graf, B., Stuttgarter Geographische Studien, 140, 85–99, 2007.

Matzarakis, A. and De Rocco, M.: Thermal bioclimate in Strasbourg – the 2003 heat wave, Theor. Appl. Climatol., 98, 209–220, 2009.

Mayer, H.: Biometeorologische Auswirkungen des Stadtklimas, VDI Berichte, 1330, 109–126, 1997.

Mearns, L. O., Katz, R. W., and Schneider, H.: Extreme high-temperature events: changes in their probabilities with changes in mean temperature, J. Clim. Appl. Meteorol., 23, 1601–1613, 1984.

Meehl, G. and Tebaldi, C.: More intense, more frequent, and longer lasting heat waves in the 21st century, Nature, 305, 994–997, 2004.

Merbitz, H.: Feinstaubkonzentration in Abhängigkeit von Witterung und Großwetterlagen im Raum Aachen, Aachener Geographische Arbeiten, 46, 121–134, 2009.

Merbitz, H.: Feinstaubbelastung in Aachen und in der Euregio Maas-Rhein, Aachener Geographische Arbeiten, 47, 48–59, 2010.

Merbitz, H., Ketzler, G., and Schneider, C.: Untersuchungen zu den Windverhältnissen im Innenstadtbereich von Aachen, Aachener Geographische Arbeiten, 47, 99–118, 2010.

Merbitz, H., Buttstädt, M., Michael, S., Dott, W., and Schneider, C.: GIS-based identification of spatial variables enhancing heat and poor air quality in urban areas, Appl. Geogr., doi:10.1016/j.apgeog.2011.06.008, 2011.

Monn, C., Carabias, V., Junker, M., Waeber, R., Karrer, M., and Wanner, H. U.: Small-scale spatial variability of particulate matter <10 μm (PM10) and nitrogen dioxide, Atmos. Environ., 31(15), 2243–2247, 1997.

Mücke, H.-G., Klasen, J. Schmoll, O., and Szewzyk, R.: Gesundheitliche Anpassung an den Klimawandel, Umweltbundesamt, Deutscher Wetterdienst (Eds.), Dessau-Roßlau, 2009.

Oke, T. R.: Boundary layer climates, Routledge, New York, 1987.

Orlowsky, B., Gerstengarbe, F.-W., and Werner, P. C.: A resampling scheme for regional climate simulations and its performance compared to a dynamical RCM, Theor. Appl. Meteorol., 92, 209–223, 2008.

Osenberg, H.: Raumordnung als Teil einer nationalen Anpassungsstrategie an den Klimawandel, in: Raumentwicklungsstrategien zum Klimawandel, Dokumentation der Fachtagung am 30. Oktober 2007 im Umweltforum Berlin, edited by: BMVBS (Bundesministerium für Verkehr, Bau und Stadtentwicklung) and BBR (Bundesamt für Bauwesen und Raumordnung), Bonn, Berlin, 7, 2007.

Pfaff, G., Koppe, C., and Jendritzky, G.: Hitzewellen und extreme Klimaereignisse – Herausforderungen für das Gesundheitswesen, Epidemiologisches Bulletin, 25, 200-204, 2004.

Ren, C., Ng, E. Y., and Katzschner, L.: Urban climatic map studies: a review, Int. J. Climatol., doi:10.1002/joc.2237, 2010.

Roberts, S.: Interactions between particulate air pollution and temperature in air pollution mortality time series studies, Environ. Res., 96(3), 328–337, 2004.

Sachsen, T., Ketzler, G., and Käsmacher, O.: Stadt- und geländeklimatologische Untersuchungen in Aachen und Umgebung, Aachener Geographische Arbeiten, 47, 39–47, 2010.

Semanza, J. C., Rubin, C. H., Falter, K. H., Selanikio, J. D., Flanders, W. D., Howe, H. L., and Wilhelm, J. L.: Heat-Related Deaths During the July 1995 Heat Wave in Chicago, New Engl. J. Med., 335(2), 84–90, 1996.

Siuda, A., Hahmann, J., Hofmeister, H., and Pfaffenbach, C.: Hitzegefährdete Bevölkerungsgruppen in Aachen, Aachener Geographische Arbeiten, 47, 69–76, 2010.

Sterns, H. L.: The Aging Worker in a Changing Environment: Organizational and Individual Issues, J. Vocat. Behav., 47, 248–268, 1995.

Taha, H.: Urban climates and heat islands. albedo, evapotranspiration, and anthropogenic heat, Energ. Buildings, 25, 99–103, 1997.

Thornburg, J, Rodes, C. E., Lawless, P. A., and Williams, R.: Spatial and temporal variability of outdoor coarse particulate matter mass concentrations measured with a new coarse particle sampler during the Detroit Exposure and Aerosol Research Study, Atmos. Environ., 43(28), 4251–4258, 2009.

UBA (Umweltbundesamt) and DWD (Deutscher Wetterdienst): Klimawandel und Gesundheit, Informationen zu gesundheitlichen Auswirkungen sommerlicher Hitzewellen und Tipps zum vorbeugenden Gesundheitsschutz, Dessau-Roßlau, 2008.

UN World Population Prospects (Population Division of the Department of Economic and Social Affairs of the United Nations Secretariat, World Population Prospects): The 2009 Revision, http://esa.un.org/unpd/wpp/unpp/p2k0data.asp, 2009.

Weber, S.: Spatio-temporal covariation of urban particle number concentration and ambient noise, Atmos. Environ., 43(34), 5518–5525, 2009.

Wigley, T. M. L.: The effect of changing climate on the frequency of absolute extreme events, Clim. Change, 97, 67–76, 2009.

Wolf-Benning, U., Draheim, T., and Endlicher, W.: Spatial and temporal differences of particulate matter in Berlin, Special Issue on Urban Air Pollution: Problems, Control Technologies and Management Practices, International Journal of Environment and Waste Management, 4(1/2), 3–16, 2009.

WHO (World Health Organisation): WHO Air quality guidelines for particulate matter, ozone, nitrogen dioxide and sulfur dioxide, 2006.

Yang, W. and Omaye, S. T.: Air pollutants, oxidative stress and human health, Mutat. Res.-Gen. Tox. En., 674, 45–54, 2009.

Zebisch, M., Grothmann, T., Schröter, D., Hasse, C., Fritsch, U., and Cramer, W.: Klimawandel in Deutschland. Vulnerabilität und Anpassungsstrategien klimasensitiver Systeme, Dessau, 2005.

7

An application of ensemble/multi model approach for wind power production forecasting

S. Alessandrini[1], P. Pinson[2], R. Hagedorn[2], G. Decimi[1], and S. Sperati[1]

[1]RSE, Ricerca Sistema Energetico, Milano, Italy
[2]ECMWF, Reading, UK

Abstract. The wind power forecasts of the 3 days ahead period are becoming always more useful and important in reducing the problem of grid integration and energy price trading due to the increasing wind power penetration. Therefore it's clear that the accuracy of this forecast is one of the most important requirements for a successful application. The wind power forecast applied in this study is based on meteorological models that provide the 3 days ahead wind data. A Model Output Statistic correction is then performed to reduce systematic error caused, for instance, by a wrong representation of surface roughness or topography in the meteorological models. For this purpose a training of a Neural Network (NN) to link directly the forecasted meteorological data and the power data has been performed. One wind farm has been examined located in a mountain area in the south of Italy (Sicily). First we compare the performances of a prediction based on meteorological data coming from a single model with those obtained by the combination of models (RAMS, ECMWF deterministic, LAMI). It is shown that the multi models approach reduces the day-ahead normalized RMSE forecast error (normalized by nominal power) of at least 1% compared to the singles models approach. Finally we have focused on the possibility of using the ensemble model system (EPS by ECMWF) to estimate the hourly, three days ahead, power forecast accuracy. Contingency diagram between RMSE of the deterministic power forecast and the ensemble members spread of wind forecast have been produced. From this first analysis it seems that ensemble spread could be used as an indicator of the forecast's accuracy at least for the first three days ahead period.

1 Introduction

The accuracy of short-term wind power forecasts is very important to facilitate the integration of wind power into the grid. Especially in area such the south part of Italy where a strong wind power production is associated with a poor development of the grid, not originally built to support such a kind of distributed input. Also in the day ahead market more accurate forecasts may increase the value of wind energy compared to that of more conventional sources. In this contest can become important an information about the prediction accuracy associated with a deterministic power forecast. In this work a multi model approach to improve the accuracy of deterministic forecast has been applied to a wind farm. Furthermore, considering that the more common wind power forecast systems for the three days ahead period are based on meteorological models and that a big part of systematic errors are removed by MOS (Model Output Statistics) techniques, it makes sense that the forecast error is related to the atmospheric predictability condition. In this work we report about usage of the member-spread of the ECMWF ensemble model system for prediction of deterministic forecasts accuracy.

2 Site and wind data description

The wind farm called Valledolmo is located in Sicily in a mountain area of the south Italy, with 9 equal turbines for a total 7.65 Mw of nominal power. Time series power data for the singles turbines are available for the year 2008 together with wind data measured at a height of 50 m a.g.l. (the same as the hubs) by an anemometer located inside the wind park. To increase the number of valid power data an average value over the working turbines has been computed for each hour and then considered for the following application.

Table 1. Statistical indexes for hourly power forecast evaluation computed using measured data in Valledolmo wind farm. RMSE and BIAS are normalized by the nominal power (NP).

	Index	Day 1	Day 2	Day 3
Best model	RMSE	14.2%	15.1%	15.6%
	Correlation	0.76	0.72	0.69
	Bias	−0.02	−0.02	−0.004
Multimodel	RMSE	13.7%	14.6%	15.0%
	Correlation	0.77	0.73	0.71
	Bias	−0.001	−0.001	−0.001

3 The wind power forecast system

The wind power forecast system is built by a meteorological model that provides hourly wind forecast for the three days ahead period. The wind field is then adjusted, in order to reduce systematic errors, by a MOS performed using a Neural Network (NN). Historical time series of measured power or wind data and forecasted wind data are used to train the NN. In this study three meteorological models have been used: RAMS (4 km resolution) (Pielke et al., 1992), LAMI (7 km resolution, http://www.arpa.emr.it/SMR/archivio/downloads/ambiente/report_lm_verif_fv.pdf), and ECMWF (25 km) deterministic forecast model. Both RAMS and LAMI use the ECMWF deterministic forecast fields as boundary conditions.

4 The multi-model application

Three independent hourly power forecasts have been performed using the three meteorological models (above mentioned) for the Valledolmo wind farm. The first 6 months have been used as a training period for three independent NN that link directly the forecasted wind with the power data. The three independent power forecasts are then regressed towards the measured power data during the training period, defining the coefficient of the linear combination to be used during the test period for the multi-model prediction. In Table 1 some statistical indexes comparing the best (between the three used) models performances with the multi model are reported. The multi-model outperforms the best model gaining 1–2% of RMSE/NP, i.e. it gains quite one day of predictability. It is also interesting to underline that, comparing the three separate power forecasts (not shown in this paper), an increase in resolution doesn't necessary assures better performances. An explanation is that, for this wind farm, the synoptic winds are more often involved in power production rather than local thermal winds, for instance. The large scale circulation is well forecasted even with a horizontal resolution of 25 km and the attempt of forecasting smaller scales flows can be seen sometimes as a "noise" signal that decreases the NN efficacy.

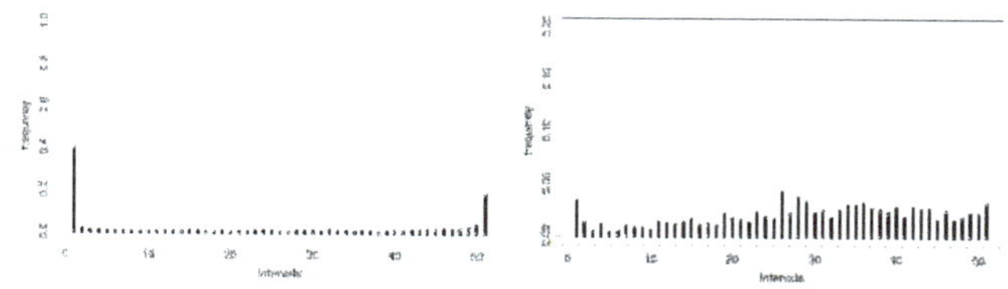

Figure 1. Rank histogram of ensemble wind forecast before (left) and after (right) the recalibration procedure. The frequency of measurements occurrences (Y-axis) between the 51 members intervals (X-axis) is reported.

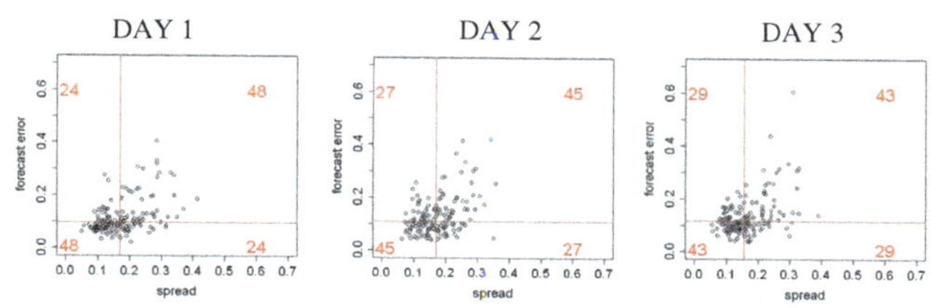

Figure 2. ECMWF deterministic power forecast error (RMSE/NP) vs ensemble power spread computed as a daily average for three lead times. The lines are plotted on the median. The red numbers indicate the points inside each square.

5 The ensemble model application

The operational initial perturbations of the ECMWF EPS are constructed using singular vectors with maximum total energy growth. The 51 ensemble wind forecasts of the ECMWF ensemble model system are processed by a MOS trained on the control run during the training period. Ensemble forecasts of wind speed are probabilistically recalibrated. The method employed consists of: (i) taking the log-transform of wind speed forecasts so that they are closer to Gaussian; (ii) adaptively correcting the mean and variance of the transformed wind speed forecasts; and (iii) transform them back by using the inverse of the log-transform. A generalization of this method is described by Pinson (2011) for the case of ensemble forecasts of (u, v)-winds. In Fig. 1 the rank histogram before the recalibration shows an overconfident model (the measurements fall quite always outside the members spread therefore the first and the last intervals are the most populated) that is improved by the recalibration procedure, obtaining a more uniform distribution among the 51 intervals. The recalibrated wind data are used to compute an ensemble power prediction using the nominal power curve of the turbine. In Fig. 2 some plots similar to those proposed by Von Bremen (2007) are shown. The daily RMSE/NP of ECMWF deterministic forecast are plotted against the power spreads. The diagonal cases are more populated than the off-diagonal cases therefore we can say that low spread of the ensemble is more often related to low forecast errors (lower left quadrant) than to large forecast errors (upper left quadrant).

6 Conclusions

In this application the use of a multi model approach produces a more accurate deterministic power forecast. At the same time it seems that ensemble power spread could be used as an indicator of the forecast's accuracy at least for the first three days ahead period.

Acknowledgements. This work has been financed by the Research Fund for the Italian Electrical System under the Contract Agreement between RSE (formerly known as ERSE) and the Ministry of Economic Development – General Directorate for Nuclear Energy, Renewable Energy and Energy Efficiency stipulated on 29 July 2009 in compliance with the Decree of 19 March 2009. This work was partly funded by the EU project SafeWind (EU-FP7, Grant Agreement No. 213740).

Edited by: S.-E. Gryning

References

Von Bremen, L.: Combination of deterministic and probabilistic meteorological models to enhance wind farm forecast, J. Phys. Conf. Ser., 75, 1–8, 2007.

Pielke, R. A., Cotton, W. R., Walko, R. L., Tremback, C. J., Lyons, W. A., Grasso, L. D., Nicholls, M. E., Moran, M. D., Wesley, D. A., Lee, T. J., and Copeland, J. H.: A Comprehensive Meteorological Modeling System – RAMS, Meteorol. Atmos. Phys., 49, 69–91, 1992.

Pinson, P.: Adaptive recalibration of ensemble forecasts of (u,v)-winds, Technical report, ECMWF, 2011.

8

Estimating urban heat island effects on near-surface air temperature records of Uccle (Brussels, Belgium): an observational and modeling study

R. Hamdi and H. Van de Vyver

Royal Meteorological Institute, 3 Avenue Circulaire, 1180 Brussels, Belgium

Abstract. In this letter, the Brussels's urban heat island (UHI) effect on the near-surface air temperature time series of Uccle (the national suburban recording station of the Royal Meteorological Institute of Belgium) was estimated between 1955 and 2006 during the summer months. The UHI of Brussels was estimated using both ground-based weather stations and remote sensing imagery combined with a land surface scheme that includes a state-of-the-art urban parameterization, the Town Energy Balance scheme. Analysis of urban warming based on the remote sensing method reveals that the urban bias on minimum air temperature is rising at a higher rate, 2.5 times (2.85 ground-based observed) more, than on maximum temperature, with a linear trend of 0.15 °C (0.19 °C ground-based observed) and 0.06 °C (0.06 °C ground-based observed) per decade respectively. The summer-mean urban bias on the mean air temperature is 0.8 °C (0.9 °C ground-based observed). The results based on remote sensing imagery are compatible with estimates of urban warming based on weather stations. Therefore, the technique presented in this work is a useful tool in estimating the urban heat island contamination in long time series, countering the drawbacks of an ground-observational approach.

1 Introduction

It is important to know whether, and to what extent, estimates of global warming trends can be explained by the growth of the urban heat island (UHI) due to increased urbanization. In fact, if observations of near-surface air temperatures in growing cities are used in the assessment of global warming trends, these trends may be overestimated. However, the change in urbanization over time is smaller for a station that originally was established in a densely built-up area than for a station originally installed in a rural or little urbanized environment that has experienced growth. Jones et al. (2008) have shown that near-surface air temperatures in central London and Vienna did not rise relative to rural locations nearby in recent decades. Nevertheless, suburban sites continue to warm relative to nearby rural areas until local urbanization is complete, as shown for London's Heathrow airport by Jones and Lister (2009).

The past observational approach compared urban near-surface air temperature records with records of a rural area. However, selective use of rural sites, requires information (metadata) about the site and its surroundings. Some forms of metadata, such as city population statistics, must be used with care because they may not be representative of the immediate vicinity of the observing site. Also, in situ observations usually suffer from inhomogeneities caused by non-climatic factors such as changes in observation time, instrumentations, location (altitude and latitude), and other local meteorological features.

With the advent of remote sensing, it has become possible to monitor local urban climate changes associated with land use changes over rapidly expanding urban areas. Specifically, the quantity of impervious surfaces is related to urban growth and urban density (Fricke and Wolff, 2002). The proportion of impervious surfaces has been reported to be a good indicator for the monitoring of the UHI. A positive correlation between the proportion of impervious surfaces and land surface temperatures was identified by Yuan and Bauer (2007) and many earlier studies, also the expansion of the built-up area was found to be the main factor in long-term changes in near-surface air temperatures (Huang et al., 2008; Shouraseni and Yuan, 2009).

Recently, Hamdi et al. (2009) presents a new technique for estimating effects of increased urbanization on near-surface air temperature trends. Specifically, they combine data from remote sensing imagery and a land surface model including state-of-the-art urban parameterization of Masson (2000), the Town Energy Balance (TEB) scheme. In their study: (i) the land surface model was coupled to a surface boundary layer scheme following the methodology described in Masson and Seity (2009) and evaluated over Basel, Switzerland in Hamdi and Masson (2008), (ii) urbanization was assessed by measuring changes of percent impervious surface areas, and (iii) the land surface scheme was run in a stand-alone mode, coupled to downscaled ERA-40 reanalysis data (Uppala et al., 2005). It should be noted that it was the first time a modeling approach was applied to assess the degree to which background near-surface air temperature trends are amplified by urbanization. This new technique was applied to the Brussels Capital Region (BCR) in Belgium, which has experienced a rapid conversion of agricultural land and native vegetation to buildings and impermeable pavements over the last century (Vanhuysse et al., 2006). However, in their study, no observed data were found to compare the results. Their model-based estimate of urban warming was based on calculating the difference between two model integrations: (i) "the rural" scenario representing a hypothetical situation with no urban areas inside the Brussels Capital Region domain and (ii) the "urban" scenario, which represented the climate in the presence of urban areas using the measured historical changes of surface cover fractions. Recently, a near-surface air temperature record (summer 1955–2006, minimum $T_{\min}$ and maximum $T_{\max}$ temperature) were found for a rural station far away about 20 km from Brussels (Van de Vyver, 2010). This station allows the increase of the UHI around Brussels to be determined. Thus, in this work, the UHI effects on the near-surface air temperature time series of Brussels during summer months is estimated using both ground-based weather stations and remote sensing imagery combined with a land surface scheme and the results compared. Therefore, the aim of this letter is to evaluate the technique presented in Hamdi et al. (2009).

2 Data and model

2.1 Study area

The focus of our study is the Brussels Capital Region (see Fig. 1), centrally located in Belgium, with a size of 161.78 km^2 and a registered population of 1 031 215 on 1 January 2007, estimated by the National Institute of Statistics (INS, 2009). As in the majority of large European cities, it is only during the 19th century that the population strongly increased, exceeding 1 million inhabitants in 2004. There has been a rapid expansion of the capital during the last 50 yr. The area covered by the BCR is circular (Fig. 1) with a diameter of 12 km and the effects of urbanization variability dominate the topographic influence (orography is beneath 150 m).

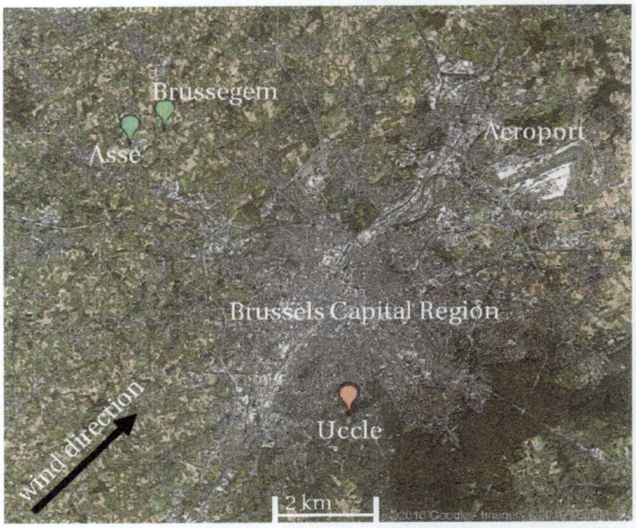

Figure 1. Study area of the Brussels Capital Region. The national recording station of the Royal Meteorological Institute of Belgium (Uccle, urban station) is marked with a red bulb, and rural stations (Asse and Brussegem) with a green bulb. The black arrow presents the direction of the prevailing wind direction in the region.

2.2 Meteorological weather station

The national recording station of the Royal Meteorological Institute (RMI) of Belgium (Uccle, World Meteorological Organization (WMO) code 06447) is situated some 6 km south of the center of the capital (Fig. 1), in a suburban area (50.80° N, 04.35° E). The series has been used to monitor climate change in Belgium and is in the world database of land-surface air temperature and in the Global Climate Observing System (GCOS) surface network (Peterson et al., 1997). In order to assess the degree to which the Uccle near-surface air temperature trends are amplified by urbanization two ground-based meteorological stations, situated 20 km far away from the center of Brussels and managed by the RMI, are used (see Fig. 1, Table 1). These rural stations are located outside the area influenced by the urban effect of Brussels (Van Weverberg et al., 2008, their Fig. 5) and are not influenced by the SW-prevailing wind direction. Therefore the impact of the city should be negligible. Though the UHI may not be zero in small villages (e.g. Fujibe, 2009), they did not experience a rapid growth during the last century and therefore the urban infrastructure should not have changed its influence on outgoing longwave radiation.

Table 1. Individual near-surface air temperature series used to estimate the urban warming of the city of Brussels, their observation hours, location and approximate elevation. No missing daily data were found in the datasets of the four climatological series.

Time span	Observation time	Location (above mean sea level)
1 June 1955–31 August 2006	Daily max- and minimum	Uccle (104 m)
1 June 1991–31 August 2006	Daily max- and minimum	Brussegem (~53 m)
1 June 1980–31 August 1990	Daily max- and minimum	Asse (~53 m)
1 June 1955–31 August 1971	Daily max- and minimum	Asse (~53 m)

2.3 Evolution of surface cover fraction

The evolution of surface cover fractions over the study region were derived from Vanhuysse et al. (2006). This study aims to assess the evolution of the fraction of impervious surfaces in the catchment basins including the BCR since the 1950s date of the acceleration of urban growth linked to widespread use of car as a new mode of transport. Two periods were studied:

1. From 1955 to 1985: Vanhuysse et al. (2006) used the MURBANDY database of topographic maps and areal photos (Fricke and Wolff, 2002) for 1955, 1970, and 1985, to estimate the fraction of impervious surfaces. The results indicate that impervious surfaces have increased from 26% in 1955 to 39% in 1985.

2. From 1985 to 2006: Vanhuysse et al. (2006), first, conducted a binary classification of land use (pervious and impervious) based on a very-high-resolution satellite image (UrbOrtho, QuickBird) for 2006. A grid 30 m resolution was used with the binary classification to extract the percentage of impervious surfaces. A simple regression model between the percentage of impervious surfaces and different spectral variables derived from high-resolution Landsat images was used to estimate the percentage of impervious surfaces for 1993 and 1986 for which only high-resolution images (Landsat Thematic Mapper with 30 m resolution) are available. The results indicate that impervious surfaces have increased from 39% in 1985 to 47% in 2006 (see Fig. 2).

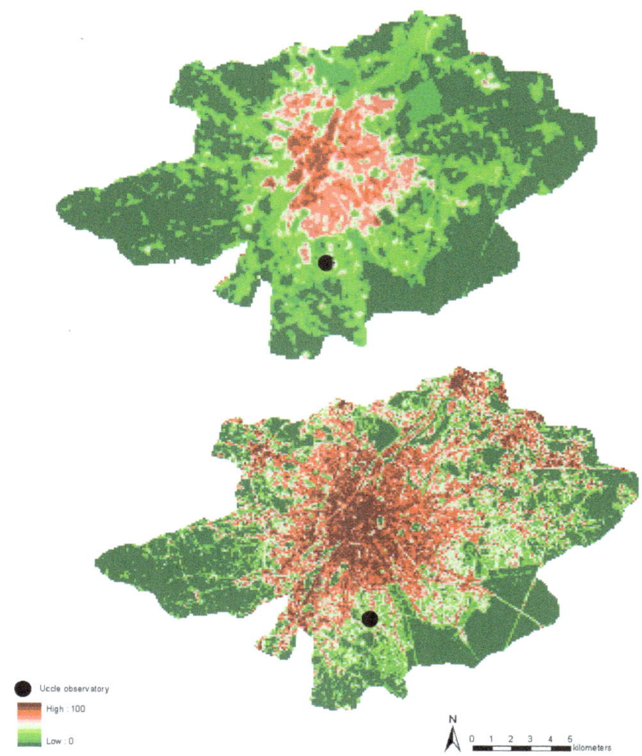

Figure 2. The percentage of impervious surfaces in the catchment basins including the Brussels Capital Region in 1955 (top) and 2006 (bottom). The black dot shows the Uccle observatory location.

2.4 Atmospheric data

The ERA-40 re-analysis (Uppala et al., 2005), produced by the European Centre for Medium-Range Weather Forecasts (ECMWF), covers the period September 1957 to August 2002 with a temporal resolution of 6 h and a spatial resolution of about 120 km for Western Europe. In order to increase the spatial resolution over Belgium, a dynamical downscaling method was applied using the numerical weather prediction limited area model ALADIN developed by the ALADIN International Team (1997).

A dynamical downscaling to 10 km resolution over Belgium is performed in two steps: (i) the ALADIN model is coupled to the ERA-40 data and run at a resolution of 40 km on a domain encompassing most of Western Europe, (ii) these results are then used as initial conditions and lateral boundary conditions for a second downscaling run at 10 km resolution on a smaller domain over Belgium (see Hamdi et al. (2009) for more details).

2.5 Land surface scheme

We use the offline surface scheme of Météo-France SURFEX (SURFace EXternalisée) (Le Moigne, 2009). In SURFEX, each grid box is made of four adjacent surfaces: nature, urban areas, sea or ocean, and lake. Horizontal interaction does not exist between the different surface area tiles. The coverage of each of these surfaces is known through the global ECO-CLIMAP database (Masson et al., 2003). During a model time step (300 s), each surface grid box receives the upper air temperature, specific humidity, wind speed, pressure, total precipitation, long-wave radiation, and short-wave radiation.

For urban surfaces, SURFEX uses the Town Energy Balance (TEB) (Masson, 2000) single-layer urban canopy module which assumes an isotropic array of street canyons. The advantage is that relatively few individual surface energy balance evaluations need to be resolved, radiation interactions are simplified, and therefore computational time is kept low. TEB simulates heat and water exchanges and climate of three generic surfaces (roof, wall, and road), where heat transfers are computed through several layers of materials, generally four. Anthropogenic heat and vapor releases from buildings, vehicles and chimneys can also be added. For vegetated tiles, Noilhan and Planton (1989) Interaction between Soil, Biosphere, and Atmosphere (ISBA) scheme is used. TEB is applied with literature-based surface thermal parameters and observed or simulated atmospheric and radiation data from above roof level. Despite the simplification, offline simulations of TEB have been shown to accurately reproduce surface energy balance, canyon air temperature, and surface temperatures observed in dense urban areas: Vancouver and Mexico City (Masson et al., 2002), Marseille (Lemonsu et al., 2004), Basel (Hamdi and Masson, 2008).

2.6 Model parameter

The domain is 10 km × 10 km over the BCR, centered on the city center of Brussels. This small heterogeneous domain is overlaid on a 1 km resolution land cover classification provided by the ECOCLIMAP database. The land cover types contained in this domain are then aggregated into 4 tiles (Sea, Lake, Vegetation, and Urban) with the corresponding fractional coverage (0%, 0%, 53%, 47%) to be used as the contemporaneous land cover setting. SURFEX is run in one offline single column mode from 1 June 1960 to 31 August 1999 (40 yr), and the forcing variables are derived from the downscaling of the ERA-40 (Sect. 2.4) for the grid point the closest to the BCR. For the vegetation tile, radiative, thermal, and soil properties (albedo, roughness length, emissivity, thermal inertia, leaf area index, etc.) are taken from the ECOCLIMAP database (Masson et al., 2003) and remain fixed through the simulation. For the urban tile, SURFEX uses only one urban land-use class as input. In this study, geometrical, thermal, and radiative properties of roofs, walls, and roads were set to values representing a typical midsize European city (see Table 1 in Hamdi et al., 2009). Another important urban-related aspect is the anthropogenic heat. This term includes all heat emitted by human activities: traffic, release from industry, and release from residential buildings. Over the area presented in this study, releases from buildings have been shown to be the dominant component of the anthropogenic heat (Van Weverberg et al., 2008). In SURFEX, to mimic space heating, a fixed minimum internal building temperature of 19 °C is specified.

3 Results and discussion

3.1 Urban warming from weather stations

3.1.1 Homogeneity testing of the rural series

In situ observations usually suffer from inhomogeneities caused by nonclimatic factors such as changes in observation time, instrumentation, location (altitude and latitude), and other meteorological features. Without assurance of homogeneity, trend estimates are unreliable and artifacts in long-term observations and rural/urban differences can be introduced and thus may bias the estimate of the UHI. For this reason, the assumption that the three rural subseries could be linked to constitute a reference rural series is tested with respect to homogeneity (details about the homogenization of the time series of Uccle can be found in Demarée et al. (2002)). Following the guidelines described in Wijngaard et al. (2003), the four methods selected to test the departure of homogeneity in the time series are: the standard normal homogeneity test for a single break, the Buishand range test, the Pettitt test, and the Von Neumann ratio test. The first three tests are location-specific tests which means that they are capable of locating the year when a break is likely. The mathematical formulation of the tests and tables with critical values for their test statistics are listed in detail in Wijngaard et al. (2003). The four tests are carried out on two testing variables which are characteristics of the diurnal near-surface air temperature range (DTR): its annual mean (mDTR) and its annual mean of the absolute day-to-day differences (vDTR). In our study, we considered the summer average. A particular test accepts the null hypothesis if and only if the two variables mDTR and vDTR pass the test at the 1% level. The results of the different tests are grouped into three classes: useful, doubtful, and suspect. The classification depends on the number of tests rejecting the null hypothesis (i.e. no breaks are present) at the 1% level: useful (at most one test), doubtful (two test), suspect (at least three tests). The critical values for the test statistics are known only for $n = 20$, 30, 40, 50, 70, 100. Since our series has 43 yr of data we need to apply the statistical tests to sliding blocks of 40 yr.

In Fig. 3, the results of the standard normal homogeneity test and the Pettitt test applied to the mDTR and vDTR series (1955–2002) of the three rural subseries are shown. While the results of the standard normal homogeneity test

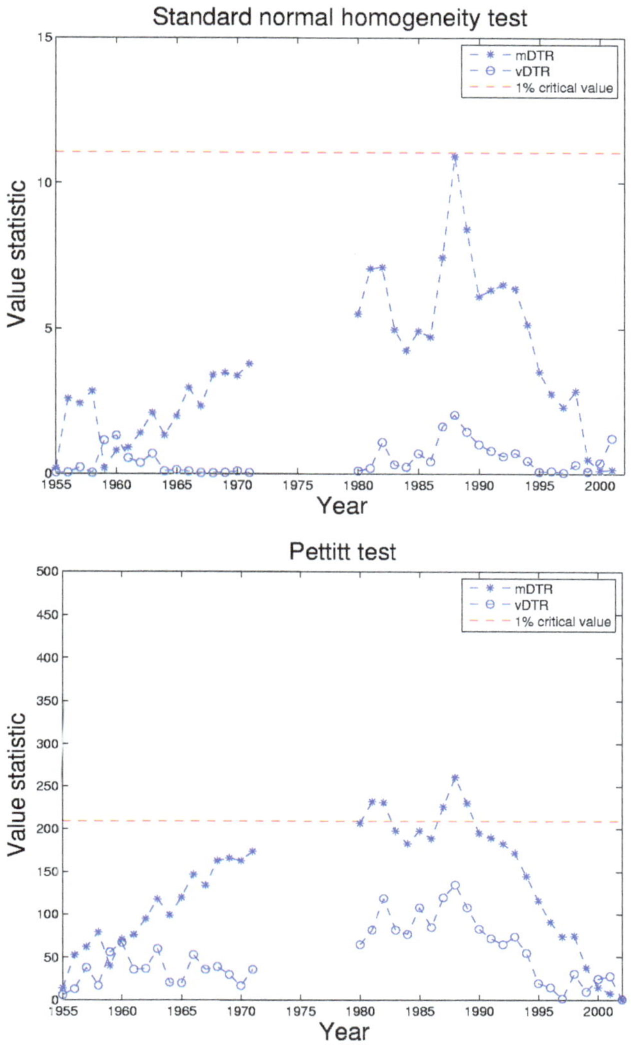

Figure 3. Results of the standard normal homogeneity test and Pettitt test applied to the 3 rural subseries (1955–2002, see Table 1).

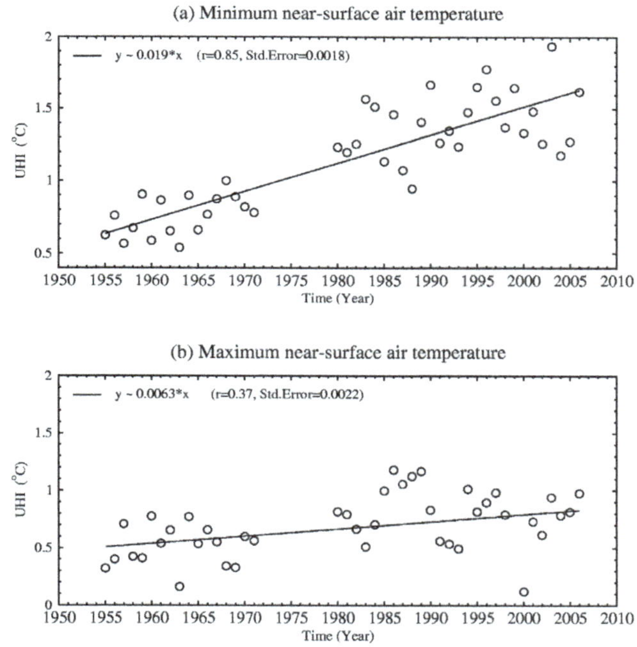

Figure 4. The summer-mean urban heat island (UHI) effect on **(a)** minimum and **(b)** maximum near-surface air temperature estimated as the difference between urban and rural ground-based weather stations, with the linear trend, 1955–2006. r is the correlation coefficient and Std. Error is the standard error of the regression coefficient (no data were found between 1972 and 1979).

did not exceed the 1% level, the null hypothesis of no breaks is not accepted by the Pettitt test for mDTR. This break point during the 80s is also observed in the long time series of Uccle (not shown) and is probably due to the fact that the last two decades of the 20th century were marked by very high yearly average near-surface air temperatures in Belgium. The statistical analysis of the Uccle temperature data (www.climat.be) distinguishes four periods. Between 1833 and 1909 the yearly average near-surface air temperature was 8.8 °C. Between 1910 and 1942 and between 1943 and 1983 the yearly average near-surface air temperature amounted to respectively 9.5 °C and 9.7 °C. From 1984 onwards, the yearly average is about 10.4 °C.

The values of the Von Neumann ratio test are 1.67 and 2.24 for mDTR and vDTR, respectively. The null hypothesis of no breaks is accepted when the test statistics is larger than the critical 1% value. For $n = 40$, the critical value is 1.29, which clearly indicates the homogeneity at the 1% level. Similarity, the Buishand range test indicates no homogeneity break in the series, the values are 1.62 and 0.85 for mDTR and vDTR, respectively. The null hypothesis of no breaks is accepted when the test statistics is lower than the critical 1% value. For $n = 40$, the critical value is 1.74, which clearly indicates the homogeneity at the 1% level. Since three tests accept the null hypothesis at the 1% level, the constructed reference rural series is assigned to class 1 "useful".

3.1.2 Urban warming

The UHI intensity is defined as the difference in near-surface air temperature between urban and rural stations. Estimates of urban bias at the Uccle recording station, on maximum and minimum near-surface air temperature, calculated during the summer of each year between 1955 and 2006, are plotted with the linear trends in Fig. 4 (unfortunately, no data were found between 1972 and 1979).

As indicated by Fig. 4, the UHI effect on minimum air temperature is shown to be rising at a higher rate (2.85 times more) than on maximum temperature, with a linear trend of 0.19 °C (±0.02 °C) and 0.06 °C (±0.02 °C) $(10\,\text{yr})^{-1}$, respectively. This result is consistent with previous work

Table 2. Performance statistics for daily maximum (T_{max}) and minimum (T_{min}) near-surface air temperature, obtained from the "urban" scenario, based on formulas by Wilmott (1982). $RMSE_{SYS}$ and $RMSE_{UNSYS}$ are the systematic and unsystematic root mean square error, respectively.

	Correlation	Index of agreement	Bias (°C)	$RMSE_{SYS}$ (°C)	$RMSE_{UNSYS}$ (°C)
T_{max}	0.96	0.97	1.12	1.38	2.33
T_{min}	0.96	0.97	−0.43	0.66	1.89

suggesting that the maximum air temperature is substantially less affected by urbanization than the minimum temperature (Landsberg, 1981; Karl et al., 1988; Kalnay and Cai, 2003; Hua et al., 2008). The increase of summer-mean urban bias on minimum temperature may be attributed to: (i) the higher thermal inertia, which, in combination with lower albedo of urban surfaces, delays the cooling of the cities at nights compared to rural areas (more explanation can be found in Hamdi and Schayes, 2008), (ii) to the limited evapotranspiration which prevents evaporative cooling of urban areas and (iii) during night hours, the contribution of anthropogenic heat can also influence long-term trend of near-surface air temperature.

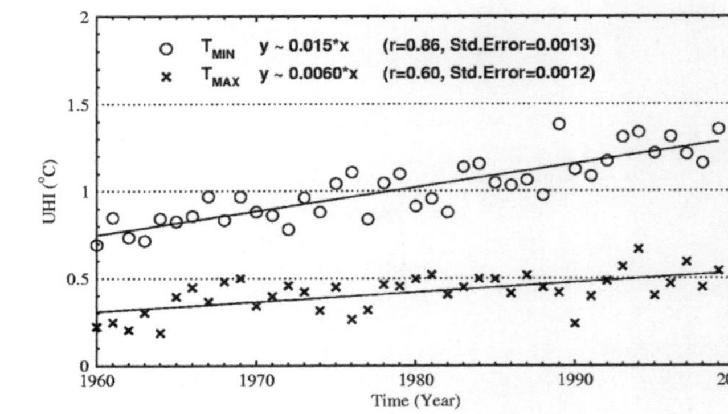

Figure 5. The summer-mean urban heat island (UHI) effect on minimum and maximum near-surface air temperature estimated as the difference between the "rural" and "urban" model integrations with the linear trend, 1960–1999. r is the correlation coefficient and Std. Error is the standard error of the regression coefficient.

3.2 Urban warming from model simulations

To isolate effects of urbanization on local near-surface climate conditions, we calculate the difference between two model integrations: (i) the "rural" scenario representing a hypothetical situation with no urban areas inside the Brussels Capital Region domain and (ii) the "urban" scenario, which represented the climate in the presence of urban areas using the measured historical changes of surface cover fractions. For this run, the surface cover fractions are updated each year using a linear interpolation. The use of the land surface model in an offline mode does not account for atmospheric feedback and therefore allows isolation of the effects of landscape difference on local near-surface climate conditions. The UHI effect is estimated as the difference between the "rural" and "urban" model integrations.

3.2.1 Evaluation of the model

To evaluate model performance, a comparison is made between the urban run and the routine observations of the Uccle ground station. Table 2 presents the statistics that quantify model performance for daily maximum (T_{max}) and minimum (T_{min}) air temperature using statistical measures (correlation coefficient, index of agreement, bias, and systematic and unsystematic root mean square error) based on formulas by Willmott (1982). The 2-m temperatures predicted by the urban run are consistent with meteorological observations, with a correlation coefficient of 0.96 and an index of agreement of 0.97 for T_{max} and T_{min}. The systematic root mean square error ($RMSE_{SYS}$) which should account for physical processes that the model does not routinely simulate well is small compared to the unsystematic root mean square error ($RMSE_{UNSYS}$). The proportions of systematic errors in the model are 26% and 11% for T_{max} and T_{min} respectively. The minimum temperature is better simulated by the model with a minor negative bias of −0.43 °C and $RMSE_{SYS}$ of 0.66 °C against 1.12 °C and 1.38 °C for maximum temperature. More details about the results of simulations in both scenarios can be found in Hamdi et al. (2009).

3.2.2 Urban warming

Model simulations (see Fig. 5) show that the UHI effect on near-surface minimum temperature is rising at higher rate (2.5 times, observed 2.85 more) than on maximum temperature, with a linear trend of 0.15 °C (±0.01 °C) (observed 0.19 °C ± 0.02 °C) and 0.06 °C (±0.01 °C) (observed 0.06 °C ± 0.02 °C) per decade, respectively. The summer-mean urban bias on the mean temperature ($T_{mean} = (T_{min} + T_{max})/2$) at Uccle is 0.8 °C (observed 0.9 °C). Based on 3-D modeling approach using the ARPS model, Van Weverberg et al. (2008) found a somewhat higher value. They estimated the summer-mean urban bias at the Uccle station to be 1.13 °C. However, the authors used a very simple parameterization for representing urban land and the model runs were restricted to a few weather episodes given the large computational costs of simulations. Vandiepenbeeck (1998) applied another observational approach to assess the average UHI contribution between 1966 and 1995. He studied the influence of urban heat advection on the near-surface air temperature time series of Uccle empirically by comparing the meteorological observation with those of a rural station far away from Brussels (St. Hubert, more than 100 km). He examined the dependence of the temperature differences on wind direction. He found the summer-mean urban heat contamination to be 0.8 °C which is close to the value found in this study.

The modeling approach used in this study is able to capture all the observed characteristics of the urban warming even if:

1. We consider BCR as a lumped urban volume (no horizontal interactions). In fact, wind direction can exacerbate the urban signal (Zhang et al., 2009). However, since we are interested in climate, the urban effects corresponding to the sum of all weather events can be represented by a circle around the city and the urban effects gradually decrease toward the outskirts as it is found by the 3-D modeling study of Van Weverberg et al. (2008) for Brussels.

2. The entire BCR is composed of one homogeneous material comprising uniform thermophysical properties, irrespective of spatial variability. In this study, only one urban land use class is used and the geometrical, thermal, and radiative properties of roofs, walls, and roads were set to values representing a typical midsize European city (Hamdi et al., 2009).

3. Urbanization was assessed only by measuring changes of the impervious surface area (percent) in the BCR.

4 Conclusions

In this study, the urban heat island (UHI) effects on the near-surface air temperature time series of Uccle during summer months was estimated using both ground-based weather stations and remote sensing imagery combined with a land surface scheme. The modeled estimate of urban warming was based on calculating the difference between two model integrations: (i) "the rural" scenario representing a hypothetical situation with no urban areas inside the Brussels Capital Region domain and the "urban" scenario, which represented the climate in the presence of urban areas using the measured changes of surface cover fractions. Results of our simulations is compatible with estimate of urban warming based on weather stations. However, exact compatibility is not to be expected because with the 10 km horizontal resolution of the climate drivers used to run the land surface scheme, we are not able to replicate the micrometeorology in the required detail.

This new technique combining data from remote sensing imagery and a land surface scheme presented in this study is a useful tool to estimate the urban heat island contamination in long time series, countering the drawbacks of an observational approach. It would be very simple and useful to applied this research method to other cities where credible rural ground weather stations do not exist. Furthermore, because of its local character, the results of this study will be particularly helpful for planners in developing scenarios for future land cover changes.

Acknowledgements. We thank S. Vanhuysse for providing us the data of the surface cover fraction. These data are obtained in the frame of the project 2008-02-AL. The authors are very grateful to G. Demarée for fruitful discussion about the history of weather stations and to two anonymous reviewers for comments that greatly improved the manuscript.

Edited by: M. Brunet-India

References

ALADIN International Team: The ALADIN project: Mesoscale modelling seen as a basic tool for weather forecasting and atmospheric research, WMO Bull., 46, 317–324, 1997.

Demarée, G. R., Lachaert, P. J., Verhoeve, T., and Thoen, E.: The long-term daily central Belgium temperature (CBT) series (1767–1998) and early instrumental meteorological observations in Belgium, Climatic Change, 53, 269–293, 2002.

Fricke, R. and Wolff, E.: The MURBANDY project: development of land use and network databases for the Brussels area (Belgium) using remote sensing and aerial photography, Int. J. Appl. Earth Observ. Geoinform., 4, 33–50, 2002.

Fujibe, F.: Detection of urban warming in recent temperature trends in Japan, Int. J. Climatol., 29, 1811–1822, 2009.

Hamdi, R. and Masson, V.: Inclusion of a drag approach in the Town Energy Balance (TEB) scheme: offline 1-D evaluation in a street canyon, J. Appl. Meteor. Clim., 47, 2627–2644, 2008.

Hamdi, R. and Schayes, G.: Sensitivity study of the urban heat island intensity to urban characteristics, Int. J. Climatol., 28, 973–982, 2008.

Hamdi, R., Deckmyn, A., Termonia, P., Demarée, G. R., Baguis, P., Vanhuysse, S., and Wolff, E.: Effects of historical urbanization in the Brussels Capital Region on surface air temperature time series: a model study, J. Appl. Meteorol. Clim., 48, 2181–2196, 2009.

Hua, L. J., Ma, Z. G., and Guo, W. D.: The impact of urbanization on air temperature across China, Theor. Appl. Climatol., 93, 179–194, 2008.

Huang, L., Lia, J., Zhaoa, D., and Zhub, J.: A fieldwork study on the diurnal changes of urban microclimate in four types of ground cover and urban heat island of Nanjing, China, Build. Environ., 43, 7–17, 2008.

Institut National de Statistique: Chiffres provisoires de la population résidente au 1er janvier, par année 1990–2009, SPF Economie – Direction Générale Statistique et Information Economique, 2009.

Jones, P. D. and Lister, D. H.: The urban heat island in central London and urban-related warming trends in Central London since 1900, Weather, 64, 323–327, 2009.

Jones, P. D., Lister, D. H., and Li, Q.: Urbanization effects in large-scale temperature records, with an emphasis on China, J. Geophys. Res., 113, D16122, doi:10.1029/2008JD009916, 2008.

Kalnay, E. and Cai, M.: Impact of urbanization and land use on climate change, Nature, 423, 528–531, 2003.

Karl, T. R., Diaz, H. F., and Kukla, G.: Urbanization: its detection and effect in the United States climate record, J. Climate, 1, 1099–1123, 1988.

Landsberg, H. E.: The urban climate, International Geophysics Series, 28, Academic Press, 275 pp., 1981.

Lemonsu, A., Grimmond, C. S. B., and Masson, V.: Modeling the surface energy balance of an old Mediterranean city core, J. Appl. Meteor., 43, 312–327, 2004.

Le Moigne, P.: SURFEX scientific documentation, Note de centre (CNRM/GMME), Météo-France, Toulouse, France, 211 pp., 2009.

Masson, V.: A physically-based scheme for the urban energy budget in atmospheric models, Bound.-Layer Meteor., 94, 357–397, 2000.

Masson, V. and Seity, Y.: Including atmospheric layers in vegetation and urban offline surface schemes, J. Appl. Meteor. Clim., 48, 1377–1397, 2009.

Masson, V., Grimmond, C. S. B., and Oke, T. R.: Evaluation of the Town Energy Balance (TEB) scheme with direct measurements from dry districts in two cities, J. Appl. Meteor., 41, 1011–1026, 2002.

Masson, V., Champeaux, J. L., Chauvin, F., Meriguet, C., and Lacaze, R.: A global database of land surface parameters at 1km resolution in meteorological and climate models, J. Climate, 16, 1261–1282, 2003.

Noilhan, J. and Planton, S.: A simple parameterization of land surface processes for meteorological models, Mon. Wea. Rev., 117, 536–549, 1989.

Peterson, T. C., Daan, H., and Jones, P. D.: Initial selection of a GCOS surface network, Bull. Amer. Meteor. Soc., 78, 2145–2152, 1997.

Shouraseni, S. R. and Yuan, F.: Trends in extreme temperatures in relation to urbanization in twin cities metropolitan area, Minnesota, J. Appl. Meteor. Clim., 48, 669–679, 2009.

Uppala, S. M., Kallberg, P. W., Simmons, A. J., Andrae, U., da Costa Bechtold, V., Fiorino, M., Gibson, J. K., Haseler, J., Hernandez, A., Kelly, G. A., Li, X., Onogi, K., Saarinen, S., Sokka, N., Allan, R. P., Andersson, E., Arpe, K., Balmaseda, M. A., Beljaars, A. C. M., van de Berg, L., Bidlot, J., Bormann, N., Caires, S., Chevallier, F., Dethof, A., Dragosavac, M., Fisher, M., Fuentes, M., Hagemann, S., Holm, E., Hoskins, B. J., Isaksen, L., Janssen, P. A. E. M., Jenne, R., McNally, A. P., Mahfouf, J.-F., Morcrette, J.-J., Rayner, N. A., Saunders, R. W., Simon, P., Sterl, A., Trenberth, K. E., Untch, A., Vasiljevic, D., Viterbo, P., and Woollen, J.: The ERA-40 re-analysis, Q. J. Roy. Meteor. Soc., 131, 2961–3012, 2005.

Van de Vyver, H.: Evolution of extreme temperatures in Belgium since the 1950s, Theor. Appl. Climatol., submitted, 2010.

Vandiepenbeeck, M.: On a recent urban effect in the time series of the temperature in different stations in Belgium, Proc. Second European Conf. on Applied Climatology, Vienna, Austria, Österreichse Beiträge zu Meteorologie und Geophysik, CD-ROM, 19, 1998.

Vanhuysse, S., Depireux, J., and Wolff, E.: Etude de l'évolution de l'imperméabilisation du sol en région de Bruxelles-Capitale, Université Libre de Bruxelles, IGEAT, Brussels, Belgium, 60 pp., 2006.

Van Weverberg, K., De Ridder, K., and Van Rompaey, A.: Modeling the contribution of the Brussels Heat Island to a long temperature time series, J. Appl. Meteor. Clim., 47, 976–990, 2008.

Wijngaard, J. B., Klein Tank, A. M. G., and Können, G. P.: Homogeneity of the 20th century European daily temperature and precipitation series, Int. J. Climatol., 23, 679–692, 2003.

Wilmott, C. T.: Some comments on the evaluation of model performance, Bull. Amer. Meteor. Soc., 63, 1309–1313, 1982.

Yuan, F. and Bauer, M. E.: Comparison of impervious surface area and normalized difference vegetation index as indicators of surface urban heat island effects in Landsat imagery, Remote Sens. Environ., 106, 375–386, 2007.

Zhang, D. L., Shou, Y. X., and Dickerson, R. R.: Upstream urbanization exacerbates urban heat island effects, Geophys. Res. Lett., 36, L24401, doi:10.1029/2009GL041082, 2009.

9

Large scale integrated hydrological modelling of the impact of climate change on the water balance with DANUBIA

M. Prasch[1], T. Marke[1,*], U. Strasser[1,*], and W. Mauser[1]

[1]Department of Geography, Ludwig-Maximilians-Universität Munich, Germany
[*]now at: Institute of Geography and Regional Science, University of Graz, Austria

Abstract. Future climate change will affect the water availability in large areas. In order to derive appropriate adaptation strategies the impact on the water balance has to be determined on a regional scale in a high spatial and temporal resolution. Within the framework of the BRAHMATWINN project the model system DANUBIA, developed within the project GLOWA Danube (GLOWA Danube, 2010; Mauser and Ludwig, 2002), was applied to calculate the water balance components under past and future climate conditions in the large-scale mountain watersheds of the Upper Danube and the Upper Brahmaputra. To use CLM model output data as meteorological drivers DANUBIA is coupled with the scaling tool SCALMET (Marke, 2008). For the determination of the impact of glacier melt water on the water balance the model SURGES (Weber et al., 2008; Prasch, 2010) is integrated into DANUBIA. In this paper we introduce the hydrological model DANUBIA with the tools SCALMET and SURGES. By means of the distributed hydrological time series for the past from 1971 to 2000 the model performance is presented. In order to determine the impact of climate change on the water balance in both catchments, time series from 2011 to 2080 according to the IPCC SRES emission scenarios A2, A1B, B2 and Commitment are analysed. Together with the socioeconomic outcomes (see Chapter 4) the DANUBIA model results provide the basis for the derivation of Integrated Water Resources Management Strategies to adapt to climate change impacts (see Chapter 9 and 10).

1 Introduction and objectives

The impact of climate change on the water resources is one of the most essential issues for the population of mountain areas and their forelands in the future (Barnett et al., 2005). In order to determine the impact of future climate change on the water balance and to identify appropriate adaptation strategies (see Chapter 9 and 10), the main objective of the hydrological modelling was to deliver distributed hydrological time series for historical as well as future climate conditions in the large mountain watersheds of the Upper Danube and the Upper Brahmaputra on a proper temporal and spatial resolution. Therefore the model DANUBIA, developed within the research project GLOWA-Danube (GLOWA Danube, 2010; Mauser and Ludwig, 2002), was applied in the framework of the project BRAHMATWINN in both basins. In order to use the outcomes of the regional climate model CLM (see Chapter 2) as meteorological drivers, a further downscaling of the climate model outcomes was carried out wit the tool SCALMET (Marke, 2008). The analysis of the model results should explicitly comprise the impact of snow and glacier melt on the water balance in respect to climate change, so that glacier dynamics had to be modelled in an appropriate way. In this paper we introduce the hydrological model DANUBIA with the tools SCALMET and SURGES. At first, the model performance is shown. Afterwards the analysis of the model results with respect to the impact of climate change on the water balance is presented.

2 Role within the integrated project

The hydrological modelling with DANUBIA enables the simulation of distributed hydrological dynamics in a high spatial and temporal resolution. Since the model is based on physical principles and is not calibrated to measured discharges it can be applied in regions with different environmental conditions for past as well as for future climatic conditions. The analysis of the time series of the water balance components for the time period from 1971 to 2080, provided by DANUBIA, enables the determination of the influence of climate change on water balance components and accordingly on the water availability at the regional scale. As four different IPCC SRES scenarios (see Chapter 2) are used as meteorological drivers, a range of possible future trends is given. Together with the outcomes of the socio-economic modelling the DANUBIA model results provide the basis for the development of the Integrated Water Resources Management Strategies IWRMS as one of the ultimate outcomes of BRAHMATWINN.

3 Scientific methods applied

Hydrological models should fulfil several requirements to determine the impact of climate change on the water balance in large-scale watersheds. The spatial heterogeneity of complex basins with mountain headwatersheds as the Alps or the Himalaya on the one hand and mountain forelands with large river valleys like the Brahmaputra or the Danube on the other hand have to be captured synchronously by the model. This includes the ability to simulate the relevant processes under different climatic conditions as well as varying hydrological regimes (Mauser and Bach, 2009). Therefore process based spatially distributed models were developed which can be divided into two groups. The first group considers hydrological homogenous areas as hydrological response units (HRU) (Leavesley et al., 1983; Flügel, 1996), whereas the second is based on spatial grids, where spatial as well as linear hydrological processes are modelled with a consistent data set (e.g. Lee et al., 2005; Mauser and Bach, 2009). The latter models have the advantage that the coupling with climate models is easier, because climate models usually are also raster-based. A further requirement for the coupling is to close the energy and mass balance within all model components. Therefore in the framework of BRAHMATWINN the integrative decision support system (DSS) DANUBIA was applied in the UDRB as well as the UBRB. The DSS consists of both socio-economic and natural process models which are coupled to consider the interactions between socio-economics and nature. The physically based, fully spatially distributed model PROMET (Processes of Radiation, Mass and Energy Transfer) (Mauser and Bach, 2009) builds the landsurface core model of DANUBIA for the simulation of the natural processes. PROMET was developed as a raster model with a spatial resolution of 1×1 km and runs on an hourly time step. It strictly conserves mass and energy within as well as throughout all its components and feedbacks. So all processes related to water fluxes of complex watersheds are considered in fully interacting components for meteorology, land surface energy and mass balance, vegetation, snow and ice, soil hydraulics and soil temperature as well as in a groundwater and channel-flow component with hydraulic structures. PROMET is not calibrated using measured discharges to fully exploit the predictive capabilities of

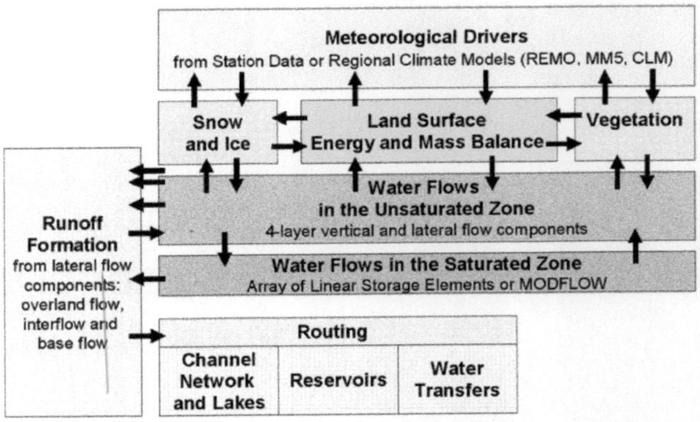

Figure 1. Schematic diagram of the components of PROMET, the DANUBIA landsurface core model, and the interfaces between them. Boxes indicate components and arrows indicate interfaces, through which data is exchanged (Mauser and Bach, 2009, p. 365).

the physically based approach. Figure 1 gives a schematic overview of the mentioned components and its interfaces for data exchange (Mauser and Bach, 2009).

Since the spatial resolution of climate models is still limited, e.g. the CLM has a resolution of 50×50 km, land surface heterogeneities cannot be considered properly, so that for example the topography especially in high mountain areas and accordingly the environmental variability is not fully captured by the RCMs. As a consequence, the impact of climate change cannot be investigated at a local scale (Gerstengarbe, 2001). In order to bridge the scales, basically two downscaling techniques are used. The dynamical technique is based on physics and enables the consideration of small scale climate processes (Gerstengarbe, 2001), whereas the statistical downscaling uses empirical relations. In most cases, both downscaling techniques are combined as it is done by the tool SCALMET (Scaling Meteorological variables) (Marke, 2008). To use the provided CLM output (see Chapter 2) at the required resolution for the process description at the landsurface (1×1 km) a further downscaling was applied to the model output with SCALMET. As shown in Fig. 2, SCALMET connects models simulating the atmosphere with those operating at the land surface. In order to reflect the natural climate gradients more realistically, SCALMET uses globally available sub-grid information (e.g. terrain elevation, topographic slope and aspect) in a spatial resolution of 1×1 km to adjust the RCM model output. A regression approach is applied to adjust those meteorological variables with respect to sub-grid topography that strongly vary with terrain elevation, e.g. air temperature. For the remapping of wind speed and long- and shortwave radiation, separate submodels have been implemented in the coupler to account for sub-grid heterogeneities within the remapping process. Furthermore, a temporal interpolation routine to disaggregate the 3 h values of the CLM output to an hourly time basis is implemented into SCALMET. To maintain the energy and

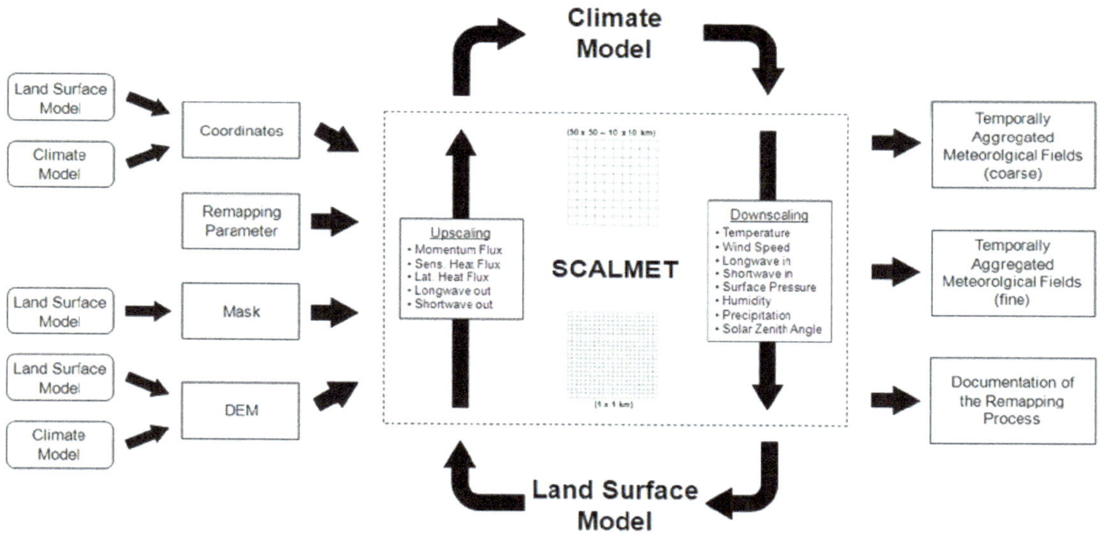

Figure 2. Schematic overview of the coupled model SCALMET (Marke, 2008, p. 41).

mass budged predetermined by the RCM outputs, special algorithms have been integrated that systematically assure the conservation of mass and energy between the model scales. As the downscaling techniques implemented in SCALMET are based on physical and statistical approaches, they are completely general and do not require any further parameterization.

In order to completely describe the hydrological processes, especially in mountain watersheds like the UDRB and the UBRB, the consideration of cryospheric processes is necessary. For the determination of the influence of glaciers under changing climate conditions, various glacier models of different complexity have been developed (Oerlemans, 2008). To numerically describe the glacier melt water release on a raster scale of 1 km^2, glacier dynamics have to be considered appropriately by finding a compromise between complex and simple models. For that purpose DANUBIA has been extended with the component SURGES (SUbscale Regional Glacier Extension Simulator) (Weber et al., 2008; Prasch, 2010), simulating glacier advance and retreat utilizing a glacier inventory which provides geometry characteristics for each single glacier. The total area of every glacier is subdivided into equidistant elevation levels with a homogenous ice thickness for all grid cells. In every time step, the hourly meteorological data, provided by DANUBIA is distributed to each subscale level. In this way mass and energy balance can be calculated for every elevation level to model accumulation and ablation and accordingly, quantify the melt water release to runoff. In a simple parameterization, snow metamorphosis and ice flow are also considered in SURGES. Hence, the individual dynamics of any single glacier and the effect of climate change on the ice resources in the Alpine or respectively Himalayan headwatersheds can be investigated. Figure 3 presents a schematic overview of SURGES.

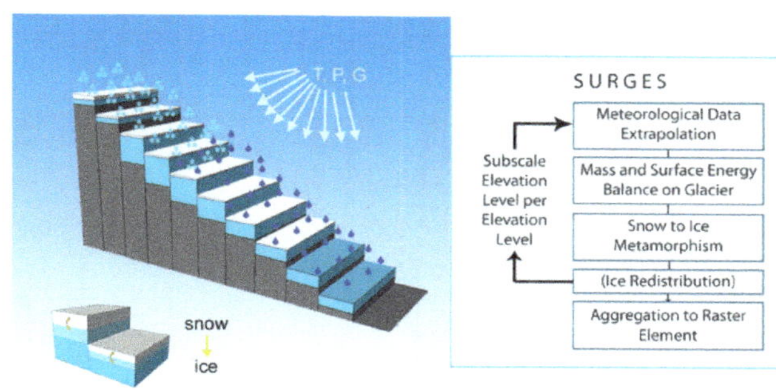

Figure 3. Scheme of the glacier model SURGES, symbolizing the subscale elevation levels of the glacierized area (Weber et al., 2009, modified).

The model framework of DANUBIA including the tools SCALMET and SURGES as explained above fulfils all requirements for the hydrological modelling within BRAHMATWINN to deliver distributed hydrological time series in both test basins for past as well as future climate conditions. Before the model performance and the model results are presented in the following section, the required input data are briefly described.

4 Results achieved and deliverables provided

In order to run DANUBIA, input data fields of topography, landuse and landcover, soil texture as well as glacier properties with a spatial resolution of 1 × 1 km are required besides meteorological drivers. Furthermore, properties describing soil and landuse/landcover classes have to be provided (Mauser and Bach, 2009). For the UDRB all data were available from the GLOWA-Danube project. For the

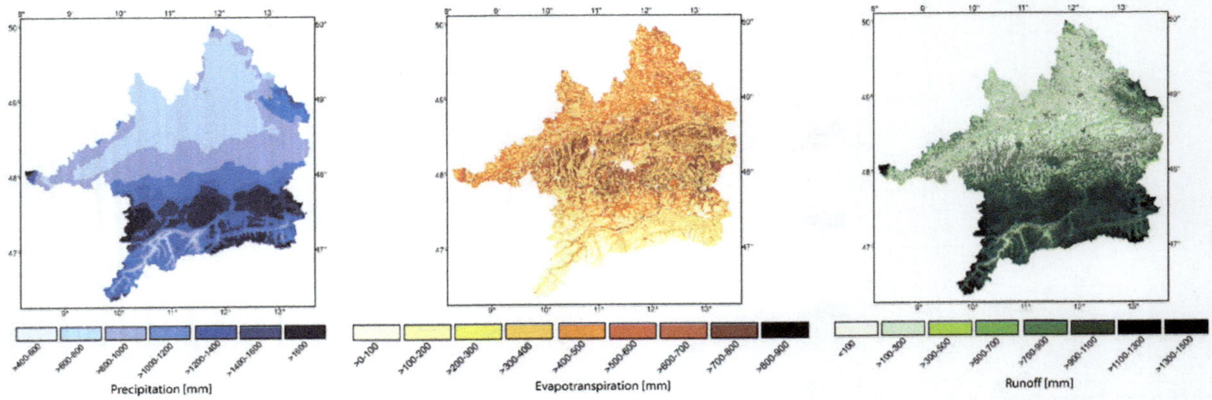

Figure 4. Modelled distributed water balance in the UDRB in using station recordings as meteorological drivers for the period 1971–2000.

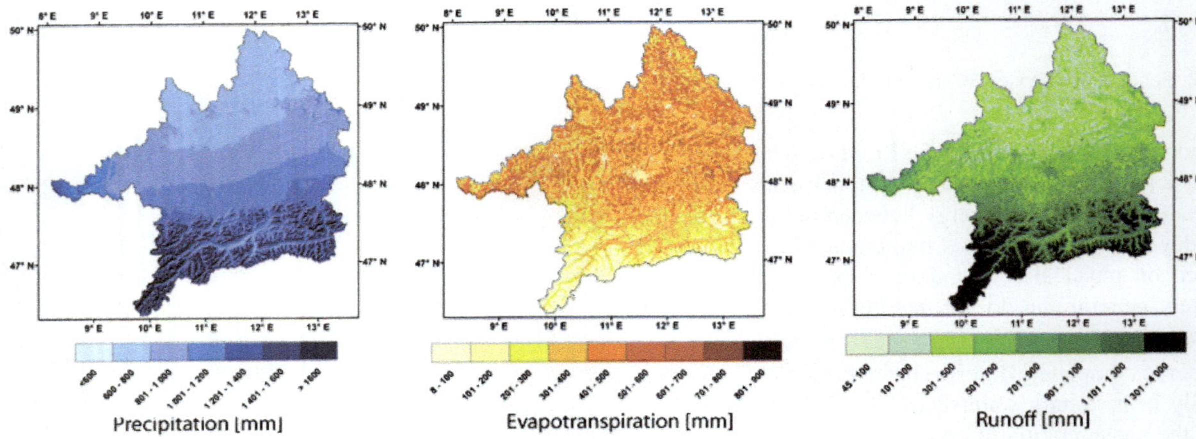

Figure 5. Modelled distributed water balance in the UDRB in using CLM ERA data as meteorological drivers for the period 1971–2000.

UBRB topography was deduced from the SRTM data (Jarvis et al., 2006). To provide a suitable landuse and landcover data set, we use the publically available classification from MODIS/TERRA (Boston University, 2004). The digital Soil Map of the FAO/UNESCO (FAO-UNESCO, 2003) represents the basis for the applied soil texture classification. For the glacier model, the data were derived of the Chinese glacier inventory (WDC, 2009). All data are stored in a raster GIS structure and initialised at the beginning of the modelling. Detailed information of soil physics and plant parameters, stored as data tables, are taken from literature and/or field campaigns. The results of the DANUBIA hydrological model consist of both a specified set of output variables for selectable raster elements and of spatially distributed fields for the catchment area, describing e.g. the water balance and runoff components.

4.1 The model performance of DANUBIA

Detailed validation studies of the model DANUBIA and its components were conducted, so that it is referred to Mauser and Bach (2009), Marke (2008), Weber et al. (2008), Prasch (2010) and GLOWA Danube (2010) for further details at this point. In order to check the quality of the hydrological model DANUBIA, driven with CLM model output data and to assess uncertainties, model runs were carried out for the period from 1971 to 2000 for both the UDRB and the UBRB. The simulated water balance of the UDRB is compared to model results driven by meteorological station recordings. In using the latter, about 1039 mm of precipitation and 430 mm of evapotranspiration occur on average per year. The difference between precipitation and evapotranspiration results in a modelled runoff of 609 mm. This value resembles the observed discharge of 590 mm at the catchment outlet gauge of the UDRB in Achleiten near the city of Passau, so that the accuracy of the simulated runoff is in the range of 3% (Fig. 4). The comparison of modelled with observed daily channel runoff values of the Danube in Achleiten is presented in Fig. 6 (left). A Nash-Sutcliffe efficiency coefficient (Nash and Sutcliffe, 1970) of 0.84 and a coefficient of determination of 0.87 allow the conclusion that DANUBIA can reproduce

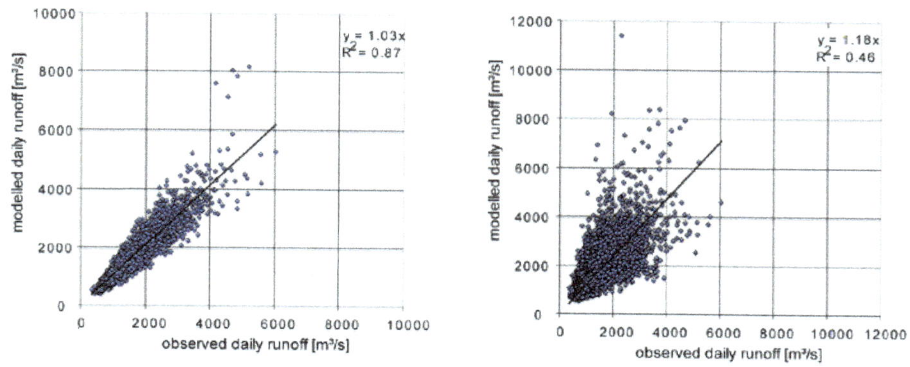

Figure 6. Comparison of observed and modelled daily runoff at the UDRB outlet gauge in Achleiten for 1971–2000, using station recordings (left) and downscaled CLM ERA output data (right).

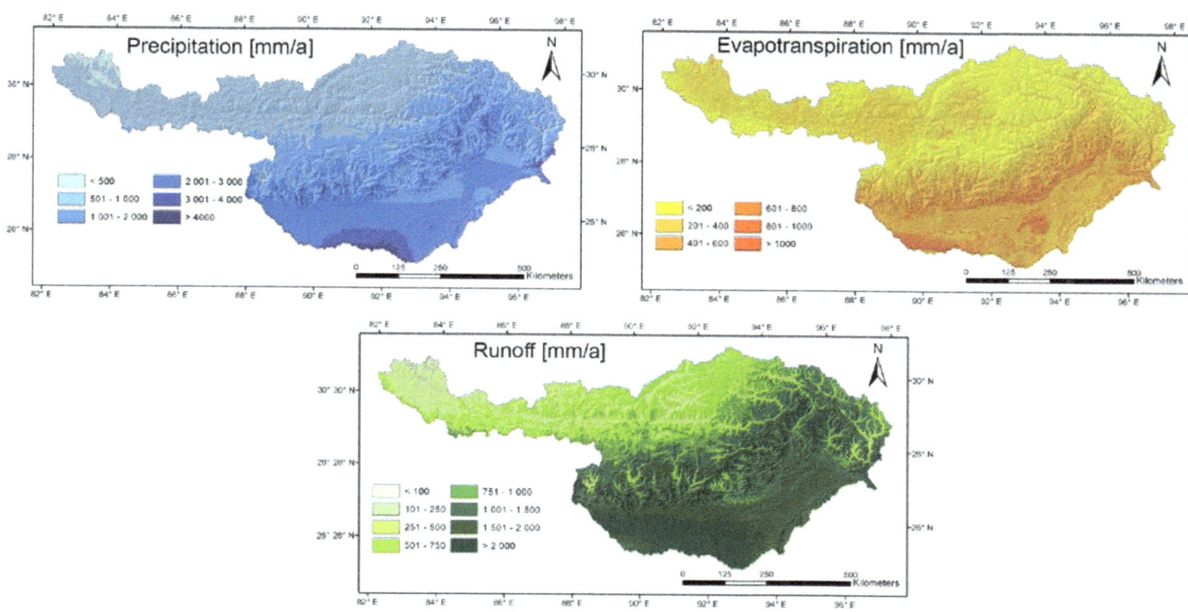

Figure 7. Modelled distributed water balance in the UBRB in using CLM ERA data as meteorological drivers for the period 1971–2000.

the water balance of the past in the UDRB, which is confirmed by the values for other gauges of subcatchments in the UDRB according to Mauser and Bach (2009). Figures 5 and 6 (right) show the similar model results for the water balance in using downscaled CLM ERA output data as meteorological drivers. Compared to the station driven results, precipitation is higher whereas evapotranspiration is lower on average. The spatial distribution of precipitation also is different. Especially the wind and leeside effects at the northern alpine border line are not reproduced by the CLM data. This causes an overestimation of the runoff of 21 percent for the period from 1971–2000 compared to the observed data. Using CLM ECHAM data as meteorological drivers leads to a further overestimation, so that the deviation of the runoff is 26 percent altogether. Although the overestimation affects the daily runoff course, it is reproduced, the coefficient of determination is 0.46 (Fig. 6, right). Reasons for the deviation can be found in the different precipitation pattern of the CLM data and the lower evapotranspiration which can be traced back to the very high air humidity in the data set.

In the UBRB the number of station recordings and the observed data are insufficient to use them as meteorological drivers, so that only CLM data can be used in the Asian catchment. Figure 7 shows the modelled water balance in the UBRB driven by downscaled CLM ERA data for 1971–2000. About 1550 mm of precipitation and 320 mm evapotranspiration occur in average per year. The difference between precipitation and evapotranspiration results in a mean modelled discharge of 19 810 $m^3\,s^{-1}$. According to Jain et al. (2007) the average discharge near the confluence of the Brahmaputra with the Ganges River is 19 200 $m^3\,s^{-1}$. This

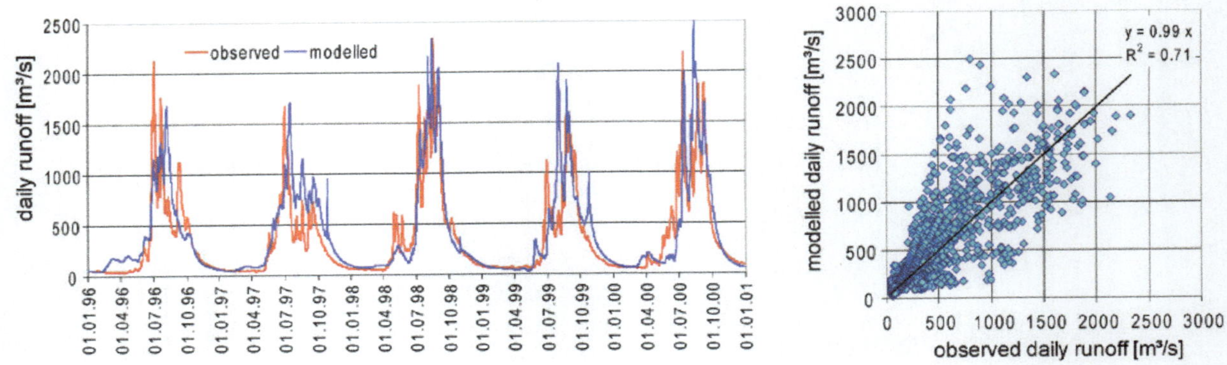

Figure 8. Modelled (CLM ERA driven) and observed monthly runoff at the gauge in Lhasa at the Lhasa River for 1996–2000.

value corresponds very well with the modelled discharge, although in general the latter is slightly overestimated by the model. The spatial distribution shows the subdivision of the UBRB in the dry Tibetan parts north of the Himalayas and the wet southern parts of India and Bhutan with high monsoon precipitation. The varying air temperature due to the different elevation and the different water availability causes low evapotranspiration rates in the north, whereas in the southern parts of the catchment values over 1500 mm per year can be reached. These values are measured by stations in this region (Singh et al., 2004).

Annual courses of runoff data are sparse in the UBRB, so that modelled and observed runoff values are compared for the subcatchment of the Lhasa River in Tibet. Figure 8 shows the modelled as well as the observed daily runoff at the gauge in Lhasa from 1996 to 2000. The simulated runoff explains 71 percent of the observed runoff, although the quality strongly varies between the years. A Nash-Sutcliffe efficiency coefficient of 0.67 confirms the result. Reasons for the deviations could be manifold. Besides changes in the observation methodology, the precipitation pattern is not fully captured by the CLM ERA model data in some years (see Chapter 2).

Although the model uncertainty increases in using climate model output as meteorological drivers in the hydrological model as shown in the validation steps, the results proof the robustness and applicability of the modelling framework DANUBIA for the project purposes within the bounds of possibility, taking into account several uncertainties of the input data as well as of the observations.

4.2 The impact of climate change on the water balance

The regional climate model data of the CLM, driven by the global climate model ECHAM 5, is provided for the four IPCC SRES emission scenarios A2, A1B, B1 and the commitment model run (see Chapter 2). They are used as meteorological drivers for DANUBIA to calculate future time series of the water balance components for the UDRB and the UBRB. In order to determine the impact of future

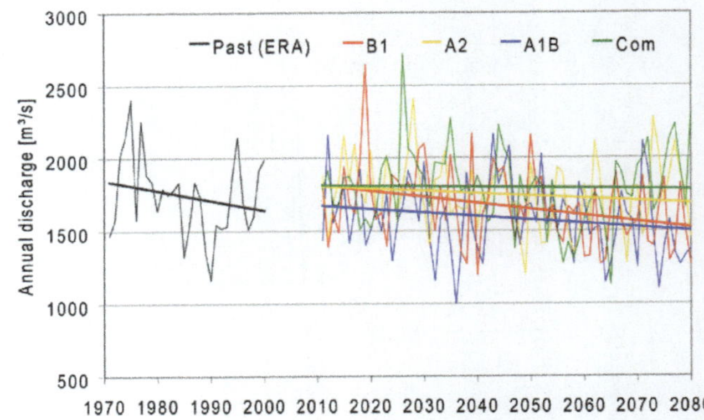

Figure 9. Development of the mean annual discharge in the UDRB for the past and the four scenarios from 1971–2080.

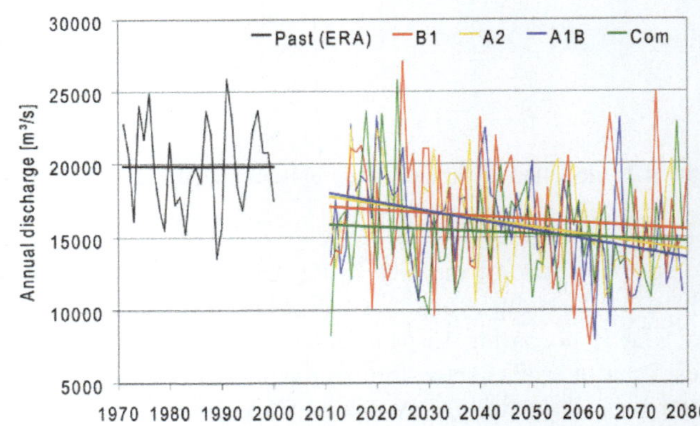

Figure 10. Development of the mean annual discharge in the UBRB for the past and the four scenarios from 1971–2080.

climate change on the water balance, the hydrological time series were analysed in detail for the period from 2011 to 2080.

In the UDRB as well as the UBRB a reduction of the mean annual runoff is simulated (Figs. 9 and 10). Hence, the water availability in both catchments is decreasing in

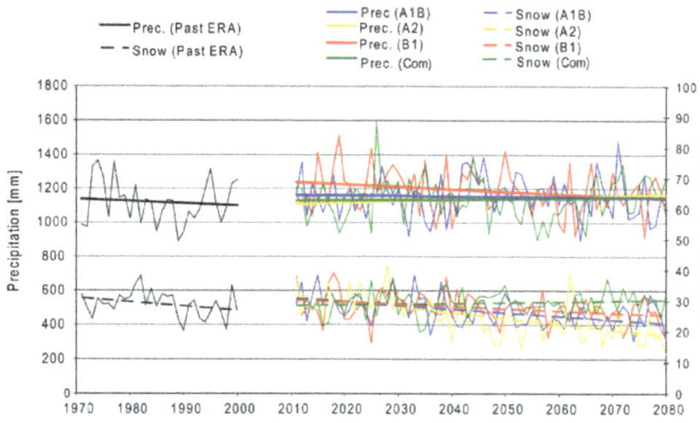

Figure 11. Development of the annual precipitation sum and the percentage of snow precipitation in the UDRB for the past and the four IPCC SRES scenarios from 1971–2080.

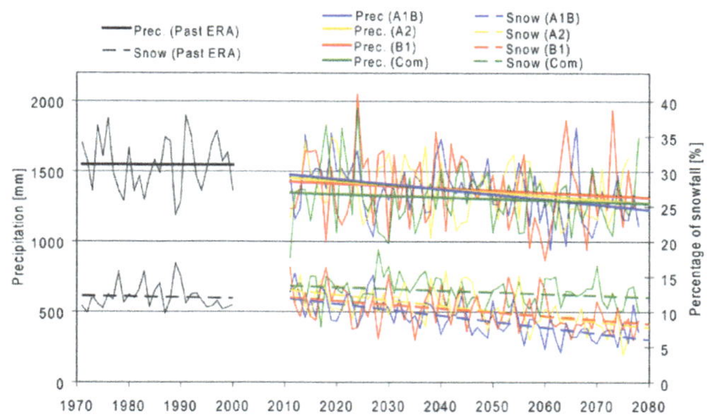

Figure 12. Development of the annual precipitation sum and the percentage of snow precipitation in the UBRB for the past and the four IPCC SRES scenarios from 1971–2080.

average according to the future simulations. The magnitude of the trends varies according to the chosen IPCC SRES scenario. While there is no significant trend for the commitment scenario, the strongest decrease is modelled for B1 in the UDRB. In the Asian test site a decrease is simulated for all four scenarios, which rises from the commitment to the A1B scenario conditions. Reasons for the reduction can be found in the precipitation sum on the one hand. Whereas in the UBRB precipitation is becoming less in all four scenarios (see Chapter 2), there is no significant trend in the UDRB except for the B1 scenario with a precipitation decrease. Not only the precipitation sum is of importance for the water availability but also the precipitation phase plays an essential role because of the temporal storage function of snow and ice. The decreasing trend of the percentage of snowfall during the past period from 1971 to 2000 continues in both basins as well as for all four scenarios except the commitment run (Figs. 11 and 12). The magnitude of the trend is related to the different conditions according to the SRES scenarios with different increasing air temperatures (see Chapter 2). Hence for the commitment model run without increasing emissions in the future a slight increase of snowfall is simulated. On the other hand, an increase in evapotranspiration also causes a reduced runoff. The increasing temperatures as well as longer vegetation periods result in an increasing evapotranspiration sum in both basins, where the trend is clear for all scenarios except the A2 scenario conditions in the UDRB.

In the next step, the average runoff pattern from the climate periods from 2011 to 2040 and from 2051 to 2080 is compared to the past period from 1971 to 2000, where the reduction of water availability described above is confirmed. Furthermore the average from 2011 to 2040 and from 2051 to 2080 of the mentioned output parameters is analysed in its spatial distribution, focussing on the B1 and the A1B scenario results. Figure 13 shows the difference of the future values to the past. In the UDRB in both scenarios in the western parts, especially in the southwest, a decrease in runoff can be seen whereas in the eastern parts an increase in runoff is simulated. Reasons are the different precipitation patterns which correspond to the runoff. Moreover the annual evapotranspiration increases in the low and high mountain ranges due to higher temperatures and accordingly longer growing seasons whereas in the lower valleys, especially in the northern parts of the watershed, there is a small decrease modelled which might be caused by water stress of the plants.

The UBRB is also divided into regions with increasing water availability in the Tibetan parts, whereas in Assam in India the runoff will decrease according to the scenarios. Again this represents the precipitation pattern. Evapotranspiration is determined by the water availability due to precipitation. With increasing precipitation in the Tibetan parts the evapotranspiration also increases a little whereas in Assam both an increasing as well as a decreasing trend is simulated.

Finally, the monthly runoff was studied to find out seasonal changes in the water availability in the future. Therefore the mean runoff of the climate periods mentioned above is compared. According to all scenarios the runoff will decrease in the UDRB (Fig. 14) especially during the summer month. This trend goes on in the last period of the scenario time. As this is the growing season it is especially of importance for agriculture but also for energy production and shipping in the Danube, not only in the Upper basin, but also in the attached regions of Austria and Hungary. In the UBRB runoff will decrease in all four scenarios in the first 30 years of the modelled future period, especially during the summer months. The decreasing trend is going on in the second period from 2051 to 2080. Regarding the annual runoff peak, no clear trend can be seen. Figure 15 shows the mean monthly runoff in the UBRB for the three periods and the four different scenarios.

Besides the decrease of snow precipitation, the influence of glacier melt on runoff was analysed for the test site of

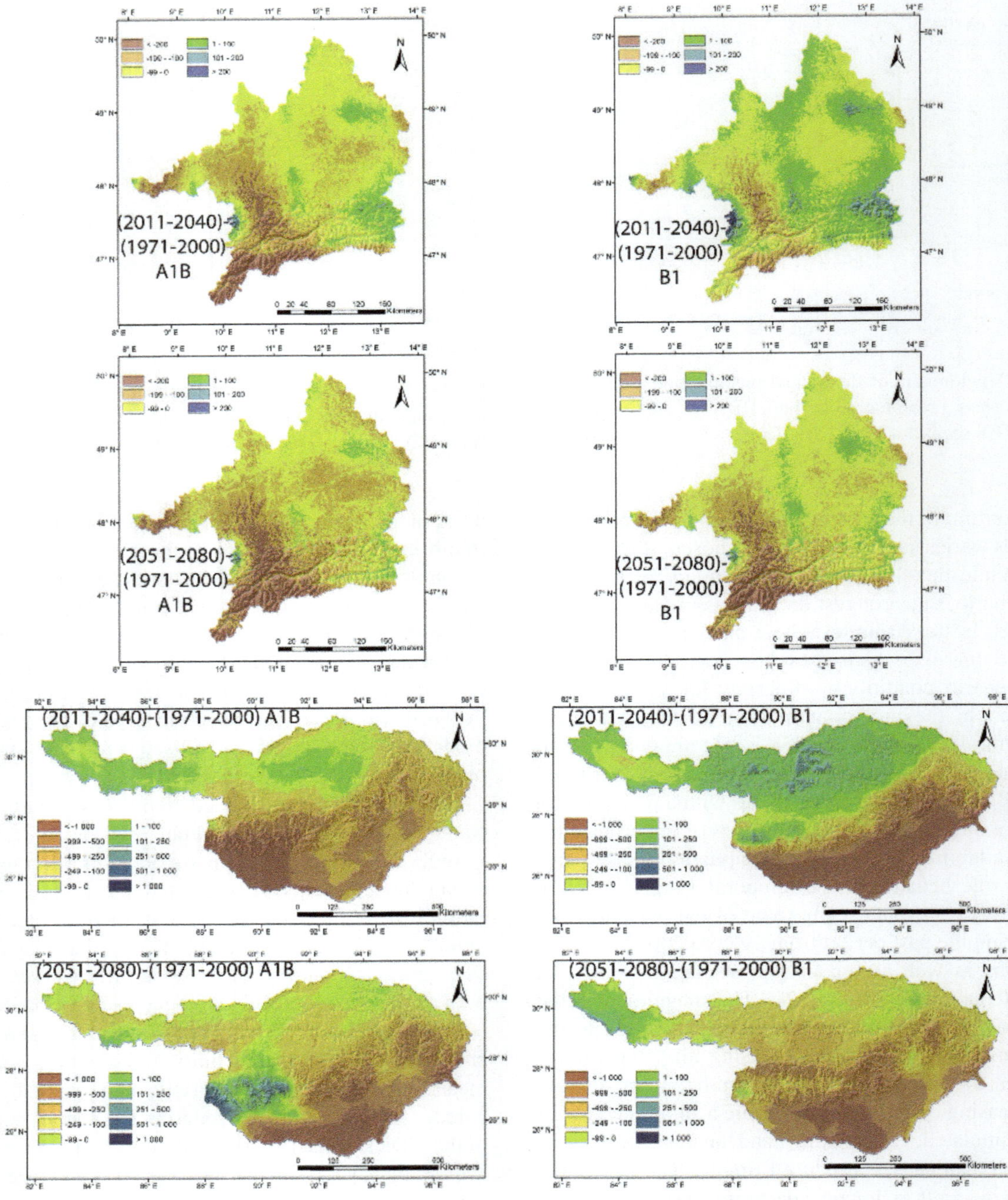

Figure 13. Difference of the mean annual runoff [mm] of the two future 30-years periods from 2011–2040 and 2051–2080 according to the A1B and B1 scenario compared to the past 30-years period from 1971 to 2000 in the UDRB (top) and the UBRB (bottom).

the Lhasa River catchment. In comparing model run outputs with model run outputs without considering glaciers, the amount of ice melt of the glaciers can be determined. Due to increasing air temperatures the melting periods will be longer in the future, so that the influence of ice melt during the future period from 2011 to 2040 increases compared to the past. Afterwards the influence goes down because of the melting away of the glaciers and their water reservoirs (Fig. 16). As the influence of glacier runoff is little compared to the influence of the monsoon precipitation this is only of importance during periods without precipitation in spring and autumn (Prasch, 2010). This confirms the results for the UDRB (GLOWA Danube, 2010) where the influence of glaciers, which is only relevant in alpine headwatersheds, will further go down because of the melt out of the glaciers.

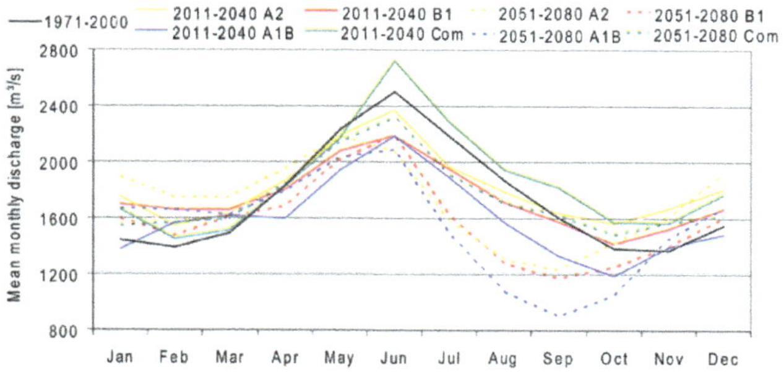

Figure 14. Developing of the mean monthly discharge in the UDRB for the past from 1971–2000 in comparison to the 30-years period from 2011–2040 and 2051–2080.

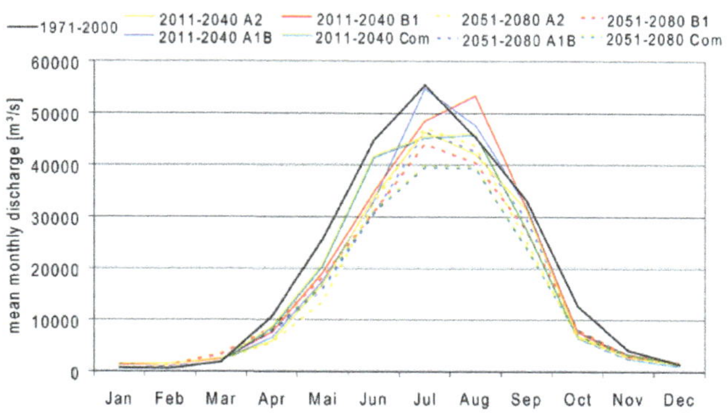

Figure 15. Developing of the mean monthly discharge in the UBRB for the past from 1971–2000 in comparison to the 30-years period from 2011–2040 and 2051–2080.

5 Contributions to sustainable IWRM

Under the assumption that climate will develop according to the CLM ECHAM 5 model output due to the chosen IPCC SRES scenarios, climate change will have a significant impact on the hydrological dynamics of the water balance in the UDRB and the UBRB. There will be less water available in both the UDRB and the UBRB. Reasons for the decrease predominantly are less precipitation and higher evapotranspiration due to rising air temperatures. Accordingly, snowfall will decrease and the glaciers will melt out in the headwatersheds. However, the glacier melt water only has a small influence on the water availability in the populated mountain forelands. Its amount will not suffice to compensate less water in the rivers, because the influence is small compared to precipitation. Besides various other impacts, there will be less water available for hydropower production, which is the main renewable power source in the UDRB at the moment. Furthermore, the changes will affect agriculture. In regions with higher elevation, the growing season will be longer due to higher temperatures than in the past. In combination with increasing precipitation as it is simulated for Tibetan parts in the UBRB as well as for the Eastern Alps in the UDRB,

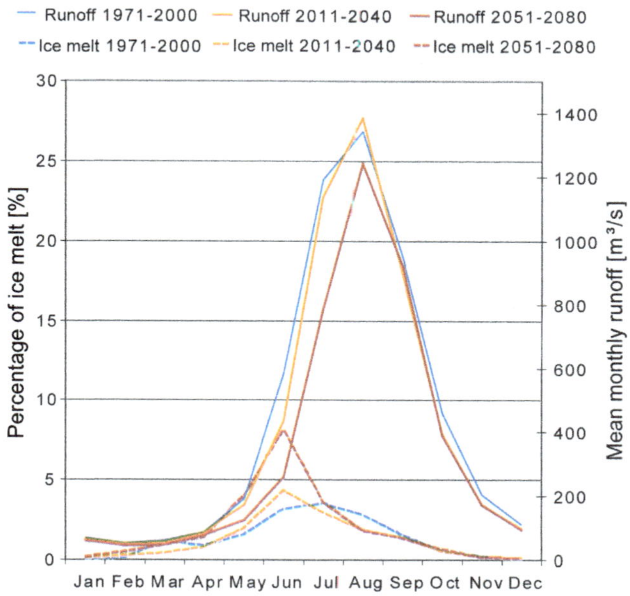

Figure 16. Development of the percentage of ice melt on the mean monthly discharge in the Lhasa River catchment for the 30-years-periods from 1971–2000, from 2011–2040 and from 2051–2080 according to the A1B scenario.

the situation for agriculture might improve. On the contrary, drought stress might increase in the valleys of the Danube or the Brahmaputra in Assam due to higher temperatures, more evapotranspiration and partly less precipitation. To be able to adapt to the impact of climate change, this has to be considered in future plans, e.g. for the construction of hydro power plants or landuse changes in the IWRMS.

6 Conclusions and recommendations

The physically based distributed hydrological model DANUBIA was applied successfully not only in the UDRB but also in the UBRB. The environmental and climatic conditions in both watersheds differ considerably. Based on the available historic data validated hydrological model results are delivered for both catchments, so that the transferability of the approach, which was developed in Europe, to Asia was successfully demonstrated. Furthermore, the model results for the past justify the application for future scenarios in both catchments, using future regional climate model output data of the CLM as meteorological drivers. Although several input data are required to apply DANUBIA, the model setting allows a distributed simulation of the impact of climate change on the water balance with a high spatial and temporal resolution. Hence, the time series of water cycle components for historical and future climatic conditions for both the UDRB and the UBRB enable the determination of the regional impact of climate change on the hydrologic cycle. In this way, the development of the Integrated Water

Resources Management System (IWRMS) as one of the ultimate outcomes of BRAHMATWINN for the adaptation to climate change impacts is based on the results of DANUBIA together with the socio-economic results (see Chapter 4). In a next step the proposed adaptation strategies should be simulated in further DANUBIA model runs to determine their effectiveness and efficiency in an integrated approach. For instance, the effect of land use changes on runoff and evapotranspiration could be assessed with a different land use classification as input.

Acknowledgements. The authors thank the European Commission for funding BRAHMATWINN within the Sixth Framework Program (FP 6). The project GLOWA-Danube, funded by the BMBF, the State of Bavaria and the Ludwig-Maximilians-Universität Munich, and especially Markus Weber is acknowledged for their cooperation.

References

Barnett, T. P., Adam, J. C., and Lettenmaier, D. P.: Potential impacts of a warming climate on water availability in snow-dominated regions, Nature, 438, 303–309, 2005.

Boston University: NASA TERRA/MODIS HDF-EOS MOD12Q1 V004 Land Cover Product Binary Data, Eurasia subset, IGBP Class Scheme, http://duckwater.bu.edu/lc/mod12q1.html (last access: 26 April 2007), 2004.

FAO-UNESCO: The Digital Soil Map of the World, Version 3.6, 2003.

Flügel, W.-A.: Hydrological Response Units (HRU) as modelling entities for hydrological river basin simulation and their methodological potential for modelling complex environmental process systems – Results from the Sieg catchment, Die Erde, 127, 42–62, 1996.

Gerstengarbe, F.-W.: Conditions and Potentials for Climate Projections, in: Climate of the 21st Century: Changes and Risks, edited by: Lozan, J. L., Graßl, H., and Hupfer, P., Hamburg, 2001.

GLOWA-Danube: http://www.glowa-danube.de (last access: 3 November 2010), 2010.

Jain, S., Agarwal, P., and Singh, V.: Hydrology and Water Resources of India, Springer, Dodrecht, 2007.

Jarvis, A., Reuter, H. I., Nelson, A., and Guevara, E.: Hole-filled seamless SRTM data V3, International Centre for Tropical Agriculture (CIAT), http://srtm.csi.cgiar.org (last access: 2 March 2007), 2006.

Leavesley, G. H., Lichty, R. W., Troutman, B. M., and Saindon, L. G.: Precipitation – runoff Modeling System – Users Manual, US Geological Survey Water-Resources Investigations Report 83, 4238, 207 pp., 1983.

Lee, H., Silvapalan, M., and Zehe, E.: Representative elementary watershed (REW) approach, a new blueprint to distributed hydrologic modelling at the catchment scale, in: Predictions in Ungauged Basins: International Perspectives of State of the Art and Pathways Forward, edited by: Franks, S. W., Silvapalan, M., Takeuchi, K., and Tachikawa, Y., IAHS Press, Wallingford, Oxen, UK, 2005.

Marke, T.: Development and Application of a Model Interface to couple Regional Climate Models with Land Surface Models for Climate Change Risk Assessment in the Upper Danube Watershed, Dissertation der Fakultät für Geowissenschaften, Digitale Hochschulschriften der LMU München, 188, available at: http://edoc.ub.uni-muenchen.de/9162/, München, 2008.

Mauser, W. and Bach, H.: PROMET – Large scale distributed hydrological modelling to study the impact of climate change on the water flows of mountain watersheds, J. Hydrol., 376, 362–377, 2009.

Mauser, W. and Ludwig, R.: GLOWA-DANUBE – a research concept to develop integrative techniques, scenarios and strategies regarding global changes of the water cycle, in: Climatic Change: Implications for the Hydrological Cycle and for Water Management, Advances in Global Change Research, edited by: Beniston, M., Vol. 10, Kluwer Academic Publishers, Dordrecht and Boston, 171–188, 2002.

Nash, J. E. and Sutcliffe, J. V.: River flow forecasting through conceptual models, a discussion of principles, J. Hydrol., 227, 33–51, 1970.

Oerlemans, J.: Minimal Glacier Models, Utrecht, 2008.

Prasch, M.: Distributed Process Oriented Modelling of the Future Impact of Glacier Melt Water on Runoff in the Lhasa River Basin in Tibet, Dissertation der Fakultät für Geowissenschaften der LMU München, 228 pp., 2010.

Singh, V. P., Sharma, N., and Ojha, C. Sh. P.: The Brahmaputra Basin Water Resources, Water Science and Technology Library, Vol. 47, Dodrecht, 2004.

WDC: Chinese Glacier Inventory of the World Data Center For Glaciology and Geocryology, Lanzhou, China, http://wdcdgg.westgis.ac.cn/DATABASE/Glacier/glacier_inventory.asp (last access: 17 August 2009), 2009.

Weber, M., Prasch, M., Strasser, U., und Mauser, W.: Embedded grid-modelling of 506 glaciers in the upper Danube basin, Geophys. Res. Abstr., 10, EGU2008-A-01029, 2008.

Weber, M., Braun, L., Mauser, W., and Prasch, M.: Die Bedeutung der Gletscherschmelze für den Abfluss der Donau gegenwärtig und in der Zukunft, The relevance of glacier melt for the Upper Danube River discharge today and in the future, Mitteilungsblatt des hydrographischen Dienstes in Österreich, No. 86, 1–29, 2009.

10

The effect of weather and its changes on emotional state – individual characteristics that make us vulnerable

Z. Spasova

National Center of Public Health and Analyses, Sofia, Bulgaria

Correspondence to: Z. Spasova (z_spassova@abv.bg)

Abstract. Given the proven effects of weather on the human organism, an attempt to examine its effects on a psychological and emotional level has been made. Emotions affect the bio tone, working ability, and concentration; hence their significance in various domains of economic life such as health care, education, transportation, and tourism.

The present pilot study was conducted in Sofia, Bulgaria over a period of eight months, using five psychological methods: Eysenck Personality Questionnaire, State-Trait Anxiety Inventory, Test for Self-assessment of the emotional state, Test for evaluation of moods and Test "Self-confidence-Activity-Mood". The Fiodorov-Chubukov's complex-climatic method was used to characterize meteorological conditions in order to include a maximal number of meteorological elements in the analysis. Sixteen weather types are defined depending on the meteorological elements values according to this method. Abrupt weather changes from one day to another, defined by the same method, were also considered.

The results obtained by t-test showed that the different categories of weather led to changes in the emotional status, which indicates a character either positive or negative for the organism. The abrupt weather changes, according to expectations, have negative effects on human emotions – but only when a transition to the cloudy weather or weather type, classified as "unfavorable", has been realized. The relationship between weather and human emotions is rather complicated since it depends on individual characteristics of people. One of these individual psychological characteristics, marked by the dimension "neuroticism", has a strong effect on emotional reactions in different weather conditions. Emotionally stable individuals are more "resistant" to the weather influence on their emotions, while those who are emotionally unstable have a stronger dependence on the impacts of weather.

1 Introduction

The weak but significant relationship between weather and mood is one of the oldest and most frequent topics in popular biometeorology. Although the statement seems intuitively obvious, the experimental verification of this relationship has been difficult and complex (Persinger, 1975). Studies on the effect of weather on human emotions are relatively low in number and in Bulgaria they are an exception. These studies are also difficult to interpret, susceptible of confounding variables, and of mixed results (Scott, 2007). Most of them focus on the relationship between the weather and mood.

Barnston (1988) performed an experiment in which 62 university student subjects in Illinois kept structured diaries of their feelings and their productivity for six weeks in early autumn. During the same period, he monitored the daily frequency of telephone calls to a crisis intervention service and collected the complete daily weather data for the vicinity from a local meteorological research facility. The results show that the weather appears to influence mood and productivity, but only to a small extent compared with the aggregate of all other controlling factors. Males experienced a relatively stronger effect than females. Psychologically troubled people generally appeared to be more affected by weather. People with mild problems tended to be stressed more when the weather is unstable, cloudy, warm and humid; and least stressed during sunny, dry, cool weather with rising barometric pressure. The crisis service clients with severe problems reacted oppositely to these two weather types.

Hribersek et al. (1987) researched 21-months of the calls to a telephone counseling service at Louvain. Contrary to Barnston (1988) they found that women are more sensitive to weather compared to men, which they explained with hormonal differences. They also explained the phenomenon women make more calls when the weather is adverse with the fact they prefer to stay at home during severe weather. This statement is supported by Trenkle (1982) and Michael and Zumpe (1983) as well.

Persinger (1975) did a detailed study on the relationship between the weather and mood. He calculated correlations between daily self-evaluated mood reports (4 reports/day) of 10 student subjects and 10 meteorological-geophysical variables over a 90-day period. The variables are mean and/or change measures of temperature, barometric pressure, relative humidity, sunshine hours, wind speeds and global geomagnetic activity. Multiple regression analyses indicate that the weather matrix used could account for not more than 35 % of the mood variance on the day of the evaluation. Lag correlations over the previous week indicates a greater number of significant correlations between mood reports and weather of the previous two days. The mean correlation coefficient is 0.27. The author concluded that "lower moods" are associated with fewer sunshine hours, higher relative humidity, and smaller humidity variations than "higher moods". It is concluded that mood reports can show weak response to antecedent weather fluctuation.

Howarth and Hoffman (1984) investigated the relationship between eight weather variables (hours of sunshine, precipitation, temperature, wind direction, wind velocity, humidity, change in barometric pressure and absolute barometric pressure) and 10 mood variables (concentration, cooperation, anxiety, potency, aggression, depression, sleepiness, skepticism, control, and optimism). Data was collected from 24 male subjects over 11 consecutive days. The results show that "*humidity, temperature and hours of sunshine have the greatest effect on mood. High levels of humidity lowered scores on concentration, while increasing reports of sleepiness. Rising temperatures lowered anxiety and skepticism mood scores*".

Palinkas (2001) compared cold temperatures with mental processes. He found that low temperatures influence the concentration of attention, memory and general cognitive processes. "*There is evidence of a dose-response relation involving decrements in cognitive performance with respect to decline in core body temperature and complexity of tasks performed. However, it is unclear whether these effects are due to distraction or increased anxiety.*"

Mood disturbances like nervousness, irritability, aggressiveness or phlegm, reduced concentration of attention are symptoms of so-called foehn illness. As many authors say, these symptoms could be registered during extreme temperature changes and during the passage of weather fronts (Tromp, 1979; Fletcher, 1988); rapid drop of atmospheric pressure followed by its rapid rise (Fletcher, 1988); strong winds blowing (Auliciems, 1978; Thomson, 1979; Sulman, 1980; Fletcher, 1988).

Some of the authors make attempts to connect the perception of weather with the residence of investigated people and with their personal characteristics. Scott (2007) hypothesize the "*relocated individuals are more susceptible to fluctuations in mood stemming from novel weather conditions than indigenous individuals*". He has conducted a research with 70 life-long Minnesota residents and 25 individuals who have spent minimum of one year living outside of Minnesota. Participants completed a mood self-report measure online for four consecutive weeks to determine positive and negative affect levels. Data was then matched with corresponding weather data for the same time period. The author found no support for the hypothesis; however, "*sunshine was identified as the crucial factor for mood adjustment*".

Denissen et al. (2008) suggest that the individual differences may have an effect on how people perceive the weather. They press on fact that differences in sensitivity to daily weather have not been studied previously, but some studies suggest a link between seasonality and personality, especially concerning the trait of neuroticism (e.g. Jang et al., 1997; Murray et al., 1995) (cit. by Denissen et al., 2008). They use the Big Five Inventory Test (BFI) to determine if the weather influences in a different way people with different personality traits. The test is measuring Extraversion, Neuroticism, Openness to Experiences, Conscientiousness, and Agreeableness. The authors examine the effects of six weather parameters (temperature, wind power, sunlight, precipitation, air pressure, and photoperiod) on mood (positive affect, negative affect, and tiredness). They analyze the results from an online diary study of 1233 people. They reveal main effects of temperature, wind power, and sunlight on negative affect. Sunlight has a main effect on tiredness and mediated the effects of precipitation and air pressure on tiredness. According to Denissen et al. (2008) the average effect of weather on mood is small, though there is significant random variation across individuals, especially regarding the effect of photoperiod. There are also no evidences that the individual differences in weather sensitivity could be explained by the Five Factor Model personality traits, gender, or age.

Several findings in general deny the relationship between weather and emotional state in spite of finding some significant coefficients. Findikyan and Sells (1964) for example investigated students from Texas, who had to fill out a diary with their everyday subjective feelings. The group as a whole did not show significant correlations between the weather parameters and mood, although the female group showed a significant negative correlation between the dimension "humidity" and positive mood.

Table 1. Distribution of the investigated persons by sex and age (%).

Sex	Age groups					Mean age
	16–20	21–30	31–40	41–50	51–60	
Males (44)	25	4	5	7	3	23.48
Females (143)	90	17	13	17	6	26.36
All (187)	115	21	18	24	9	25.68

Watson's (2000) research is of diary reports of eight different samples of students from Texas (total number of 478 people) for 8-yr period – 1985–1993. Analyzing the amount of sunshine and rain, he fails to find consistent effect on any of the daily mood variables. Interested if the extreme weather conditions have an effect on people's mood, Watson (2000) also compares days with 0 % sunshine with days with 100 % sunshine. He finds that the sunshine only influenced the overall intensity of participants' mood reports, not the valence of these reports.

Driscoll and Stillman (2002) analyzed the calls to telephone counseling services ("hotlines") each serving communities in a major metropolitan area of the United States (Detroit, Washington DC, Dallas and Seattle) for the period 1997–1998. They investigated associations with subjectively derived weather types for all cities except Seattle, as well as with individual weather elements as cloudiness, precipitation, wind speed, and interdiurnal temperature change for all four cities. The authors reveal that despite the statistically significant results "*obtained in scattered instances, the total number was within that expected by chance*", and that "*there was little in the way of consistency with these associations*". One exception is the increase of call frequency during severe weather, "*when there is obvious concern about the damage done by it*".

Keller et al. (2005) investigate the effect of temperature and barometric pressure on mood and cognition (memory and cognitive style) of 97 North American participants. The authors do not find consistent main effect of weather on mood, though they observe that pleasant weather (with higher temperature or barometric pressure) is "*related to higher mood, better memory and "broadened" cognitive style during the spring*" (only this season) "*as time spent outside increased*". On hot summer days, conversely, spending more time outside is associated with deteriorated mood. "*These results are consistent with findings on seasonal affective disorder, and suggest that pleasant weather improves mood and broadens cognition in spring because people have been deprived of such weather during the winter*".

The contradictions in the results of the listed studies reveal that the issue requires an additional consideration. We hope our study, in spite being a pilot limited in time, would contribute in revealing some aspects of the weather's influence on human emotional state.

2 Data and methods

The present research took place in Sofia City during 25 days within a period of 8 months (November 1998–June 1999). A group of 187 volunteers was examined. Their number varied for different dates. The group included males and females aged between 16 and 60 yr, without serious disturbances in their physical and mental status. Among the persons examined, there were secondary and high school students, teachers and university lecturers, workers and pensioners. Detailed information about the gender and age distribution of the tested people is given in Table 1.

Five psychological methods were used – four of them intended for emotional state analysis, while the fifth determined the human temperament.

One of the most frequently used questionnaires for assessing anxiety as an emotional state is the State-Trait Anxiety Inventory (STAI) (Spielberger et al., 1970). The test has been adopted for Bulgarian population by Shtetinski and Paspalanov (1989). STAI is a self-evaluating test, which comprises of two separate scales, which measure the S-anxiety (Y-1 form) and T-anxiety (Y-2 form). In the S-scale, the main qualities being under scrutiny are related to the so-called "gloomy presentiment" (i.e. the sensation of danger) of the tension, nervousness and inconvenience type. This scale measures the situative anxiety. In the T-scale, the assessed qualities are related to the persistent sensation of danger, tension and nervousness (i.e. it includes the extent of proneness to anxiety). The evaluation of the test results is done individually for each researched individual following the key to the STAI test. The numerical values obtained for all items on each scale are collected and the result is compared with test norms for the scale. The marks on both scales can vary in the range between 20 and 80 points.

The Test for Self-assessment of the emotional state (SAES) was developed by the American psychologists A. Wessman and D. Ricks. It is comprised of four subscales: "calmness-anxiety", "vigor-tiredness", "high spirits-low spirits" and "self-confidence – lack of self-confidence" (Karelin, 1997).

The test "Evaluation of moods" investigates the most prolonged emotional state, which determines one's attitude toward the environment (Karelin, 1997). The test diagnoses three states of the mood: normal, euphoric mood, and asthenic state (it shows the negative moods of the subjects at the moment of taking the test).

The "Self-confidence-Activity-Mood" (SCAM) Questionnaire was developed by specialists from the Military Academy in Saint Petersburg, and is designed for operational evaluation of self-confidence, activity and mood of subjects (Rimskii and Rimskii, 1995).

The Eysenck Personality Questionnaire (EPQ) (Paspalanov et al., 1984) does not measure the emotional state of the subjects. It was included in the study because it is the one which has a scale of "lie", serving as an indicator of the reliability of responses. That also makes it possible to recognize the impact of weather on people having major basic personality traits such as extraversion, psychoticism, and especially neuroticism which is directly connected with human emotional states and processes.

Neuroticism and emotional stability are basic personal characteristics that refer to emotionality as a personality trait. On physiological level they are connected with the inflammability level of the autonomic nervous system and visceral brain. Individuals who score high on neuroticism are more likely than the average to experience feelings as anxiety, guilt, anger, and depressed mood (Matthews and Deary, 1998). They respond more poorly to environmental stress, they are often self-conscious and shy, and they may have trouble controlling urges and delaying gratification (Goleman, 1997). They are highly emotional and impressionable, with strong reactions to the all kind of stimulus and return slower to the normal state. Emotionally stable individuals have higher thresholds of emotional reaction, they return faster to the normal state after emotionally excitement. They are usually calm, steady, and control themselves well. They have higher adaptive abilities especially in stressful situations (Eysenck and Eysenck, 1975).

The Fiodorov-Chubukov's complex-climatic method (see Tishkov, 1985) was used to characterize meteorological conditions for the purpose of including in the analysis a maximal number of meteorological elements. The method considers cloudiness ass well, an important factor for purposes of our study as it has been found the sunlight has a significant effect on mood. Lambert et al. (2002) states that "*the rate of production of serotonin by the brain was directly related to the prevailing duration of bright sunlight and rose rapidly with increased luminosity*". Another advantage of the complex-climatic method is that it has been applied for characterizing the climate of Sofia which makes the results of our study applicable.

According to Fiodorov-Chubukov's complex-climatic method there are three major types of weather: warm weather, weather with transition of temperature through 0 °C, and frosty weather. These weather groups include 16 weather types/classes (see Figs. 1 and 2).

Abrupt weather changes from one day to another, defined by the same method (Fig. 3) were considered as well, with the presumption they could be a source of additional environmental stress on the tested people.

Y. Ajizkii (cited by Boksha and Bogutskii, 1980) classified the 16 weather types according their impact on human biocomfort. He differentiated three groups:

1. Favorable weather types (II, III, IV, V, IX, X, XI types);
2. Relatively favorable weather types (I, VI, VIII, XII);
3. Unfavorable weather types (VII, XVI, XIII, XVI and XV), as well as weather types from the second group with strong wind blowing (over $9\,m\,s^{-1}$), storm, fog, hailstorm, blizzard, or sand storm.

By recurrence of testing in the same weather types with the same persons, we managed to eliminate the suspicion that it was weather that leads to certain emotional states. We also conducted a brief inquiry which helped us to eliminate responses of those people whose emotions at the time of survey were strongly influenced by external factors. The statistical processing of the results included calculation of the average values and standard deviation for grouped data and comparative analysis through the t-test.

3 Hypotheses of the study

3.1 First hypothesis

It can be assumed that different weather classes will affect the emotional state of people and that influence will show statistically significant differences among the different weather groups (favorable, relative favorable and unfavorable).

3.2 Second hypothesis

It is expected that abrupt weather changes will provoke negative emotions in subjects.

3.3 Third hypothesis

It is believed that changes in emotional state affected by weather will depend on individual differences of people; they will be stronger in people with high levels of neuroticism.

4 Results and discussion

During the 25 days of testing, seven of the 16 weather types were observed – both sunny and cloudy in some of the types from the groups, with transition of temperature through 0 °C and frosty weather. The following weather types have been observed: III (slightly cloudy) – in 7 days; VII (rainy) – in 1 day, VIII (weather with transition of the temperature through 0 °C – warm spell) – in 4 days, IX (weather with transition of the temperature through 0 °C – cold spell) – 6 days, X (slightly frosty) – 2 days, XI (moderate frosty) – 4 days and

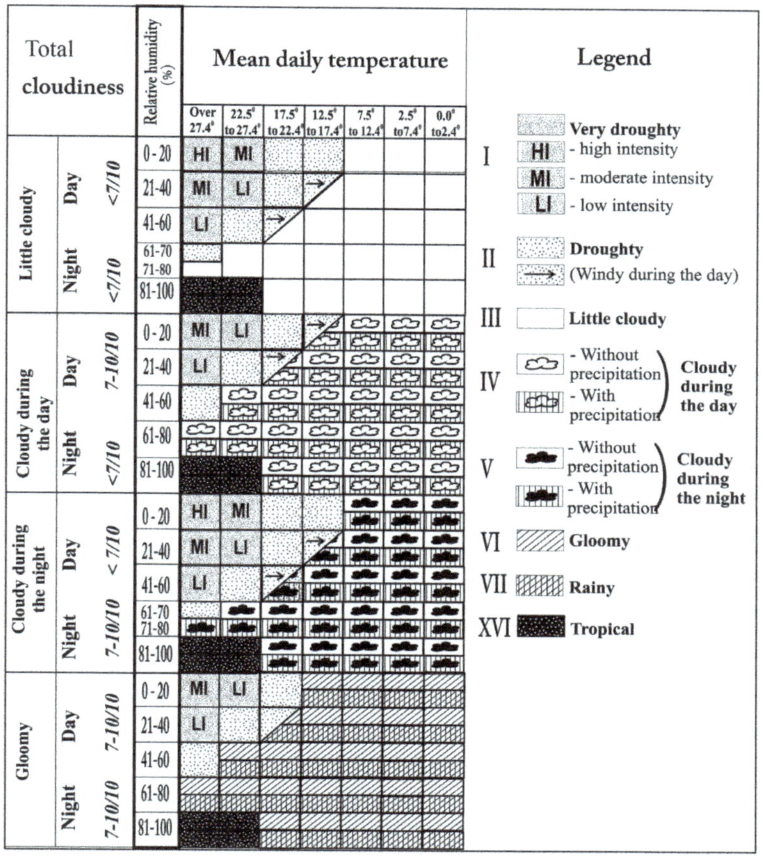

Figure 1. Weather classes during the warm part of the year (source: Tishkov, 1985).

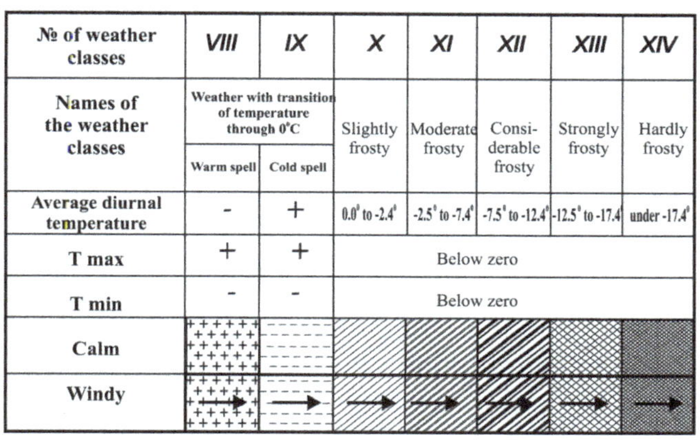

Figure 2. Weather classes during the cold part of the year (source: Tishkov, 1985).

XVI (tropical) – 1 day. Here we provide a short description of the observed weather types during the study period:

Class III (partly cloudy) is characterized with cloud cover less than 7/10 during the day and night. Temperature and humidity may vary broadly, but usually the class is characterized with moderate values of these meteorological elements.

Class VIII (rainy) is observed when cloud cover is more than 7/10 during the day and night, with precipitation of rain. In most cases this weather type is connected with atmospheric frontal passage. "*Such type of weather often provoke depressive mood in people*." (Hristov and Tanev, 1978).

Class VIII and class IX (weather with transition of temperature through 0 °C) is characterized by temperatures above and below 0 °C – during class VIII the maximum temperature is above 0 °C, while the minimum and diurnal average temperatures are below 0 °C; and during class IX the maximum and diurnal mean temperatures are above 0 °C, while the minimum temperature is negative.

Frosty weather is characterized by negative mean, maximum and minimum temperatures and is divided accordingly its severity in five classes (see Fig. 2).

On four of the days of the study period abrupt changes of weather occurred (see Table 2).

4.1 Results of the first hypothesis

The results of all tests and their sub-scales showed an increase of negative emotions during adverse weather type occurrences – although in the group of favorable weather types

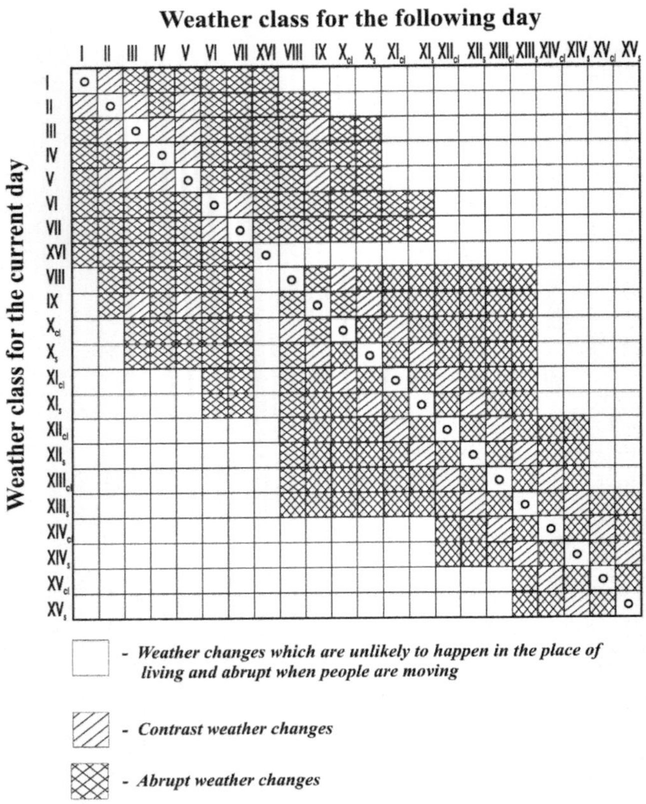

Figure 3. Scheme for determination the contradistinction of the weather changes (source: Baibakova et al., 1966).

there were contradictory results. It should be concluded that weather types defined as "favorable" do not always have a favorable effect on human emotions. For example, we had investigated weather types VII (rainy) and XVI (tropical) from the group of unfavorable weather classes with tests STAI and SAES. The statistical relation of class XVI (tropical) and classes III (little cloudy), VIII (with transition of temperature through 0 °C – warm spell), IX (with transition of temperature through 0 °C – cold spell) and X (slightly frosty) revealed that there are significant differences only with III and IX weather type. This can be interpreted as low favorability of classes VIII and X in terms of emotions, and higher for classes III and IX. This finding is confirmed by the relationship between the above mentioned four classes. In all methods used in almost all cases (except two) between class III and IX, on the one hand, and class VIII and X, on the other, significant differences are realized. Mean average values in different sub-scales showed that the first couple of weather types have a positive impact on human emotions and the second pair – a negative impact.

Significant differences have been observed between IX (weather with transition of temperature through 0 °C – cold spell) and III (little cloudy) types at the one hand, and type XI (moderate frosty) on the other. The latter also belongs to the group of favorable classes. These differences and the mean average values of the test SCAM showed that the XI type (moderate frosty) has less beneficial effect on human emotional processes and states compared with III and IX classes, although it is also a part of the favorable group.

4.2 Results of the second hypothesis

As we mentioned above, during the period of the research abrupt weather changes on four days were observed. In two of the cases a favorable weather class is changed with another favorable weather class (X cloudy turns to IX and IX turns to XI cloudy). One of the abrupt weather changes is when weather turns from unfavorable to favorable class (class VII has been changed with class IX). We have also observed one transition from favorable weather class to unfavorable weather class (when type IX has been changed with VII) (see Table 2).

The results of the hypothesis show that we could speak of a tendency of negativity of emotional states during abrupt weather changes, but not for objective law. Moreover, the presence of normal emotional state indicates that abrupt changes do not have a consistent influence on the human emotional state – but are internally differentiated. Probably there are transitions from one weather class to another which alter the effects of contrasting weather changes depending on the direction of the occurring transitions.

This is confirmed by analysis of weather changes during the dates of the study. Normal emotional state is observed when the weather is changing to classes from the "favorable" group, or when weather turns from cloudy to sunny. In cases of transition from favorable weather class to unfavorable weather class we observe negative emotional state, in our case – heightened anxiety (see Table 2).

4.3 Results of the third hypothesis

The all 187 persons involved in the study have been tested once with EPQ. Women prevail forming about 3/4 of the group (see Table 1). The share of neurotic people is 58.3 % and emotionally stable are 35.8 % (Fig. 4). The mean value of neuroticism scores is 13.06 for the group.

In our third hypothesis we expected that people with higher neuroticism scores would be more affected by the weather and its changes. The results entirely confirmed our assumption. We had investigated the relationship between emotionally stable people and neurotic people (ambiverts, the individuals characterized with the same degree of emotional stability and neuroticism had been excluded from the analysis). The obtained values by t-test are high – not only in its statistical significance, but also as a numerical score.

The biggest differences in the sub-scale for high mood are observed in the SAES test (Table 3). Obviously the greater sensitivity of the neurotics affects their emotional reaction to changes in environment, including weather changes. As a

Table 2. Qualitative assessment of the emotions of the volunteers by mean average values in different sub-scales during abrupt weather changes.

Abrupt weather changes	SAES Test			
	I	II	III	IV
Xcl → IXs	4.40	4.70	4.36	4.80
/Slightly frosty (cloudy) → Weather with transition of temperature through 0 °C – cold spell (sunny)/	Moderately negative, close to normal state			
VII → IXcl	5.33	5.63	5.72	5.81
/Rainy → Weather with transition of temperature through 0 °C – cold spell (cloudy)/	Moderately negative	Normal emotional state		
	Test Evaluation of Moods			
IXs → XIcl	I	II		III
/Weather with transition of temperature through 0 °C	4.03	4.50		4.91
– cold spell (sunny) → Moderate frosty (cloudy)/	Moderately negative			Normal emotional state
IXcl → VII	STAI Test – S-anxiety			
/Weather with transition of temperature through 0 °C – cold spell (cloudy) → Rainy/	44.18 – Heightened anxiety			

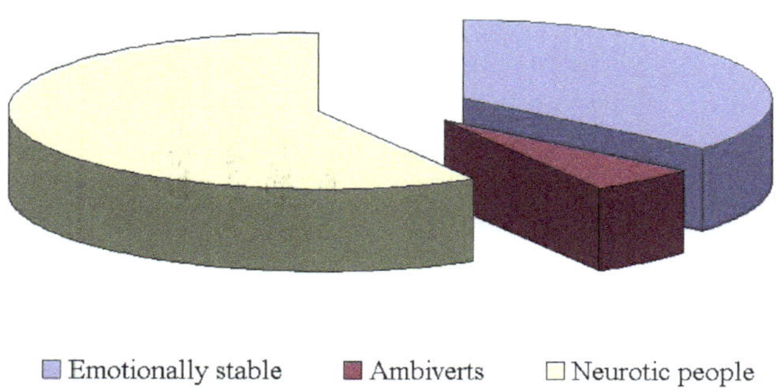

Figure 4. Distribution of the investigated people according to dimention Neuroticism/Stability.

matter of principle, the same thing is observed for the other three sub-scales – which measure other kinds of emotional states. Results on these sub-scales are lower but remain statistically significant. In smallest (but statistically significant) degree, weather affects self-confidence of people with different emotional stability. The reasons for this have already been discussed.

The results of the test "Evaluation of moods" show some specifics of meaningful and quantitative nature. The values of sub-scales which measure mood and asthenic state are not only significant, but also high; while the euphoria result is statistically insignificant. This test shows that the individual differences in neuroticism dimension affects strongly the mood – and in lesser extent – the asthenic state of people. Euphoria does not show significant changes in this comparison as well (Table 3).

The results of SCAM test show that all t-test values are statistically significant, but it is interesting that the highest value is of the sub-scale measuring self-confidence, followed by those measuring mood and activity (Table 3). It seems that one of the peculiarities of neurotics lies in the fact that the increased excitability of their nervous system has a strong impact of their self-feeling. Probably this reflects on their entire perception of the world and strongly affects their self-confidence. Different number of tested people on both tests (SCAM and SAES) could also change the ratio. In the same direction works the different weather types during the testing procedure of both groups. Finally, another important fact is that the SCAM test has a slightly higher level of reliability of the results, which inevitably has given its impact on the ratio of results obtained by the different sub-scales.

The results of STAI test showed significant differences between neurotics and emotionally stable people, although they are minimal ($t_{critical} = 1.98$) (see Table 3). STAI has a maximum level of reliability, which makes its results highly accurate. Moreover it measures only S-anxiety while the SAES test does not differentiate the personal and situational anxiety. This is the main reason for the quantitative differences in measurement of anxiety between these tests, and which did not affect the quality of the value of benchmarking which remains significant in both cases.

Table 3. Comparative analysis by t-quotient of the results of the used psychological methods for the impact of the weather on people with different neuroticism scores.

SAES Test				
Sub-scales	Calmness	Energy	High mood	Self-confidence
N (neurotic people) – 135 persons	S (stable) – 93 persons 4.07*	S 3.71*	S 4.52*	S 3.53*
Test Evaluation of Moods				
Sub-scales	Mood	Asthenic state	Euphoria	
N – 92 persons	S – 71 persons 5.90*	S 4.57*	S 1.23	
SCAM Test				
Sub-scales	Self-confidence	Activity	Mood	
N – 95 persons	S – 69 persons 4.57*	S 2.94*	S 2.97*	
STAI Test				
S-anxiety N – 109 persons			S – 67 persons 1.99*	

* The quotients marked with an asterisk are statistically significant.

5 Discussion

The aim of our research was to reveal the effect of different weather types and abrupt weather changes on people's emotions and to find if there are differences in perception of weather for people with different neuroticism scores.

As a pilot study the present research had some limitations that should be discussed. It did not include the effect of the whole number of 16 weather types of the used classification for several reasons. Despite the limited number of dates of testing, the described weather types have also a different frequency during the year. Some of their manifestations are exceptional for the climate of Sofia, like XVI (tropical), XIV (hardly frosty) and XV class (severe frosty). The XV class has been even excluded from the adjusted classification for Bulgaria because it's not typical for the country. At the same time, weather class number III (little cloudy) is observed during all months of the year, varying from 1 % of the days in January to almost half of the days in September (46 %). The same is valid for rainy weather (class VII) which is observed throughout the year, ranging from 3 % in January and February to almost a quarter of the days in April, May and June. Weather with transition of temperature trough 0 °C (VIII and IX class) is typical for the spring and autumn but is often observed during the winter as well (Hristov and Tanev, 1978). This fact requires more precise choosing of the dates of testing people in which the weather forecast could play an important role. As we found some contradictions in terms of favourableness of the weather types given by Ajizkii, the results of a more detailed study could be used for determination the effect of weather types on human emotions and even classifying the weather in a new way accordingly its positive or negative effect on the emotional state.

Another problem we strike concerned testing of people with different psychological methods. It has been found that if a particular group is tested frequently with the same tests that might lead to inaccurate results because many people will give similar answers. That requires testing a bigger number of people for a longer period.

Even describing the effect of a limited number of weather types on human emotions we could notice that the negative effect of the unfavorable weather classes is clearly manifested. This result confirmed the findings of Denissen et al. (2008) who state that the weather considerably affects negative emotions measured by items "afraid", "scared", "nervous", "jittery", irritable", "hostile", "guilty", "ashamed", "upset", and "distressed". Our findings although state the weather can affect in a different extend the positive emotions as well. Actually we found that all of the emotional states excluding euphoria are affected by the weather. It is understandable because euphoria is an extreme emotion influenced by very strong positive factors for the person. The rest of emotional states are affected in different extend by the weather. The strongest is the effect of the weather on mood, which varies widely according to the weather characteristics. To a smaller extent, weather affects activity, asthenic state, vigor, anxiety and high mood. Self-confidence is lesser influenced by weather, which could be considered normal because it depends mostly on individual experience, way of life, social status, education, etc. It could be concluded from the present study that weather influences the simple and short-lived emotional states, while in the complicated and long-lived emotions its effect is lower or non-existent.

We also supposed that the abrupt weather change is a stressor for the human organism and therefore should negatively affect people's physiological and respectively emotional state. This assumption although was not fully confirmed by the results of the study (see Sect. 4.2). One of the possible explanations could be the age of the tested volunteers. They were young and they may have higher potential to adapt to changing weather; while for older people, especially those with chronic diseases, this would require physiological tension which could lead to negative emotions. Another explanation is the character of abrupt weather changes, as when the weather "improves" (turns from unfavorable to favorable types or from cloudy to sunny) the emotional state improves as well. The problem requires drawing a deeper attention on it in the future as the abrupt weather changes are expected to increase in frequency with climate change.

We also found that the perception of weather and its changes strongly depends on individual psychological characteristics of people. Our assumption does not confirm the findings of Denissen et al. (2008) who were not able to prove that the individual differences in weather sensitivity could be explained by the Five Factor Model personality traits (one of its sub-scales is measuring neuroticism), gender, or age. Our findings reveal that neurotic people have higher sensitivity to the weather while the emotionally stable individuals are more "protected" against the weather influence on their emotions. Although not dividing the studied group by gender we could in indirect way state that women are more vulnerable to the weather because they have higher scores of neuroticism. The norms of the dimension neuroticism are 9.69 points for men and 13.25 for women (statistically significant difference). The mean value of neuroticism scores is 13.06 for the group investigated by us which could be explained by the higher percent of women.

In conclusion, we would like to emphasize that the thesis of this work deserves further development, since its theoretical and applied effects are not negligible. Emotions affect the bio tone, working ability and concentration; hence their significance in various domains of economic life, such as health care, education, transportation, and tourism. We believe the relation between climatology and psychology has a future in theoretical and practical aspects, revealing interesting dependences between climatic phenomena and psychical reactions of humans.

Acknowledgements. The author would like to thank psychologist Petar Zahariev for providing methodological help for this paper.

Edited by: T. Cegnar

References

Auliciems, A.: Mood dependency on low intensity atmospheric variability, Int. J. Biometeorol., 22, 20–32, 1978.

Baibakova, E., Butyeva, I., Ilyicheva, E., and Nevraev, G.: Weather change and its assessment for the medical characterization of climate. Issues of culturology, physiotherapy and medical physical exercise, 145–150, 1966.

Barnston, A.: The effect of weather on mood, productivity, and frequency of emotional crisis in a temperate continental climate, Int. J. Biometeorol., 32, 134–143, 1988.

Boksha, V. and Bogutskii, B.: Medical climatology and climatotherapy, 1980 (in Russian).

Denissen, J., Butalid, L., Penke, L., and van Aken, M.: The effects of weather on daily mood: A multilevel approach, Emotion, 8, 662–667, 2008.

Driscoll, D. and Stillman, D.: Weather and emotional state: a search for associations between weather and calls to telephone counseling services, Int. J. Biometeorol., 47, 21–34, 2002.

Eysenck, H. and Eysenck, S.: Manual of the Eysenck Personality Questionnaire (Junior and Adult), Hodder and Stoughton, 1975.

Findikyan, N. and Sells, S.: Some relations of meteorological variables to day-to-day fluctuations in subjective feeling, in: Dimensions of stimulus situations which account for behavior variance, Contract Nour-3436(00), Group Psychology Bronch, Office of Naval Research, Research Note No. 1, 1964.

Fletcher, R.: "Föhn illness" and human biometeorology in the Chinook area of Canada, Int J Biometeorol, 32, 168–175, 1988.

Goleman, D.: Emotional Intelligence, New York, Bantam, 1997.

Howarth, E. and Hoffman M.: A multidimensional approach to the relationship between mood and weather, Brit. J. Psychol., 75, 15–23, 1984.

Hribersek, E., Van de Voorde, H., Poppe, H., and Casselman, J.: Influence of the day of the week and the weather on people using a telephone support system, Brit. J. Psychiat., 150, 189–192, 1987.

Hristov, P. and Tanev, A.: Climate of Sofia, Sofia, 1978.

Jang, K., Lam, R., Livesley, W., and Vernon, P.: The relationship between seasonal mood change and personality: More apparent then real?, Acta Psychiat. Scand., 95, 539–543, 1997.

Karelin, A. (Ed.): Encyclopaedia of psychological tests. Personality, motivation, necessities, Moscow, 1997 (in Russian).

Keller, M., Fredrickson, B., Ybarra, O., Cote, S., Johnson, K., Mikels, J., Conway, A., and Wagner, T.: A warm heart and a clear head: The contingent effects of weather on mood and cognition, Psychol. Sci., 16, 724–731, 2005.

Lambert, G., Reid. C., Kaye, D., Jennings, G., and Esler, M.: Effect of sunlight and season on serotonin turnover in the brain, Lancet, 7, 360(9348):1840-2, 2002.

Matthews, G. and Deary, I.: Personality traits. Cambridge, UK, Cambridge University Press, 1998.

Michael, R. and Zumpe, D.: Sexual violence in the United States and the role of season, Am. J. Psychiat., 140, 883–886, 1983.

Murray, G., Hay, D., and Armstrong, S.: Personality factors in seasonal affective disorder: Is seasonality and aspect of neuroticism?, Pers. Indiv. Differ., 19, 613–617, 1995.

Palinkas, L.: Mental and cognitive performance in the cold, Int. J. Cireumpolar Health, 60, 430–439, 2001.

Paspalanov, I., Shtetinski, D., and Eysenck, S.: Bulgarian adaptation of the Eysenck Personality Questionnaire, Psychology, 5, 279–292, 1984 (in Bulgarian).

Persinger, M.: Lag responses in mood reports to changes in the weather matrix, Int. J. Biometeor., 19, 108–114, 1975.

Rimskii, R. R. and Rimskii, S. A. (compilers): Almanac of psychological tests, Moscow, 1995 (in Russian).

Scott, J.: Impact of weather conditions on mood. The impact of weather conditions on mood variability in geographically relocated versus non-relocated individuals, Minnesota State University, Mankato, available at: http://www.mnsu.edu/urc/journal/2007/scott.pdf, 2007.

Shtetinski, D. and Paspalanov I.: Methodical guide for work with the Bulgarian form of the C. Spielberger's Anxiety Inventory (STAI – form Y), Sofia, 1989 (in Bulgarian).

Spielberger, C., Gorsuch, R., and Lushene, R.: The State-Trait Anxiety Inventory: Test manual, Palo Alto, CA, Consulting Psychologist Press, 1970.

Sulman, F.: Keine Angst mehr vor dem Föhn: die Wetterfühligkeit und ihre Behandlung, Umsch. Frankf. a/M, 80, 291–292, 294, 295, 1980.

Thomson, W.: A change of air, Charles Scribner, New York, 1979.

Tishkov, H.: Method of complex-climatic analysis – General climatology of D. Dimitrov, (3rd edn.), 1985 (in Bulgarian).

Trenkle, H.: Wetter und Mensch. Medizinische Welt, 33, 1285–128, 1982.

Tromp, S.: Studies on the origin and biological effects of the Chinook in western Canada, in: Biometeorological survey, vol. 1, part A: Human Biometeorology, edited by: Tromp, S. and Bouma, J., Heyden, London, 191–194, 1979.

Watson, D.: Mood and temperament, New York, Guilford Press, 2000.

11

MEGAPOLI: concept of multi-scale modelling of megacity impact on air quality and climate

A. Baklanov[1], M. Lawrence[2], S. Pandis[3], A. Mahura[1], S. Finardi[4], N. Moussiopoulos[5], M. Beekmann[6,7], P. Laj[6], L. Gomes[6], J.-L. Jaffrezo[6], A. Borbon[7], I. Coll[7], V. Gros[8], J. Sciare[8], J. Kukkonen[9], S. Galmarini[10], F. Giorgi[11], S. Grimmond[12], I. Esau[13], A. Stohl[14], B. Denby[14], T. Wagner[2], T. Butler[2], U. Baltensperger[15], P. Builtjes[16], D. van den Hout[16], H. D. van der Gon[16], B. Collins[17], H. Schluenzen[18], M. Kulmala[19], S. Zilitinkevich[19], R. Sokhi[20], R. Friedrich[21], J. Theloke[21], U. Kummer[21], L. Jalkinen[22], T. Halenka[23], A. Wiedensholer[24], J. Pyle[25], and W. B. Rossow[25]

[1]Danish Meteorological Institute (DMI), Lyngbyvej 100, 2100, Copenhagen, Denmark
[2]Max Planck Institute for Chemistry (MPIC), Germany
[3]Foundation for Research and Technology (FORTH), Hellas, University of Patras, Greece
[4]ARIANET Consulting, Italy
[5]Aristotle University Thessaloniki (AUTH), Greece
[6]Centre National de Recherche Scientifique (CNRS), France
[7]Laboratory Inter-universitaire des Systèmes Atmosphériques (LISA), France
[8]Laboratoire des Sciences du Climat et de l'Environnement (LSCE), France
[9]Finnish Meteorological Institute (FMI), Finland
[10]Joint Research Center (JRC), Ispra, Italy
[11]International Centre for Theoretical Physics (ICTP), Italy
[12]King's College London (KCL), UK
[13]Nansen Environmental and Remote Sensing Center (NERSC), Norway
[14]Norwegian Institute for Air Research (NILU), Norway
[15]Paul Scherrer Institute (PSI), Switzerland
[16]TNO-Built Environment and Geosciences, The Netherlands
[17]UK MetOffice (MetO), UK
[18]University of Hamburg (UHam), Hamburg, Germany
[19]University of Helsinki (UHel), Finland
[20]University of Hertfordshire – Centre for Atmospheric & Instrumentation Research (UH-CAIR), UK
[21]University of Stuttgart (UStutt), Germany
[22]World Meteorological Organization (WMO), Switzerland
[23]Charles University Prague (CUNI), Czech Republic
[24]Institute of Tropospheric Research (IfT), Germany
[25]Centre for Atmospheric Science, University of Cambridge (UCam), UK

Abstract. The EU FP7 Project MEGAPOLI: "Megacities: Emissions, urban, regional and Global Atmospheric POLlution and climate effects, and Integrated tools for assessment and mitigation" (http://megapoli.info) brings together leading European research groups, state-of-the-art scientific tools and key players from non-European countries to investigate the interactions among megacities, air quality and climate. MEGAPOLI bridges the spatial and temporal scales that connect local emissions, air quality and weather with global atmospheric chemistry and climate. The suggested concept of multi-scale integrated modelling of megacity impact on air quality and climate and vice versa is discussed in the paper. It requires considering different spatial and temporal dimensions: time scales from seconds and hours (to understand the interaction mechanisms) up to years and decades (to consider the climate effects); spatial resolutions: with model down- and up-scaling from street- to global-scale; and two-way interactions between meteorological and chemical processes.

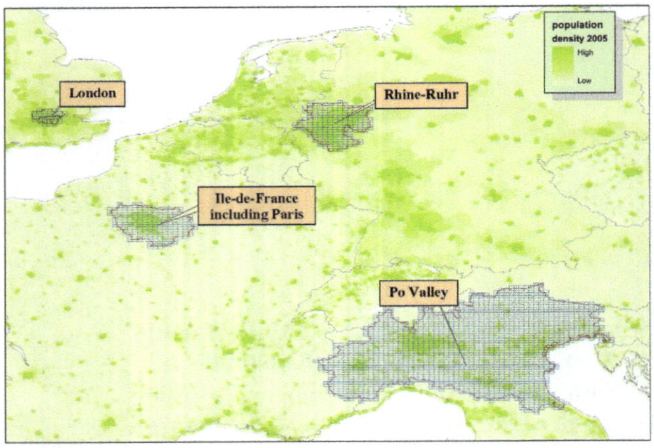

Figure 1. Domain of selected 1st level Megacities in MEGAPOLI (van der Gon et al., 2009).

1 Introduction

For the past few hundred years, human populations have been clustering in increasingly large settlements. In 2007, for the first time in history, the world's urban population exceeded the rural population. At present, there are about 20 cities worldwide with a population of 10 mln inhabitants or greater, and 30 with a population of exceeding 7 mln inhabitants. These numbers are expected to grow considerably in the near future. Such coherent urban areas with more than about 5 mln people are usually called megacities (although there is no formal definition of a megacity at present). In Europe there are 6 major population centres that clearly qualified as megacities: London, Paris, the Rhine-Ruhr region, the Po Valley, Moscow, and Istanbul. The first four of them are considered in the project as the 1st level megacities or urban agglomerations in focus (Fig. 1).

2 Concept and project objectives

The new EU FP7 Project MEGAPOLI: "Megacities: Emissions, urban, regional and Global Atmospheric POLlution and climate effects, and Integrated tools for assessment and mitigation" (http://megapoli.info) started in October 2008, to investigate the interactions among megacities, air quality and climate. The consortium consists of 23 full main partners from 11 European countries, 12 international research non-funded partners and multiple end-users from Canada, Chile, China, Egypt, France, India, Mexico, Russia, Thailand, Turkey, UK, Ukraine, and USA. The project includes both basic and applied research, and bridges the spatial and temporal scales that connect local emissions, air quality and weather conditions with global atmospheric chemistry and climate. The main objectives are (i) to assess impacts of megacities and large air-pollution hot-spots on local, regional and global air quality, (ii) to quantify feedbacks among megacity air quality, local and regional climate, and global climate change, (iii) to develop improved integrated tools for prediction of air pollution in megacities.

In order to fulfil the objectives the following scientific questions are addressed:

- What is the change of exposure of the overall population to the major air pollutants as people move into megacities? What are the health impacts of this exposure?
- How do megacities affect air quality on regional and global scales? What is the range of influence for major air pollutants (ozone, particulate matter, etc.)?
- What are the major physical and chemical transformations of air pollutants as they are moving away from megacities? What happens to the organic particulate matter, volatile organic compounds, etc.?
- How accurate are the current emission inventories for megacities in Europe and around the world? What are the major gaps?
- How large is the current impact of megacities on regional and global climate?
- How will the growth of megacities affect future climate at global and regional scales?
- What is the impact of large-scale dynamic processes on air pollution from megacities?
- What are the key feedbacks between air quality, local climate and global climate change relevant to megacities? For example, how will climate change affect air quality in megacities?
- How should megacities (emissions, processing inside megacities, meteorology) be parameterised in regional and global models?
- What type of modelling tools should be used for the simulation of multi-scale megacity air quality – climate interactions?
- Which policy options are available to influence the emissions of air pollutants and greenhouse gases in megacities and how can these options be assessed?

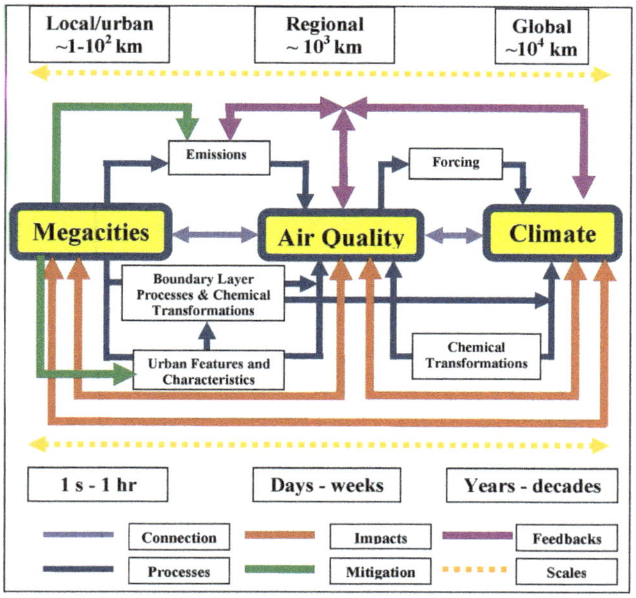

Figure 2. Schematic showing the main linkages between megacities, air quality and climate. The connections and processes will be the focus of MEGAPOLI. In addition to the overall connections between megacities, air quality and climate, the figure shows the main feedbacks, ecosystem, health and weather impact pathways, and mitigation routes which will be investigated. The relevant temporal and spatial scales are additionally included.

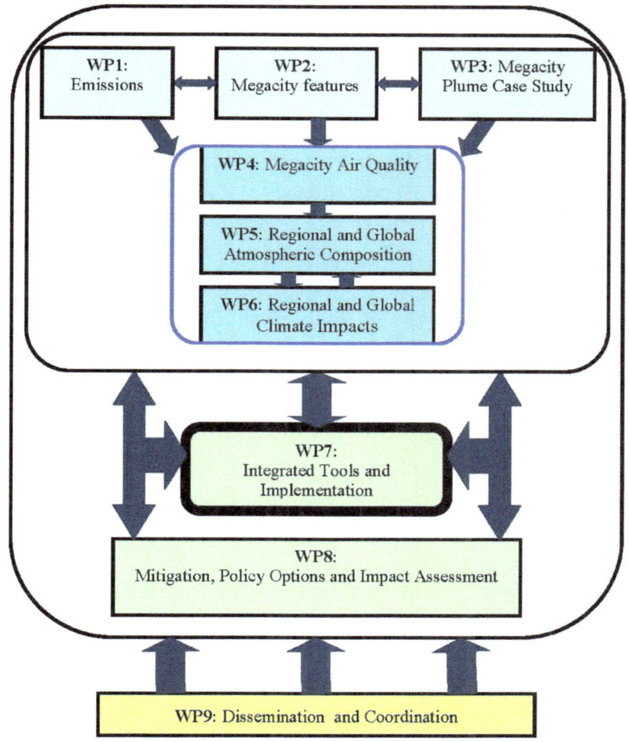

Figure 3. MEGAPOLI Work Packages (WPs) structure and linkage.

3 Methodology and results

The project focuses on research spanning the range from emissions to air quality, effects on regional and global climate, and feedbacks and mitigation potentials. The project takes into account different features and growing trends that characterize cities located in developed and developing countries to highlight their present and future effects on local to global air quality and climate (Baklanov and Mahura, 2009).

The proposed hypothesis is that megacities around the world have an impact on air quality not only locally, but also regionally and globally and therefore can also influence the climate of our planet. In Fig. 2 a schematic description of how megacities, air quality and climate interact is presented. Although some of the links shown have already been considered, a complete quantitative picture of these interactions is still missing. Understanding and quantifying of these missing links is the focus of MEGAPOLI.

The first critical step in improving our understanding of how megacities impact air quality, atmospheric composition and climate on different scales is the development of high-quality inventories of the emissions of relevant gases and aerosols and their precursors, and determining how these are anticipated to change in the mid-term future, as well as how these change under various scenarios (e.g., relocation of 10% of the population out of a megacity and into the surrounding countryside). This forms the key input to the remaining components of the study, and will also, through an understanding of the sensitivity of emissions of different compounds and from different sectors, form the basis for sensible approaches to mitigation strategies. This task is used as a starting point with the corresponding emission inventories developed by local administrations for major urban areas. These will be improved when necessary, adjusted to the appropriate model scale and integrated into larger scale datasets for their use in regional and global scale atmospheric composition modelling. This approach allows the exploitation of former investments and available information, and builds connections between local air quality management authorities and the air quality and climate research community. The first version of the MEGAPOLI European gridded emission inventory for the base year 2005 has the resolution of 6×6 km with nesting up to 1×1 km inventories for the 1st level megacities (Paris, London, Rhine-Ruhr area, Po Valley) (van der Gon et al., 2009). Additionally to the pollutant emission inventory an anthropogenic heat flux (AHF) model (0.25×0.25 arc-minute resolution) was developed and used to compute the AHF inventories for Europe and London (Allen et al., 2010). Emissions are the focus of WP1, which links to the other aspects of the project as shown in Fig. 3.

WP2 is focusing on the megacity features (e.g. morphology, AHF), along with processes taking place in the urban canopy and boundary layer, which are responsible for the airborne transport and transformation of pollutants and urban climate effects. This WP is aimed at the development of databases of morphology/landuse classifications for megacities, as well as developing databases and sub-grid parameterisations of urban layer processes for megacity, regional and global scale models. A morphology database for Paris has been developed (Sievinen et al., 2010), along with a hierarchy of urban canopy and energy budget parameterisations for different scale models (Grimmond et al., 2010; Mahura and Baklanov, 2010), which are being used to evaluate the surface flux balance modelling and urban features needed for climate and air quality models.

Pollutant emissions impact the chemical composition of the atmosphere on different scales. This in turn influences the climate through radiative transfer and effects on clouds and the hydrological cycle (precipitation, etc.). These issues form the core of the project, and are analysed by WPs 3–6 (Fig. 3).

WP3 focuses on the characterization of aerosols and relevant precursors at urban, suburban, and surrounding non-urban areas. Two intensive measurement campaigns were performed around Paris during 1–31 July 2009 and 15 January–15 February 2010. The campaigns aimed at examining the evolution of aerosols and gas-aerosol interactions in the urban outflow and specifically at better quantifying primary and secondary organic aerosol (SOA) sources for a European megacity. They included 3 primary and 7 secondary fixed ground measurement sites, an aircraft and 5 mobile vans (Beekmann et al., 2010). More than 25 research laboratories participated. It was found that the pollution plume was still well defined at more than 100 km downwind from the agglomeration, which gives a clear framework for later studying SOA build-up in the plume. Significant new particle formation events were frequently observed during the campaigns. Paris has been chosen mainly because it is a very concentrated European urban pollution hot spot surrounded by rural areas.

The objective of WP4 is to improve understanding and modelling of local and urban-scale impacts of megacity emissions, on the urban and the surrounding area air quality on examples of the four key megacities (Fig. 1). In particular, new physical and chemical parameterisations and zooming approaches (Moussiopoulos et al., 2010; Koraj and Pandis, 2010) have been implemented and are being tested for several megacities, providing information about the relative importance of the various parameterisations when examining megacity air quality and especially its relation to meteorology. Online coupled chemistry-meteorology models with two-way feedbacks (e.g., Korsholm, 2009) were used to analyse effects of urban aerosols on urban microclimate and to classify meteorological patterns favouring development of urban air pollution episodes in European megacities (Baklanov and Mahura, 2010). Urban aerosols were found to significantly affect several meteorological variables (temperature, inversion layers, radiation budget, cloud processes, precipitation, fog, etc.) in and far from the megacities due to the direct and indirect effects.

Continental and global scale impacts of megacities on atmospheric composition and climate using a combination of monitoring and modelling tools are considered in WPs 5–6. These WPs consider also the effects of future climate and emission scenarios. Each WP activity comprises basic research concerning the individual processes critical for understanding the impacts analysed. Additionally, applied research is devoted to building connections between the scales and aspects introduced, and towards developing integrated assessment tools to define impacts and mitigation strategies. Progress was made in developing and evaluating the satellite-based methods for the measurement of tropospheric gases and aerosols, especially NO_2, in and around megacities (Wagner et al., 2010). The radiative forcing from megacity emissions on the global scale was examined (Folberth et al., 2010). Generally, megacities contribute about 2–5% of the total global annual anthropogenic emission fluxes for various compounds. Megacity pollutants were found to contribute a radiative forcing of $+6.3 \pm 0.4\,\mathrm{mW/m^2}$ from an increase in the ozone burden, while the impact on CH_4 contributes a forcing of $-1.0 \pm 0.5\,\mathrm{mW/m^2}$. The aerosol forcing from megacity pollutants amounts to $-8.0 \pm 1.6\,\mathrm{mW/m^2}$ under present-day conditions.

The information from WPs 1–6 is brought together in WPs 7–8. In WP7 the information and model developments are used to develop integrated tools for prediction of megacity impacts on air quality. Progress has been made on producing a European framework for online and offline coupling of meteorological and atmospheric chemical transport models (Baklanov, 2010). The integrated methodology will be implemented to assess the air quality and climate impacts of selected world megacities by employing improved multi-scale air quality and climate models (urbanized, online and off-line coupled, one- and two-way nested). In addition, the results of the atmospheric and climate modelling are used to estimate and assess (in monetary terms) impacts on human health and ecosystems and climate change impacts of megacities with a methodology developed in WP8. Finally, the information from the integrated assessments will be input into WP8, focusing on mitigation options, which will be assessed by creating scenarios of possible future developments of megacities, in which these options are implemented. The emissions for these scenarios are calculated in WP1 and used as input for the integrated modelling. By comparing the results, the different scenarios and policy options are assessed.

MEGAPOLI follows a pyramid strategy (Baklanov et al., 2008) of undertaking detailed measurements in one European major city (Paris), performing detailed analysis for the four 1st level megacities (Fig. 1) and twelve 2nd level megacities with existing air quality datasets, and investigating the effects of all megacities on climate. The results will be disseminated to authorities, policy community, researchers and the other stakeholders in the corresponding megacities with aim to improve predicting megacity environment for the future. Results and instruments will support the definition of areas and scales of effective measures to improve present and future air quality in large urban agglomerations.

4 Summary and expected impacts

The MEGAPOLI project (http://megapoli.info) is contributing to the strategic goal of promoting sustainable management of the environment and its resources. It will do this by advancing our knowledge on possible interactions between air quality, climate and human activities related to large urban centres and hotspots. Megacities constitute major sources of anthropogenic air pollution and hence, affect the lives of hundreds of millions of people in the world directly by the quality of air that they breathe and through complex interactions resulting in climate change. Research within this project will lead to improved modelling and assessment tools. In particular, MEGAPOLI will formulate a European methodology for integrated air quality and climate assessment over multiple scales (urban to global).

MEGAPOLI places emphasis on the interactions between air quality and climate change impacts resulting from megacities on regional to global scales and potential mitigation options. It will further lead to an integrated methodology and corresponding tools to assess these impacts not only in Europe but also elsewhere.

MEGAPOLI will lead to significant scientific innovations including:

i. Integration of the interactions and processes affecting air quality and climate change on regional to global scales coupled with the capability of estimating the human, ecosystem and economic impact of air pollution resulting from megacities;

ii. Development of an integrated European methodology and tools to assess the impacts within and from megacities on city to global scales;

iii. Integration of ground-based, aircraft and satellite technologies with state-of-the-art modelling tools;

iv. Integrated approaches for addressing the feedbacks and interlinkages between climate change and regional air quality related to megacities;

v. Integration of knowledge and practical implementation of improved tools according to level of complexity to the main megacities and hotspots in focus;

vi. Improved current and future emission estimates for natural and anthropogenic sources of air pollutants;

vii. Development of an integrated assessment methodology for supporting EU and global policy frameworks. This will be achieved through the assessment of mitigation options and the quantification of impacts from polluted air-masses on larger scale atmospheric dynamics;

viii. Examination of the important feedbacks among air quality, climate and climate change;

ix. A robust, global information dissemination gateway on air quality, climate change and mitigation and policy options for European stakeholders strengthening the European Research Area.

MEGAPOLI will significantly extend the current state-of-the-art in the assessment capabilities by developing and implementing reliable integrated tools on multiple scales and for multiple pollutants. These will be applied to assess directly the impact of the largest urban centres and hotspots in Europe and globally by employing highly advanced as well as simpler tools. The project brings together current off-line approaches as well as new on-line methods enabling feedbacks to be quantified on multiple scales enabling mitigation options to be examined more effectively.

Acknowledgements. The research leading to these results has received funding from the European Union's Seventh Framework Programme FP/2007-2011 within the project MEGAPOLI, grant agreement no. 212520.

Edited by: M. Piringer

References

Allen, L., Beevers, S., Lindberg, F., Iamarino, M., Kitiwiroon, N., and Grimmond, C. S. B.: Global to City Scale Urban Anthropogenic Heat Flux: Model and Variability, Deliverable 1.4, MEGAPOLI Scientific Report 10-01, MEGAPOLI-04-REP-2010-03, ISBN 978-87-992924-4-8, 87 pp., 2010.

Baklanov, A. (Ed.): Framework for Integrating Tools, Deliverable D7.1, MEGAPOLI Scientific Report 10-11, MEGAPOLI-14-REP-2010-03, ISBN 978-87-993898-4-1, 68 pp., 2010.

Baklanov, A. and Mahura, A. (Eds.): First Year MEGAPOLI Dissemination Report, Deliverable 9.4.1, MEGAPOLI Scientific Report 09-03, MEGAPOLI-03-REP-2009-12, ISBN 978-87-992924-3-1, 57 pp., 2009.

Baklanov, A. and Mahura, A. (Eds.): Interactions between Air Quality and Meteorology, Deliverable D4.3, MEGAPOLI Scientific Report 10-10, MEGAPOLI-13-REP-2010-03, ISBN 978-87-993898-3-4, 48 pp., 2010.

Baklanov, A., Lawrence, M., and Pandis, S.: MEGAPOLI Description of Work (2008–2011), Copenhagen, ISBN 978-87-992924-0-0, 150 pp., 2008.

Beekmann, M., Baltensperger, U., and the MEGAPOLI campaign team: Database of Chemical Composition, Size Distribution and Optical Parameters of Urban and Suburban PM and its Temporal Variability (Hourly to Seasonal), Deliverable D3.1, MEGAPOLI Scientific Report 10-15, MEGAPOLI-18-REP-2010-10, http://megapoli.dmi.dk/publ/MEGAPOLI_sr10-15.pdf, 21 pp., 2010.

Folberth, G., Rumbold, S., Collins, W. J., and Butler, T.: Determination of Radiative Forcing from Megacity Emissions on the Global Scale, Deliverable D6.2, MEGAPOLI Scientific Report 10-08, MEGAPOLI-11-REP-2010-03, ISBN 978-87-993898-1-0, 19 pp., 2010.

Grimmond, C. S. B., Blackett, M., and Best, M. J.: Urban Energy Balance Models Comparison, Deliverable D2.3, MEGAPOLI Scientific Report 10-07, MEGAPOLI-10-REP-2010-03, ISBN 978-87-993898-0-3, 72 pp., 2010.

Koraj, D. and Pandis, S. N.: Evaluation of Zooming Approaches Describing Multiscale Chemical Transformations, Deliverable D4.2, MEGAPOLI Scientific Report 10-05, MEGAPOLI-08-REP-2010-01, ISBN 978-87-992924-8-6, 29 pp., 2010.

Korsholm, U.: Integrated modeling of aerosol indirect effects – development and application of a chemical weather model, PhD thesis University of Copenhagen, Niels Bohr Institute and DMI, Research Department, http://www.dmi.dk/dmi/sr09-01.pdf, 2009.

Mahura, A. and Baklanov, A. (Eds.): Hierarchy of Urban Canopy Parameterisations for Different Scale Models, Deliverable D2.2, MEGAPOLI Scientific Report 10-04, MEGAPOLI-07-REP-2010-03, ISBN 978-87-992924-7-9, 50 pp., 2010.

Moussiopoulos, N., Douros, J., and Tsegas, G. (Eds.): Evaluation of Zooming Approaches Describing Multiscale Physical Processes, Deliverable D4.1, MEGAPOLI Scientific Report 10-03, MEGAPOLI-06-REP-2010-01, ISBN 978-87-992924-6-2, 41 pp., 2010.

Sievinen, P., Hellsten, A., Praks, J., Koskinen, J., and Kukkonen, J.: Urban Morphological Database for Paris, France, Deliverable D2.1, MEGAPOLI Scientific Report 10-02, MEGAPOLI-05-REP-2010-03, ISBN 978-87-992924-5-5, 13 pp., 2010.

van der Gon Denier, H. A. C., Visschedijk, A. J. H., van der Brugh, H., Dröge, R., and Kuenen, J.: A base year (2005) MEGAPOLI European gridded emission inventory (1st version), Deliverable 1.2, MEGAPOLI Scientific Report 09-02, MEGAPOLI-02-REP-2009-10, ISBN 978-87-992924-2-4, 17 pp., 2009.

Wagner, T., Beirle, S., and Shaiganfar, R.: Characterization of Megacity Impact on Regional and Global Scales Using Satellite Data, Deliverable D5.1, MEGAPOLI Scientific Report 10-09, MEGAPOLI-12-REP-2010-03, ISBN 978-87-993898-2-7, 25 pp., 2010.

High resolution modelling of wind fields for optimization of empirical storm flood predictions

B. Brecht and H. Frank

Deutscher Wetterdienst, Offenbach, Germany

Correspondence to: B. Brecht (benedict-manuel.brecht@dwd.de)

Abstract. High resolution wind fields are necessary to predict the occurrence of storm flood events and their magnitude. Deutscher Wetterdienst (DWD) created a catalogue of detailed wind fields of 39 historical storms at the German North Sea coast from the years 1962 to 2011. The catalogue is used by the Niedersächsisches Landesamt für Wasser-, Küsten- und Naturschutz (NLWKN) coastal research center to improve their flood alert service.

The computation of wind fields and other meteorological parameters is based on the model chain of the DWD going from the global model GME via the limited-area model COSMO with 7 km mesh size down to a COSMO model with 2.2 km. To obtain an improved analysis COSMO runs are nudged against observations for the historical storms. The global model GME is initialised from the ERA reanalysis data of the European Centre for Medium-Range Weather Forecasts (ECMWF).

As expected, we got better congruency with observations of the model for the nudging runs than the normal forecast runs for most storms. We also found during the verification process that different land use data sets could influence the results considerably.

1 Introduction

Storm floods at coastal regions are natural phenomena occuring at irregular intervals and variable intensity. The three biggest storm floods at the German North Sea coast in the 20th century occurred in the years 1962, 1976 and 1999. The flood in 1976 reached the highest water levels at almost all German North Sea stations. Nevertheless the flood of Hamburg in 1962 caused much more damage and 315 fatal casualties (Jensen et al., 2006). In the 21th century three very severe storm floods, with a water level of at least 3.5 m over the mean high water level (MHW), occurred in the years 2006, 2007 and 2013 (e.g. Wiegand, 1990; BSH, 2014).

The central target of the project "OptempS-MohoWif" (High resolution modelling of wind fields for optimization of empirical storm flood predictions) is an improvement of the regional storm flood alert service at the North Sea coast of Lower Saxony. The project is divided into two parts. The first part of the project concerning the wind fields is done by Deutscher Wetterdienst (DWD). The second part concerning water level modeling is done by the flood alert service of Niedersächsisches Landesamt für Wasser-, Küsten- und Naturschutz (NLWKN). To improve their flood model, highly-resolved spatial and temporal wind fields of severe storms are necessary, which are provided by the DWD.

Therefore, a catalogue of 39 historical storms, which lead to storm floods at the German North Sea coast from the years 1962 to 2011, was created (see Table 1). Each storm event was simulated for about 4–5 days, except for two events with 7 and 11 days respectively, when several storms are connected. The forecast of an event starts three days before the highest water level occurres. It ends one or two low tides after the maximum water level. The simulations of the meteorological fields are computed by the model chain of DWD, which is started from reanalysis data of ECMWF (European Centre for Medium-Range Weather Forecasts). To verify the simulations the meteorological fields are compared with observational data of up to 25 stations at the German, Dutch and Danish North Sea coast, or stations in the North Sea like platforms or fire ships.

In 2005 there was a similar project for 22 storms from 1962 until 2002 (Frank and Majewski, 2006). The major improvements in the new project are the nudging of the COSMO simulations to observations as well as newer model versions, higher resolution, greater model domain, and more frequent reanalysis data used by GME.

In the following section the model setup and the nudging process are described. Then the verification procedure is shown, in presenting the results of one selected storm, followed by the demonstration of the results for the set of the storms. Finally, the most important information are summarized.

2 Model chain

The model chain of the DWD consists of the global model GME (Majewski et al., 2002) and the limited-area non-hydrostatical model COSMO (COSMO model documentation, 2014), which is run in the two versions COSMO-eu and COSMO-de. GME is a hydrostatic model operating on an icosahedral grid. For these simulations a mesh width of 30 km and 60 vertical layers is used whereas, operationally, GME runs on a 20 km grid. Here, GME is initialized four times a day (00:00, 06:00, 12:00, and 18:00 UTC) from reanalysis data of ECMWF. For storms in and after 1979 the ERAInterim reanalysis data (Dee et al., 2011) is used. Before 1979 we use ERA-40 reanalysis data (Uppala et al., 2005). GME provides initial and boundary data for COSMO-eu with a mesh width of 7 km and 40 vertical layers. COSMO-eu in turn drives COSMO-de which has a mesh width of 2.2 km and 65 vertical layers. At this resolution deep convection can be explicitly resolved (Baldauf et al., 2011). Therefore, only shallow convection is parametrized in COSMO-de. Compared to the operational COSMO-EU at the DWD, the domain of COSMO-eu is decreased in the south and the east (see Fig. 1), as these regions are not important for storm flood predictions at the North Sea coast. COSMO-de is increased in the north, west and south compared to the operational COSMO-DE (see Fig. 1). The vertical resolution is increased from operationally 50 to 65 layers. COSMO-de serves as test bed for a possible new operational setup of COSMO-DE.

In a first series of experiments the model chain was started by a new reanalysis field every 6 h and run to 18 h. 6-hourly segments of the 18 h forecasts were connected to get a continuous time series with minimal jumps from one forecast to the next. Tests were run, in order to determine which segments yielded the minimal root mean square difference of mean sea level pressure (p.m.s.l.) from the end of a segment to the start of the segments of the next forecast, for example "p.m.s.l.(00:00 UTC + 8 h)–p.m.s.l.(06:00 UTC + 2 h)". This test was made for 3 complete storms. It was found that the RMSE at "X UTC + 10 h–(X+6) UTC + 4 h" was minimal. Therefore the hours 5–10 of every 18 h forecast are used and connected to a continuous time series to obtain the forecast with the minimal jump of p.m.s.l. between two consecutive forecasts of one storm.

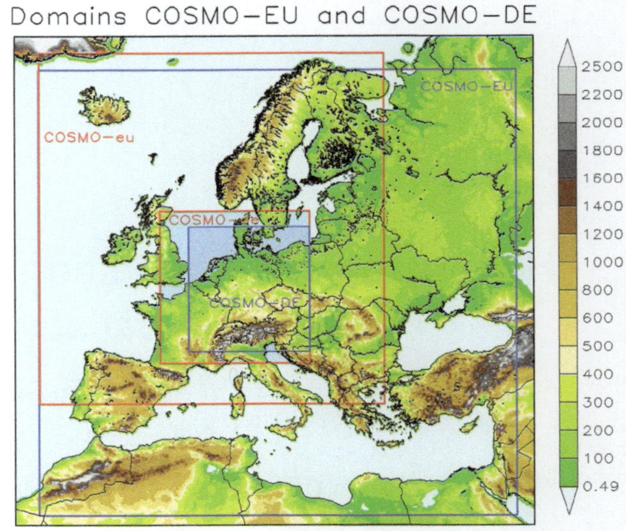

Figure 1. Domains of COSMO: the red lines show the domains of the 7 km COSMO-eu, and 2.2 km COSMO-de. The blue lines show the domains of the operational versions COSMO-EU and COSMO-DE.

2.1 Nudging runs

In order to get an improved analysis at high resolution, the simulations with COSMO have additionally been carried out with the Newtonian relaxation or nudging technique (Stauffer et al., 1990) to obtain one continuous analysis without jumps. With this method the prognostic variables of the model are relaxed towards prescribed values within a period of time (Schraff, 1997). In this case the prescribed values are direct observations, which is advantageous for asynoptic observations (Stauffer and Seaman, 1994). The relaxation against the observations is done via additional terms in the prognostic equations. In practical applications, the nudging term should and usually does remain smaller than the largest term of the dynamics. This situation is related to the basic idea of the method that the model fields are to be relaxed towards the observed values without significantly disturbing the dynamic balance of the model (Schraff and Hess, 2002).

Observations are used from radiosondes (horizontal wind, temperature, humidity and pressure), aircraft measurements (horizontal wind and temperature), wind profiler (horizontal wind) and surface observations (10 m wind, only used for stations below 100 m above sea level to select only stations in flat terrain).

The observations are obtained from the DWD archives, which date back until 1966. Hence, for the first storm in 1962 no nudging run is available. For the nudging runs the

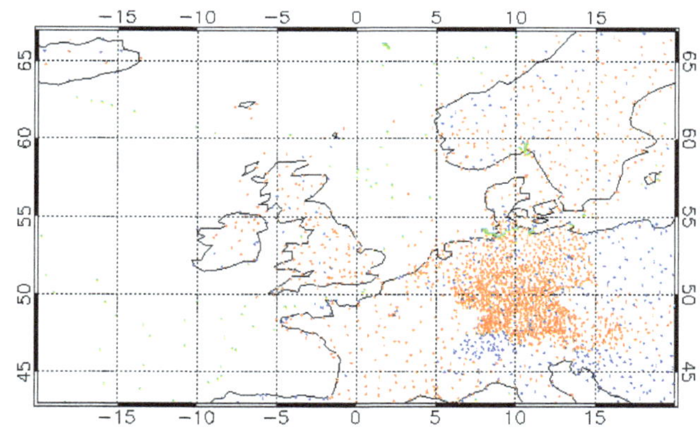

Figure 2. Data coverage of Synop land and ship observations on 9 November 2007. The blue dots show the manned land stations, the red dots the automatic land stations and the green dots the ships.

observations are processed with the same quality control procedure as the operational runs.

A more detailed description of the nudging in COSMO is given by Schraff and Hess (2002).

The nudging runs of COSMO were made in segments of 6 h. However, a new segment is started from the last analysis of the previous segment. Hence, each storm analysis is simulated in a continuous analysis cycle, which is initialized from the reanalysis data only at the very start of the storm event.

2.1.1 Availability of Nudging data

The data availability increases from older to more recent storms. The total available observations per 6 h for the COSMO-eu area are 2000–3000 Synop land and ship (relation 10 : 1) observations for events from 1966–1976. In 1977 it increased to approximately 5000 per 6 h mainly due to an increased frequency of Synop measurements (to 3 hourly or hourly observations), and due to the first use of drifting buoys and aircraft observations. Until the year 2000 it slowly increased up to 15 000 observations per year. Radiosonde data have been available since 1991, and wind profiler data have been available since 2000. Since 2006 over 40 000 observations in 6 h are available. As an example, Fig. 2 shows the coverage of available Synop land and ship observations on 9 November 2007 during storm "Tilo".

The used data are less than the available data caused by the quality control, the availability of meta information like station height, or differences between real station height and model orography. Approximately 40 % to 90 % of the available data are used for the nudging.

For the storms from 1966–1976 only 40–50 % of the data are used by the model, mostly due to missing station altitudes of Synop land stations. From 1977–2000 80–90 % of the measurements are accepted. For later runs this ratio decreases again to 60–65 %.

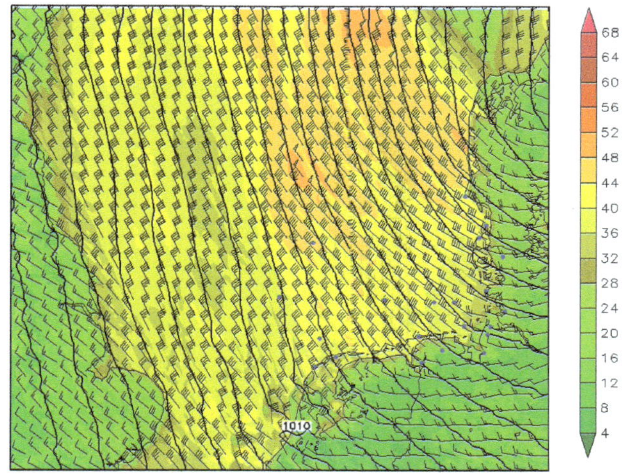

Figure 3. Wind field over the North Sea of storm "Tilo" at 9 November 2007, 06:00 UTC, 3 h before the water reached the highest level at Norderney. Shown is a COSMO-de nudging run. The wind 10 m above ground is shown in knots. As contour lines p.m.s.l. is shown in 2 hPa intervals. The blue dots show the locations of the stations which were used for verification.

3 Results

The COSMO model data were verified with observations of up to 25 Synop stations at or near the North Sea coast of Denmark, Germany and the Netherlands, as well as some stations in the North Sea (platforms, fire ships). Except for one station (Scharhörn, external data), these stations were also used in the nudging runs we made. However, not all stations were available for all storms. The verification was made for sea level pressure, p.m.s.l., wind speed at 10 m, v_10m, and wind direction at 10 m, dd_10m. We calculated the bias of the model (model-obs), and the RMSE of the model values vs. the observation values. In the next section we present results for storm "Tilo" in 2007, accordingly we present the general results for all 39 simulated storms.

3.1 Verification of storm "Tilo" 2007

Storm "Tilo" passed from the south of Iceland to the west coast of Norway and to the North of the Skagerrak. For many hours the water was pushed by northwesterly winds into the German Bight, see Fig. 3, which resulted in the high water level. As is typical for Northern Hemisphere mid-latitude storms the strongest winds of "Tilo" occurred on the southwest side of the low pressure centre behind the cold front.

For the verification of storm "Tilo" we verified data of 21 stations. We can compare both COSMO models as well as nudging runs with forecast runs (see Fig. 4). For COSMO-eu the RMSE and biases for all 3 variables, v_10m, dd_10m, and p.m.s.l., are less for the nudging analysis than for the forecast runs. The biases are $1.1 \pm 1.1\,\mathrm{m\,s^{-1}}$ compared to

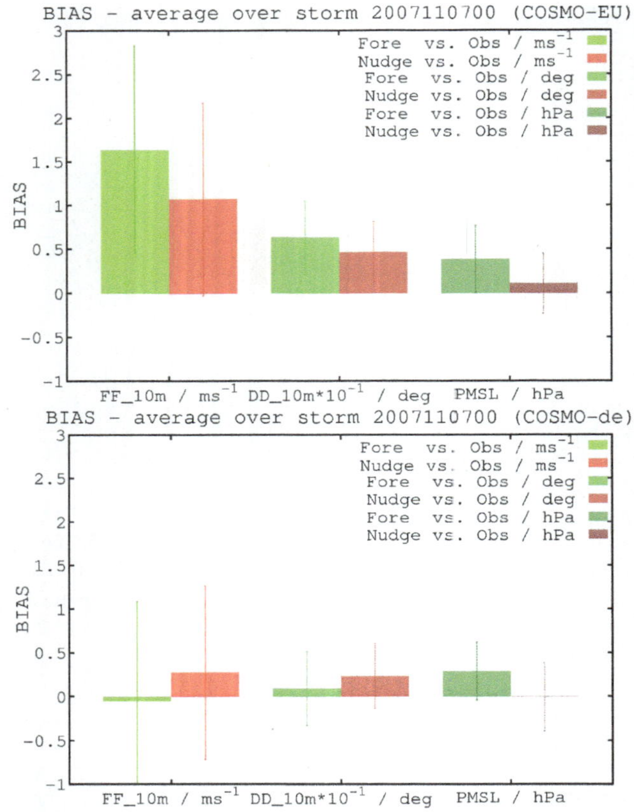

Figure 4. The biases and uncertainties of the biases for storm "Tilo" for the three variables v_10m, dd_10m and p.m.s.l. for COSMO-eu and COSMO-de and both for nudging runs (red) and runs without nudging (green). At the top the biases for COSMO-eu and at the bottom the biases for COSMO-de are shown.

$1.6 \pm 1.2\,\mathrm{m\,s^{-1}}$, $4.6 \pm 3.4°$ vs. $6.3 \pm 4.2°$, and 0.1 ± 0.3 hPa vs. 0.4 ± 0.3 hPa (see Fig. 4, top). The uncertainties of the values presented here are the standard deviations over all observation stations.

The RMSE values are $2.2 \pm 0.6\,\mathrm{m\,s^{-1}}$ vs. $2.7 \pm 0.8\,\mathrm{m\,s^{-1}}$, $13.0 \pm 4.2°$ vs. $15.6 \pm 4.5°$, and 0.5 ± 0.2 hPa vs. 0.8 ± 0.2 hPa. For this storm, the results of the nudging run were better than for the forecast run for all variables. The results for the biases also show that the variances of the observations are not negligible. The variances for v_10m and dd_10m can arise from the difference of land and sea observations and if their nearest grid points of the model are land or sea points.

COSMO-de showed qualitatively similar behaviour for the RMSE, but differences in the bias. The RMSE of the nudging runs versus the normal forecast runs were $2.0 \pm 0.5\,\mathrm{m\,s^{-1}}$ vs. $2.2 \pm 0.6\,\mathrm{m\,s^{-1}}$, $13.4 \pm 5.0°$ vs. $14.5 \pm 4.3°$, and 0.5 ± 0.2 hPa vs. 0.8 ± 0.2 hPa for p.m.s.l. Hence, the nudging run was slightly better than the run without nudging. But, the bias showed no unambiguous tendency between the two modes (see Fig. 4, bottom).

Table 1. Catalogue of the 39 historical storm floods and the RMSE and bias of the wind speed (v10, $\mathrm{m\,s^{-1}}$) for COSMO-eu (eu) and COSMO-de (de) forecast runs (fo) and nudging runs (an). The first three columns show the storm number (No), the start date (UTC) and the duration (Dur) of a storm (in days, d). An event lasts from approximately 3 days before the main maximum water level until one or two low tides later. Simulations start at 00:00 UTC of the first day and end at 24:00 UTC of the last day.

No	Start	Dur	RMSE eu v10 an/fo	bias eu v10 an/fo	RMSE de v10 an/fo	bias de v10 an/fo
1	1962-02-14	4 d	–	–	–	–
2	1966-11-28	4 d	2.9/2.6	0.5/0.7	2.9/2.6	−0.1/−0.1
3	1967-02-21	4 d	3.0/3.4	−0.3/−0.3	3.6/3.9	−0.5/−0.8
4	1967-02-26	4 d	2.2/2.5	0.2/0.2	2.3/2.6	−0.5/−0.4
5	1973-11-10	11 d	3.4/3.4	0.7/1.0	3.9/3.7	0.6/0.1
6	1973-12-04	4 d	2.6/2.8	0.7/1.4	3.1/3.1	0.7/0.8
7	1973-12-11	4 d	3.0/2.8	1.3/1.4	3.2/3.2	1.0/1.0
8	1976-01-01	4 d	3.4/3.2	−0.2/1.0	3.8/3.3	−0.7/0.1
9	1976-01-18	5 d	2.7/2.6	0.1/0.9	3.5/3.3	−0.6/−0.3
10	1977-12-28	4 d	2.4/2.9	0.1/1.0	2.8/3.3	−0.4/0.1
11	1981-11-21	5 d	2.4/2.6	0.1/1.0	2.8/3.0	−0.6/−0.2
12	1983-01-30	4 d	3.2/3.2	−0.1/0.4	2.1/2.3	−0.8/−0.6
13	1988-12-02	4 d	1.8/2.0	0.6/0.9	1.8/1.7	0.0/0.0
14	1989-02-11	5 d	2.1/2.2	0.3/0.6	2.2/2.3	−0.7/−0.5
15	1990-01-23	5 d	2.4/2.4	0.6/0.7	2.8/2.6	−0.7/−0.8
16	1990-02-24	5 d	2.2/2.5	0.0/0.1	2.6/2.8	−1.0/−1.1
17	1990-08-18	4 d	1.6/2.0	0.0/0.0	1.9/2.2	−0.2/−0.9
18	1991-12-17	5 d	1.9/2.2	0.1/0.3	2.1/2.2	−0.4/−0.5
19	1993-01-20	4 d	2.0/2.2	0.1/0.1	2.9/2.8	−1.5/−1.6
20	1993-12-07	4 d	2.0/2.3	0.0/0.4	2.3/2.4	−0.5/−0.7
21	1993-12-17	4 d	1.8/1.9	0.0/0.3	2.2/2.3	−0.7/−0.9
22	1994-01-25	7 d	2.1/2.2	−0.4/−0.1	2.4/2.5	−0.9/−1.0
23	1994-03-11	4 d	1.7/1.8	−0.3/0.3	2.3/2.2	−1.0/−0.9
24	1994-12-30	4 d	2.3/2.6	0.1/0.6	2.5/2.7	−0.3/−0.6
25	1995-01-07	5 d	1.8/2.4	0.4/1.0	1.9/2.2	0.1/0.3
26	1996-10-27	4 d	1.9/2.3	0.4/1.1	2.0/2.1	−0.2/−0.2
27	1999-02-02	5 d	1.9/2.0	0.2/0.3	2.0/2.1	−0.4/−0.7
28	1999-11-30	5 d	2.4/2.5	0.9/1.1	2.4/2.4	−0.1/−0.6
29	2000-01-27	4 d	1.8/1.9	0.5/0.8	2.0/2.0	−0.5/−0.3
30	2002-10-25	4 d	2.2/2.4	0.4/0.7	2.2/2.4	−0.2/−0.4
31	2006-10-29	4 d	2.1/2.3	0.9/1.2	1.9/2.0	−0.1/−0.3
32	2007-01-09	4 d	2.4/2.8	1.0/1.5	2.2/2.3	−0.6/−0.5
33	2007-01-16	4 d	2.2/2.5	1.1/1.4	2.0/2.3	0.0/0.0
34	2007-03-16	5 d	2.0/2.2	0.4/0.6	1.9/2.1	0.2/0.4
35	2007-11-07	4 d	2.2/2.7	1.1/1.6	2.0/2.2	0.3/−0.1
36	2007-11-23	4 d	2.0/2.4	0.7/1.2	1.8/2.0	0.0/−0.1
37	2008-03-10	4 d	1.9/2.2	0.8/1.1	1.8/1.8	−0.2/−0.3
38	2010-11-09	5 d	1.7/2.2	0.6/1.2	1.6/1.7	−0.2/−0.2
39	2011-02-02	4 d	1.7/1.8	0.4/0.5	1.7/1.8	−0.3/−0.6

3.2 General results

In general we saw the largest improvement by the nudging results for the variable p.m.s.l. For every time series of the nudging runs at nearly all verified stations the RMSE and biases of p.m.s.l. were better than those of the normal forecast runs. There were only a few exceptions at storm events before 1980. For wind speed and direction the forecast without nudging was better for a few stations compared to the nudging runs, but not averaged over all stations.

Comparing all storms (except number one from 1962, where we have no observation data) we see that the RMSE of both COSMO-eu and COSMO-de for v_10m is between $1.6\,\mathrm{m\,s^{-1}}$ and $3.9\,\mathrm{m\,s^{-1}}$, while the higher values mainly occur at older events (before and in 1983, see Table 1). This is also the case for the variable p.m.s.l. For dd_10m no tendency could be observed. If we compare COSMO-eu and COSMO-de we have no significant difference for the RMSE of v_10m of COSMO-eu and COSMO-de averaged over all storms. The same is valid for p.m.s.l. and dd_10m. The results of the RMSE of v_10m for the nudging runs are better or equal compared to the forecast runs for nearly all storms, except for the storms 2, 7, 8 and 9 for COSMO-eu and 2, 5, 8, 9, 13, 15, 19 and 23 for COSMO-de (see Table 1). For dd_10m we qualitatively get the same result, but for p.m.s.l. all nudging runs yield better or equal results as the normal forecast runs.

One reason for the improved results of newer events can be the better quality and higher resolution of the ERAInterim data for storms from 1979 on. Another reason for the improved results of the nudging runs is surely the rapid increase of observation data during the time of the occurrence of the storm events.

Comparing the biases of p.m.s.l. we obtain nearly the same values for COSMO-eu and COSMO-de. We have a small positive bias for normal forecast runs and values around zero for the nudging runs. The standard deviations of the biases are in the range as the biases themselves as we have seen for storm "Tilo". For dd_10m COSMO-eu has a slightly positive bias, which is lower for the nudging runs compared to the forecast runs. COSMO-de has a minimally lower bias than COSMO-eu, averaged over all storms. The standard deviation of the biases of dd_10m is in the range of the biases themselves or even a little bit larger due to the different characteristics of the observation points. The wind speed biases of COSMO-eu and COSMO-de have a similar magnitude, but different sign, positive for COSMO-eu and negative for COSMO-de. Averaged over all storms the positive biases for COSMO-eu are $0.4 \pm 1.4\,\mathrm{m\,s^{-1}}$ for runs with nudging and $0.7 \pm 1.5\,\mathrm{m\,s^{-1}}$ without nudging. The negative biases for COSMO-de are $-0.3 \pm 1.6\,\mathrm{m\,s^{-1}}$ for runs with nudging and $-0.4 \pm 1.6\,\mathrm{m\,s^{-1}}$ without nudging.

Where does this big difference of wind speed bias ($0.7\,\mathrm{m\,s^{-1}}$ for nudging runs respectively $1.1\,\mathrm{m\,s^{-1}}$ without nudging) of both COSMO models comes from? There could be two reasons where the differences result from. One answer is the different mesh width, which will influence the wind speed at grid points near the coast. However, looking at a map of the mean wind speed difference of COSMO-eu and COSMO-de (see Fig. 5) we see the main difference over land outside the Alps which is the result of different roughness length, z_0. The differences in z_0 comes from the usage of different land use classifications. For COSMO-eu z_0 was derived from the GLC2000 data set (Bartholomé and Belward, 2005), whereas for COSMO-de the newer GlobCover

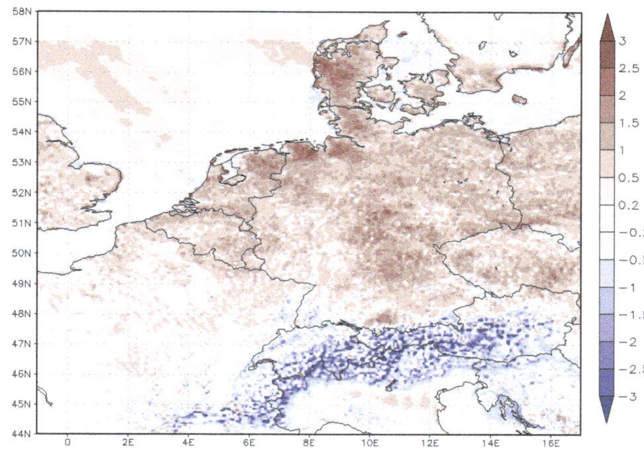

Figure 5. Difference of v_10m for COSMO-eu minus COSMO-de (nudging runs), averaged about the storm event (96 h) of "Tilo" 2007.

data set (Arino et al., 2008) has been employed. Outside the Alps z_0 is higher in the data derived from GlobCover which yields lower near surface winds, which are in better accordance with the observations. In the Alps z_0 for COSMO-eu contains an additional contribution from the orographic variability. We ran COSMO-eu with z_0 derived from GlobCover for three storms ("Tilo" 2007, "Kyrill" 2007, and "Britta" 2006). The wind speed bias of COSMO-eu decreased by $0.2\,\mathrm{m\,s^{-1}}$ averaged over all stations, $0.1\,\mathrm{m\,s^{-1}}$ for sea stations and $0.35\,\mathrm{m\,s^{-1}}$ for land stations which is an improvement for the storm. We conclude GlobCover yields better roughness values for COSMO-eu than GLC2000.

4 Summary

To improve the storm flood model of the NLWKN, DWD produced high-resolution spatial and temporal wind fields of 39 historical storms in the North Sea. The data are available for COSMO-eu and COSMO-de, both, in forecast and in nudging mode. I.e. we have a mini ensemble with 4 members for each storm event.

As expected we got better results for the nudging runs than for the normal forecasts for the RMSE as well as for the biases, with exception of the COSMO-de wind biases, where no clear tendency can be seen. Comparing COSMO-eu and COSMO-de we see no significant difference in the RMSE. The biases of COSMO-eu and COSMO-de differ significantly for the 10 m wind where COSMO-eu overestimates both values. The biases of COSMO-eu and COSMO-de for p.m.s.l. differ only marginally and are both very low.

Looking for the reason of the significant difference of the 10 m wind between COSMO-eu and COSMO-de, we detected big difference in the roughness length z_0 because it was derived from different land use data sets. The wind speed bias of COSMO-eu over land could be reduced by $0.35\,\mathrm{m\,s^{-1}}$ if z_0

was derived from GlobCover as has been done for COSMO-de. A second reason could be the different mesh width of both models.

Further experiments will be made changing tuning parameters in the turbulence scheme of the COSMO model, and with wind speed dependent Charnock constant to calculate the sea surface roughness.

Edited by: M. M. Miglietta

References

Arino, O., Bicheron, P., Achard, F., Latham, J., Witt, R., and Weber, J.-L.: GlobCover the most detailed portrait of Earth, ESA Bulletin, 136, 25–31, 2008.

Baldauf, M., Seifert, A., Förstner, J., Majewski, D., Raschendorfer, M., and Reinhardt, T.: Operational Convective-Scale Numerical Weather Prediction with the COSMO model: Description and Sensitivities, Mon. Weather Rev., 139, 3887–3905, 2011.

Bartholomé, E. and Belward, A. S.: GLC2000: a new approach to global land cover mapping from Earth observation data, Int. J. Remote Sens., 26, 1959–1977, 2005.

Bundesamt für Seeschifffahrt und Hydrographie (BSH), Storm flood classification: http://www.bsh.de, last access: 7 January 2014.

Consortium for Small-scale Modelling, COSMO documentation: http://www.cosmo-model.org, last access: 7 January 2014.

Dee, D. P., Uppala, S. M., Simmons, A. J., Berrisford, P., Poli, P., Kobayashi, S., Andrae, U., Balmaseda, M. A., Balsamo, G., Bauer, P., Bechtold, P., Beljaars, A. C. M., van de Berg, L., Bidlot, J., Bormann, N., Delsol, C., Dragani, R., Fuentes, M., Geer, A. J., Haimberger, L., Healy, S. B., Hersbach, H., Hólm, E. V., Isaksen, L., Kållberg, P., Köhler, M., Matricardi, M., McNally, A. P., Monge-Sanz, B. M., Morcrette, J.-J., Park, B.-K., Peubey, C., de Rosnay, P., Tavolato, C., Thépaut, J.-N., and Vitart, F.: The ERA-Interim reanalysis: configuration and performance of the data assimilation system, Q. J. Roy. Meteorol. Soc., 137, 553–597, 2011.

Frank, H. and Majewski, D.: Hindcasts of historic storms with the DWD models GME, LMQ and LMK using ERA-40 reanalyses, 2006, ECMWF Newsletter, 109, 16–21, 2006.

Jensen, J., Mudersbach, C., Müller-Navarra, S. H., Bork, I., Koziar, C., and Renner, V.: Modellgestützte Untersuchungen zu Sturmfluten mit sehr geringen Eintrittswahrscheinlichkeiten an der deutschen Nordseeküste, Die Küste, 71, 123–167, 2006.

Majewski, D., Liermann, D., Prohl, D., Ritter, B., Buchhold, B., Hanisch, T., Paul, G., Wergen, W., and Baumgardner, J.: The Operational Global Icosahedral-Hexagonal Gridpoint Model GME: Description and High-Resolution Tests, Mon. Weather Rev., 130, 319–338, 2002.

Schraff, C. H.: Mesoscale Data Assimilation and Prediction of Low Stratus in the Alpine Region, Meteorol. Atmos. Phys., 64, 21–50, 1997.

Schraff, C. and Hess, R.: Datenassimilation für das LM. promet, Jahrgang 27, Nr. 3/4, 156–164, 2002.

Stauffer, D. R. and Seaman, N. L.: Use of Four-Dimensional Data Assimilation in a Limited-Area Mesoscale Model Part I: Experiments with synoptic scale data, Mon. Weather Rev., 118, 1250–1277, 1990.

Stauffer, D. R. and Seaman, N. L.: Multiscale four-dimensional data assimilation, J. Appl. Meteorol., 33, 416–434, 1994.

Uppala, S. M., Kållberg, P., Simmons, A., Andrae, U., da Costa Bechtold, V., Fiorino, M., Gibson, J., Haseler, J., Hernandez, A., Kelly, G., Li, X., Onogi, K., Saarinen, S., Sokka, N., Allan, R., Andersson, E., Arpe, K., Balmaseda, M., Beljaars, A., van de Berg, L., Bidlot, J., Bormann, N., Caires, S., Chevallier, F., Dethof, A., Dragosavac, M., Fisher, M., Fuentes, M., Hagemann, S., Hólm, E., Hoskins, B., Isaksen, L., Janssen, P., Jenne, R., McNally, A., Mahfouf, J.-F., Morcrette, J.-J., Rayner, N., Saunders, R., Simon, P., Sterl, A., Trenberth, K., Untch, A., Vasiljevic, D., Viterbo, P., and Woollen, J.: The ERA-40 re-analysis, Q. J. Roy. Meteorol. Soc., 131, 2961–3012, 2005.

Wiegand, P.: Küstenfibel – Ein Abc der Nordseeküste, Westholsteinische Verlagsgesellschaft Boyens & Co., Heide in Holstein, 1990.

Monitoring of climate change in Germany – data, products and services of Germany's National Climate Data Centre

F. Kaspar, G. Müller-Westermeier, E. Penda, H. Mächel, K. Zimmermann, A. Kaiser-Weiss, and T. Deutschländer

Deutscher Wetterdienst, Frankfurter Str. 135, 63067 Offenbach, Germany

Correspondence to: F. Kaspar (frank.kaspar@dwd.de)

Abstract. Germany's national meteorological service (Deutscher Wetterdienst, DWD) is the responsible authority for monitoring climate change in Germany. To fulfill this task it operates the National Climate Data Centre ("Nationales KlimaDatenZentrum, NKDZ"). The historical and current instrumental measurements and visual observations of DWD's station network are archived, quality-controlled and used to provide aggregated products, as for example daily and monthly means or climate normals. Gridded data are generated and used to derive time series of national and regional averages. Phenological observations and radiosonde data are also part of the data base.

In recent years, additional historical data have been digitized to expand the data base. The products are used for informing the public, e.g. as an element of the German climate atlas (http://www.deutscher-klimaatlas.de). One major recent activity was the provision of information for the new climatological reference interval 1981–2010 and an updated climatological analysis based on the newly digitized data.

1 Introduction

Climate is the synthesis of weather conditions in a given area, characterized by long-term statistics (mean values, variances, probability of extreme values, etc.) of the meteorological elements in that area (World Meteorological Organisation, 1992). Climate shows great variability in space and time and the knowledge about observed variability and trends is an important information for several societal sectors. To characterize long-term climate change, especially in extremes, observations over many decades are required.

Ground-based meteorological observations have been performed by National Meteorological and Hydrological Services and their predecessors for more than a century and therefore provide the basis for long-term analysis of regional climate change. However, over this interval several changes have been applied to the observing systems. Therefore the data need to be accurately documented.

Deutscher Wetterdienst (DWD) operates a dense measurement and observation network[1] for Germany. The data from this network is the basis for Germany's National Climate Data Centre (Nationales KlimaDatenZentrum, NKDZ). In the following, we give a short overview of the historic development of the German observation network and the NKDZ. We also describe typical derived products and how they are used to inform the public about the status of German climate.

[1]The term "measurement" refers to determining the value of a quantity, typically based on an instrument; whereas the term "observation" is used in a more general sense and does not necessarily involve the use of a measurement instrument (i.e. in the case of cloud cover observations of human observers or phenological observations).

Table 1. The primary network of Deutscher Wetterdienst (as of 8 December 2012).

Type of station	Number of stations
main meteorological watch offices and automatic weather stations,	179
… manned around the clock	38
… manned part-time	28
… fully automatic weather stations	113
main upper-air stations	9
… fully automatic stations (auto launcher)	5
… integrate ozone soundings	2
stations for measuring radiation	119
… with measurements of global radiation	119
… with measurements of diffuse illumination	117
… with measurements of atmospheric thermal radiation	9
stations measuring radioactivity	48
weather radar sites	17
surface weather stations run by the Bundeswehr Geoinformation Service (as part of the joint network)	35

2 The development of the measurement and observation networks in Germany

In the year 1847 the Prussian Meteorological Institute was established (the predecessor of DWD), the first national meteorological service in the northern part of Germany that collected, analyzed and archived the meteorological observations of the voluntary observers. These data were published in an annual book together with a standard observer instruction. An additional, denser precipitation network was built up in the 1880s where only precipitation was measured by volunteers once a day.

Over the years further 8 national meteorological services were founded in other parts of Germany which had their own instruments and instructions. After the establishment of the International Meteorological Organization (IMO) in 1873 the different standards were continuously adjusted to this specification. In 1934 these different services were united and had the same instrumentation and observer instruction (see Wege, 2002 for a detailed description of the development of meteorological services in Germany).

After World War II, in the divided Germany, different instrumentation and observation practices have been developed with different observational times. After the reunification in 1990 these practices were again harmonized.

At the beginning in 1848 the climate network included about 24 stations and increased up to 200 stations in 1900 and about 400 stations during World War II. The peak was reached in the 1980s with about 600 stations. With the beginning of the automatization of the network the number of stations decreased (Fig. 1). On the other hand, the automatic

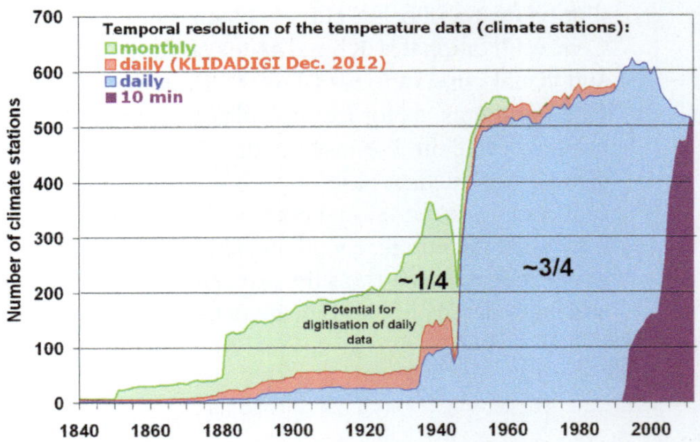

Figure 1. Number of temperature stations in the data base with specific temporal resolution in the database. The blue area indicates the amount of data that was already digitally available at daily resolution prior to the recent digitization efforts. The red area indicates the data that have already been digitized at daily resolution within the KLIDADIGI project. The green area indicates the amount of data with monthly resolution. This data can potentially be completed with daily resolution by digitizing further original documents.

stations provided measurements at higher temporal resolution.

The precipitation network increased rapidly to more than 1400 station in the year 1901 and reached the peak in the 1980s with about 4500 stations.

The current network (as of 8 December 2012) of DWD is based on a so-called "primary network", which has mainly been planned for the needs of daily weather forecasting. The components of that network are summarized in Table 1.

Table 2. Number of stations of the total network of DWD (the 35 stations of the Bundeswehr Geoinformation Service included) measuring specific parameters.

Parameter	Number of stations
Temperature and humidity	507
Air pressure	212
Precipitation	1925
Wind	297 (68 as part of the storm warning network)
Sunshine duration	302

As for spatial climatological investigations a higher density is desirable, a secondary "climatological" network is supported. It consists of 1787 stations run by voluntary weather observers. The densest network exists for precipitation, as this parameters shows an especially large spatial variability. Table 2 shows the total number of stations measuring selected parameters.

In addition, DWD operates a phenological network which consists of 1304 secondary phenological observation sites, largely based on voluntary observers.

Selected stations are operated as so-called "climate reference stations". For these stations long time series are already available and manual observations will be continued with automatic weather stations in parallel. This will allow a long-term comparison of both techniques and thereby also supports the homogenization for those stations where the old technique is not continued.

The climate reference stations are: Helgoland, Hamburg, Schleswig, Potsdam, Görlitz, Lindenberg, Brocken, Aachen, Fichtelberg, Frankfurt (Main) and Hohenpeißenberg.

An up-to-date description of the components of the German climate observing system can be found in Deutscher Wetterdienst (2013).

3 Data base of Germany's National Climate Data Centre (NKDZ)

The data base of NKDZ mainly comprises "conventional" in situ observations that are made within the station network of DWD. The archive contains historical German climate data and current data are continuously added. In addition to the classical meteorological data taken near the Earth's surface, the archive also comprises data from the free atmosphere and the soil, as well as phenological data and data on the classification of weather conditions. The number of stations available has varied in the course of time and depends on the parameter. The longest time series goes back to the 18th century (Hohenpeißenberg, since 1781 with measurements of temperature, pressure and precipitation), but many do not start before ~ 1945.

The main aim of the NKDZ is the sustained development of a comprehensive data base of high quality, which is available to applied climatology and climate research. A major focus of the activity is to guarantee a defined data quality by means of validation and correction of the data (or at least a description of errors), as well as a detailed description of the data base, including provision of detailed records of station metadata, as e.g. change of instrumentation (see Table 3 for an example), location or observer instructions.

4 Quality assurance

Continuous quality assurance and documentation is necessary to provide a high quality data base for research purposes. Quality assurance of NKDZ comprises the identification and correction or description of errors in the historical data. Documentation comprises the description of the historical time series, including the changes in the coding and observation regulations.

Data from the current observation network of DWD pass through several steps of quality control. The first step is a quality control directly at the automatic stations. As this is an automatic test, rather broad limits are applied. It includes tests for completeness, thresholds, temporal and internal consistency. Based on these tests, a quality flag is assigned to the data. The data are then transferred to the data base. At this stage, a second test (with thresholds of the data base) is applied. Based on the values in the data base, a third level of quality control is performed with smaller limits than during the first step. It is based on a software (QualiMet, see Spengler, 2002) that tests for completeness and climatological, temporal, spatial and internal consistency. Doubtful values are manually examined and the quality flag is adapted. A final step of quality control is applied after all data of a month is available, with focus on the aggregated values. The quality flags are stored in the data base and are also provided to the users, e.g. when providing the data in the internet.

The introduction of near real-time automatic quality control in the current observation network allowed shifting the focus of the NKDZ to the quality control of previously insufficiently verified historical data. Original quality control was only subjectively and undocumented. There has not been a complete, systematic and well documented quality assurance of the data that was observed before the introduction of the computer-based verification methods (i.e. prior to 1979). For the quality control of these historical data, suitable procedures have been developed and are currently applied to the data. The focus is on daily values. In general, the quality of these values can be considered as reasonably good, but there are still doubtful values, especially in data prior to 1979 (in the order of about 0.1–1 %).

It is important for the user to note that data might be affected by certain none-climatic effects, e.g. changes in instrumentation or observation time. Except for a few exceptions,

Table 3. An example for the provision of supplementary metadata on station/instrumentation history: instruments used to measure precipitation at station Bremen. Further metadata, e.g. observation times, are provided in additional tables (table is shown as provided with the data, i.e. in German language).

ID	Stations-name	Geo. Länge [Grad]	Geo. Breite [Grad]	Stations-höhe [m]	Von Datum	Bis Datum	Gerätetyp Name	Messverfahren
691	Bremen	8.79	53.05	3	01.02.1949	31.05.1978	Niederschlagsmesser	Niederschlagsmessung, Hellmann
691	Bremen	8.79	53.05	3	01.01.1961	31.05.1978	Niederschlagsschreiber (beheizt)	Niederschlagsmessung, Hellmann
691	Bremen	8.8	53.05	4	01.06.1978	31.08.1990	Niederschlagsmesser	Niederschlagsmessung, Hellmann
691	Bremen	8.8	53.05	4	01.06.1978	30.10.2008	Niederschlagsschreiber (beheizt)	Niederschlagsmessung, Hellmann
691	Bremen	8.8	53.05	4	20.06.1991	30.10.2008	Ombrometer Lambrecht 1518H3	Niederschlagsmessung, Hellmann, elektr.
691	Bremen	8.8	53.05	4	14.10.2008		PLUVIO-OTT	Niederschlagsmessung, Hellmann, elektr.
691	Bremen	8.8	53.05	4	19.06.1991		Niederschlagsgeber Kroneis	Niederschlagsmessung, elektr.

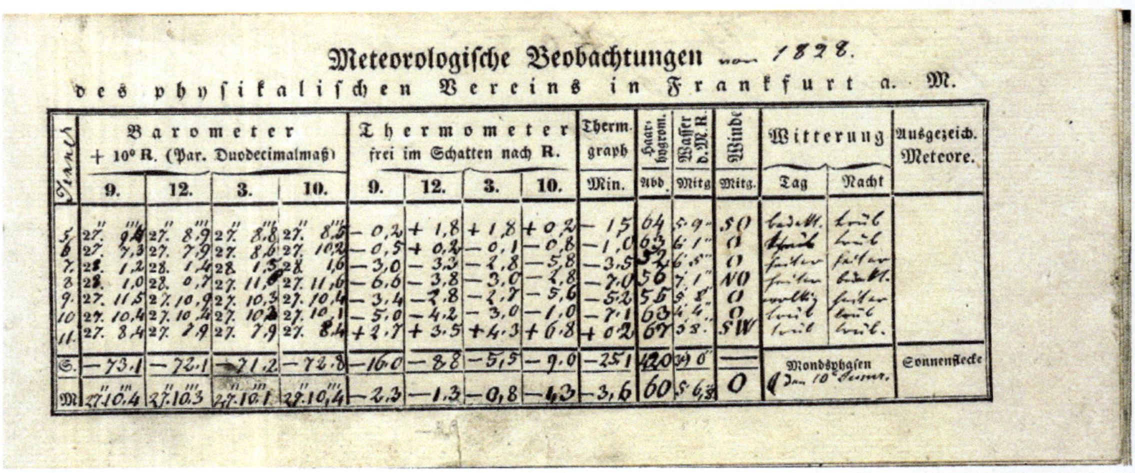

Figure 2. An example of handwritten historical documents with weather observations: Frankfurt, 1828 (5 to 11 January).

data are provided "as observed", e.g. no homogenization procedure has been applied. Information on these changes is therefore provided together with the data. A homogenization procedure has only been applied to selected parameters at monthly resolution (Herzog and Müller-Westermeier, 1998). These monthly data are available as an alternative version, but are not continuously updated.

A systematic verification of the upper-air and phenological data has also not yet been carried out.

5 Digitisation of historical climate data

Apart from the integration of current data, the data base is also expanded by adding additional historical time series. Such data are still available as hand-written documents (Fig. 2 shows an example). Therefore the digitization can only be done manually and is a time-consuming and costly process, which can only be carried out as a long-term activity on a best effort basis.

In particular for the time prior to 1951 (in the new federal states prior to 1969) the availability of daily data (or data of a higher temporal resolution) is still insufficient. A comparison of the number of monthly values in the data base to the daily data illustrates the missing information in the data base at that temporal resolution (see Fig. 1).

The project KLIDADIGI was initiated to digitize additional data, but also to rescue the original documents, esp. by taking photographs of the paper documents. A focus is the expansion and completion of data for stations with long time series (Mächel and Kapala, 2013).

During the last 7 yr, daily records from 97 climate stations or 2932 station-years (about 12.7 % of the total potential) were digitized (Fig. 1). Records from 514 precipitation stations were digitized which equals to 19 300 station-years or 11.7 % of the total potential. 492 000 photos of paper documents have been taken.

The greatest deficiencies exist in East Germany before 1969, where only 50 % of the precipitation stations are digitally available. The filling of these gaps and digitizing all available precipitation stations back to the end of World War II have the highest priority in the next few years.

6 Climate analysis

The NKDZ data base can be used to describe the climate in Germany over a period of more than 130 yr. For the provision of aggregated information the following two steps are performed: (a) calculation of gridded fields and (b) aggregation of regional averages based on these gridded data.

Gridded data at a resolution of 1 km × 1 km are produced for the monthly mean of temperature, monthly sums of precipitation and sunshine duration, grass reference evaporation, aridity index and the climatic water balance.

The calculation of the gridded fields is done in three steps: first, the monthly values at the station locations are reduced to a reference level (sea level). Due to the high station density, linear regression between topographic height and climatological parameter is sufficient within a specific region. For precipitation and other parameters with a very dense network, the regression is calculated within regions of 1° × 1°. For other parameters, 10 larger regions are used (for details see Müller-Westermeier, 1995). The regression coefficients are interpolated between the center locations of the regions and are calculated for each of the 12 months, based on the period 1961–1990. With the interpolated coefficients (covering the whole area), the climatological values at each individual station are reduced to sea level.

In the second step, the values at reference level are horizontally interpolated (inverse squared-distance weighted).

In the third step, the values are transferred from the reference level to the real height of the grid point. A grid with 1 km resolution is used (Gauss-Krüger projection with central meridian at 9° E; EPSG-code: 31467). The grids are produced for each month. More details can be found in Müller-Westermeier et al. (2005). Gridded fields for reference intervals (e.g. 1961–1990) are produced with the same methodology based on the station-based mean values for the respective interval. Grids for the new reference interval 1981–2010 have recently been calculated.

The monthly climate monitoring products are produced within a few days after the end of a month in order to provide near real time information on the status of German climate. At that stage, not all steps of the quality control procedures have been applied to the data. Therefore, the gridded data are reproduced once again after a few months. Figure 3 shows an example.

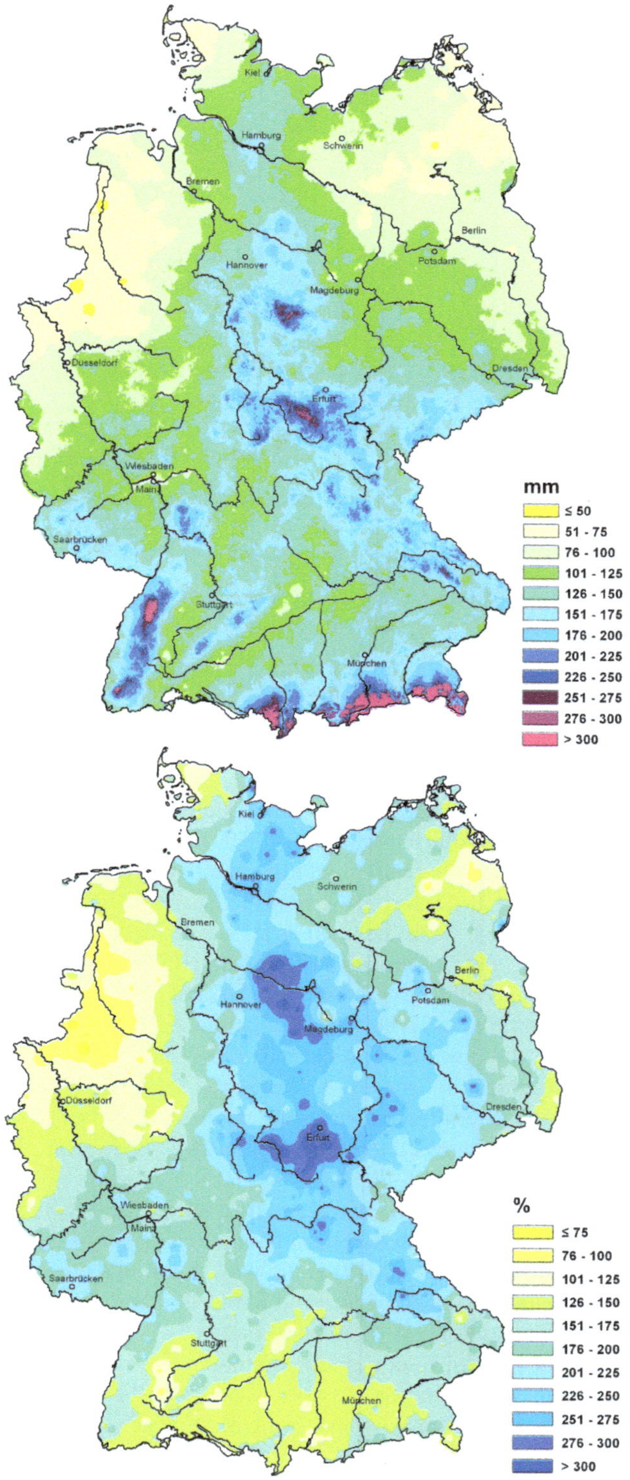

Figure 3. An example for the gridded products: precipitation in May 2013, which was the 2nd wettest May in Germany since 1881; (top): absolute values in mm. (bottom): in percent of the long-term mean 1961–1990.

Table 4. Trend of the annual mean temperature in Germany's federal states from 1881 to 2012 (i.e. change in areal mean temperature). For small states values are combined.

Region	Trend in degree Celsius from 1881 to 2012
Schleswig-Holstein	1.1
Lower Saxony (Niedersachsen), Hamburg, Bremen	1.3
North Rhine-Westphalia (Nordrhein-Westfalen)	1.3
Rhineland Palatinate (Rheinland-Pfalz), Saarland	1.4
Baden-Württemberg	1.2
Hesse (Hessen)	1.2
Bavaria (Bayern)	1.3
Mecklenburg-Vorpommern	1.1
Brandenburg, Berlin	1.0
Saxony (Sachsen)	1.1
Saxony-Anhalt (Sachsen-Anhalt), Thuringia (Thüringen)	1.2
Germany (Deutschland)	1.2

The gridded data are used to calculate means for administrative areas (esp. for Germany and the federal states) which are then combined to make time series. The gridding procedure considers location and height of the stations. It can therefore be assumed that the time series of the spatial means are less affected by such changes in the measurement network than the time series of the individual stations.

In 2012, the gridded temperature fields have been reprocessed for the full interval 1881–2011 based on the updated and completed data base, including the newly digitized data. To improve the interpolation, also data from neighboring countries have been included (stations within a distance of ~ 50 km to the border; taken from the HISTALP, CRU and GHCN data collections as they are available from the archive of the International Surface Temperature Initiative). Especially in the north-east of Germany, the station density has been significantly improved. Compared to previous results, the north-eastern federal states are now slightly more consistent. Table 4 shows the updated trends for the federal states for the interval 1881–2012.

The interpolation method has been verified and compared with other (GIS-based) interpolation methods (Maier and Müller-Westermeier, 2010) and it was shown that other standard interpolation procedures do not lead to better results.

For climatological reference periods (so-called "climate normals"), mean values of temperature, precipitation amount and sunshine duration are provided at station locations. These products are now also available for the period 1981–2010.

7 The German climate atlas

The monthly maps are one element of the German climate atlas (Deutscher Klimaatlas; http://www.deutscher-klimaatlas.de). The German climate atlas is an interactive online tool to provide comprehensive information about past, current and projected future climate in Germany, together with sector-specific information (currently agriculture, forestry, soil). The atlas offers a combined view on observation-based information from the past (back to 1881) and results of regional climate model projections (until 2100). The scenario information is based on an multi-model ensemble (from the EU-project ENSEMBLES and additional sources). The maps are visualized in combination with time series of the national mean. This allows a direct visual comparison of observed to projected data. Figure 4 shows an example of the atlas.

8 Data access

Relevant parts of the data are accessible via internet-based access tools. The Climate Data Centre (CDC) of DWD provides an entry point to all access tools (http://www.dwd.de/cdc). Due to legal restrictions only a selection of the data can be provided free of restrictions. Additional data are available for science, research and education, but registration is required. The set of free data currently comprises the time series from 78 selected stations, the long-term average values for all stations, grid fields for the climatological (30 yr) reference periods and the phenological data. The data can be found on the internet pages of NKDZ (http://www.dwd.de/nkdz, http://www.dwd.de/phaenologie). The time series of the stations are provided with metadata that describes the history of the station, including all relevant modifications in the instrumentation. General information on the status of climate in Germany is also provided at http://www.dwd.de/klimamonitoring. Data from all German stations and monthly grid fields are available for science, research and education after registration via the *Web-based Weather Request and Distribution System* (WebWerdis; http://www.dwd.de/webwerdis). The phenological data are also provided to the PEP725-project (Pan-European Phenology; http://www.pep725.eu) and to the plant-phenological

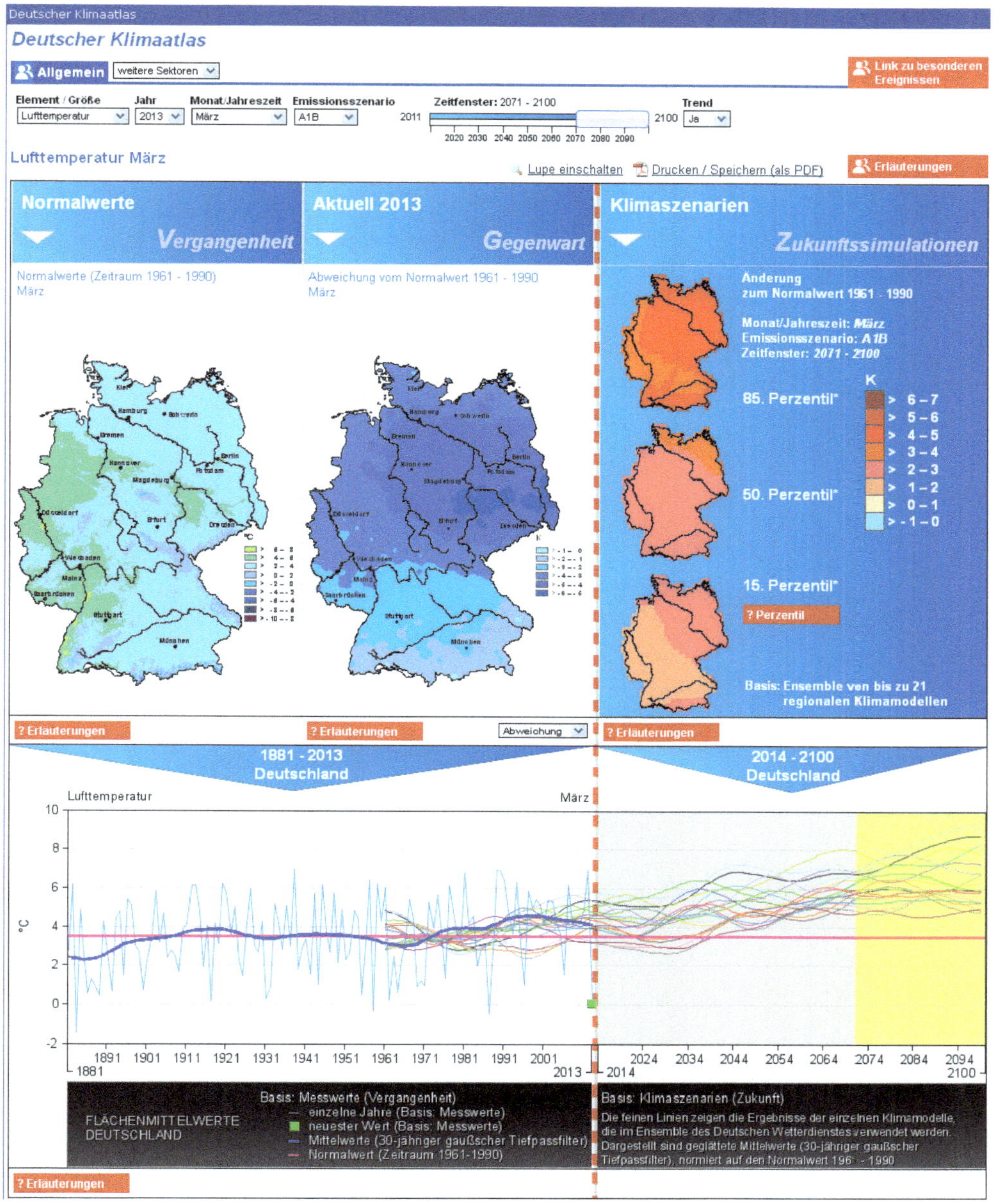

Figure 4. The German climate atlas provides a comprehensive overview of climate in Germany. Here, the parameter "temperature" of "March 2013" is selected (which was the 5th coldest since 1881). The left map shows the average values for March during the reference period 1961–1990. The middle map shows the anomaly in March 2013. The right maps show results of regional climate models (also for March; interval 2071–2100 is selected). At the bottom, the time series of March temperatures is shown. The observation-based information (black and pale blue; left part) can directly be compared with the climate projections (assorted colours; right).

online database (PPODB; www.ppodb.de) and are accessible at these portals. The aim of PEP725 is to establish a pan European open access database with plant phenology data sets for science, research and education. PPODB also contains phenological data from additional sources and supplements the DWD-data especially for periods prior to 1951.

Edited by: M. Brunet-India

References

Deutscher Wetterdienst: Die deutschen Klimabeobachtungssysteme. Inventarbericht zum Global Climate Observing System (GCOS), Selbstverlag des Deutschen Wetterdienstes, Offenbach am Main, 130 pp., http://www.gcos.de/inventarbericht, 2013.

Herzog, J. and Müller-Westermeier, G.: Homogenitätsprüfung und Homogenisierung klimatologischer Messreihen im Deutschen Wetterdienst, Berichte des Deutschen Wetterdienstes 202, Selbstverlag des Deutschen Wetterdienstes, Offenbach am Main, 1998.

Mächel, H. and Kapala, A.: Multivariate testing of spatio-temporal consistence of daily precipitation records, Adv. Sci. Res., 10, 85–90, doi:10.5194/asr-10-85-2013, 2013.

Maier, U. und Müller-Westermeier, G.: Verifikation klimatologischer Rasterfelder. Berichte des Deutschen Wetterdienstes 235, Selbstverlag des Deutschen Wetterdienstes, Offenbach am Main, 2010.

Müller-Westermeier, G.: Numerisches Verfahren zu Erstellung klimatologischer Karten. Berichte des Deutschen Wetterdienstes 193, Selbstverlag des Deutschen Wetterdienstes, Offenbach am Main, 1995.

Müller-Westermeier, G., Walter, A., and Dittmann, E.: Klimaatlas Bundesrepublik Deutschland, Teil 1–4, Selbstverlag des Deutschen Wetterdienstes, Offenbach am Main, 2005.

Spengler, R.: The new Quality Control- and Monitoring System of the Deutscher Wetterdienst. Proceedings of the WMO Technical Conference on Meteorological and Environmental Instruments and Methods of Observation, Bratislava, 2002.

Wege, K.: Die Entwicklung der meteorologischen Dienste in Deutschland. Geschichte der Meteorologie in Deutschland – Band 5. Selbstverlag des Deutschen Wetterdienstes, Offenbach am Main, 2002.

World Meteorological Organisation: International Meteorological Vocabulary, 2nd Edn., WMO No. 182, 1992.

14

Measurement procedures for characterization of wind turbine wakes with scanning Doppler wind LiDARs

G. V. Iungo and F. Porté-Agel

Wind Engineering and Renewable Energy Laboratory (WIRE), École Polytechnique Fédérale de Lausanne (EPFL), Switzerland

Correspondence to: G. V. Iungo (valerio.iungo@epfl.ch)

Abstract. The wake flow produced from an Enercon E-70 wind turbine is investigated through three scanning Doppler wind LiDARs. One LiDAR is deployed upwind to characterize the incoming wind, while the other two LiDARs are located downstream to carry out wake measurements. The main challenge in performing measurements of wind turbine wakes is represented by the varying wind conditions, and by the consequent adjustments of the turbine yaw angle needed to maximize power production. Consequently, taking into account possible variations of the relative position between the LiDAR measurement volume and wake location, different measuring techniques were carried out in order to perform 2-D and 3-D characterizations of the mean wake velocity field. However, larger measurement volumes and higher spatial resolution require longer sampling periods; thus, to investigate wake turbulence tests were also performed by staring the LiDAR laser beam over fixed directions and with the maximum sampling frequency. The characterization of the wake recovery along the downwind direction is performed. Moreover, wake turbulence peaks are detected at turbine top-tip height, which can represent increased fatigue loads for downstream wind turbines within a wind farm.

1 Introduction

The evaluation of wind turbine performance, and design and optimization of wind farms are typically performed through a synergistical interaction of different tools. Indeed, results obtained from basic analytical models are compared to the ones of CFD numerical simulations of the wake flow produced from wind turbines, or to wind tunnel tests of downscaled wind turbine models. However, all these tools need to be assessed through field measurements of wakes produced by real wind turbines. Current industry standards for field measurements in wind energy consist in tests preformed with instrumented masts, which could be insufficient for the characterization of wind turbine wakes, because of the limited number of measurement locations. Therefore, in order to investigate over larger measurement volumes, the use of remote sensing techniques is growing rapidly, in particular Doppler wind Light Detection and Ranging (LiDAR). Wind turbine wake measurements can be performed with ground-based LiDARs, as for Iungo et al. (2013) and for the present work, or through nacelle-mounted LiDARs in order to keep the instrument always aligned with the mean wind direction for varying wind conditions (Bingöl et al., 2010; Trujillo et al., 2011).

A Doppler wind LiDAR measures wind velocity through the evaluation of the Doppler shift undergone by a laser beam emitted into the atmosphere and back-scattered due to the presence of aerosol. Therefore, a Doppler wind LiDAR is only sensitive to the wind component parallel to the laser beam, the so-called radial velocity or line-of-sight velocity. LiDAR tests can be performed by staring the laser beam at a fixed direction, i.e. producing 1-D velocity profiles carried out with the maximum sampling frequency. These tests can be useful for the characterization of wake turbulence. 2-D measurements with a longer sampling period can be performed over vertical planes, denoted as Range Height Indicator (RHI) scans, or over conical surfaces, denoted as Plan Position Indicator (PPI) scans (see Käsler et al., 2010). 3-D scans can be also performed as a combination of consecutive RHI or PPI scans. Multi-component wake velocity field

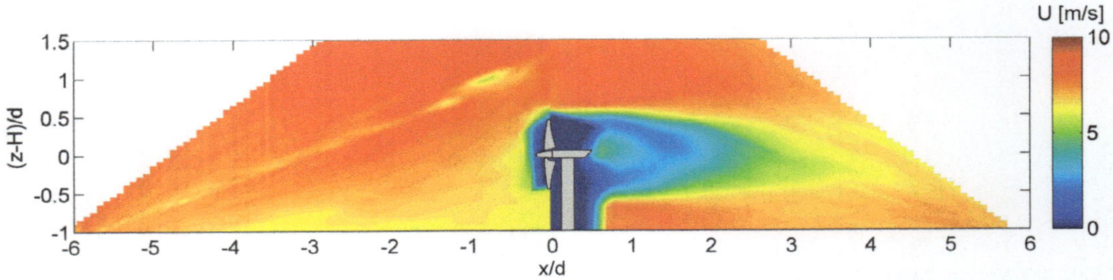

Figure 1. Velocity field measured over the wake vertical symmetry plane through simultaneous RHI scans with two LiDARs.

can be retrieved by simultaneously staring different LiDARs at the same measurement volume, see Mann et al. (2012) and Iungo et al. (2013). A review on remote sensing techniques for wind energy applications is reported in Emeis et al. (2007).

The paper is organized as follows: the wind turbine site and the experimental setup are described in Sect. 2, whereas the used LiDAR measurement techniques are reported in Sect. 3. Finally, conclusions are drawn in Sect. 4.

2 Wind turbine site and experimental setup

The presented LiDAR measurements were performed in June 2012. The tested wind turbine is a pitch-regulated 2.3 MW Enercon E-70, located in Collonges, Switzerland, in the central part of a narrow valley. The turbine rotor diameter, d, is 71 m, while hub height, H, is 98 m.

Three Streamline Doppler wind LiDARs, produced by Halo Photonics, were used for this campaign. The LiDAR laser is characterized by a wavelength of 15 μm and a pulse repetition rate of 15 kHz. The highest spatial resolution is 18 m, with the first measurement point at a distance of 40 m from the LiDAR location, and maximum range of 3 km. The maximum sampling frequency available to obtain a single velocity profile along the direction of the laser beam is 0.77 Hz.

The wind turbine location, the LiDAR locations and azimuthal displacements of the LiDAR laser beams were evaluated through a GPS system. The used reference frame has its origin at the the base of the turbine tower. The x-axis corresponds to the mean wind direction, positive in the downstream direction. The z-axis is along the vertical direction, positive from the bottom to the top, while the y-axis is along the spanwise direction. Throughout this paper the approximated streamwise velocity is always reported, which is the projection of the measured line-of-sight velocity on the x direction. This calculation is performed by considering geometrical relations that take location, azimuthal and elevation angles of the LiDAR into account (Machefaux et al., 2012). More details about the site, the LiDARs and the setup can be found in Iungo et al. (2013).

3 LiDAR measurements

The first LiDAR measurement technique described in the present paper is the simultaneous RHI performed with two LiDARs over the vertical symmetry plane of the wind turbine wake. One LiDAR was located upwind at $x/d = -6$ and a second LiDAR at $x/d = 6$. The two LiDARs measured simultaneously by varying the elevation angle of the laser beams from 1° up to 38° with an angular step of 1°. The sampling period required for a single RHI scan was 77 s, and 30 consecutive scans were performed with roughly constant mean wind conditions.

The mean velocity field obtained as average of all the selected RHI scans is reported in Fig. 1. Velocity values reported for the locations upwind to the wind turbine position, i.e. with $x/d < 0$, are obtained from the LiDAR located upwind at $x/d = -6$. Conversely, the wake velocity measurements for $x/d > 0$ were performed by the downwind LiDAR at $x/d = 6$. A large error on the velocity measurements is observed for the locations $-0.5 < x/d < 0.5$ and $(z-H)/D < 0.5$, which is due to the back-scattering produced from the wind turbine and affecting wind measurements at those locations.

The vertical heterogeneity of the atmospheric boundary layer (ABL) flow investing the wind turbine is observed, as well as the wake flow produced from the wind turbine. The typical velocity deficit induced by the blade rotation, which is strictly connected to the power production, is clearly detected. Moving downstream the wake recovers, and the velocity deficit is reduced.

More quantitative information can be gained if the vertical profiles of the velocity field are extracted, as reported in Fig. 2. For instance, the vertical profile of the incoming wind, evaluated for the location $x/d = -2$, clearly represents an ABL flow. This velocity profile is fitted through a power law with an exponent of 1.085, which is typical for neutral stability conditions of the ABL. The LiDAR measurements performed within the wake confirm the typical velocity profiles of wind turbine wakes, which have been observed through wind tunnel tests (e.g. Chamorro and Porté-Agel, 2010) and numerical simulations (e.g. Wu and Porté-Agel, 2011). The largest velocity deficit is observed in proximity of

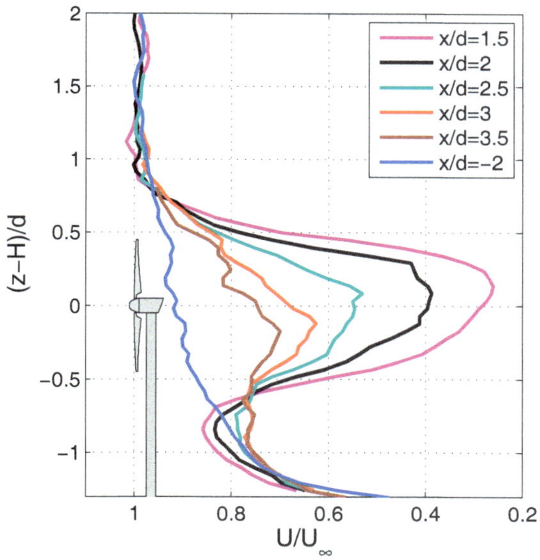

Figure 2. Vertical velocity profiles obtained through simultaneous RHI scans.

the hub height, then moving downstream the wake recovers approaching the vertical profile of the incoming ABL flow.

A limitation of the simultaneous RHI technique is that the two LiDARs must be deployed in such a way to perform scans over the wake vertical symmetry plane. Thus, by considering fixed LiDAR locations, this setup can be ensured only for a roughly constant wind direction, and with negligible adjustments of the turbine yaw angle. For the presented test the standard deviation of the wind direction was 0.7°. In case a proper setup is not ensured, RHI scans are performed over planes not parallel to the mean wind direction, thus misleading wake measurements can be carried out.

To overcome this drawback, wind turbine wake flow can be characterized through simultaneous PPI scans, which are performed with a fixed elevation angle of the LiDAR laser beams and varying the azimuthal angle. By performing PPI scans, variations of the wind direction, thus of the turbine yaw angle, are observed through changes of the wake location. For these tests one LiDAR was located upwind and one LiDAR downwind, both at a streamwise distance of 6 d from the turbine location. The LiDARs measured over an azimuthal range of 16°, centered with the azimuthal location of the wind turbine, and an angular step of 0.5°. The sampling period for each PPI scan was 70 s.

The upwind LiDAR measured with an elevation angle of the laser beam of 15.4°, in order to intersect hub height at the location $x/d = -1$, which is convenient for the characterization of the incoming wind. The downwind LiDAR measured with different elevation angles in order to intersect hub-height at different downstream locations. In Fig. 3 the mean velocity field obtained by scanning the downwind LiDAR at hub-height at the location $x/d = 2$ is reported. This plot was obtained by averaging 30 consecutive PPI scans performed with roughly constant wind conditions. For this tested wind direction, the downwind LiDAR is not aligned with the mean wind direction, but located at the transversal location $y/d = 1$ for logistical reasons.

The wind turbine wake is characterized through the simultaneous PPI scans, as well as the incoming wind. Besides the possibility of performing wake measurements with varying wind direction and turbine yaw angle, the physical interpretation of the wind data obtained through simultaneous PPI scans is not straightforward as for the simultaneous RHI scans. Indeed, each transversal velocity profile, i.e. measurements with a fixed streamwise position x/d, corresponds to different heights. In other words, in Fig. 3 velocity variations are due to changes both in streamwise and vertical positions.

To overcome possible issues related to the variability of wind direction and turbine yaw angle, for the RHI scans, or to physical interpretation of the wind data, for the PPI scans, volumetric scans of the wind turbine wake flow field were also performed. These measurements were carried out by varying both elevation and azimuthal angles of the LiDAR laser beam, thus they can be considered as a combination of consecutive PPI or RHI scans. Measurements were performed by varying the azimuthal angle of the laser beam within a range of 20°, centered with the turbine location, and an angular step of 2°. The elevation angle was varied from 2° up to 16° with an angular step of 2°. The measurement volume was chosen big enough to include the wake and considering possible variations of the turbine yaw angle. Consequently, a limited spatial resolution was used, i.e. angular step of 2°, in order to ensure a reasonable sampling period. Each volumetric scan was performed with a sampling period of 220 s, if all the measurements were acquired with one LiDAR; otherwise the sampling period was halved if the measurements were simultaneously carried out with two LiDARs, thus the measurement volume split in two parts. A third LiDAR was used to characterize the incoming wind.

A wake velocity field measured through a volumetric scan is reported in Fig. 4. For this test the LiDAR was deployed at the downstream location $x/d = 12$. Transversal planes are reported for different downstream locations, which enable to characterize cross-sections of the wind turbine wake, in particular its shape, location and wake velocity deficit. It is evident the clarity for the physical interpretation of the wind data obtained through a volumetric scan; however, the significant sampling period required does not enable to investigate possible time variations of the wake and wake turbulence.

To this end, the maximum sampling frequency of 0.77 Hz was achieved by scanning the LiDAR at fixed directions, by setting the laser beam over four different elevation angles, and using 110 gates. Consecutive measurements were performed for a sampling time of 11 min for each tested laser direction. In Fig. 5 vertical profiles of standard deviation show peaks for different elevation angles in correspondence of the turbine top-tip height for downstream distances lower than

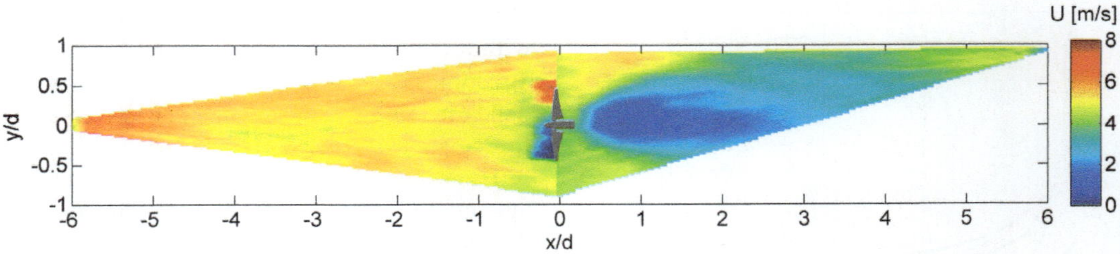

Figure 3. Velocity field measured through simultaneous PPI scans.

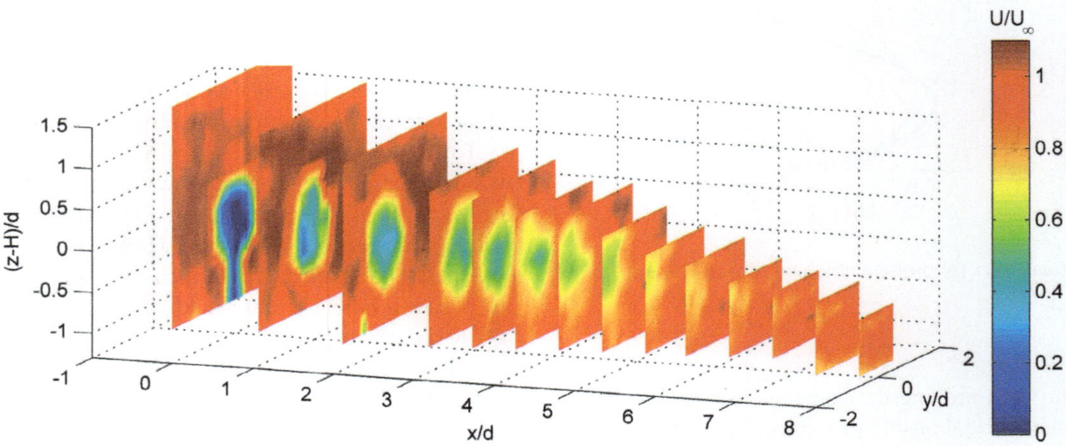

Figure 4. Volumetric scan of the wind turbine wake.

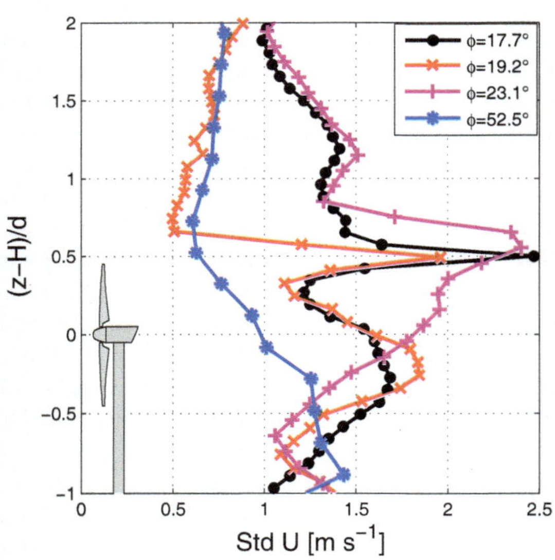

Figure 5. Velocity standard deviation obtained by staring a LiDAR with different elevation angles.

5 *d*. This increased turbulence could represent a wake region with an increased turbulence and a hazard for downstream turbines, because of the increased fatigue loads. This wake feature confirms previous results presented in Chamorro and Porté-Agel (2010) and Wu and Porté-Agel (2011) through wind tunnel tests and numerical simulations, respectively.

4 Conclusions

Several LiDAR measurement procedures have been presented for the characterization of the wake flow produced from an Enercon E-70 wind turbine. Simultaneous tests were performed with three scanning Doppler wind LiDARs in order to characterize incoming wind and wake flow. The presented measurement techniques are characterized by different size of the measurement volume, spatial resolution and sampling period, depending if they were optimized for the characterization of the mean wake flow or of the wake dynamics. 2-D scans were performed over the wake vertical symmetry plane, in order to characterize the wake recovery, or over conical surfaces to detect variations of the wake position due to changes of the wind direction and turbine yaw angle. 3-D volumetric scans were also performed for a global characterization of the wake flow; however, the significant sampling period required does not enable to investigate wake dynamics. Finally, tests with fixed directions of the LiDAR laser beam were performed with the maximum sampling frequency in order to characterize wake turbulence. Turbulence peaks have been observed in correspondence of the turbine top-tip height, which can represent dangerous fatigue loads for downstream wind turbines within a wind farm.

Edited by: S.-E. Gryning

References

Bingöl, F., Mann, J., and Larsen, G. C.: Light detection and ranging measurements of wake dynamics; Part I: One-dimensional scanning, Wind Energy, 13, 51–61, 2010.

Chamorro, L. and Porté-Agel, F.: Effects of thermal stability and incoming boundary-layer flow characteristics on wind-turbine wakes: a wind-tunnel study, Bound.-Lay. Meteorol., 136, 515–533, 2010.

Emeis, S., Harris, M., and Banta, R.: Boundary-layer anemometry by optical remote sensing for wind energy applications, Meteorol. Z., 16, 337–347, 2007.

Iungo, G.V., Wu, Y., and Porté-Agel, F.: Field measurements of wind turbine wakes with lidars, J. Atmos. Ocean. Technol., 30, 274–287, 2013.

Käsler, Y., Rahm, S., Simmet, R., and Kühn, M.: Wake measurements of a multi-MW wind turbine with coherent long-range pulsed Doppler wind lidar, J. Atmos. Ocean. Tech., 27, 1529–1532, 2010.

Machefaux, E., Troldborg, N., Larsen, G. C., Mann, J., and Madsen, H. A.: Experimental and numerical study of wake to wake interaction in wind farms, EWEC 2012, Copenhagen, 2012.

Mann, J., Cariou, J.-P., Courtney, M. S., Parmentier, R., Mikkelsen, T., Wagner, R., Lindelöw, P., Sjöholm, M., and Enevoldsen, K.: Comparison of 3D turbulence measurements using three staring wind lidars and a sonic anemometer, IOP Conf. Series: Earth Environ. Science 1, 012012, 2012.

Trujillo, J., Bingöl, F., Larsen, G. C., and Mann, J.: Light detection and ranging measurements of wake dynamics; Part II: Two-dimensional scanning, Wind Energy, 14, 61–75, 2011.

Wu, Y.-T. and Porté-Agel, F.: Large-eddy simulation of wind-turbine wakes: evaluation of turbine parametrisations, Bound.-Lay. Meteorol., 138, 345–366, 2011.

Preliminary step towards comparing large scale features in surface wind field between ERA-Interim re-analysis and a higher resolution simulation

A. Boilley[1], **T. Ranchin**[1], **and A. Ghennioui**[2]

[1]MINES ParisTech, O.I.E. – Centre Observation, Impacts, Energie, CS 10207 rue Claude Daunesse, 06904 Sophia Antipolis Cedex, France

[2]Institut de Recherche en Energie Solaire et Energies Nouvelles, Rabat, Morocco

Correspondence to: A. Boilley (alexandre.boilley@mines-paristech.fr)

Abstract. The weather research and forecasting model (WRF) is initialized with ERA-Interim re-analysis to produce a higher resolution dataset over a one month period. WRF is supposed to introduce small scale information and maintain the mean large scale features. A detailed spectral analysis is performed to verify this statement. It reveals that consistency in large-scale features is not achieved for wavelengths greater than or equal to 6° even with the activation of spectral nudging.

1 Introduction

Morocco has engaged itself in an appraisal of the wind energy resources. The first step is a wind atlas that will depict the mean wind speed at 100 m height above land and coastal areas. The wind atlas is created by averaging instantaneous wind fields during several years.

High resolution wind fields needed for the atlas are currently unavailable. The dynamical downscaling method is an approach to create them. Dynamical downscaling consists in initializing a numerical meteorological model with a coarse resolution analysis and run the model at a higher resolution over the area of interest. Most of the time a re-analysis field is used (Castro et al., 2005; Heikkilä et al., 2011) since it is considered as the best field a numerical model can produce at the resolution of the re-analysis. In our case, no local measurements are assimilated in the high resolution model.

Traditionally the atlas is compared to several independent measurements to estimate its reliability. It determines its ability to represent some particular local flows but do not validate the atlas over the entire region. To overcome this challenge, one may think of comparing the re-analysis and the high resolution fields once remapped at the coarse resolution. Ideally, the difference should be small and the consistency of large scale features between both data sets should be large. This is particularly true for variables independent from orographic effects.

This paper presents a method to perform such a comparison and to study the amount of large scale patterns from re-analysis maintained by a numerical downscaling process with and without spectral nudging. It aims at evaluating the magnitude of the consistency error between a re-analysis dataset and a higher resolution simulation for wind speed at 100 m. In the case when the consistency error is high, a further effort should be carried out to modify the large scale features of the wind atlas through data fusion for example (Fichaux et al., 2005).

2 Method

2.1 Numerical simulation setup

We use the European Centre for Medium-range Weather Forecast (ECMWF) ERA-Interim re-analysis (Dee et al., 2011) as input and the Weather Research and Forecasting (WRF, Skamarock et al., 2008) as high resolution model. WRF is initialized once with ERA-Interim and the boundary conditions are then updated each 6 h with ERA-Interim during the month of January 2004; output files are produced each hour. The simulation is carried out on a domain of 256×256

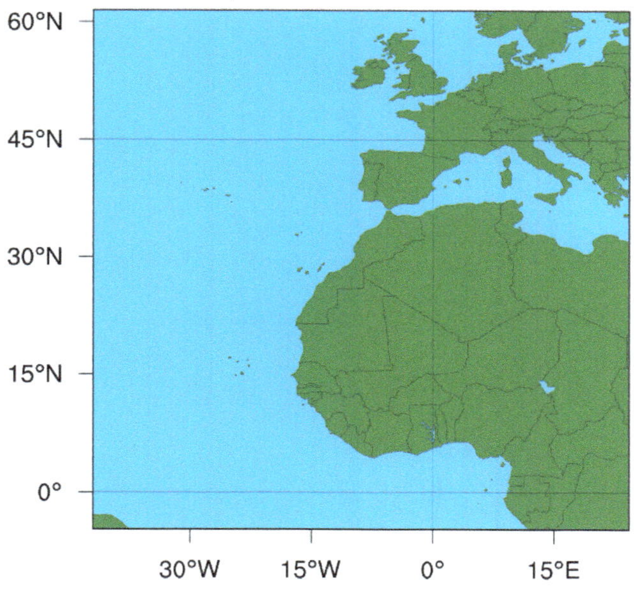

Figure 1. WRF simulation domain at 0.25° resolution.

nodes at 0.25° resolution containing Europe and a part of the African continent (Fig. 1). We designed 40 vertical levels to fit the 40 first vertical levels of ERA-Interim at initial time. The domain of study can represent the coarser domain for a dynamical downscaling method to produce a wind atlas over Morocco. No higher resolution domains are nested in this paper.

The fully compressible non-hydrostatic equations are integrated on a 60 s time step using the 3rd order Runge Kutta scheme. Land and lakes surface temperature and moisture as well as snow cover and frozen soil physics are computed by the Noah Land Surface Model (Chen et al., 1996) in four layers every time steps. The sea-surface temperature and the deep soil layer are updated every 6 h with ERA-Interim data. Cumulus are parameterized using the Kain-Fritsch scheme (Kain and Fritsch, 1990) and since the final goal is to obtain a high resolution atlas a sophisticated 6-class microphysics scheme (Hong and Lim, 2006) is used. The other parameters are set to the default value.

A simulation (NODA) is run without data assimilation. The interior part of the domain, far from the lateral boundary conditions is left alone and only the physical and dynamical processes take place. This is the worst case configuration; the low predictability of the atmosphere implies an expected low consistency in that case. A second simulation (SPECNUD) is conducted with spectral nudging turned on above the 10th vertical level (approximately 1200 m a.g.l.). The nudging coefficient linearly increases to full strength ($5.10^{-4}\,s^{-1}$) for the next 3 vertical levels. The wave number considered for nudging is set to 3 in each directions corresponding to a resolution of 21° approximately. The configuration with spectral nudging activated should lead to a better agreement between WRF and ERA-Interim large scales and could be used for the computation of a wind atlas.

2.2 Consistency error

ERA-Interim dataset is the reference field in the following. In order to avoid the introduction of discrepancies in the reference through the re-sampling and interpolation processes (Wald et al., 1997), the ERA-Interim wind fields are only re-sampled from the 0.75° original grid points onto a 0.25° grid, each 6 h. The WRF high resolution fields are then interpolated horizontally and vertically on the same grid. Even if the two fields are gridded the same way, the information scale contained in each field is different. Obviously, the WRF simulation exhibits more fine structures than ERA-Interim. Due to the resolution difference and since the simulation NODA is run freely, a direct comparison between reference and simulations results cannot be considered.

The physical processes scale represented by a gridded domain can be at most twice the resolution of the grid as stated by the Nyquist-Shannon sampling theorem. Hence ERA-Interim, which resolution is 0.75°, represents processes at most 1.5° resolution. A theoretical anti-aliasing low-pass filter (H_{theo}) has then been designed to remove the information contained in scales less than 1.5° (twice the resolution of ERA-Interim) from the WRF simulations, noted RUN in the following equations. H_{theo} is a sinus cardinal function in the spatial domain apodized with a Blackman function (Blackman and Tukey, 1958). By construction, the remapped re-analysis noted ANA in the following equations, does not contain information at scales less than 1.5°. The difference between the one month mean analysis $\overline{\mathrm{ANA}}$ and simulation $\overline{\mathrm{RUN}}$ convoluted to the theoretical filter is called the consistency error (CE) and is defined as follows:

$$\mathrm{CE} = \overline{\mathrm{ANA}} - H_{\mathrm{theo}} \cdot \overline{\mathrm{RUN}} \qquad (1)$$

The absolute value of monthly mean consistency error is presented in Fig. 2a for the NODA simulation and in Fig. 2b for the SPECNUD simulation. Large errors up to $5\,\mathrm{m\,s^{-1}}$ appear in complex terrain for the NODA simulation since the 100 m wind is not independent of orographic effects. Even if diminished of 25 % in mean, the error remains large for the SPECNUD simulation. The error over the ocean and in coastal area is however high.

For both NODA and SPECNUD simulations the consistency error exhibits a fine scale structured field. It means that the theoretical filter did not remove enough scales from the WRF field to be coherent with the features from ERA-Interim. The smaller scale theoretically represented by ERA-Interim are modified by WRF.

2.3 Can the consistency error be decreased?

We have further investigated the change in consistency error with scale. As stated previously the theoretical filter is

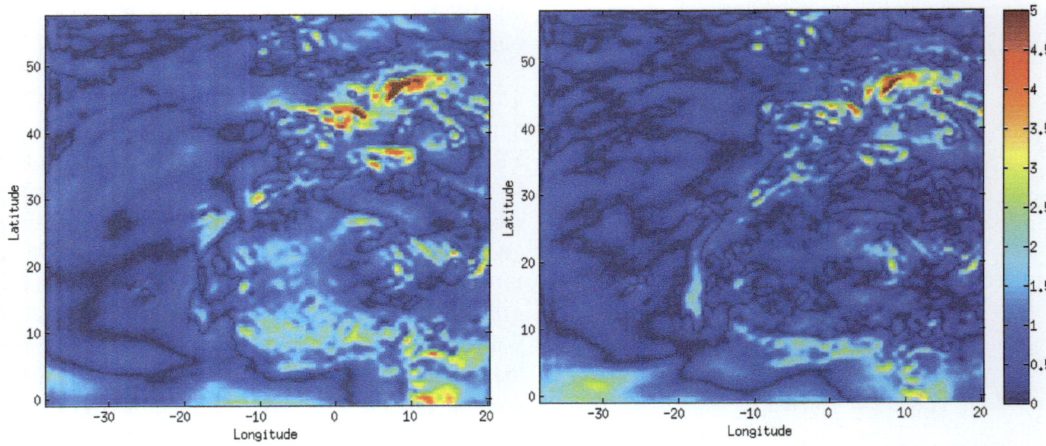

(a) (b)

Figure 2. Absolute value of the consistency error for the monthly mean of the NODA simulation **(a)** and SPECNUD simulation **(b)** obtain with the theoretical filter.

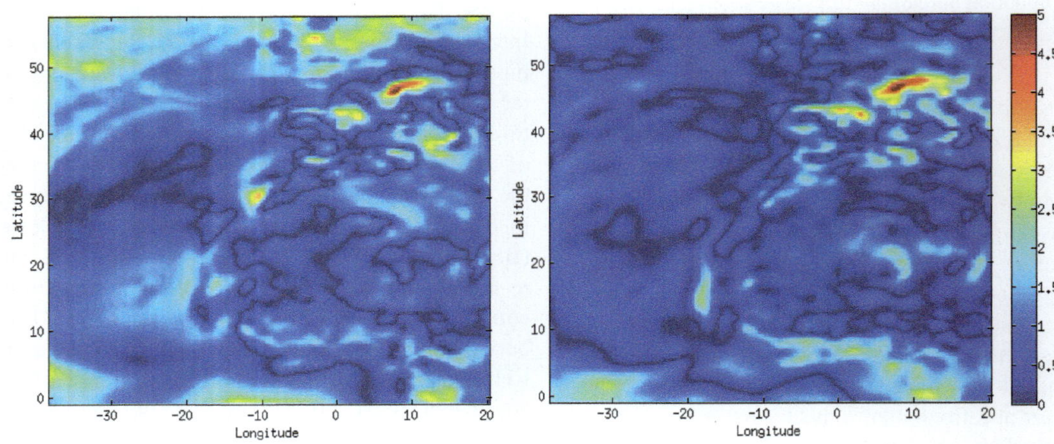

(a) (b)

Figure 3. Absolute value of the consistency error for the monthly mean of the NODA simulation **(a)** and SPECNUD simulation **(b)** obtain with the optimal filter.

not adequate to obtain a great consistency between WRF and ERA-Interim. The objective is yet to assess the existence and the value of a threshold scale above which the error is small or acceptable. We expect the spectral nudging to modify this threshold since the WRF large scales are constrained towards large scales from ERA-Interim. A filter H_{opt} is designed such as to minimize the consistency. It is constructed from the Fourier transform of the auto-correlation function (Γ):

$$S(f) = \int_{-\infty}^{\infty} \Gamma(\tau) e^{-2j\pi f \tau} \mathrm{d}\tau \qquad (2)$$

so that $E[(H_{opt} \cdot \mathrm{RUN} - \mathrm{ANA})^2]$ is minimal. Assuming 2nd order stationary processes and applying the perpendicularity theorem, it can be proven that the re-analysis auto-correlation is related to the cross-correlation:

$$H_{opt} = \mathrm{TF}^{-1}\left(\frac{S_{\mathrm{ANA}}}{S_{\mathrm{RUN,ANA}}}\right) \qquad (3)$$

Using Eq. (1), H_{opt} is applied to the one month period mean wind speed. The obtain filter provides the threshold above which the large scale structures are maintained by the high resolution model in the mean wind field and its amplitude determine the amount of information maintained.

By definition, this is the smallest consistency error that can be obtained by removing information from the high resolution simulation. The mean error magnitude is reduced by about 20 % for the NODA simulation and by about 15 % for

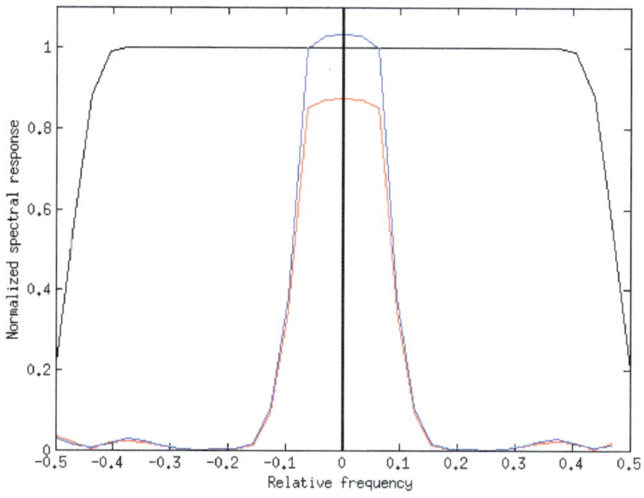

Figure 4. Latitudinal profile of the optimal filter for the one month mean of the 100 m wind speed plotted on relative frequency range (0.5 corresponds to half the sampling rate of the re-analysis) for the NODA simulation (in red) and for the SPECNUD simulation (in blue). The theoretical filter is represented as the thin black line.

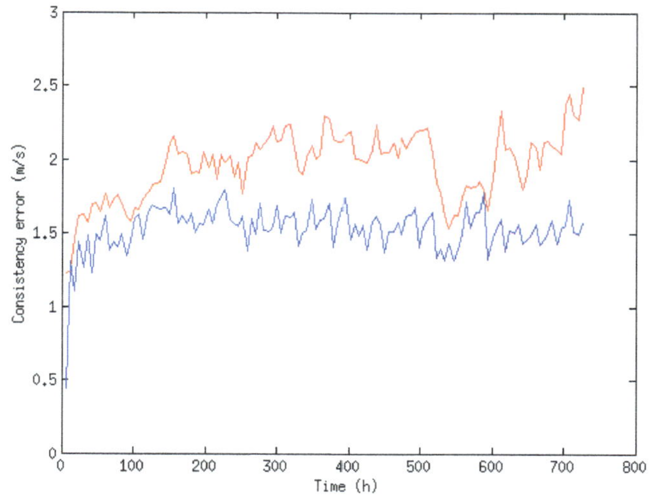

Figure 5. 6 h-time series of the consistency error over the one month period for the NODA simulation in red and for the SPECNUD simulation in blue.

the SPECNUD simulation (Fig. 3) compared to the consistency error in the theoretical case (Fig. 2). Many previously observed fine scales structures have been removed. The error is however slightly increased over ocean and reduced over flat terrain.

The obtained filter shapes are isotropic which means that there is no preferential treatment along one direction upon the other. The latitudinal profile of the optimal filter for both simulations is presented in Fig. 4 and superimposed on the transect of the theoretical filter. In Fig. 4 the frequency axis is graduated relatively to the sampling frequency of the re-analysis: a value of 0.5 corresponds to a wavelength of 1.5°. The optimal filters are mostly rectangular from −0.12 to 0.12. It corresponds to a low-pass filter with a low-pass threshold of 6.25°; structures from ERA-Interim less than 6.25° are strongly modified by the WRF model. This threshold is almost the same for both simulations.

It can be noted that the maximum spectral response of the filter is not equal to 1. When the model is run freely the large scale mean wind speed is increasing along with the resolution. When the spectral nudging is turned on the amount of large scale structures is slightly lesser in WRF than in ERA-Interim. To perfectly match the ERA-Interim large scales a smaller nudging coefficient ($4.10^{-4}\,s^{-1}$ for example) should be used. Hence this comparison technique can be used to determine the best WRF nudging configuration.

Once H_{opt} is obtained for the monthly mean it is applied to the WRF output every 6 h and the consistency error time series is computed. The time evolution of the error is presented in Fig. 5 for the month of January 2004. The time series demonstrates that the error is growing along with the duration of the simulation when the model is run freely but remains quite constant when spectral nudging is turned on. It confirms that the model has to be constraint regularly towards a re-analysis to perform a long lasting wind atlas simulation. The rapid increase at the beginning of the time series corresponds to the spin-up period of the model that can be estimated to 24 h.

3 Conclusions

The method presented in this paper is a possible approach to quantify the spatial representativeness of high resolution numerical model compared to re-analysis. It is able to determine the scales from re-analysis, considered as the best information, kept in the high resolution wind field. In the present study, two WRF simulations with and without data assimilation are compared to ERA-Interim re-analysis. After removing the theoretical range of information from high resolution simulations to be on the exact same theoretical range scale than re-analysis, the consistency error remains large and includes still fine scale structures. An optimal filter is then computed to minimize the consistency error. The obtained filter is a thinner low-pass filter compared to the theoretical filter. It means that WRF modifies the information up to 5 times the size of the grid cell resolution of the re-analysis. Turning on the spectral nudging does not modify this threshold but maintains the magnitude of large scale under this threshold. It leads to a higher consistency between WRF and ERA-Interim and this consistency remains stable over the one month period simulation whereas it increases without constraint.

Simulations with different spectral nudging configurations could be run to determine the effect of the nudging coefficient or of the wave number for example. A longer simulation is also needed to verify that the WRF model remains coherent with the re-analysis over long time periods such as years needed for a wind atlas. The difference between the NODA and SPECNUD simulations raises the question on the impact of the misrepresentation of the large scale features. Indeed the range of the large scale from re-analysis maintained in both high resolution simulations is similar. The amount is however superior over these scales when spectral nudging is activated. The obtained wind atlas (here over a one month period) is more reliable for the SPECNUD than for the NODA simulation. Would the wind atlas be even more reliable with a stronger constraint towards re-analysis?

Acknowledgements. The authors would like to acknowledge Lucien Wald for his remark about the paper and Philippe Blanc for his help in mathematics and programming. This work was achieved with the financial support of IRESEN and in collaboration with Transvalor Innovation.

Edited by: E. Batchvarova

References

Blackman, R. B. and Tukey, J. W.: The measurement of power spectra, from the point of view of communications engineering, Dover Publications, 1958.

Castro, C., Pielke, R., and Leoncini, G.: Dynamical downscaling: Assessment of value retained and added using the regional atmospheric modeling system (RAMS) RID A-5015-2009, J. Geophys. Res.-Atmos., 110, D05108, doi:10.1029/2004JD004721, 2005.

Chen, F., Mitchell, K., Schaake, J., Xue, Y. K., Pan, H. L., Koren, V., Duan, Q. Y., Ek, M., and Betts, A.: Modeling of land surface evaporation by four schemes and comparison with FIFE observations, J. Geophys. Res.-Atmos., 101, 7251–7268, doi:10.1029/95JD02165, 1996.

Dee, D. P., Uppala, S. M., Simmons, A. J., Berrisford, P., Poli, P., Kobayashi, S., Andrae, U., Balmaseda, M. A., Balsamo, G., Bauer, P., Bechtold, P., Beljaars, A. C. M., van de Berg, L., Bidlot, J., Bormann, N., Delsol, C., Dragani, R., Fuentes, M., Geer, A. J., Haimberger, L., Healy, S. B., Hersbach, H., Hólm, E. V., Isaksen, L., Kållberg, P., Köhler, M., Matricardi, M., McNally, A. P., Monge-Sanz, B. M., Morcrette, J.-J., Park, B.-K., Peubey, C., de Rosnay, P., Tavolato, C., Thépaut, J.-N., and Vitart, F.: The ERA-Interim reanalysis: configuration and performance of the data assimilation system, Q. J. Roy. Meteor. Soc., 137, 553–597, doi:10.1002/qj.828, 2011.

Fichaux, N., Poglio, T., and Ranchin, T.: Mapping offshore wind resources: Synergetic potential of SAR and scatterometer data, IEEE J. Ocean. Eng., 30, 516–525, doi:10.1109/JOE.2005.857504, 2005.

Heikkilä, U., Sandvik, A., and Sorteberg, A.: Dynamical downscaling of ERA-40 in complex terrain using the WRF regional climate model, Clim. Dynam., 37, 1551–1564, doi:10.1007/s00382-010-0928-6, 2011.

Hong, S.-Y. and Lim, J.-O. J.: The WRF Single-Moment 6-Class Microphysics Scheme (WSM6), J. Korean Meteor. Soc., 42, 129–151, 2006.

Kain, J. and Fritsch, J.: A One-Dimensional Entraining Detraining Plume Model and Its Application, J. Atmos. Sci., 47, 2784–2802, doi:10.1175/1520-0469(1990)047<2784:AODEPM>2.0.CO;2, 1990.

Skamarock, W. C., Klemp, J. B., Dudhia, J., Gill, D. O., Barker, D. M., Duda, M. G., Huang, X.-Y., Wang, W., and Powers, J. G.: A Description of the Advanced Research WRF Version 3, NCAR technical note, 2008.

Wald, L., Ranchin, T., and Mangolini, M.: Fusion of satellite images of different spatial resolutions: Assessing the quality of resulting images, Photogramm. Eng. Remote Sens., 63, 691–699, 1997.

The 27 July 2002 tornado event in Athens, Greece

J. T. Matsangouras and P. T. Nastos

Laboratory of Climatology and Atmospheric Environment, Faculty of Geology and Geoenvironment,
University of Athens, Panepistimiopolis, 157 84 Athens, Greece

Abstract. This study analyzes the tornado event on 27 July 2002 at Athens International Airport (Eleutherios Venizelos), a suburban area located at the east of Athens. The tornado was formed approximately at 10:20 UTC and characterized as T4 (Torro Scale). A synoptic discussion of the ECMWF analysis charts from the surface up to the 500 hPa geopotential height level is presented along with the daily composite anomaly (reference period: 1968–1996) of the relative meteorological parameters from NCEP/NCAR reanalysis dataset. The vertical profile of the atmosphere is also presented, and derived from the operational sounding of the nearest upper air meteorological station to the tornado incidence site. The dynamic indices of the atmosphere revealed unstable atmospheric conditions capable for thunderstorm formation. The tornado caused injuries to a woman due to the shift of a parked airplane during the disembarkation procedure and also several damage at the airport facilities.

1 Introduction

Tornadoes, as results of extreme convective weather at local or large scale, are associated with strong winds, which cause extended damage and in many cases loss of life. The increased atmospheric moisture content due to global warming may force an increase in severe weather and tornado activity. These fury phenomena are rare in Greece and in most of the cases appear in the sea without causing significant damage, until they take place over an urban region with remarkable consequences.

Tornadoes occur in many parts of the world (Fujita, 1973) and several publications during the last two decades indicate the occurrence of tornadoes in many European countries (Dessens, 1984; Dessens and Snow, 1987; Paul, 1999; Reynolds, 1999; Dotzek, 2001; Holzer, 2001; Bechini et al., 2001; Sioutas, 2003; Bertato et al., 2003). The tornado damage path is considered to extend a few hundred meters in width causing though considerable damage in buildings and in many cases posing a serious threat for human life, especially when tornadoes develop near urban areas or hit international airports. Bech et al. (2007) analyzed a tornado outbreak on 7 September 2005 in the Llobregat delta river, affecting a densely populated and urbanised area and the Barcelona International airport (NE Spain). At least five short-lived tornadoes were confirmed, four of which were weak (F0, F1) and the other one was significant (F2 on the Fujita scale).

Greece has experienced a mean number of six tornadoes per year (Sioutas, 2003), as its complex geomorphology creates a favorable environment for the development of tornado activity. This tornado activity causes several damages in buildings and in a few cases lethal effects. During the last century whirlwind phenomena (33 events occurred in 28 days) caused the loss of 4 lifes, the injury of 40 people and numerous damage on human constructions and cultivations (Nastos and Matsangouras, 2009). Although that the number of 33 cases is greater than the number of 17 cases presented in previous research by Sioutas (2003), this database is assumed incomplete as more cases are likely to be found in the future. The tornado reporting in Greece within the 20th century revealed an increasing trend, mainly over the west and south coastal areas, while the temporal variability showed that the maximum of tornadoes activity dominated within the cold period of the year (autumn and winter) during the warm hours of the day of the day (09:00–15:00 UTC) (Nastos and Matsangouras, 2009). It should be noted that historical reviews contribute to a significant climatological information (Paul,1999; Tyrrell, 2001).

Figure 1. Tornado at Athens International Airport "Eleftherios Venizelos" at 10:20 UTC on 27 July 2002.

This study analyzes the tornado event on 27 July 2002 at Athens International Airport (Eletherios Venizelos), a suburban area located at the east of Athens. The tornado was formed approximately at 10:20 UTC after the passage of a thunderstorm and was characterized as T4 (Torro Scale). Torro scale was developed by Meaden (1976) of the Tornado and Storm Research Organisation (TORRO), a meteorological organization in the United Kingdom, as an extension of the Beaufort scale. The International Tornado Intensity Scale categorizes wind speeds of tornadoes. The scale is directly related to the Beaufort scale and is the only true tornado intensity scale with a sound scientific base. Tornadoes of strength T0, T1, T2, T3 are termed weak tornadoes, those reaching T4, T5, T6, T7 are strong tornadoes and T8, T9, T10, T11 are violent tornadoes.

2 Tornado event analysis

The tornado formed at 10:20 UTC (13:20 Greek Local Time) on 27 July 2002, approximately 8 km SW from Athens International Airport (Eletherios Venizelos) shortly after the passage of a thunderstorm (Fig. 1). According to the authors' interview from a weather officer, who was on duty that day, the tornado moved NE straight to the airport area and when it reached the Athens International Cargo Station it was dissipated and reformed after that point. The tornado reinforced as it was crossing the airports runways and moved another 2 km to NE causing several damage. In the process, the tornado weakened and a rope funnel cloud was formed on the sky suggesting the dissipation of tornado.

Although the tornado event lasted only 10 min, it caused several damage on light constructions which was on its path and it was characterized as a T4-tornado (Sioutas, 2003). Damage was also reported within the airport's area as debris flow and other light objects hit the planes that were parked causing significant damage to their frames. The strong winds resulted in shifting a parked airplane, hurting a passenger, who was in the disembarkation procedure, the time when she was using the landing stairs. The existence of the tornado at the airport was the reason that many departures and arrivals were cancelled because of the significant possibility of having an accident. Besides, pilots during the landing procedure experienced severe turbulence as the tornado came closer to the airport. On that day another tornado event (T3, Torro scale) was noticed in Peraia, Northen Greece (Sioutas, 2003).

3 Synoptic discussion

As the event took place at 10:20 UTC, we used the 06:00 UTC reanalysis from the European Center for Medium-Range Weather Forecasts (ECMWF) for the synoptic discussion of several levels of the atmosphere. Additionally, the daily composite anomalies (reference period: 1968–1996) for relevant meteorological parameters were calculated from the reanalysis datasets of the National Centers for Environmental Prediction/National Center for Atmospheric Research (NCEP/NCAR). Finally, the vertical profile of the atmosphere was presented using the available upper air data at 00:00 UTC on 27 July 2002, in Athens.

3.1 ECMWF reanalysis

The synoptic conditions of the event day were characterized by a closed low at the isobaric level of 500 hPa associated with low temperatures over the area of interest causing a SW upper air flow (Fig. 2a). This synoptic type is characterized as the dominant type for tornado formation in Greece (Sioutas, 2003). The cyclonic circulation also existed at lower levels such as 700 hPa, 850 hPa and at Mean Sea Level (MSL) causing S-SW air stream over the studied area (Fig. 2b, c, d). The reanalysis of the divergence-convergence revealed convergence values equal to $1.2\,s^{-1}$ along with vertical velocity values equal to -1.12 Pa/s at the lower levels of the atmosphere (850–925 hpa) (not shown). These values were greater than the respective mean monthly values of July 2002 (divergence at 850–925 hPa: $0.2\,s^{-1}$ and vertical velocity at 850–925 hPa: +0.06 Pa/s) (not shown).

3.2 NCEP/NCAR reanalysis

Studying the NCEP/NCAR reanalysis datasets (Kalnay et al., 1996), relevant conclusions were extracted. Daily composite anomalies (reference period: 1968–1996) of the geopotential heights occurred from the middle to the lower atmospheric levels (500 hPa – mean sea level) on 27 July 2002 (Fig. 3a–c). Moreover, a significant daily composite anomaly (reference period: 1968–1996) of the horizontal wind shear extended over the area, from 700 hpa to 925 hpa (Fig. 3d–f).

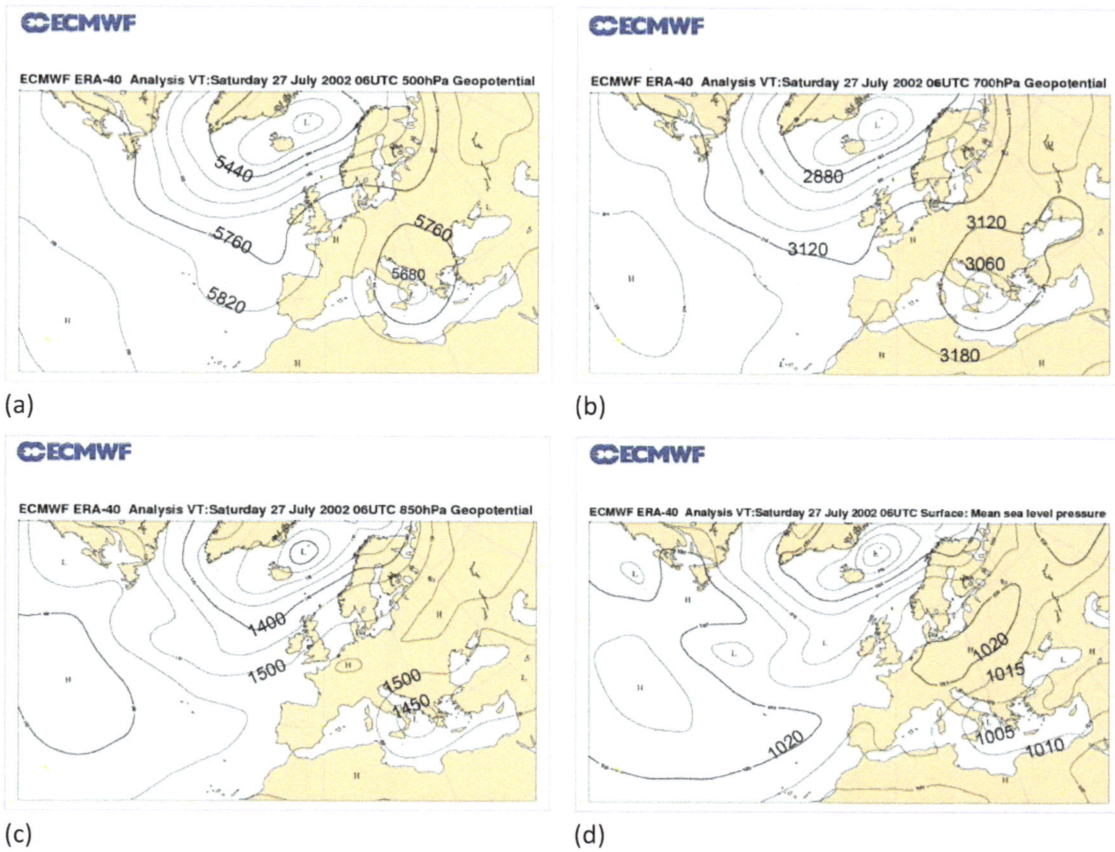

Figure 2. ECMWF reanalysis charts of geopotential heights at: **(a)** 500 hPa, **(b)** 700 hPa, **(c)** 850 hPa and **(d)** reanalysis of MSL pressure, at 06:00 UTC on 27 July 2002.

3.3 Upper air data

The vertical profile plot of the atmosphere over Athens at 00:00 UTC (Fig. 4) was a significant contribution to study the vertical dynamic structure of the atmosphere. In this point, it is useful to notify that the Upper Air Station (37°52 N, 23°44 E), from which the vertical profile of the atmosphere was taken, is located about 20km southwest from the Athens International Airport (37°56 N, 23°57 E). The Skew-T plot analysis was downloaded from Wyoming's University database (http://weather.uwyo.edu/upperair/europe.html). The dynamic indices of the atmosphere revealed unstable atmospheric conditions capable for thunderstorm formation; K Index = 23.3 °C, Showalter Index (SI) = 3.7 °C, Severe Weather Threat Index (SWEAT) = 93.6, Convective Available Potential Energy (CAPE) = 1993 ($J\,Kg^{-1}$) and Lifted Index (LIFT) = −6.5 °C. Furthermore, the above dynamic indices extracted from the Skew-T plot analysis at 12:00 UTC (1.5 h after the tornado formation-dissipation) revealed further unstable atmospheric conditions: K Index = 21.2 °C, SI = 1 °C, SWEAT = 315.6, CAPE = 1545 ($J\,Kg^{-1}$) and LI = −5.8 °C.

3.4 Satellite images

The tornado formed exactly after the pass of a severe thunderstorm over Athens International Airport at 10:20 UTC. The storm was spotted on the visual spectrum (VIS 0.6 μm) of Meteosat 7, at a sequence of satellite images from 09:30 UTC to 11:00 UTC (Fig. 5). Studying the sequence of the images at visual spectrum, we found out that the storm was spotted firstly at 09:30 UTC (Fig. 5a) over Saronikos Gulf (south of Athens). As the storm was propagating NE, it was further more developed on vertical structure until it was dissipated over the Southern Evoikos Gulf (eastwards of Athens international Airport) (Fig. 5d).

4 Conclusions

On 27 July 2002 at 10:20 UTC, a tornado (T4-Torro Scale) hit the Athens International Airport, a suburban area located at the east of Athens. The tornado was formed SW from the airport and propagated NE with a trajectory of approximately 10 km, causing significant damage to the airport facilities and at parked airplanes. The analysis of the ECMWF charts

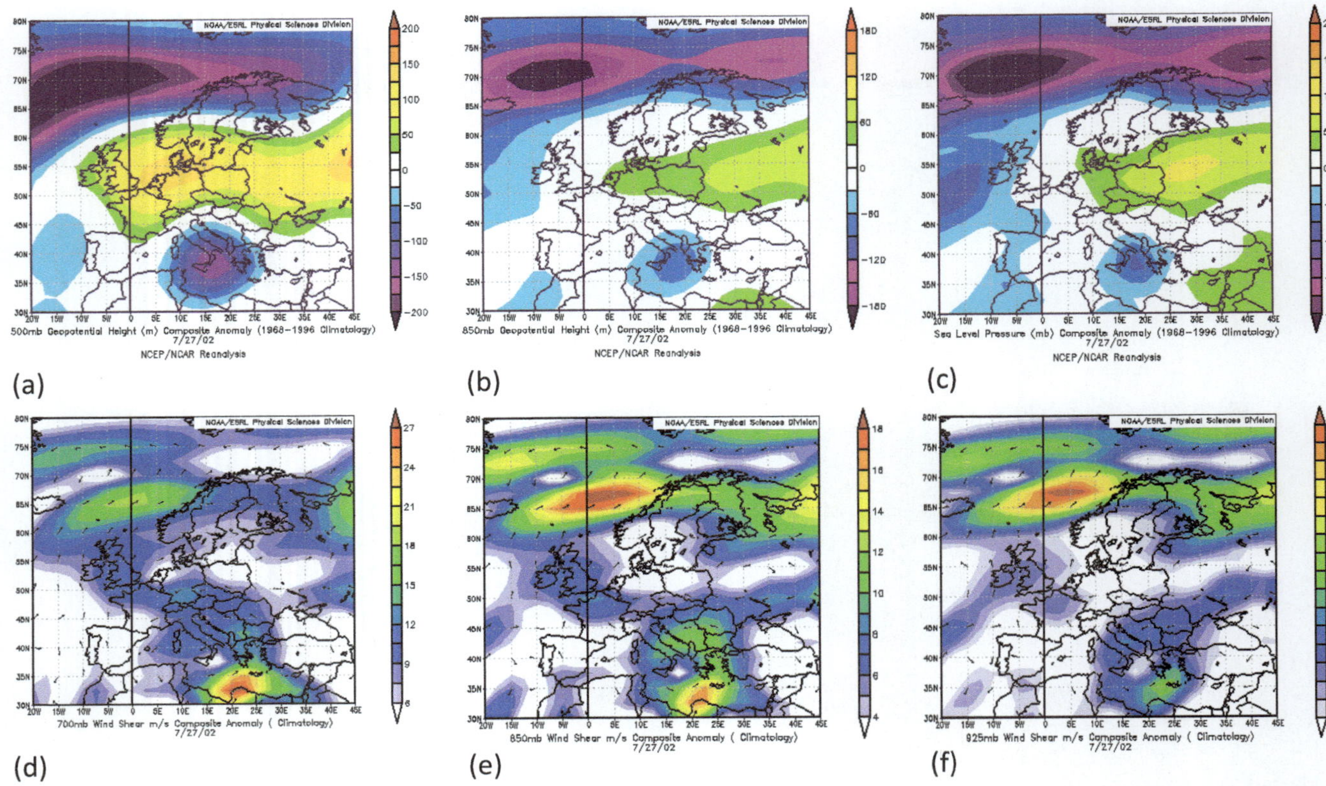

Figure 3. Daily composite anomaly (reference period: 1968-1996) of geopotential heights at the isobaric levels of 500 hPa (a) and 850 hPa (b), of sea-level pressure (c) and daily composite anomaly (reference period: 1968-1996) of wind shear at 700 hPa (d), 850 hPa (e) and 925 hPa (f), on July, 27 2002, from NCEP/NCAR reanalysis datasets.

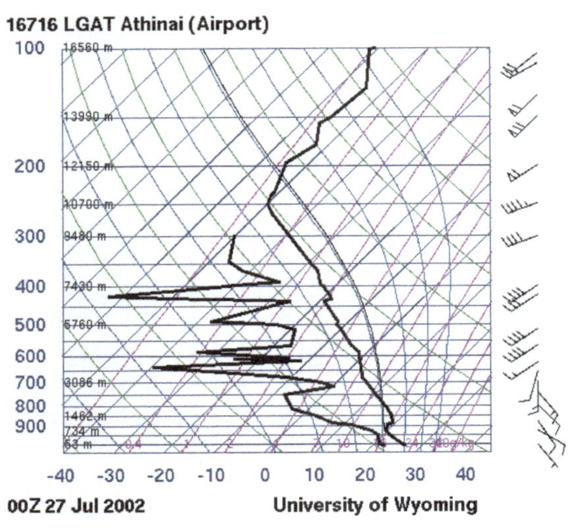

Figure 4. Skew-T plot and indices analysis for Athens at 00:00 UTC, 27 July 2002, retrieved from the database of Wyoming University USA.

revealed a cyclonic circulation from the middle to the lower levels of the atmosphere, while daily composite anomalies of the wind shear from NCEP/NCAR reanalysis datasets appeared at the lower levels of the atmosphere. The vertical profile of the atmosphere showed instability indices typical for the development of severe thunderstorm. A woman was hurt by the tornado activity due to the shift of an airplane during the disembarkation procedure, while several damage at the airport facilities was reported.

The gain of understanding the extracted results from the performed analysis is the possible use of the specific instability indices' thresholds in order to establish an early warning system for short range prognosis of tornadoes' incidence, especially in urban areas and international airports. Future work on local scale is needed in order to verify the atmospheric conditions favorable for tornadoes' development.

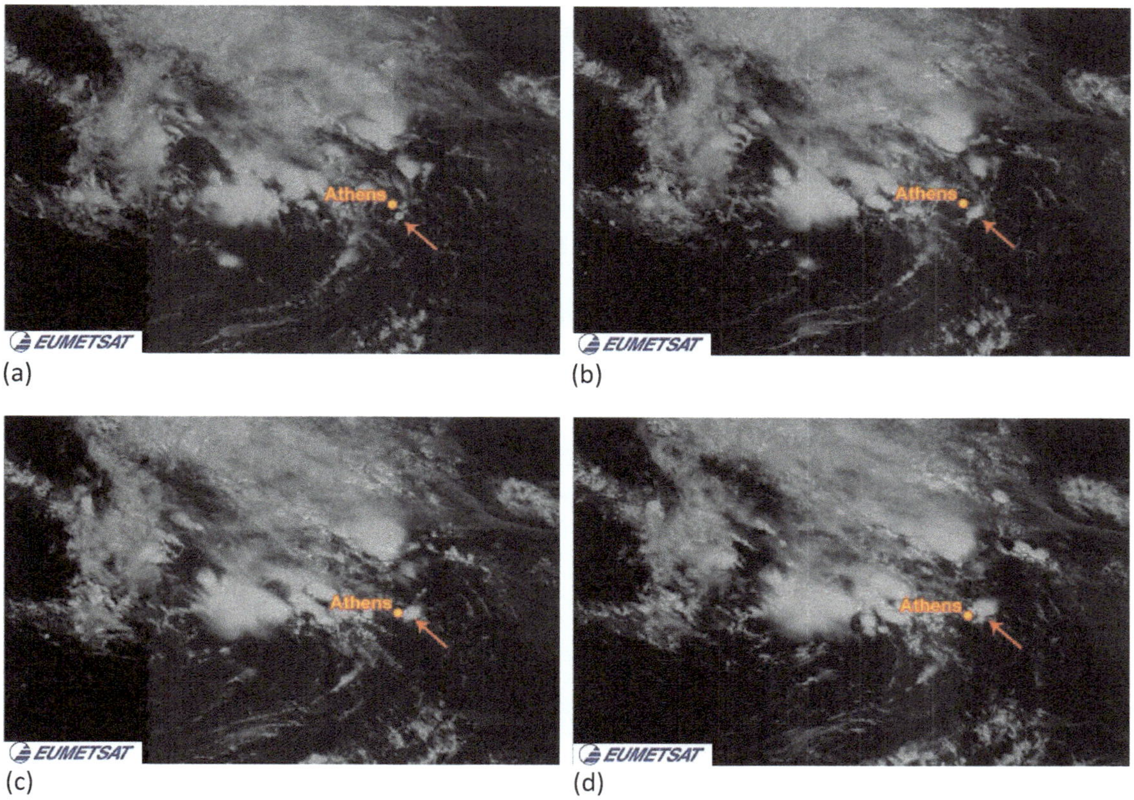

Figure 5. Sequence of satellite images at: **(a)** 09:30 UTC, **(b)** 10:00 UTC, **(c)** 10:30 UTC and **(d)** 11:00 UTC on 27 July 2002 from visual spectrum (VIS 0.6 μm) of Meteosat 7 (Eumetsat). Red arrow indicates the associated storm that formed the tornado over the Athens International Airport.

Acknowledgements. The authors are grateful to the European Center for Medium range Weather Forecasts for the use of reanalysis archive datasets at several levels. Moreover, we thank the National Centers for Environmental Prediction/National Center for Atmospheric Research, the Eumetsat and the Hellenic National Meteorological Service. Special thanks to the HNMS Meteo Officer 1st Lieutenant Mr. Litras for the detailed information about the tornado event.

Edited by: D. Giaiotti

References

Bech, J., Pascual, R., Rigo, T., Pineda, N., López, J. M., Arús, J., and Gayà, M.: An observational study of the 7 September 2005 Barcelona tornado outbreak, Nat. Hazards Earth Syst. Sci., 7, 129–139, 2007, http://www.nat-hazards-earth-syst-sci.net/7/129/2007/.

Bechini, R., Giaiotti, D., Manzato, A., Stel, F. and Micheletti, S.: The June 4th 1999 severe weather episode in san Quirino, Italy: a tornado event?, Atmos. Res., 56, 213–232, 2001.

Bertato, M., Giaiotti, D. B., Manzato, A., and Stel, F.: An interesting case of tornado in Friuli-Norteastern Italy, Atmos. Res., 67–68, 3–21, 2003.

Dessens, J.: Les Trombes en France. Climatologie et Caracteristiques Physiques, Lannemezan, France, 1984..

Dessens, J. and Snow, J. T.: Tornadoes in France, Weather Forecast., 4, 110–132, 1987.

Dotzek, N.: Tornadoes in Germany, Atmos. Res., 56, 233–251, 2001.

Fujita, T. T.: Tornadoes around the world, Weatherwise, 26, 56–83, 1973.

Holzer, A. M.: Tornado climatology of Austria, Atmos. Res., 56, 203–211, 2001.

Kalnay, E. and Coauthors: The NCEP/NCAR 40-Year Reanalysis Project, B. Am. Meteorol. Soc., 77, 437–471, 1996.

Meaden, G. T.: Tornadoes in Britain: Their intensities and distribution in space and time, J. Meteorol., 1(8), 242–251, 1976.

Nastos, P. T. and Matsangouras J. T.: Tornado activity in Greece within the 20th Century, 11th Plinius Conference on Mediterranean storms, 7–11 September, Barcelona, Spain, Plinius Conference Abstracts, Vol. 11, 11–57, 2009.

Paul, J. F.: An inventory of tornadoes in France, Weather, 54, 217–219, 1999.

Reynolds, D. J.: European tornado climatology 1960–1989, J. Meteorol., 24, 376–403, 1999.

Sioutas, M. V.: Tornadoes and waterspouts in Greece, Atmos. Res., 67–68, 645–656, 2003.

Tyrrell, J.: Tornadoes in Ireland: an historical and empirical approach, Atmos. Res., 56, 281–290, 2001.

Seasonality in onshore normalized wind profiles above the surface layer

J. N. Nissen and S.-E. Gryning

Risø National Laboratory for Sustainable Energy, Technical University of Denmark, Frederiksborgvej 399, P.O. Box 49, 4000 Roskilde, Denmark

Abstract. This work aims to study the seasonal difference in normalized wind speed above the surface layer as it is observed at the 160 m high mast at the coastal site Høvsøre at winds from the sea (westerly).

Normalized and stability averaged wind speeds above the surface layer are observed to be 20 to 50% larger in the winter/spring seasons compared to the summer/autumn seasons at winds from west within the same atmospheric stability class.

A method combining the mesoscale model, COAMPS, and observations of the surface stability of the marine boundary layer is presented. The objective of the method is to reconstruct the seasonal signal in normalized wind speed and identify the physical process behind. The method proved reasonably successful in capturing the relative difference in wind speed between seasons, indicating that the simulated physical processes are likely candidates to the observed seasonal signal in normalized wind speed.

1 Introduction

An increasing demand for the ability to estimate the wind climate above the surface layer in coastal regions has been generated by the wind power industry because of the steady increase of modern wind turbines with hub heights well above the surface layer height. To choose the optimal site for a wind turbine, information about the wind energy potential needs to be properly assessed. The wind energy potential is proportional to the cube of the wind speed at hub height, implying that even a small error in estimating the wind speed can have a large impact on the wind energy assessment. Conventional methods for assessing wind speed at heights of up to 200 m, such as that of Troen and Lundtang (1989), are derived from wind measurement taken close to the surface and extrapolated to greater heights by assuming a logarithmic increase of the wind speed with height, with a correction for stability effects. The derivation of the logarithmic wind profile assumes homogeneous surface properties and a constant momentum flux and is therefore confined to the surface layer. Little is known about boundary layer wind profiles above the surface layer, but recent attempts to address this problem have been put forward by Högström et al. (2006) and Gryning et al. (2007), where extension to the Monin-Obukhov surface similarity scaling is suggested with some success. In a coastal region, the description of the wind profile is however complicated by non-homogenous upstream surface conditions, orography, change of roughness and heat capacity, all of which influence the wind profile and are difficult to describe using simple, universal formulas. An observational study is undertaken in order to quantify the effects on the wind profile from the upstream surface conditions. The observed effects are caused by multiple boundary layer processes at play in the coastal boundary layer and in order to identify the most significant ones, we apply a well-validated mesoscale model in an idealized setup. The selected boundary layer processes in the idealized modelling are kept under easy control and we are therefore able to test the models ability to reproduce the observed wind profile in an idealized atmosphere where only a few boundary layer processes are at play.

2 Observations

The observational basis for this work consists of measurements from the Høvsøre National Test Station for Large Wind Turbines situated in the northwestern part of Denmark (Fig. 1). The data is gathered during the period from

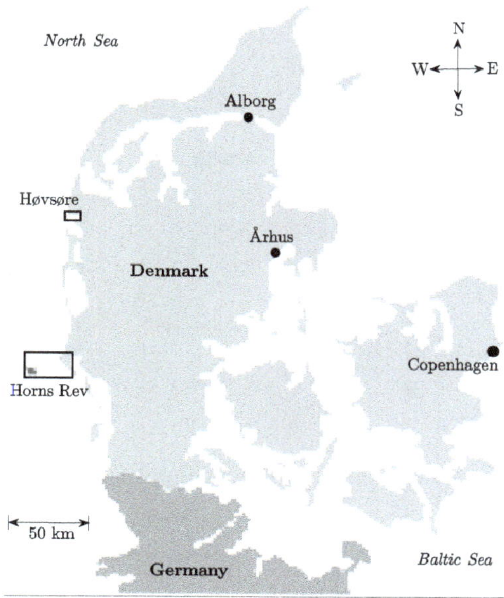

Figure 1. Høvsøre and Horns Rev position.

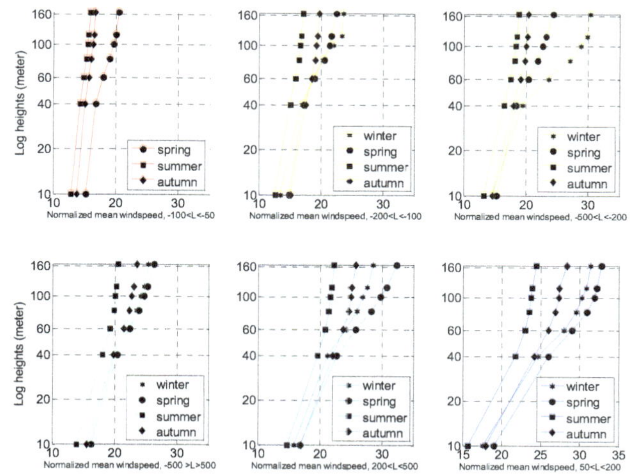

Figure 2. Observed stability-bin-averaged and normalized wind profiles for each season at Høvsøre for the onshore wind direction sector (260–280 degree west.)

February 2004 to May 2008 for the onshore wind direction sector 260°–280°.

Upstream observations of the stability distribution in the marine boundary layer were found to be of importance in describing the seasonal differences in the normalized wind profiles at Høvsøre and measurements from offshore wind park Horns Rev were therefore used. To classify the stability for the Høvsøre observations, the Obukhov length, L, defined in Eq. (1), is evaluated from the Høvsøre sonic anemometer mounted at 20 m height. The observations are subsequently sorted into 7 stability bins following Gryning et al. (2007) and normalized by the friction velocity at 20 m as defined in Eq. (2) and finally averaged according to season.

$$L = \frac{-u_{*20\mathrm{m}}^3}{\kappa(g/T_{20\mathrm{m}})\left(\overline{w'T'}\right)_{20\mathrm{m}}} \tag{1}$$

κ = von Karman constant
g = Acceleration of gravity
T = Temperature

$$u_{*20\mathrm{m}}^2 = \left[\overline{u'w'_{20\mathrm{m}}}^2 + \overline{v'w'_{20\mathrm{m}}}^2\right]^{0.5} \tag{2}$$

The results from this procedure can be seen in Fig. 2. It is here observed how the seasons wind speed systematically splits at heights above 40 m. The spring season is always the most windy (except for bin $-500 < L < -200$ which is poorly represented for the winter season) and the summer is the less windy season. Winter and autumn falls systematically between the spring and summer seasons for all stability bins.

3 Method and physical process

It was during this work observed, that the annual stability distribution between the coastal station Høvsøre (Fig. 3) and the upstream marine station Horns Rev (Fig. 4) was phase shifted. It is posed here that the associated boundary layer processes from the observed phase shift in annual stability distribution are responsible for the observed seasonal signal in the normalized wind profiles in the following way.

When a majority of observations reflect unstable marine atmospheric boundary layers (hereafter mabl), which is advected over a neutral coastal internal boundary layer (cibl), then the upper normalized wind profile will bend to the left (autumn and summer scenario, neutral case). The opposite will happen when the majority of observations reflect stable mabl's, which are advected over a neutral cibl and make the upper normalized wind profiles bend to the right (spring and winter scenarios, neutral case). The method presented here was designed to test this hypnosis's accordingly.

At Horns Rev L is calculated from the bulk Richardson number, following Grachev and Fairall (1996) in Eqs. (3) and (4) respectively, where $C_1 = C_2 = 10$ and $C_3 = 5$.

$$Ri_b = \frac{gz(\Delta\theta)_{13\mathrm{m}-(-4\mathrm{m})}}{\theta_{13\mathrm{m}} u_{15\mathrm{m}}^2} \tag{3}$$

$$\frac{z}{L} = C_1 Ri_b \qquad \text{Unstable}$$
$$\frac{z}{L} = \frac{C_2 Ri_b}{1 - C_3 Ri_b} \qquad \text{Stable} \tag{4}$$

Modeling and result: The numerical method consists of 2 sets of simulations, denoted step 1 and step 2, and are conceptually outlined in Fig. 5.

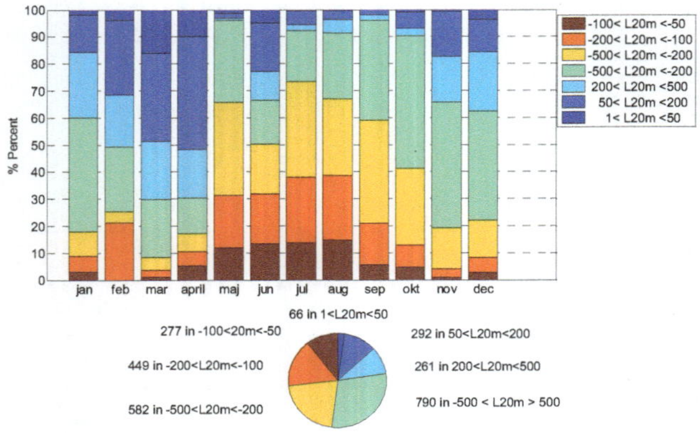

Figure 3. Høvsøre test station: monthly stability distribution (bar graph, upper) and yearly distribution of the stability distribution (pie chart, lower). Left hand integers denote number of observation in stability class (legend, lower).

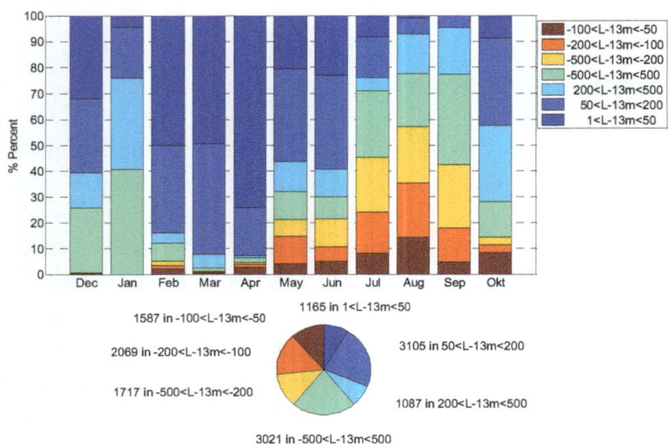

Figure 4. Horns Rev wind farm: monthly stability distribution (bar graph, upper) and yearly distribution of the stability distribution (pie chart, lower). Left hand integers denote number of observation in stability class (legend, lower). The data is from the period December 2003 to November 2004, for the 225°–315° wind direction sector.

Step 1. Generation of 18 marine atmospheric boundary layers (mabl).

Procedure: In this step, air is allowed 6 different initial temperatures and is swept across a sea surface, which is kept at temperature 277 K. The initial speed of the wind is varied from 10 m/s over 15 to 20 m/s and is initially constant with height. The marine atmospheric boundary layer is then spun up by the model and profiles of wind and temperature in equilibrium with the sea surface is generated during the 6 h of integration time.

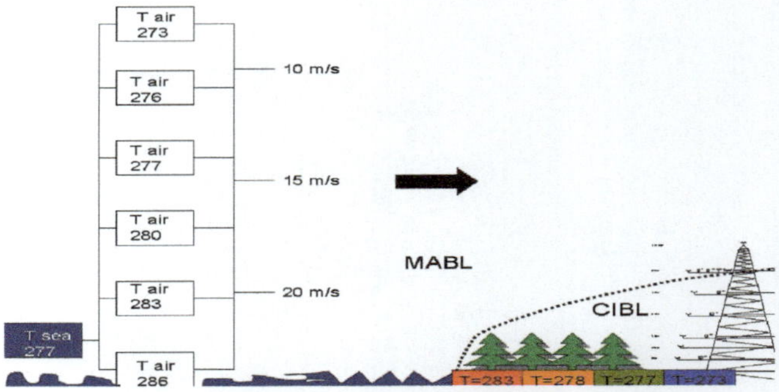

Figure 5. Conceptual model of the numerical method.

Step 2. Generation of 72 coastal wind profiles.

Procedure: The 18 mabl are hereafter exposed to a cost line with 4 different temperatures. The end result of these two simulation steps is 72 simulations, where wind profiles are extracted from a position in the model domain relating to Høvsøre test station in terms of distance from the coast.

To make the extracted profiles comparable to the normalized and seasonal averaged observations from Høvsøre, an averaging procedure reflecting the particular season is needed. The observed phase shift in annual stability distribution over the North Sea is used to average the profiles in the following way. Each extracted wind profile is assigned a weight for each season reflecting how often the mabl it originates from occurs in that season. The seasonal weights are attained by assuming that a mabl is fully characterised by the computed surface L. The probability/weight of the mabl in a given season is then the same as the probability of the computed L to occur in that season divided by the number of simulated mabl with the same L. The weight of each mabl for each season, depicted in Fig. 6, is then used to average the normalized wind profiles within each stability bin. This result is one wind profiles for each season and stability bin, as depicted in (Fig. 7) which is then immediately comparable to the observed wind profiles from Høvsøre (Fig. 2).

4 Results and conclusions

The observed seasonal pattern as seen in Fig. 2 is to some extent reproduced by the method as seen in Fig. 7. The relatively larger wind speeds observed in spring and winter is clearly detectable in all 4 simulated stability bins. Also the trend towards higher wind speed at the top of the seasonal averaged profile with increasing stability is clearly seen. The height of the equilibrium layer is also seen to be captured quite well to 40 m. The experiment proved skill in predicting the seasonality in the normalized wind profiles for Høvsøre when incorporating information on the upstream stability distribution. Using the time lag in stability distribution in

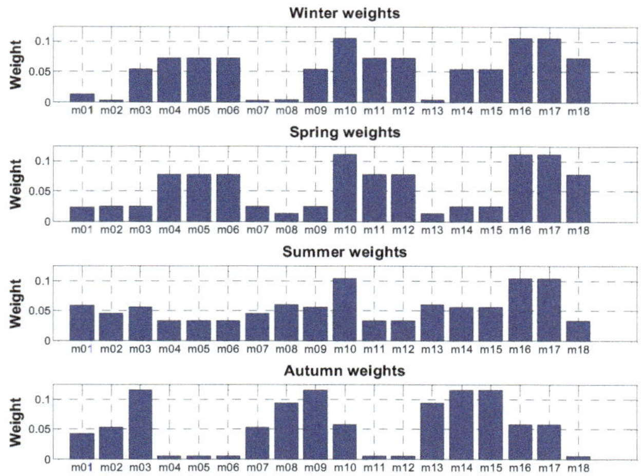

Figure 6. Seasonal weighs for the generated mabl. The syntax for the mabl is listed in Table A2.

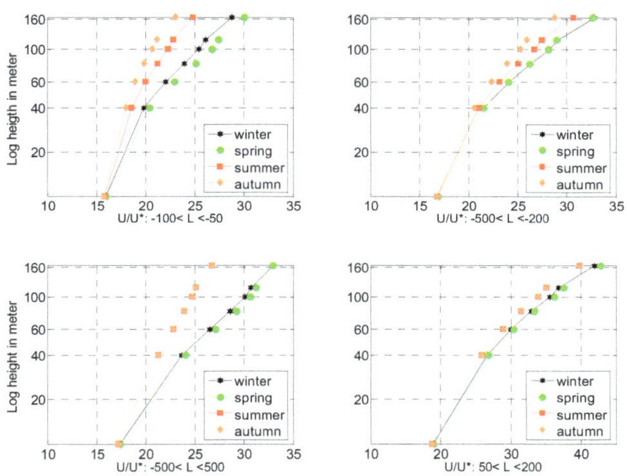

Figure 7. Simulated stability-bin-averaged and normalized wind profiles for each season at a position in model domain relating to Høvsøre.

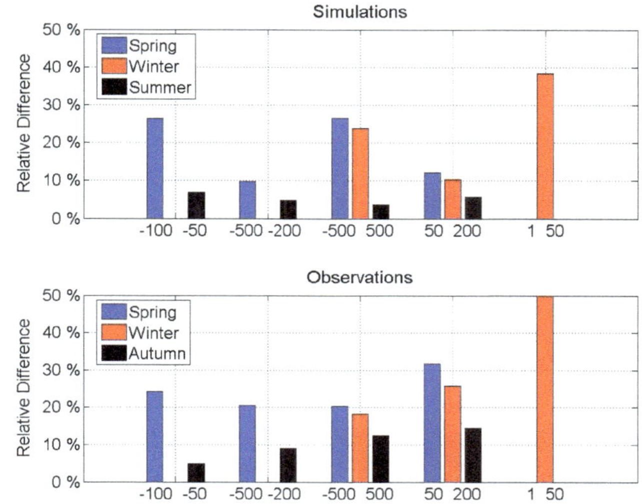

Figure 8. Simulated and observed relative difference between seasons at 100 m found according to Eqs. (5) and (6).

$$\text{RD-winter}=\frac{[\text{Winter\_Wspd}_{100\text{m}}-\text{Autumn\_Wspd}_{100\text{m}}]}{\text{Autumn\_Wspd}_{100\text{m}}}100$$

Simulated profiles (5)

$$\text{RD-winter}=\frac{[\text{Winter\_Wspd}_{100\text{m}}-\text{Summer\_Wspd}_{100\text{m}}]}{\text{Summer\_Wspd}_{100\text{m}}}100$$

Observed profiles (6)

terms of L between the North Sea and Høvsøre made it possible to partially reproduce the observed seasonality indicating that some of the important physical processes behind the observations are captured in the idealized setup presented here. The observed normalized bin averaged wind profiles are, to a reasonable degree, seen to follow the conceptual model suggested by Sempreviva et al. (1990); Jensen and Peterson (1977). The wind profiles from 100 m and up are characterized by a smaller slope and according to Sempreviva et al. (1990) and Jensen and Peterson (1977) can the smaller slope be interpreted as an equilibrium with the upstream smoother sea surface. This feature is not captured in the numerical setup and the upper part of the simulated bin averaged wind profile is seen to be over predicted for all seasons and bins. The difference between the simulations and the observations for wind speed above 100 m clearly shows that the highly idealized simulations utilities here does not capture correctly the boundary layer processes responsible for the mabl structures over the North Sea, as seen in the observations above 100 m.The most likely explanation for this mismatch between the observations and the simulations is a possible too low simulated height of the averaged upstream constant flux layer. The conceptual model suggested by Sempreviva et al. (1990) and Jensen and Peterson (1977) assumes logarithmic upwind wind profile dictating an upwind average constant flux layer height of at least 160 m in order to diagnose the characteristically kink feature in the wind profile as seen in the observations from Høvsøre.

5 Model skills relative comparison

Better agreement between the observations and the simulations is found when comparing the relative differences, denoted RD, calculated as outlined in Eqs. (5), (6) and depicted in Fig. 8. The 100 m height is chosen for a number of reasons. The most important one is that it close to the height of the internal boundary layer at the position of the observations mast and therefore reflects both local and upstream properties.

Table A1. Model details.

Grid information		Resolution		Boundary condition
Nr. Horizontal grid points 120 (E–W) 60 (N–S) Step 1 120*60 sea points Step 2: 60*60 land points 60*60 sea points	Vertical 62 levels Top: 3825 m	Horizontal (50*50) m	Vertical: 10 m below 250 m hereafter stretched gradually to 200 m above 1000 m	Step 1: Periodic Step 2 : Western: Rayleight damping on the inflow Eastern, southern and northern: Radiation condition.

Initial profile	Surface	ABL physics
Profiles of wind speed and temperature extracted from the mabl simulations in step 1 are used as initial profiles in step 2.	Step 1: 120*60 grid points of constant sea roughness = 0.0001 m Step 2: 60*60 grid points of land roughness = 0.01 m 60*60 grid points of constant sea roughness 0.0001 m	No moisture, no radiation PBL scheme: 1.5 order, level 2.5 (Mellor and Yamada, 1982) Surface layer : Louis (1979) Mixing length: Blackadar (1962)

Table A2. Syntax for the simulated mabl listed in Fig. 6.

mabl nr	m01	m02	m03	m04	m05	m06	m07	m08	m09
Wspd	10	10	10	10	10	10	15	15	15
Air tmp (K)	273	276	277	280	283	286	273	276	277
mabl nr	m10	m11	m12	m13	m14	m15	m16	m17	m18
Wspd m/s	15	15	15	20	20	20	20	20	20
Airtmp (K)	280	283	286	273	276	277	280	283	286

Good agreement is found between simulations and observations for the most convective bin 1 as seen in Fig. 8. The observed relative difference between the high wind season (winter) and the low wind season (summer) is seen to be 24% while the simulations gave 26%. The same good agreement is evident when comparing the second lowest to the lowest wind season, in the same stability class. Here observations read 4.8% and simulations 7.8%. Similar good agreements are found in the neutral stability class. A general pattern was that we found good agreement for the stability bins with numerous observations and simulations, and not so good agreements were found for the bins with only a few observations and simulations. This is encouraging and indicates that the method captures the relevant physical processes behind the seasonality in the normalized wind profile and we can therefore conclude that the phase shift in annual offshore/onshore stability distribution plays a major role in the observed seasonality in the normalized onshore wind profiles.

6 Summary

We can summarize the findings as follows. The observational analysis showed that onshore normalized wind profiles at Høvsøre had a strong seasonal dependence and the modelling strongly indicated that this was caused by boundary layer processes, associated with the offshore/onshore phase shift in annual stability distribution. We expect the seasonality to be present on the entire west coast of Jutland and therefore recommend including offshore stability condition when addressing the wind resource climate at west coast stations in Jutland.

Appendix A

Model details

COAMPS is a finite-difference approximation to the nonhydrostatic fully compressible equations of motion following Klemp and Wilhemson (1977) with a suite of physical parameterizations of surface fluxes, boundary layer physics and moist processes described in Hodur (1997) and Hodur and Doyle (1998). The physical processes behind the parameterizations can be switched on and off in order to meet the complexity of the area of interest in the numerical experiment. Below is listed information on the model-setup during simulations carried out in this study.

Acknowledgements. This work was supported in part by the Danish Council for Strategic Research sagsnr. 2104-08-0025.

Edited by: A. M. Sempreviva

References

Blackadar, A. K.: The vertical distribution of wind and turbulent exchange in a neutral boundary atmosphere, J. Geophys. Res., 67, 3095–3102, 1962.

Grachev, A. and Fairall, C.: Dependence of the Monin-Obukhov stability parameter on the Bulk Richardson number over the Ocean, J. Appl. Meteorol., 36, 406–414, 1996.

Gryning, S. E., Batcvarova, E., Brümmer, B., Jørgensen, H. E., and Larsen, S. E.: On the extension of the wind profile over homogeneous terrain beyond the surface layer, Bound.-Lay. Meteorol., 124, 251–268, 2007.

Hodur, R. M.: The Naval Research Laboratory's coupled Ocean/atmosphere meso scale predition system (COAMPS), Mon. Weather Rev., 125, 1414–1430, 1997.

Hodur, R. M. and Doyle, J. D.: The coupled ocean/atmosphere model prediction system COAMPS ". Coastal ocean predict, Coast. Estuar. Stud., 56, 125–155, 1998.

Högström, U., Smedman, A., and Bergström, H.: Calculation of the wind speed variation with height over the sea, Wind Engineering, 30, 269–286, 2006.

Jensen, N. O. and Peterson, E. W.: Wind flow near the surface of non uniform terrain, Report of progress for the year 1976 Progress report for US army grant available at Risø National Laboratory, 1976.

Klemp, J. B. and Wilhelmson, R. B.:The simulation of theredimensional convective storm dynamics, J. Atoms. Sci., 35, 1070–1096, 1977.

Mellor, G. L. and Yamada, T.: Development of a turbulence closure model for geophysical fluid problems, Rev. Geophys., 20, 851–875, 1982.

Sempreviva, A. M., Larsen, S. E., Mortensen, N. G., and Troen I.: Response of neutral boundary layers to change in Roughness, Bound.-Lay. Meteorol., 50, 205–226, 1990.

Troen, I. and Lundtang Petersen, E.: European Wind Atlas, Risø National Laboratory, Risø, Denmark, 1989.

18

WRF model and ASAR-retrieved 10 m wind field comparison in a case study over Eastern Mediterranean Sea

M. M. Miglietta, S. Zecchetto, and F. De Biasio

Institute of Atmospheric Sciences and Climate, Italian National Research Council (ISAC-CNR), Section of Padova, Italy

Abstract. In Synthetic Aperture Radar (SAR) imagery, signatures of coherent atmospheric structures, due to sea surface roughness modulation by surface winds, are usually well detected. In the present study, the wind field derived from the Envisat Advanced SAR (ASAR) sensor has been analyzed and compared with those simulated with a regional atmospheric model in a case study over an area located in the eastern Mediterranean Sea, extending southward and eastward of Crete island. This is a region subject to complex wind patterns, due to the interaction of the almost steady northerly Etesian wind with the orography of the islands in the region.

The ASAR Wide Swath Mode images provide datasets at resolutions exceptionally high compared to model data, appropriate for investigating the mesoscale phenomena on the marine atmospheric boundary layer and to retrieve the surface wind field. The latter has been obtained with a methodology based on the 2-D Continuous Wavelet Transform, suitable to isolate the backscatter patterns on the base of energy and scale considerations.

Numerical simulations with the Weather Research and Forecasting (WRF) model have been performed using three 2-way nested domains, the inner one covering the area of interest with a resolution of 1 km. Several simulations, using different diffusion and boundary layer parameterization schemes, have been performed in a case study corresponding to mountain lee waves detected in the ASAR image. The 10 m winds resulting from the numerical experiments have been compared to those retrieved from the ASAR, both quantitatively and qualitatively, in order to analyze the correspondence of observed and simulated wind structures.

1 Introduction

The spatial and temporal scales typical of the atmosphere-ocean interaction range from the micro-scale (O(1) m, O(1) s) to the synoptic scale (O(1000) km, O(10) days). While the interaction processes at synoptic scale are fairly well known, those at meso-β and meso-γ scales, relative respectively to atmospheric processes from 20 km to 200 km and from 2 km to 20 km (Orlanski, 1975) are still to be fully understood, as the amount of available data is generally insufficient to completely resolve these scales and the representation of sub-grid processes in the parametrization schemes of the atmospheric models is only approximated.

Significant modulations of the wind field may occur locally, especially in regions of surrounding steep orography. For example, when a fluid element is displaced vertically by mountains, gravity tries to restore the parcel toward equilibrium resulting in an oscillation called gravity wave. Stationary gravity waves on the downstream side of an obstacle are called lee waves (see Smith, 1979; Atkinson, 1981). As wind field leaves imprints at the sea surface detectable by radars, the presence of atmospheric waves can be identified by such sensors. Satellite-borne scatterometers are able to measure the wind blowing inside the marine atmospheric boundary layer, providing information about the spatial structure of the wind in the marine surface layer over large areas. However, they have a resolution too coarse for coastal applications, where the wind needs to be determined as close to the coast as possible. Thus, at present, the Synthetic Aperture Radar

(SAR), due to its very fine spatial resolution, is the appropriate sensor for a spatial high resolution wind determination (see Zecchetto and De Biasio, 2008 and the references quoted hereafter).

Global models cannot simulate accurately the strength and the spatial pattern of the winds in regional basins in the lowest levels. This is mainly a consequence of the wavenumber truncation of the spectral resolution of these models, corresponding in the best cases to some tens of km. Such a low resolution prevents from the detailed representation of the coastal morphology (and, in particular, of the land orography), and from the explicit representation of important processes, such as turbulence, convection, ..., that thus need to be parameterized. The task of simulating the weather fields with better local accuracy is usually pursued by using limited area models, which integrate the model equations only on limited geographical domains, but with finer horizontal resolution. The application of these models permits to better represent the small scale forcings, e.g. topographic features, and the internal dynamics and physics, so that a consistent downscaling of the large scale forecasts is produced. This is particularly relevant for those fields with low spatial correlation, such as precipitation, or strongly related to local terrain features, as the 10 m wind field.

The generation of gravity waves and the sheltering effects due to interaction of intense winds with the orography have been studied theoretically (Scorer and Wilkinson, 1956; Gjevik and Marthinsen, 1978) and are well documented from satellite imagery in the literature (Thomson et al, 1992; Vachon et al., 1994; Vachon et al., 1995). In the present paper, intense atmospheric lee waves downwind Crete and Rhodos have been analyzed. The 10 m wind fields simulated with the Weather Research and Forecasting (WRF) atmospheric model (Skamarock et al., 2005) are compared with the sea surface winds derived from an Envisat ASAR wide swath image, obtained using the two-dimensional Continuous Wavelet Transform (2D-CWT) analysis. The purpose of the paper is to test the ability of the WRF model to reproduce the features shown in the ASAR data and to identify a model configuration optimal for wind field simulation. The methodology used for the retrieval of wind from the ASAR is outlined in Sect. 2; in Sect. 3, the model implementation is briefly presented; in Sect. 4, the wind patterns obtained in simulations with different parameterization schemes and different treatments of diffusion are compared to each other and with the ASAR retrieved wind fields. Conclusions are given in Sect. 5.

2 ASAR

The satellite image used here (Fig. 1) is an Envisat ASAR Wide Swath Mode image in VV polarization (ESA, 2002). It covers an area of about 400 km by 400 km (with pixel size of 75 m by 75 m) in the Aegean Sea, close to Crete and Rhodos islands, at 08:20 UTC, 15 May 2008. The methodology of wind extraction, explained in detail in Zecchetto and De Biasio (2002, 2008), is able to extract the radar backscatter signatures related to the wind field, from which it is possible to derive the de-aliased wind direction. With the wind direction information, it is possible to evaluate the wind speed from the CMOD-5 model developed for C-band scatterometers (Hersbach et al., 2007). This model requires the knowledge of the radar incidence angle, of the normalized radar cross section and of the angle between the radar looking and the wind direction. Since the wind directions are retrieved from the backscatter signatures detected by the analysis, which are more structured and dense when the wind speed is high, the spatial resolution of the resulting wind field depends on the wind speed regime.

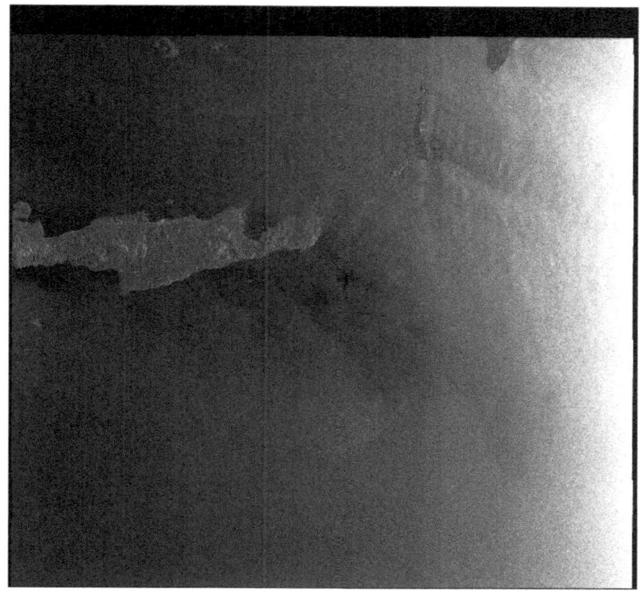

Figure 1. Detail of the Envisat ASAR Wide Swath image at 08:20 UTC, 15 May 2008 in the vicinity of Crete. Pixel size is 75 m by 75 m.

3 WRF model

In the present study, the Weather Research and Forecasting (WRF) model (see http://www.wrf-model.org) has been implemented. For the present study, the model version 3.0 has been implemented in a 3 grid configuration: the largest domain, with a horizontal resolution of 16 km, covers approximately the eastern Mediterranean basin (109×109 grid points); the middle domain covers the area of the Mediterranean sea around Crete with a horizontal resolution of 4 km (161×161 grid points); the inner domain, with a horizontal resolution $\Delta x = \Delta y = 1$ km (321×321 grid points), covers the area (33.6° N–36.4° N, 25° E–28.4° E), that corresponds approximately to the eastern part of Crete, Rhodos island

and the surrounding seas. The coarser grid reproduces the large scale features which force the local dynamics in grid 3, while the intermediate grid is needed as grid 1 and grid 3 have too different horizontal resolutions to be nested directly one into the other. The simulations, starting at 12:00 UTC, 14 May 2008 and lasting 24 h, are performed using a two-way nesting technique. Forty vertical levels are employed, more closely distributed in the lower levels. Initial and boundary conditions every three hours are provided respectively with the European Center for Medium range Weather Forecast (ECMWF) analysis and forecasts. Thus the runs are performed in an operational-like configuration.

A full set of parameterization schemes is included in the model for microphysics, convection, turbulence, soil processes, boundary layer processes, and radiation. In the model configuration used here, the following parameterization schemes have been selected: the Thompson et al. (2004) microphysics, which includes six classes of moisture species plus ice crystal number concentration as prognostic variables; the Kain (2004) cumulus parameterization on the coarser grid (no parameterization is used on the two inner grids); the Rapid Radiative Transfer Model (RRTM) for longwave radiation, based on Mlawer et al. (1997); the Dudhia (1989) scheme for shortwave radiation; a five-layer thermal-diffusion soil scheme (Skamarock et al., 2005). Two different parameterization schemes have been used for the planetary boundary layer (PBL): the Yonsei University (YSU) scheme with a nonlocal turbulent mixing coefficient in the PBL (Hong and Pan, 1996); the Mellor-Yamada-Janjic (MYJ) prognostic scheme that calculates the turbulent kinetic energy (Janjic, 2001).

4 Model versus ASAR wind

The case study is characterized by intense wind impinging on Crete and on nearby islands in the eastern Mediterranean Sea. Such a region is subject to complex wind patterns, due to the interaction of the almost steady northerly Etesian wind with the orography (Koletsis et al., 2009). In order to compare fields with different spatial sampling, both WRF and ASAR-derived winds have been re-sampled on a $0.02° \times 0.02°$ geographical grid, chosen because it permits to preserve the spatial features of both the SAR derived wind (resolution of about $0.001°$) and the WRF model (resolution of about $0.01°$). The SAR and model-derived wind-vector fields can then be cross-validated, obtaining information on the spatial extent and variability of wind features, and in particular of the atmospheric gravity wave patterns, frequently observed in the region.

The correspondence of the observed with the simulated wind structures has been analyzed for different WRF model simulations performed using two different PBL parameterization schemes (YSU vs. MYJ) and, only for the YSU scheme, different treatments of diffusion. Really, adequate modeling of atmospheric flow features of O(1 km) using limited area models with comparable horizontal grid spacings suffers from the fact that these features are within the maximum energy-containing spectral interval, and they are neither explicitly resolved on the grid nor represented statistically as subgrid motions. We are close to the "no man's land" separating classical Planetary Boundary Layer (PBL) from Large Eddy Simulation (Weisman et al., 2005) schemes. As a result, the ability of mesoscale models to accurately reproduce atmospheric phenomena on such fine spatial scales is questionable, particularly in terms of near-surface flow parameters (Gibbs and Fedorovich, 2009). At the moment, classical PBL schemes are generally employed for grid scales larger than 1 km, although there is no consensus on how to best configure the WRF for these scales (van der Velde et al., 2009; Loughner et al., 2009; Green, 2009; see also http://forum.wrfforum.com).

Figure 2 shows the wind field reproduced by the WRF model using: (a) the YSU scheme without explicit diffusion (hereafter CTL run)[1], (b) the YSU scheme with a second order diffusion based on constant eddy viscosities evaluated on model coordinate surfaces (constant coefficients $k_x = k_y = \Delta x$; hereafter DIFF run), (c) the YSU scheme with a horizontal "Smagorinsky" closure (Smagorinsky, 1963; hereafter SMG run) and finally (d) the MYJ scheme without diffusion (hereafter MYJ run). By comparing Fig. 2a with Fig. 2b, we see that the small-scale details of the wind pattern are progressively lost when larger diffusion coefficients are used.

Figure 2a, c and d shows similar gravity wave structures on the east side, generated by the flow impinging on Rhodos, Crete and the small islands of Kassos and Karpathos in between. The interaction of the wave trains at the lee side of the different islands yields a complex interference pattern, extending more than 80 km downstream. The simultaneous presence of orographic gravity waves and a low-speed island wake southward of Crete suggests that different Froude number regimes occur (Smith, 1989), depending on the local orographic features the wind impinges on.

Some relevant differences among the simulations can be identified close to Crete on the west side of the figure, especially downstream, as local wind maxima occur close to the island in Fig. 2c, d, while they are much more confined in Fig. 2a and missing in Fig. 2b. These maxima are associated with gap flows (Koletsis, 2009) or low level vortices generated downstream (Smolarkiewicz and Rotunno, 1989). Since the structure of the atmospheric boundary layer flow is more complex at the land-sea transition zone, due to the formation of coherent mesoscale circulation triggered by abrupt changes in the surface roughness, thermal forcing, land use, or orography, it is expected that the strongest differences

[1]WRF model includes an implicit scale-selective filter for advection, thus the model can run stably without artificial diffusion; however, it may sometimes be insufficient to remove noise.

Table 1. Bias (WRF-ASAR), standard deviation σ and correlation coefficients R of wind speed (ws) and direction (wd) for the different experiments after resampling at 0.05° by 0.05°. Section 4 explains the characteristics of the different experiments.

Exp.	$bias_{ws}$ ($m\,s^{-1}$)	$bias_{wd}$ (°)	σ_{ws} ($m\,s^{-1}$)	σ_{wd} (°)	R_{ws}	R_{wd}
CTL	−0.4	−10	1.9	27	0.6	0.5
SMG	−0.9	−4	2.3	33	0.3	0.3
MYJ	−0.9	−2	2.3	38	0.4	0.3

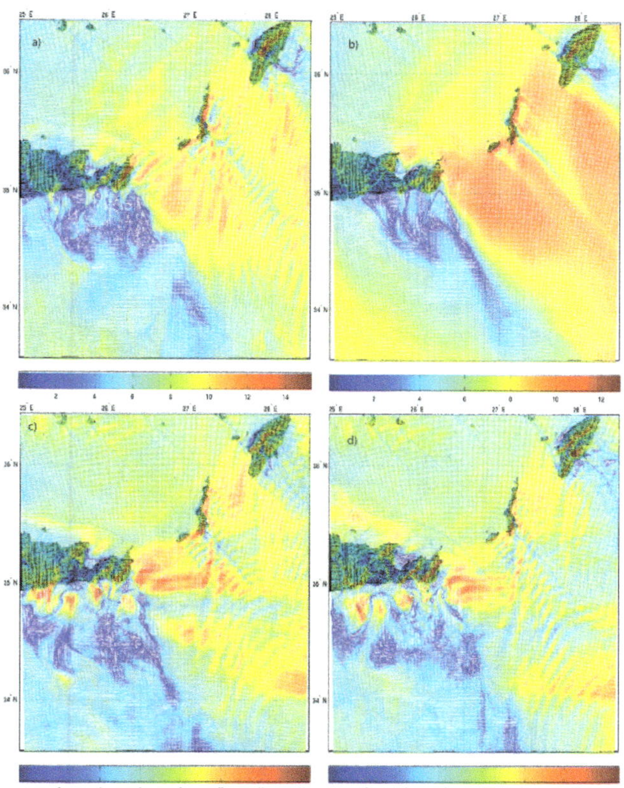

Figure 2. 10 m wind speed (colors) simulated by the WRF model (inner domain) in the CTL run (**a**, left top), the DIFF run (**b**, right top), the SMG run (**c**, left bottom), and the MYJ run (**d**, right bottom) at 08:00 UTC, 15 May 2008. The dark regions correspond to the island location. The maps cover the area delimited with the black border in Fig. 3. Section 4 explains the characteristics of the different experiments.

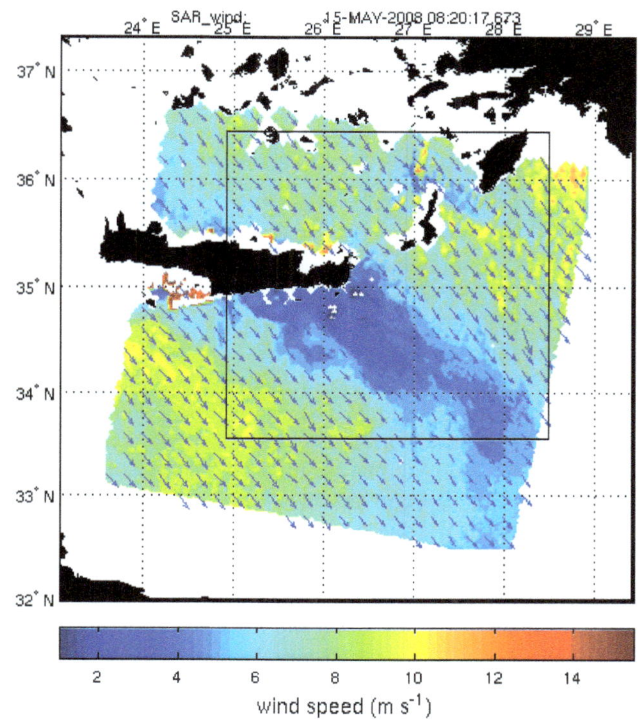

Figure 3. ASAR derived sea surface wind field (speed: colors; direction: arrows) at 08:20 UTC, 15 May 2008. The position of the WRF inner domain is represented with the black border.

among the parameterization schemes may occur in this region (Sood and Bange, 2009).

The ASAR derived wind, derived from the wind direction interpolated on a 0.02° by 0.02° grid, is shown in Fig. 3. The wind direction has been reported with a resolution ten times coarser than the original. The signature of the atmospheric gravity waves is still present. Comparing it with the WRF model fields, a closer resemblance occurs with the CTL run (Fig. 2a) and with the DIFF run (Fig. 2b) in the west side, as the low speed wake downstream of Crete is more uniform, without local wind maxima. In the east side of the domain, all simulations exhibit a wave pattern similar to ASAR, apart from the DIFF experiment, that completely misses the gravity wave structure; as a consequence, this simulation is no longer considered in the following discussion.

In order to provide a more quantitative comparison, the bias, the standard deviation and the correlation coefficients for wind speed and direction between the model simulations and the ASAR data have been calculated. Table 1 shows that the CTL run is the closest to the ASAR data for all the parameters, apart from the wind direction bias.

Figure 4 compares the normalized frequency distributions of wind speed and direction for the CTL run and the ASAR data; for a more comprehensive analysis, also the

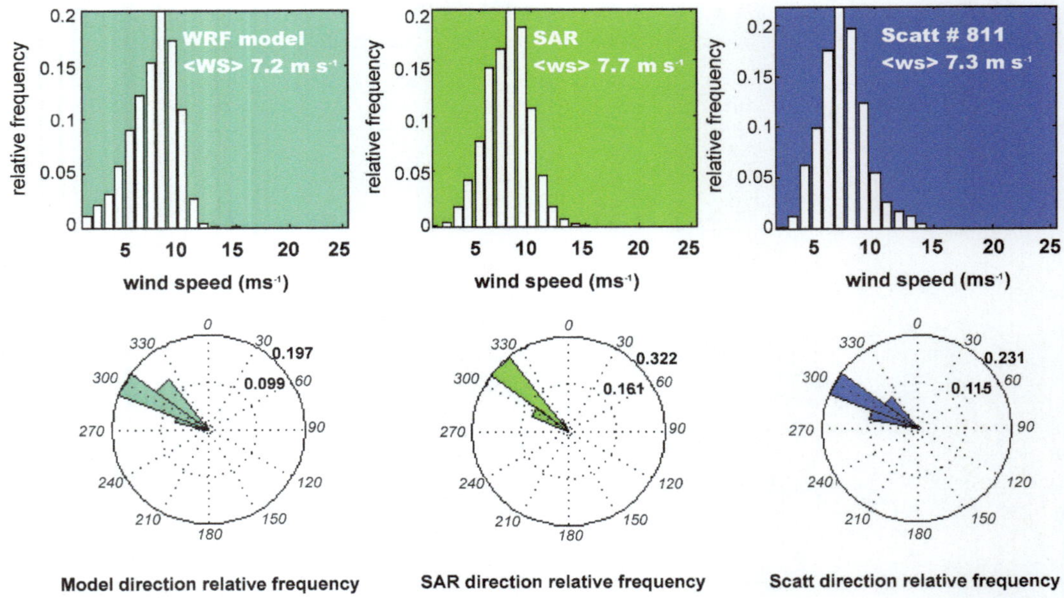

Figure 4. Frequency distributions of wind speeds (top) and wind directions (bottom) from WRF (left), ASAR (middle) and QuikSCAT (right) after resampling at 0.05° by 0.05°.

distributions obtained from the scatterometer winds derived from the NASA QuikSCAT satellite (JPL, 2006) are shown. The three patterns appear similar; however, a shift of the model distribution toward lower speeds and a threshold of $12\,\mathrm{m\,s^{-1}}$, not present in ASAR and QuikSCAT winds, can be identified. About the direction, a prevailing NW component is present in all cases, although the distribution is less unidirectional and slightly rotated anticlockwise for the WRF model and the scatterometer.

Figure 5 shows a wind speed scatterplot for the CTL run and the ASAR data, with the colors indicating the frequency of wind speed in given bins of $1\,\mathrm{m\,s^{-1}}$; it indicates a pretty good agreement between the two datasets. Again, the figure shows a general underestimation by the model, that is mainly due to the winds in the south-west side of the domain, where the model is not able to reproduce the relative maximum.

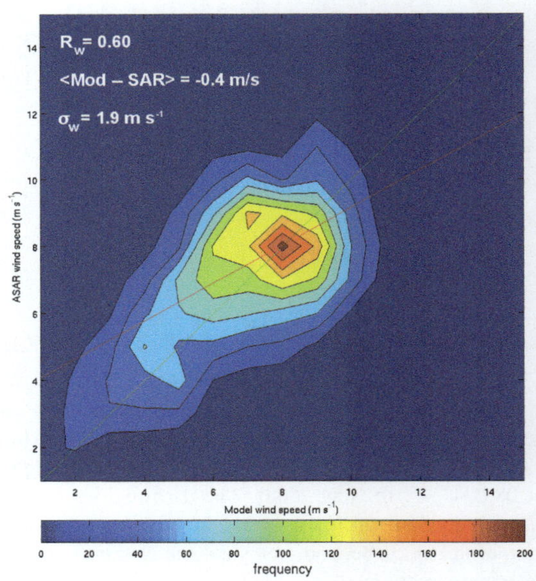

Figure 5. Scatterplot between the CTL run and the ASAR data after resampling at 0.05° by 0.05°. Colors indicate the frequency of occurrence in the $1\,\mathrm{m\,s^{-1}}$ bins. R_m is the correlation coefficient.

5 Conclusions

In the present study, an event of intense lee waves has been analyzed using the ENVISAT ASAR satellite imagery and the WRF model in the area of the Mediterranean sea localized south of Crete and Rhodos islands; this region is subject to intense and persistent low level flows, thus is favorable to the generation of atmospheric waves. Although most PBL parameterizations have demonstrated the capability to represent many meteorological phenomena, up to now little attention has been paid to the precise prediction of winds at the lowest level. Here, simulations with different diffusion and boundary layer parameterization schemes have been compared. Most of the WRF model simulations were able to forecast the existence of near-surface wind speed variability attributable to gravity waves, showing lee wave features similar to those emerging from the ASAR-retrieved wind field and interference patterns between the waves generated by different islands. The simulations with the YSU scheme and no diffusion is the closest to the ASAR data both qualitatively and quantitatively. In order to conclude that the configuration with the YSU scheme is the best for wind field simulation, it is proposed, in the future, to compare the simulations also with other types of observations (buoys, surface stations, ...) and in different atmospheric conditions.

The analysis of further cases in the same area, showing other features (funnelling effects, island sheltering effects, ...) due to the interaction of high winds with the local orography, is in progress and will be the subject of a forthcoming paper.

Acknowledgements. The Envisat ASAR Wide Swath image data have been provided by the European Space Agency (on the framework of the Project Start Up C1P.5404 of the European Space Agency). Scatterometer QuikSCAT data have been downloaded from the Physical Oceanography Distributed Active Archive Center (PODAAC) of the Jet Propulsion Laboratory, Pasadena, USA.

Edited by: A. M. Sempreviva

References

Atkinson, B. W.: Meso-scale atmospheric circulations, Academic Press Ltd., London, 1981.

Dudhia, J.: Numerical study of convection observed during the Winter Monsoon Experiment using a mesoscale two-dimensional model, J. Atmos. Sci., 46, 3077–3107, 1989.

ESA: ASAR Product Handbook, European Space Agency, Paris, France, 2002.

Gibbs, J. A. and Fedorovich, E.: Sensitivity of near-surface meteorological fields in WRF to boundary/surface-layer parameterizations in conjunction with horizontal grid spacing, 10th Users' Workshop, Boulder, CO, 23–26 June, 2009.

Gjevik, B. and Marthinsen, T.: Three-dimensional lee-wave pattern, Q. J. Roy. Meteor. Soc., 104, 947–957, 1978.

Green, M.: Performance of WRF for wind energy applications over complex terrain over complex sites in New Zealand and Australia, 10th Users' Workshop, Boulder, CO, 23–26 June, 2009.

Hersbach, H., Stoffelen, A., and de Haan, S.: An improved scatterometer ocean geophysical model function: Cmod5, J. Geophys. Res., 11, 5767–5780, 2007.

Hong, S.-Y. and Pan, H. L.: Nonlocal boundary layer vertical diffusion in a medium range forecast model, Mon. Weather Rev., 124, 2322–2339, 1996.

Janjic, Z. I.: Nonsingular implementation of the MellorYamada Level 2.5 Scheme in the NCEP Meso model. National Centers for Environmental Prediction Tech. Rep. 437, 61 pp., 2001.

JPL: QuikSCAT Science Data Product User Manual: QuikSCAT science data product user manual: overview and geophysical data products, Version 3.0, September 2006, D-18053-RevA, Jet Propulsion Laboratory (JPL), California Institute of Technology, 91 pp., 2006.

Kain, J. S.: The Kain-Fritsch convective parameterization: An update, J. Appl. Meteorol., 43, 170–181, 2004.

Koletsis, I., Lagouvardos, K., Kotroni, V., and Bartzokas, A.: The interaction of northern wind flow with the complex topography of Crete Island – Part 1: Observational study, Nat. Hazards Earth Syst. Sci., 9, 1845–1855, doi:10.5194/nhess-9-1845-2009, 2009.

Loughner, C. P., Allen, D. J., Dickerson, R. R., Zhang, D.-L., Shou, Y.-X., and Pickering, K. E.: Investigating the Use of a High Resolution WRF-URBAN CANOPY Model Simulation With CMAQ, 8th Annual CMAS Conference, Chapel Hill, NC, 19–21 October, 2009.

Mlawer, E. J., Taubman, S. J., Brown, P. D., Iacono, M. J., and Clough, S. A.: Radiative transfer for inhomogeneous atmosphere: RRTM, a validated correlated-k model for the longwave, J. Geophys. Res., 102, 16663–16682, 1997.

Orlanski, I.: A rational subdivision of scales for atmospheric processes, B. Am. Meteorol. Soc., 56, 527–530, 1975.

Scorer, R. S. and Wilkinson, M.: Waves in the lee of an isolated hill, Q. J. Roy. Meteor. Soc., 82, 419-427,1956.

Skamarock, W. C., Klemp, J. B., Dudhia, J., Gill, D. O., Barker, D. M., Wang, W., and Powers, J. G.: A description of the Advanced Research WRF Version 2, NCAR Technical Note, 468STR, 88 pp., 2005.

Smagorinsky, J.: General circulation experiments with the primitive equations. Part I. The basic experiment, J. Meteorol., 14, 184–185, 1963.

Smith, R. B.: The influence of mountains on the atmosphere, Adv. Geophys., 21, 87–230, 1979.

Smith, R. B.: Hydrostatic airflow over mountains, Adv. Geophys., 31, 1-41, 1989.

Smolarkiewicz, P. K. and Rotunno, R.: Low Froude number flow past three-dimensional obstacles. Part I: Baroclinically generated lee vortices, J. Atmos. Sci., 46, 1154-1164, 1989.

Sood, A. and Bange, J.: Validating coastal, near and far offshore boundary layer parameterizations with airborne helipod turbulence probe, EMS Annual Meeting Abstracts, Vol. 6, EMS2009-372, 2009.

Thompson, G., Rasmussen, R. M., and Manning, K.: Explicit forecasts of winter precipitation using an improved bulk microphysics scheme. Part I: Description and sensitivity analysis, Mon. Weather Rev., 132, 519–544, 2004.

Thomson, R. E., Vachon, P. W., and Borstad, G. A.: Airbome synthetic aperture radar imagery of atmospheric gravity waves, J. Geophys. Res., 97, 14249–14257, 1992.

Vachon, P. W., Johannessen, O. M., and Johannessen, J. A.: An ERS 1 synthetic aperture radar image of atmospheric lee waves, J. Geophys. Res., 99, 22483–22490, 1994.

Vachon, P. W, Johannessen, J. A., and Browne, D. P.: ERS-1 images of atmospheric gravity waves, IEEE T. Geosci. Remote, 33, 1014–1025, 1995.

Van der Velde, I. R., Steeneveld, G. J., Wichers Schreurs, B. G. J., and Holtslag, A. A. M.: Modeling and forecasting the onset and duration of a severe Dutch fog event, EMS Annual Meeting Abstracts, Vol. 6, EMS2009-13, 2009.

Weisman, M., Davis, C., and Wang, W.: Explicit convective forecasting with the WRF model, WRF-MM5 Workshop, Boulder, CO, 30 June, 2005.

Zecchetto, S. and De Biasio, F.: On shape, orientation and structure of atmosheric cells inside wind rolls in two SAR images, IEEE T. Geosci. Remote, 40, 2257–2262, 2002.

Zecchetto, S. and De Biasio, F.: A Wavelet Based Technique for Sea Wind Extraction from SAR Images, IEEE T. Geosci. Remote, 46, 2983–2989, 2008.

Estimating future air-quality due to climate change: the Athens case study

K. V. Varotsos[1,2], C. Giannakopoulos[1], and M. Tombrou[2]

[1]Institute for Environmental Research and Sustainable Development, National Observatory of Athens, Greece
[2]Division of Environmental Physics and Meteorology, National and Kapodistrian University of Athens, Greece

Abstract. The aim of this study is to investigate the development of an empirical-statistical model in order to examine the potential impact of increasing future temperatures on ozone exceedance days in the Greater Athens Area. It is based on the concept that temperature is a capable predictor for the ozone concentrations and that in a future climate change world, the likelihood of ozone pollution episodes may increase.

1 Introduction

It has become widely recognized that climate change could affect air quality, through the assumption that air quality is strongly dependent on weather systems and therefore sensitive to climate change. A thorough review of recent studies and methods used to provide estimates of the climate change effect on air quality can be found in Jacob and Winner (2009). In their study, one of the methods used seeks to determine predictive relationships between air pollution concentrations and individual meteorological parameters. A major task of research in the above framework has been dedicated to the systematic study of surface ozone correlations with meteorological variables identifying temperature, morning solar radiation, number of days since the last frontal passage, humidity and the frequency of summertime mid latitude cyclones as the closest associated with ozone (Ordóñez et al., 2005; Wise and Comrie, 2005; Camalier et al., 2007; Leibensperger et al., 2008). The ozone-temperature relationship has been studied over the years indicating that temperature is strongly associated with ozone concentrations and therefore can be used as a predictor for ozone concentrations. In particular, the strong correlation of ozone with temperature is associated with surface air ventilation, since high temperatures occur under the presence of light winds, high levels of isolation and stagnant circulation conditions (Jacob et al., 1993; Sillman and Samson, 1995). In addition, the correlation of ozone with temperature also depends on local ozone production chemistry such as thermal decomposition of PAN, on the temperature dependent biogenic emissions such as isoprene as well as to correlations of temperature with anthropogenic emissions (NO_x and VOC) (Sillman and Samson, 1995). Recently, Bloomer et al. (2009) defined as a climate penalty factor the ozone temperature slope in the rural eastern US, suggesting that the present day empirical relation between ozone and temperature, although not including non linear effects on ozone production as in state of the art 3-D photochemical models, if combined with future temperature projections could provide estimates of the future air-quality.

In addition, one of the main conclusions of the 2003 European Environmental Agency's report (EEA, 2003) after the European summer heatwave episode (Beniston, 2004; Schär et al., 2004) is that the potential summer temperature increase in Europe, due to climate change, could lead to more frequent exceedances of the ozone information threshold (hourly average concentrations of $180\,\mu g\,m^{-3}$) (EU Directive, 2002/3/EC, 2002) at the current emission levels. Although uncertainty exists, current climate change modelling studies in Europe suggest that higher temperatures like those occurred in 2003 could become the norm in the mid of the current century (Fischer and Schär, 2010; Giannakopoulos et al., 2009). As a result temperature becomes of particular interest as a predictor of ozone quality in Europe due to the ozone temperature relationship. The aim of this study is to investigate the development of an empirical-statistical model in order to examine the potential impact of increasing future temperatures on ozone exceedance days in the Greater Athens Area (GAA) using daily maximum temperatures and hourly ozone observations from two measuring stations within the GAA domain as well as daily maximum temperatures from a regional climate model.

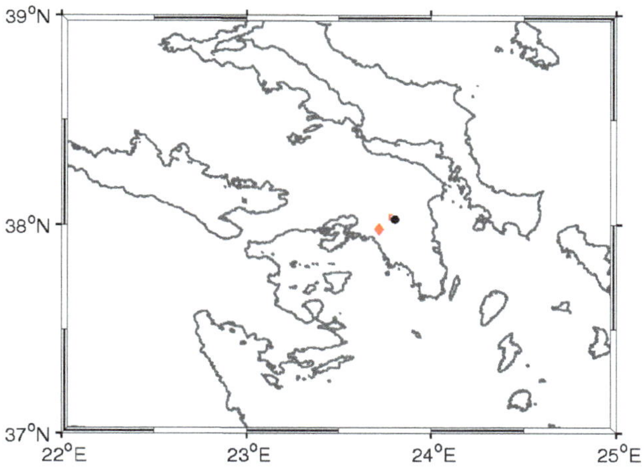

Figure 1. Greater Athens Area including the locations of the Marousi ozone station (red square), the NOA temperature station (red diamond) as well as the location of the RCM closest grid point to the ozone station location (black circle).

2 Data sources and methodology

2.1 Surface ozone and temperature observations-model data

Hourly ozone concentrations data from the suburban station of Marousi, located at the North-East of the urban core of Athens (Fig. 1), are retrieved from the Greek Ministry of Environment, Energy and Climate Change database for the 1990–1999 period. For the purpose of this study the observations are filtered choosing the daily maximum 8-h average that is calculating 8-h moving averages, indexed by the first hour, for each 8-h interval with at least 6 h of data. Local daily maximum surface temperatures are taken from the National Observatory of Athens station (NOA) located about 10 km to the South-West of the ozone station location. We extract 2 periods of daily maximum temperatures (Table 1). The period 1961–1990 is extracted for validation purposes with the regional climate model RACMO2, while a second shorter period is also extracted to correspond with the years range used in the ozone site. In addition, daily maximum temperatures, from the RACMO2 regional climate model, developed at KNMI in the Netherlands (Lenderink et al., 2003, 2007) are used. RACMO2 simulations are dynamically downscaled regional climate simulations of 25 km ($0.22° \times 0.22°$) horizontal resolution using initial and boundary conditions from the General Circulation Model (GCM) ECHAM5. For the purposes of our study 3 simulated periods of the closest grid point to the ozone station location are considered. The control period (1961–1990) for evaluating the model behavior in comparison with the observed maximum temperatures and two future projections for the periods 2021–2050 and 2071–2100. The two future period simulations of the model are based on the IPCC SRES A1B scenario (Nakicenovic et al., 2000).

Table 1. List of the data used in the analysis.

Data used in the analysis	lon	lat	year range
Ozone (Marousi)	23.78	38.03	1990–1999
Observed T_{max} (NOA)	23.71	37.97	1990–1999
			1961–1990
RCM T_{max}	23.80	38.02	1961–1990
			2021–2050
			2071–2100

2.2 Methods of analysis

To investigate the relationship of ozone exceedance days (defined here as days with maximum 8-h average $\geq$60 ppb) (EU Directive 2008/50/EC, 2008) with daily maximum temperature, we construct the probability distribution of ozone exceedance days. This is performed by calculating the number of ozone exccedance days in temperature bins of 1 °C divided by the total number of days whose temperature fell in that bin (Lin et al., 2001). By computing these probabilities we utililize the hypothesis that the total derivative of ozone with daily maximum temperature includes the partial derivatives of ozone with temperature-sensitive variables such as stagnation, chemistry, biogenic emissions, clear skies, weak winds (Mickley et al., 2004). In addition, the threshold temperature associated with the appearance of ozone exceedance days is derived which is applied with the future model simulations.

Subsequently we construct our empirical-statistical model by calculating the probability distribution of maximum 8-h average ozone concentration with daily maximum temperature, using bins of 1 °C above the pre-calculated temperature threshold and bins of 5 ppb for ozone concentration. If m are the number of ozone bins and n the number of temperature bins above the temperature threshold, we can derive an $\mathbf{m \times n}$ matrix where each n bin describes the contribution of each temperature bin to the ozone m bins in terms of probabilities. The sum of the probabilities in each ozone m bin is the total contribution of temperatures in ozone at the same bin.

Finally the calculated probabilities of the present day from the observational dataset are used to estimate the probability distributions of the future model projections.

3 Results

3.1 Relationship of ozone exceedance days with temperature

The maximum 8-h average ozone concentration values range between 4 and 143 ppb while 28% of the days exhibit maximum 8-h average concentration above 60 ppb. The daily maximum temperature varies from 1.4 to 42 °C. The correlation coefficient between the two variables is 0.66 with the confidence limits of the 95% confidence interval calculated

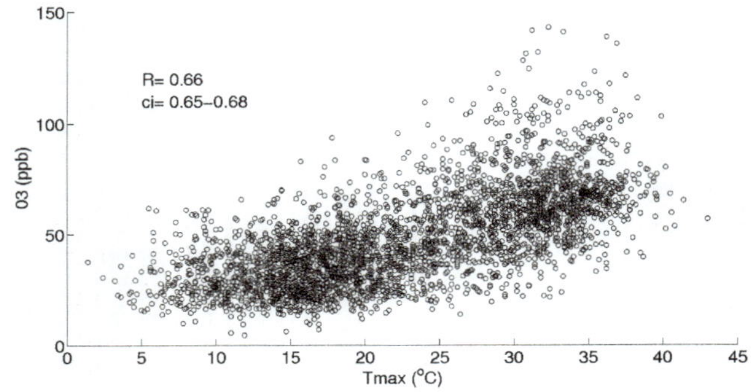

Figure 2. Correlation coefficient R and limits of the 95% confidence interval, ci, as calculated by bootstrap between maximum 8-h average ozone concentration and daily maximum temperature.

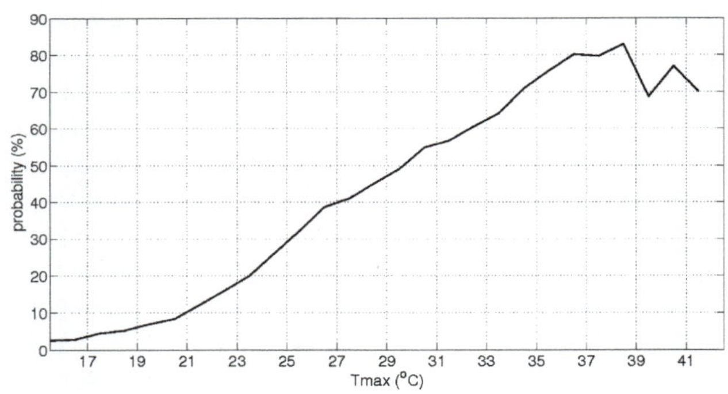

Figure 3. Probability that the daily maximum 8-h average ozone will exceed 60 ppb for a given daily maximum temperature ($T_{\max}$) for the Marousi ozone station. Values are plotted at the mid-point temperature of the 1 °C temperature bin.

by bootstrap (Efron, 1987) varying from 0.65 to 0.68 (Fig. 2). The high values of maximum 8-h average ozone are mainly attributed to O_3 precursors, due to the traffic. The station is located downwind of the urban core, when the atmospheric conditions favor the development of sea breeze circulation over the Athens basin and it is also sited close to the forestry area of Parnitha (Bossioli et al., 2007, 2009).

From Fig. 3 it is evident that the probability of ozone exceedance days with daily maximum temperature increases over the temperature range 17 to 38 °C mainly due to high activity of photo-chemical processes. On the contrary the decline of the ozone exccedance days probability when temperature is above 38 °C can be attributed to the high sensitivity of isoprene emissions to temperature. More specifically, isoprene emissions tend to increase with temperature until a certain temperature and then decline due to the impact of high temperatures in physiological processes (Guenther et al., 1993). Due to the variability of isoprene emissions to different tree species it is found that temperatures responsible of isoprene emission decrease vary from 35 to 44 °C (Singsaas and Sharkey, 2000). Finally, the ozone exceedance days re-

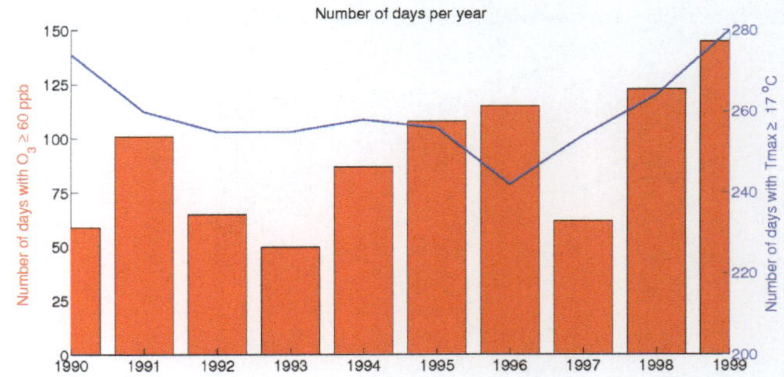

Figure 4. Number of days where daily maximum 8-h average ozone exceed 60 (red bars) and number of days where Tmax is above 17 °C (blue line) for the 1990–1999 period.

lationship with daily maximum temperature above 17 °C is reinforced by examining the number of ozone exceedance days with the number of days where temperature is above the aforementioned temperature threshold per year respectively (Fig. 4). From the figure it is apparent that the number of ozone exceedance days varies from about 50 to about 150 days.

3.2 Estimation of the future ozone exceedance days

Before presenting the results for the future simulations, the RACMO2 performance and accuracy was evaluated on an annual basis with the observed daily maximum temperature for the "recent past" period (1961–1990), using correlation coefficient and the corresponding confidence intervals for the 95-percentile as they are derived from the bootstrap percentile method. A statistically significant correlation coefficient of 0.839 (0.835–0.844) was found between the two timeseries. The analysis of the daily differences of the model and observations indicated that the model is capable of reproducing the temperature seasonal cycle. Furthermore, the 2021–2050 average annual maximum temperature in the GAA is higher by 1.5 °C than the control period whereas for the 2071–2100 the difference reaches 3.7 °C.

In Fig. 5 the probability distributions of ozone with temperature above 17 °C are shown. These distributions were calculated for the 3 periods and under the assumption of no changes in emissions for both the future periods. In general the future projections of the probability distributions follow the shape pattern of the observed. Another common feature of the two future distributions is that both indicate lower probabilities compared to the observed in the lower and higher centiles of the selected average 8-h maximum daily concentrations whereas higher than the observed probabilities are evident in the average centiles for both periods. The latter two results are in line with findings of previous climate change-air quality interactions studies (Vautard and Hauglustaine, 2007). Nevertheless, the 2021–2050 period distributions indicate smaller differences than the 2071–2100

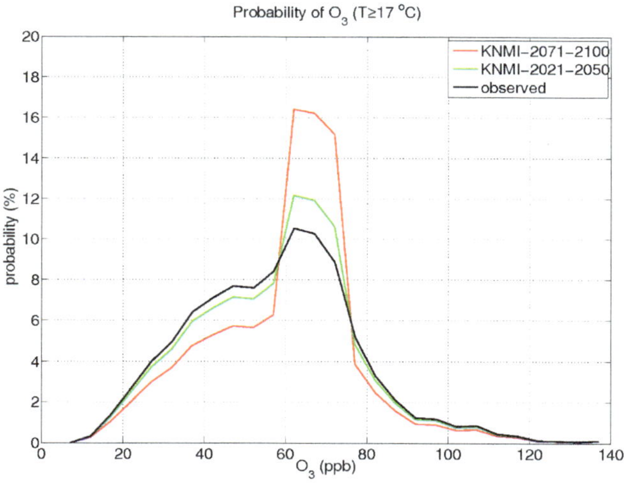

Figure 5. Ozone probability distributions for temperature above 17 °C, for three periods, the observed (1990–1999) (black), the 2021–2050 (green) and the 2071–2100 (red). Values are plotted at the mid-point concentration of the 5ppb ozone concentration bin.

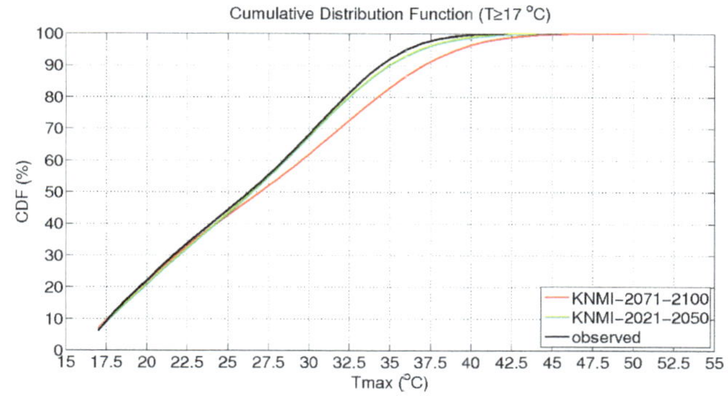

Figure 6. Cumulative distributions functions of temperature above 17 °C for the three periods, the observed (1990–1999) (black), the 2021–2050 (green) and the 2071–2100 (red).

distributions compared to the observed, due to the smaller differences in the daily maximum temperatures. As far as the number of ozone exceedance days is concerned, the results of the empirical-statistical model indicate an increase, compared to the corresponding observed distribution. In particular for the 2021–2050 simulation the percentage increase is about 8% which leads to about 7 extra ozone exceedance days per year. For the end of the 21st century (2071–2100) the percentage increase is higher, about 30% which leads to about a month (~27 days) extra ozone exceedance days per year. Since our empirical-statistical model is temperature sensitive the cumulative distributions functions of temperature above the threshold temperature for the three periods are also examined (Fig. 6). It is evident, that there is about a 70% probability that daily maximum temperatures for the first future period are not significant higher than the observed in the 17–32 °C temperature range, whereas for the 2071–2100 the probability is only about 40% and for a smaller temperature range (17–25 °C). Thus, the increases in the number of ozone exceedance days for both future periods could be explained in a similar way from the differences between the two future periods and the observed in the higher temperature centiles. More specifically, it is very likely that the difference will be less than 0.65 °C and less than 2.86 °C between the 2021–2050 period and the observed and the between the 2071–2100 and the observed period respectively. Comparison of the results of the ozone exccedance days, especially for the near future period 2021–2050, with results from recently published studies using coupled climate-chemical models at the regional scale in Europe is encouraging for our statistical results. For instance, Forkel and Knoche (2007) using a varied-resolution climate-chemical model system emission scenario, found that the impact of climate change on its own could lead to an increase of about 16 extra ozone exceedance days in Central Europe in the 2030s compared to the 1990s.

4 Summary

This study describes the development and application of an empirical-statistical model to provide estimates of the future ozone exceedance days under the impact of climate change. It is based on the concept that daily maximum temperature is closely associated with the average 8-h maximum ozone concentrations. Our results indicate that a probable increase in the higher centiles of temperature less than 0.65 °C in the 2021–2050 and less than 2.86 °C in the 2071–2100 period compared to the observed could lead to an increase in the ozone exccedance days of about 8 and about 30 extra ozone exccedance days per year respectively. Finally, although it can be used as a quick indicator of ozone sensitivity to climate change it demonstrates two caveats. Primarily it assumes constant emissions of ozone precursors and secondly it may not capture the exceedance days seasonality (variability).

Acknowledgements. This research has been co-financed by the European Union (European Social Fund – ESF) and Greek national funds through the Operational Program "Education and Lifelong Learning" of the National Strategic Reference Framework (NSRF) - Research Funding Program: Heracleitus II. Investing in knowledge society through the European Social Fund.

C. Giannakopoulos greatly acknowledges support from EU projects CIRCE (http://www.circeproject.eu/) under contract number GOCE-036961 and ENSEMBLES (http://www.ensembles-eu.org/) under contract number GOCE- 505539.

Edited by: A. Baklanov

References

Beniston, M.: The 2003 heat wave in Europe: A shape of things to come? An analysis based on Swiss climatological data and model simulations, Geophys. Res. Lett., 31, L02202, doi:10.1029/2003GL018857, 2004.

Bloomer, B. J., Stehr, J. W., Piety, C. A., Salawitch, R. J., and Dickerson, R. R.: Observed relationships of ozone air pollution with temperature and emissions, Geophys. Res. Lett., 36, L09803, doi:10.1029/2009GL037308, 2009.

Bossioli, E., Tombrou, M., Dandou, A., and Soulakellis, N.: Simulation of the effects of critical factors on ozone formation and accumulation in the greater Athens area, J. Geophys. Res., 112, D02309, doi:10.1029/2006JD007185, 2007.

Bossioli, E., Tombrou, M., Dandou, A., Athanasopoulou, E., and Varotsos, K.: The Role of Planetary Boundary-Layer Parameterizations in the Air Quality of an Urban Area with Complex Topography, Bound.-Lay. Meteorol., 131,53–72, 2009.

Camalier, L., Cox, W., and Dolwick, P.: The effects of meteorology on ozone in urban areas and their use in assessing ozone trends, Atmos. Environ., 41(33), 7127–7137, 2007.

Directive 2002/3/ EC of the European Parliament and of the Council of 12 February 2002 relating to ozone in ambient air, http://www.eea.europa.eu/policy-documents/directive-2002-3-ec...ozone-in, 2002.

Directive 2008/50/EC of the European Parliament and of the Council of 21 May 2008 on ambient air quality and cleaner air for Europe, http://www.eea.europa.eu/policy-documents/directive-2008-50-ec-o, 2008.

EEA: Air pollution by ozone in Europe in summer 2003, Topic report No. 3/2003, European Environment Agency 2003, Copenhagen, 2003.

Efron, B.: Better Bootstrap Confidence Intervals, J. Am. Stat. Assoc., 82, 171–185, 1987.

Fischer, E. M. and Schär, C.: Consistent geographical patterns of changes in high-impact European heatwaves, Nature Geosci., 3, 398–403, 2010.

Forkel, R. and Knoche, R.: Nested regional climate-chemistry simulations for central Europe, C. R. Geosci., 339(11–12), 734–746, 2007.

Giannakopoulos, C., Le Sager, P., Bindi, M., Moriondo, M., Kostopoulou, E., and Goodess, C.: Climatic changes and associated impacts in the Mediterranean resulting from a 2 °C global warming, Global Planet. Change, 68, 209–224, 2009.

Guenther, A. B., Zimmerman, P. R., Harley, P. C., Monson, R. K., and Fall, R.: Isoprene and Monoterpene Emission Rate Variability: Model Evaluations and Sensitivity Analyses, J. Geophys. Res., 98, 12609–12617, 1993.

Jacob, D. J. and Winner, D. A.: Effect of climate change on air quality, Atmos. Environ., 43, 51–63, doi:10.1016/j.atmosenv.2008.09.051, 2009.

Jacob, D. J., Logan, J. A., Gardner, G. M., Yevich, R. M., Spivakovsky, C. M., Wofsy, S. C., Sillman, S., and Prather, M. J.: Factors Regulating Ozone Over the United States and Its Export to the Global Atmosphere, J. Geophys. Res., 98, 14817–14826, 1993.

Leibensperger, E. M., Mickley, L. J., and Jacob, D. J.: Sensitivity of US air quality to mid-latitude cyclone frequency and implications of 1980–2006 climate change, Atmos. Chem. Phys., 8, 7075–7086, doi:10.5194/acp-8-7075-2008, 2008.

Lenderink, G., van den Hurk, B., van Meijgaard, E., van Ulden, A., and Cuijpers, H.: Simulations of present day climate in RACMO2: first results and model developments, Technical report TR252, Royal Netherlands Meteorological Institute, 2003.

Lenderink, G., Ulden, A., Hurk, B., and Meijgaard, E.: Summertime inter-annual temperature variability in an ensemble of regional model simulations: analysis of the surface energy budget, Climatic Change, 81(S1), 233–247, 2007.

Lin, C.-C., Jacob, D. J., and Fiore, A. M.: Trends in exceedances of the ozone air quality standard in the continental United States, 1980–1998, Atmos. Environ., 35, 3217–3228, 2001.

Mickley, L. J., Jacob, D. J., Field, B. D., and Rind, D.: Effects of future climate change on regional air pollution episodes in the United States, Geophys. Res. Lett., 31, L24103, doi:10.1029/2004GL021216, 2004.

Nakicenovic, N., Alcamo, J., Davis, G., de Vries, B., Fenhann, J., Gaffin, S., Gregory, K., Grübler, A., Jung, T. Y., Kram, T., Lebre La Rovere, E., Michaelis, L., Mori, S., Morita, T., Pepper, W., Pitcher, H., Price, L., Riahi, K., Roehrl, A., Rogner, H-H., Sankovski, A., Schlesinger, M., Shukla, P., Smith, S., Swart, R., van Rooijen, S., Victor, N., and Dadi, Z.: Special Report on Emission Scenarios, Working Group III of the Intergovernmental Panel on Climate Change (IPCC), Cambridge University Press, Cambridge, 595 pp., 2000.

Ordóñez, C., Mathis, H., Furger, M., Henne, S., Hüglin, C., Staehelin, J., and Prévôt, A. S. H.: Changes of daily surface ozone maxima in Switzerland in all seasons from 1992 to 2002 and discussion of summer 2003, Atmos. Chem. Phys., 5, 1187–1203, doi:10.5194/acp-5-1187-2005, 2005.

Schär, C., Vidale, P. L., Luthi, D., Frei, C., Haberli, C., Liniger, M. A., and Appenzeller, C.: The role of increasing temperature variability in European summer heatwaves, Nature, 427, 332–336, 2004.

Sillman, S. and Samson, P. J.: Impact of temperature on oxidant photochemistry in urban, polluted rural and remote environments, J. Geophys. Res., 100(D6), 11497-11508, 1995.

Singsaas, E. L. and Sharkey, T. D.: The effects of high temperature on isoprene synthesis in oak leaves, Plant Cell Environ., 23, 751–757, 2000.

Vautard, R. and Hauglustaine, D.: Impact of global climate change on regional air quality: Introduction to the thematic issue, C. R. Geosci., 339, 703–708, 2007.

Wise, E. K. and Comrie, A. C.: Meteorologically adjusted urban air quality trends in the Southwestern United States, Atmos. Environ., 39(16), 2969–2980, 2005.

Tornado-type stationary vortex with nonlinear term due to moisture transport

P. B. Rutkevich[1] **and P. P. Rutkevych**[2]
[1]Space Research Institute (IKI), RAS, Moscow, Russia
[2]Institute of High Performance Computing, A*STAR, Singapore

Abstract. Tornado vortex is believed to be essentially nonlinear phenomenon; and the puzzle to choose the nonlinear term(s) responsible for its formation is still unresolved. In the present work we consider the nonlinear term associated with atmosphere humidity, by introducing variable temperature gradient depending on the vertical velocity of the fluid. Such term is able to yield energy to the system and is very suitable for such a problem. Other nonlinear terms are neglected, assuming slow rotation, or in other words a "weak" tornado approximation. We consider one-dimensional radial boundary problem, and use a modificaiton of shooting method to satisfy boundary conditions at large radii. Obtained numerical solutions of the nonlinear differential equation qualitatively agree with the observed atmosphere vortices (tornados, tropical cyclones). The obtained results show general possibility of existence of unstable motion even in convectively stable atmosphere stratification.

1 Introduction

Formation mechanisms and internal structure of severe atmospheric vortices have been important questions over the past decades. In spite of their importance for both theory and applications, there is no satisfactory theory of tornado structures (Doswell and Burgess, 1993) and recently there is a tendency towards collection of observational data or numerical simulation of the phenomenon (Bluestein et al., 1997; Bluestein and Pazmany, 2000; Bluestein and Weisman, 2000; Lehmiller et al., 2001).

The problem of tornado formation appears to be complicated from both principal and technical points of view. Tornado descends from a long-living rotating thunderstorm cloud. Having appeared from the cloud the funnel narrows in its upper part and dynamically stretches in the bottom part, however the middle part of the tornado column remains cylindrical and almost unchanged during the entire lifetime of the funnel. Such structure is especially pronounced for weak tornadoes (Davies-Jones, 2001), i.e. for tornadoes with high ratio of the height to the diameter. Mathematical model of the middle part of tornado should have a steady-state solution or solution with very low growth rates.

It is commonly accepted that tornado solution cannot be derived from linear system of equations (continuity, Navier-Stokes, entropy) and therefore it is believed that tornado is described by a set of nonlinear hydrodynamic equations. However, it is still unclear which non-linear processes are responsible for its formation. Indeed, the inertia non-linear terms $(\boldsymbol{v}\nabla)\boldsymbol{v}$ lead to energy dissipation, therefore cannot be the energy source of the system. Other non-linear terms have the same problem, since they do not introduce energy in the system.

Renno and Ingersoll (1996) described convection in the atmosphere as a heat engine between warm surface (due to solar radiation) and cold troposhpere, this assumption implies heat accumulation and release by means of black-body radiation. However, continuous and powerful source of energy observed in tornadoes and cyclones has to come from a source much more efficient than the black body radiation or heat conductivity. One of possibilities is to use latent heat of water vapor as the source, since vapor content in the air is widely spread in the atmosphere, phase transitions occur fast, and the specific condensation energy is high. In this work we imply a mechanism of the latent heat release by varying the vertical temperature profiles in different parts of the tornado structure. Unlike a heat engine, the process is not reversible. Indeed, at different points the tornado has either upward or downward air flows. These flows naturally contain saturated vapor from the surface or dry air from the troposphere, respectively. The upwards fluid flow transports the vapor, which constantly condenses and releases the heat at all levels. Therefore the vertical temperature profiles in these two flows are essentially different. As it will be shown later, such system has a stable configuration with strong vertical and azimuthal velocity components.

2 Formulation of the problem

We start from the full system of hydrodynamic equations for compressible fluid, consisting of continuity equation, Navier-Stokes equations, and entropy balance (Landau and Lifshitz, 1987):

$$\frac{\partial \rho}{\partial t}+\nabla(\rho \boldsymbol{v})=0, \tag{1}$$

$$\frac{\partial \boldsymbol{v}}{\partial t}-\nu\nabla^2\boldsymbol{v}+(\boldsymbol{v}\cdot\nabla)\boldsymbol{v}=-\frac{\nabla P}{\rho}+2\Omega\boldsymbol{e}_z\times\boldsymbol{v}+\boldsymbol{e}_z g, \tag{2}$$

$$\frac{\partial s}{\partial t}+(\boldsymbol{v}\cdot\nabla)s=\frac{\chi c_P}{T}\nabla^2 T, \tag{3}$$

where thermal diffusivity coefficient $\chi=k/(c_P\rho)$ is used instead of the thermal conductivity k, and c_P is specific heat capacity of air at constant pressure; ν is the kinematic turbulent viscosity, $\boldsymbol{e}_z\Omega$ is the vector of angular velocity of the rotating mothercloud, and $\boldsymbol{v}$ is the fluid velocity. Although we describe compressible fluid, we have neglected in Eq. (2) the term with bulk viscosity for simplicity.

We introduce cylindrical coordinates with unit vector $\boldsymbol{e}_z$ directed upward. Following the standard linearization procedure (Landau and Lifshitz, 1987), density, pressure and temperature are represented as $\rho(r,z)=\rho_0(z)+\rho_1(r,z)$, $P(r,z)=P_0(z)+P_1(r,z)$ and $T(r,z)=T_0(z)+T_1(r,z)$, respectively. Here the steady-state adiabatic profiles of thermodynamic functions are denoted by index 0, and their small perturbations by index 1. The linearization procedure requires perturbations to be small compared to the steady-state values i.e. "weak" tornado.

The steady-state profiles are determined in an assumption of zero velocities, implying the temperature profiles are close to adiabatic. The corresponding set of equations contain steady-state Navier-Stokes equations $\nabla P_0+\boldsymbol{e}_z g\rho_0=0$, the perfect gas law $P_0=\rho_0 R T_0$, and the steady-state entropy balance equation: $\nabla^2 T_0=0$. Solution of the latter equation is a linear temperature profile, which we postulate as

$$T_0(z)=T_{00}+\gamma z=T_{00}+(\gamma_a+\gamma')z, \tag{4}$$

where $T_{00}=T_0(0)$ is the temperature value near the surface, $\gamma_a=-g/c_P$ is a well known neutral adiabatic gradient of dry atmosphere, and γ' is the deviation of the chosen steady-state gradient γ from the neutral profile γ_a, g is the gravity acceleration. According to our steady-state assumption the deviation is small: $\gamma'\ll\gamma_a$, and positive value of γ' corresponds to convectively stable atmosphere. Substituting temperature profile (Eq. 4) into steady-state Navier-Stokes equations one can derive the pressure $P_0(z)=P_{00}[T_0(z)/T_{00}]^{-g/(R\gamma)}\approx P_{00}[T_0(z)/T_{00}]^{\kappa/(\kappa-1)}$, and density $\rho_0(z)=\rho_{00}[T_0(z)/T_{00}]^{-g/(R\gamma)-1}\approx\rho_{00}[T_0(z)/T_{00}]^{1/(\kappa-1)}$, profiles where $\kappa=c_P/c_V$ is the ratio of specific heats. Notice the density derivative

$$\frac{1}{\rho_0}\frac{\partial\rho_0}{\partial z}=-\frac{g}{c^2}, \tag{5}$$

where $c(z)=\sqrt{\kappa R T_0(z)}$ is the sound velocity.

Linearization with the steady state Eqs. (4)–(5) leads to the following linearized system of hydrodynamic equations:

$$\frac{\partial\rho_1}{\partial t}+\mathrm{v}_z\frac{\partial\rho_0}{\partial z}+\rho_0\nabla\cdot\boldsymbol{v}=0, \tag{6}$$

$$\frac{\partial\boldsymbol{v}}{\partial t}-\nu\nabla^2\boldsymbol{v}+\frac{\nabla P_1}{\rho_0}+\frac{g\boldsymbol{e}_z\rho_1}{\rho_0}+2\Omega\boldsymbol{e}_z\times\boldsymbol{v}=0, \tag{7}$$

$$\frac{\partial P_1}{\partial t}-c^2\frac{\partial\rho_1}{\partial t}+\kappa R\rho_0\gamma'\mathrm{v}_z=\chi\kappa\nabla^2 P_1-c^2\chi\nabla^2\rho_1-2R\kappa\chi\gamma_a\frac{\partial}{\partial z}\frac{\rho_1}{\rho_0}. \tag{8}$$

To derive the last equation we have used the perfect gas entropy as $s=c_P\ln(T/T_{00})-R\ln(P/P_{00})$, substituted here the pressure and temperature profiles, expressed the temperature through the pressure and density using the perfect gas law, and finally substituted this entropy into Eq. (3).

Now we are constructing a steady-state ($\partial/\partial t=0$), isotropic ($\partial/\partial\theta=0$), and vertically homogeneous ($\partial/\partial z=0$) solution of the linearized system (Eqs. 6–8). It is convenient to express velocity in terms of potential, toroidal, and poloidal fields (Rutkevich and Rutkevych, 2009):

$$\boldsymbol{v}=\nabla\Phi+\nabla\times(\boldsymbol{e}_z\psi)+\nabla\times(\nabla\times(\boldsymbol{e}_z\varphi)), \tag{9}$$

where $\Phi(t,r)$, $\psi(t,r)$ and $\varphi(t,r)$ are the potentials of the above mentioned velocity components. In terms of these variables the set of Eqs. (6)–(8) becomes

$$\Delta\Phi+\frac{g}{c^2}\Delta\varphi=0, \tag{10}$$

$$-\nu\Delta\Delta\Phi+\frac{\Delta P_1}{\rho_0}+\frac{g^2}{c^2}\frac{\rho_1}{\rho_0}+2\Omega\Delta\psi=0, \tag{11}$$

$$-\nu\Delta\Delta\psi-2\Omega\Delta\Phi=0, \tag{12}$$

$$-\nu\Delta\Delta\Delta\varphi+\frac{g}{c^2}\frac{\Delta P_1}{\rho_0}-\frac{g}{\rho_0}\Delta\rho_1=0, \tag{13}$$

$$-\kappa R\gamma'(\mathrm{v}_z)\rho_0\Delta\varphi-\chi\kappa\Delta P_1+\chi c^2\Delta\rho_1=0, \tag{14}$$

where $\Delta=\nabla_r^2=\frac{1}{r}\frac{\partial}{\partial r}\left(r\frac{\partial}{\partial r}\right)$ is the radial component of the Laplace operator. The system of ordinary differential Eqs. (10)–(14) describes the radial profiles of the thermodynamic parameters and velocities at any given height $z=z_0$, i.e. at given density $\rho_0(z_0)$ and sound velocity $c(z_0)$.

As it has been already mentioned in the introduction, only one non-linear term is considered, namely the dependence of the temperature profile on the vertical velocity of the fluid. In this work we use a simple approximation

$$\gamma'(\mathrm{v}_z) = \gamma_0 + \gamma_1 \frac{\mathrm{v}_z}{\mathrm{v}_T}, \tag{15}$$

where $\gamma_0>0$ is the temperature gradient of stable dry atmosphere, coefficient $\gamma_1<0$ accounts for the temperature profile in moist air, and v_T is the velocity normalization coefficient. For sufficiently high vertical velocity v_z the total temperature gradient (Eq. 15) becomes negative (i.e. convectively unstable).

3 Convection without rotation

In non-rotational ($\Omega\rightarrow 0$) and incompressible ($c\rightarrow\infty$) fluid the system Eqs. (10)–(14) reduces to one equation for the poloidal velocity potential φ:

$$\Delta^2\varphi + \frac{g\gamma'}{T_0(z_0)\nu\chi}\varphi = 0. \tag{16}$$

Other velocity components, pressure and density can be derived from the solution of Eq. (16). Now we substitute the temperature gradient (Eq. 15) in Eq. (16), express v_z in terms of φ by means of Eq. (9), and obtain the following nonlinear equation

$$\Delta^2\varphi + \frac{g\gamma_0}{T_0(z_0)\nu\chi}\left(\varphi - \frac{\gamma_1}{\gamma_0}\frac{\varphi\Delta\varphi}{\mathrm{v}_T}\right) = 0, \tag{17}$$

or in dimensionless variables

$$\Delta_x^2 f + f - f\Delta_x f = 0, \tag{18}$$

where $r=\lambda x$ and $\varphi = Cf$, and

$$\lambda = \left(\frac{T_0(z_0)\chi\nu}{g\gamma_0}\right)^{1/4}, \quad C = \frac{\gamma_0}{\gamma_1}\lambda^2\mathrm{v}_T, \quad \Delta_x = \frac{1}{x}\frac{\partial}{\partial x}x\frac{\partial}{\partial x}. \tag{19}$$

Note, that according to this definition the coefficient C is negative. The unknown function $f(x)$ has to be a regular function of its argument, therefore solution of Eq. (18) can be expanded in a Taylor series near x=0:

$$f(x,A,a_2) = A\left[1 + a_2x^2 + \sum_{j=2}^{\infty} a_{2j}x^{2j}\right], \tag{20}$$

where only two coefficients are independent. Indeed, one can easily check that coefficients a_4, a_6, etc, can be recursively expressed in terms of parameters A and a_2 using Eq. (18). Construction of the solution (Eq. 20) for large x is computationally intensive due to fast divergence of the series, therefore we use 4-th order Runge-Kutta method to build a numerical solution of Cauchy problem for the 4-th order ordinary Eq. (18), using the values of function $f(x)$ and its derivatives at point x=0.1, as calculated from the Taylor expansion (Eq. 20). It appears impossible to apply Runge-Kutta method from x=0 due to singularity of the Eq. (18).

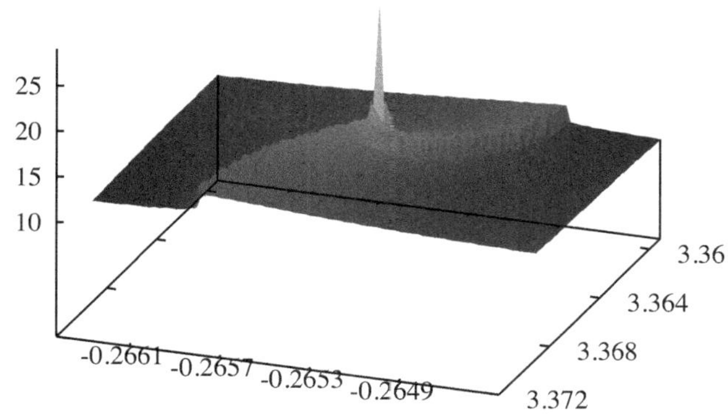

Figure 1. Profile of the x_m versus the parameters a_2 and A. The peak position indicates the best approximation of the solution of the nonlinear Eq. (18) in non-rotational fluid.

For arbitrary values of parameters A and a_2 solution (Eq. 20) diverges even for moderate x. However, the localized solution of our interest here should have a non-zero value near the axis, gradually vanish at medium distance from the axis, and uncontrollably grow far away from the axis due to numerical errors. We denote $x_m(A,a_2)$ the maximum value of dimensionless radius in the far-away range $x>10$, where the solution is small enough $|f(x_m,A,a_2)|<b|A|$. Here b is a fixed number in the range $0<b<1$, and the "far-away" radius $x=10$ corresponds to $r=700$ m, which we suppose sufficiently large. Let us denote A_0 and $a_{2,0}$ the values of parameters A and a_2, corresponding to the solution (Eq. 20) vanishing at $x\rightarrow\infty$. It is clear that function $x_m(A,a_2)$ defined above should go to $+\infty$, as $A\rightarrow A_0$, $a_2\rightarrow a_{2,0}$.

Equation (18) has to be solved numerically, therefore the exact values (A_0,$a_{2,0}$) are hardly obtainable, however, one can approximate the solution by choosing parameters (A,a_2), which give large enough value for the function $x_m(A,a_2)$. In this case the best solution would be the solution with the biggest value of x_m. In our numerical scheme the coefficients A and a_2 were varied simultaneously (see Fig. 1) to obtain maximum value of $x_m(A,a_2)$. The sharp peak on the plot corresponds to the best approximation of solution $f(x,A_0,a_{2,0})$ of the nonlinear equation (18). The position of the peak is at $A\approx -3.3641719$ and $a_2\approx -0.265424$, giving the maximum dimensionless radius $x_m\approx 25$. The corresponding solution of dimensionless poloidal potential $f(x)$ is shown in Fig. 2. Corresponding velocity profiles (v_r,v_θ,v_z) are plotted in Fig. 3 for the following values of parameters: $g=9.8\,\mathrm{ms}^{-2}$, $\nu=\chi=30\,\mathrm{m}^2\,\mathrm{s}^{-1}$, $\gamma_0=10^{-3}\,\mathrm{km}^{-1}$, $\gamma_1=-10^{-3}\,\mathrm{km}^{-1}$, $\mathrm{v}_T=1\,\mathrm{ms}^{-1}$, $\kappa=1.4$, $T_0=280\,\mathrm{K}$, $c=335\,\mathrm{ms}^{-1}$.

It should be noted here, that strictly speaking the radial and azimuthal velocity components (v_r and v_θ) are not solutions of the full system of hydrodynamic Eqs. (10)–(14), since in non-rotational case the system significantly simplifies. However, one can formally derive the profiles of all

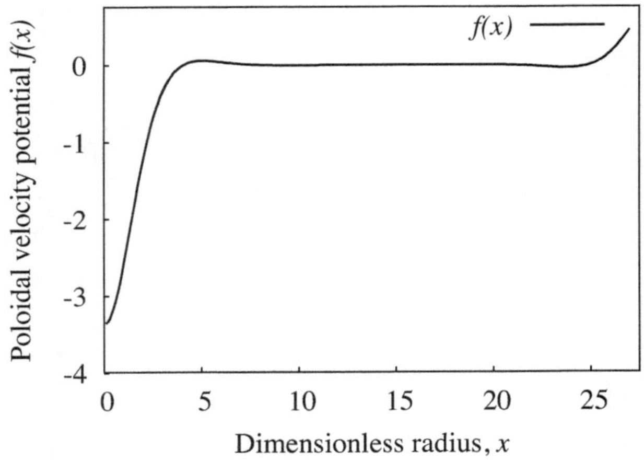

Figure 2. The best approximation of dimensionless poloidal velocity profile $f(x)$. The values of the parameters A and a_2 in the Taylor expansion (Eq. 20) correspond to the position of the peak in Fig. 1.

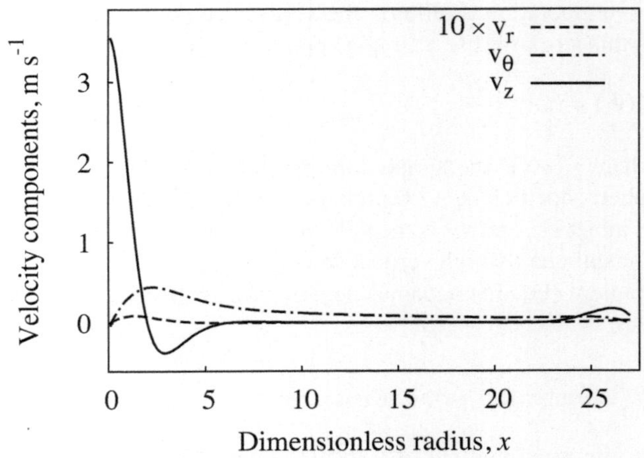

Figure 3. Profiles of velocity components in a non-rotational approximation. Values of parameters are listed at the end of Sect. 3. Notice a factor in front of the radial velocity component v_r.

velocity components from the known profile of $f(x)$, using definition (Eq. 9) and system (Eqs. 10–14) as:

$$\boldsymbol{v} = \boldsymbol{e}_r \frac{g|C|}{c^2\lambda}\frac{\partial f}{\partial x} - \boldsymbol{e}_\theta \frac{2\Omega g\lambda|C|}{\nu c^2 x}\int\limits_0^x x' f(x')dx' + \boldsymbol{e}_z \frac{|C|}{\lambda^2}\Delta_x f. \quad (21)$$

Azimuthal component of the velocity is proportional to the rotation speed of the mothercloud. The profile of v_θ in Fig. 3 has been plotted for $\Omega = 0.1\,\mathrm{s}^{-1}$. For the chosen values of parameters the radial and azimuthal components are significantly smaller than the radial component. This fact is in accordance with the initial assumption of non-rotational fluid ($\Omega \to 0$). Noteworthy, that the radial velocity of the fluid v_r is positive near the axis, indicating continuous expansion of the gas during its upwards movement.

Another noteworthy conclusion is that the dimensional parameter λ decreases with decreasing temperature (Eq. 19). This means that for other constant conditions the radius of the tornado column decreases with hight. However, this effect is of the order of 1–2% for temperature variation of ±5% (i.e. tornado height below 2 km).

4 Convection with rotation

Taking into account compressibility of the gas (i.e. $c \neq \infty$) and its rotation ($\Omega \neq 0$), one can reduce the linearized system (10)–(14) to a single high-order ordinary differential equation for the poloidal field φ. Again we assume vertically homogeneous ($\partial/\partial z = 0$), isotropic ($\partial/\partial\theta = 0$), and stationary ($\partial/\partial t = 0$) system. Using the same dimensionless units (Eq. 19) the equation reads as

$$\Delta_x[\Delta_x\Delta_x f(x) + f(x) - f(x)\Delta_x f(x)] - Qf(x) = 0, \quad (22)$$

where dimensionless parameter $Q = (\kappa - 1)(2\Omega g\lambda^3)^2\nu^{-2}c^{-4}$ characterizes the Coriolis force and compressibility. Clearly, in the case Q=0 one obtains Eq. (18).

Due to 6-th order of Eq. (22), the expansion coefficients a_6, a_8 etc. can be derived from given A, a_2, a_4, therefore the solutions now form a three-parameters set $f(x;A,a_2,a_4)$, making the numerical search procedure significantly more complicated. To solve Eq. (22) we apply the procedure described in Sect. 3, including Taylor series expansion in the vicinity of x=0 and Runge-Kutta numerical solution from x=0.1. Due to higher noise in the function, the values of the three independent parameters have been varied to achieve the smallest possible value of integral $\int_{x_1}^{x_2} f(x)^2 dx$, instead of the maximum x_m described earlier. This choice of the convergency condition reflects the fact, that the absolute value of the function f should be the smallest in the medium range of radii. Figure 4 shows the best numerical solution, which we have found in such a way, using x_1=8, x_2=14, and given Q=0.1. The solution corresponds to the following set of parameters: $A \approx 11.29547$, $a_2 \approx -5.454639$, $a_4 \approx -2.179057$. One can see, that the found solution diverges at $x_m \approx 13$.

The radial and vertical velocity components can be obtained from the potential profile $f(x)$ using Eq. (21), while for the azimuthal profile one has to solve differential equation

$$\frac{\partial v_\theta(r)}{\partial r} + \frac{v_\theta(r)}{r} + \frac{2\Omega g}{\nu c^2}\varphi(r) = 0. \quad (23)$$

Velocity profiles are plotted in Fig. 5. The air motion has several important features, namely updrift in the central area, relatively small radial velocity, and azimuthal velocity has maximum at approximately $x \approx 8$. The azimuthal velocity in axially-symmetrical system is the parameter with the slowest radial convergency, because it is limited only by the viscosity

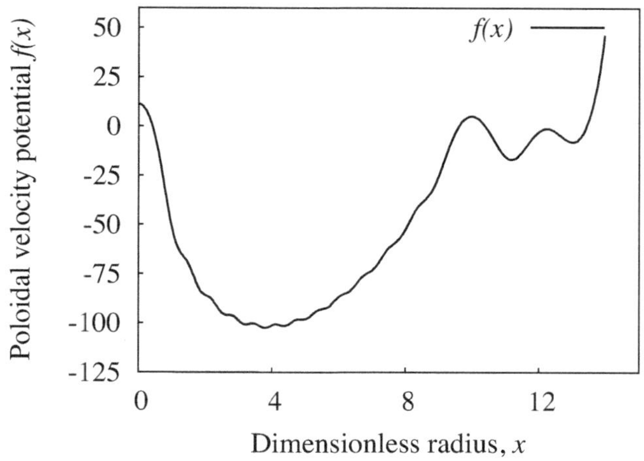

Figure 4. Profile of dimensionless potential field $f(x)$ with rotation. The values of parameters are the same as the non-rotational case, except the Coriolis parameter Ω=0.23 s^{-1}.

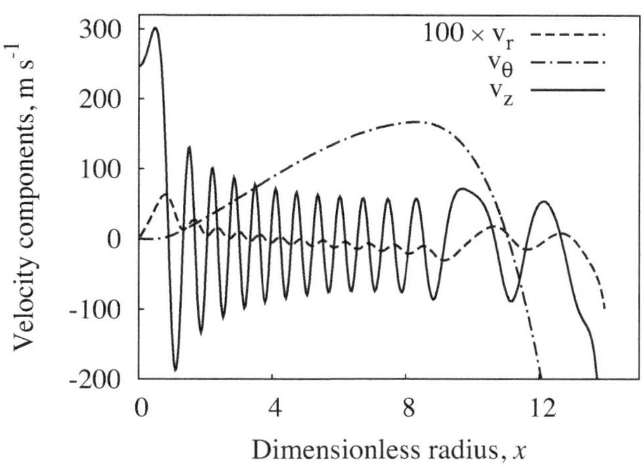

Figure 5. Profiles of velocity components with rotation, parameter values are the same as in Fig. 4.

of the fluid. In our simulation the azimuthal velocity changes the sign and infinitely increases, however we suppose this is an artifact of the numerical method. The expected convergency region of the azimuthal velocity is in the range $x \geq x_2$, where the profile $f(x)$ used in Eq. (23) has large errors. The actual profile of azimuthal velocity, which can be regarded as fine-tuning of the solution, is a subject of further study.

The coefficient λ used for dimensionless radii (Eq. 19) equals to 70 m for the chosen values of parameters. Therefore one can conclude, that the main upward air stream in Fig. 3 has radius slightly more than 100 m. Although this value is relatively large for a tornado, the model does not include the rotation, which is likely to sharpen the structure. Indeed, solution with rotation in Fig. 5 has the radius of the upwards flow less than 70 m, which is very close to a typical tornado size (Davies-Jones, 2001).

The absolute value of velocity of 3 ms^{-1} in non-rotational case is reasonable, since effectively this is the velocity of a free localized convection in atmosphere. Account of rotation in the system leads to dramatic growth of velocity, especially its vertical component. This suggests that a model of tornado must include other non-linear dissipating terms to suppress the fluid flow.

5 Conclusions

The proposed one-dimensional vertically-homogeneous model cannot be applied to the tornado funnel near the mothercloud, or to the turbulent area near the ground surface. The model is applicable at the middle level of a weak tornado, where the shape of the column changes slowly in space and time. The model shows that the system of linearized hydrodynamic equations can have a stable solution with only one non-linear term. Since the full system of equations has several important non-linear terms (including viscosity and centrifugal force) this paper cannot be regarded as the final solution of the hydrodynamic system or as a complete theory of tornado. However, the results show, that a steady-state configuration is possible even in convectively-stable atmosphere.

Proposed non-linear effect related to asymmetric vapor transport in a "weak" tornado-like structure introduces the source of energy, which can be the true mechanism in natural intensive vortices. Obtained structures have all features of real tornadoes/cyclones, however estimations show that the size of the solution is closer to the size of tornado. We expect, that accounting other non-linear terms (which are usually dissipating) in the full system of equations will lead to stabilization of the solution and especially far from the axis.

Edited by: D. Giaiotti

References

Bluestein, H. B., Gaddy, S. G., DowelL, D. C., Pazmany, A. L., Galloway, J. C., McIntosh, R. E., and Stein, H.: Doppler Radar Observations of Substorm-Scale Vortices in a Supercell, Mon. Weather Rev., 125, 1046–1059, 1997.

Bluestein, H. B. and Pazmany, A. L.: Observations of Tornadoes and Other Convective Phenomena with a Mobile, 3-mm Wavelength, Doppler Radar: The Spring 1999 Field Experiment, B. Am. Meteorol. Soc., 81, 2939–2952, 2000.

Bluestein, H. B. and Weisman, M. L.: The Interaction of Numerically Simulated Supercells Initiated along Lines, Mon. Weather Rev., 128, 3128–3149, 2000.

Davies-Jones, R.: Severe convective storm, Vol. 28, Meteorological Monographs, 2001.

Doswell III, C. A. and Burgess, D. W.: Tornadoes and tornadic storms: A review of conceptual models, in: The Tornado: Its Structure, Dynamics, Prediction and Hazards, edited by:

Church, C., Burgess, D., Doswell, C., Davies-Jones, R., Geophys. Monogr, Vol. 79, Amer. Geophys. Union, 161–172, 1993.

Landau, L. D. and Lifshitz, E. M.: Fluid Mechanics, 2nd Edn., 1987.

Lehmiller, G. S., Bluestein, H. B., Neiman, P. J., Ralph, F. M., and Feltz, W. F.: Wind Structure in a Supercell Thunderstorm as Measured by a UHF Wind Profiler, Mon. Weather Rev., 129, 1968–1986, 2001.

Renno, N. O. and Ingersoll, A. P.: Natural convection as a heat engine: a theory for CAPE, J. Atmos. Sci., 53, 572–585, 1996.

Rutkevich, P. B. and Rutkevych, P. P.: Model of oscillatory instability in vertically-homogeneous atmosphere, Adv. Geosci., 15, 57–63, doi:10.5194/adgeo-15-57-2009, 2009.

Development and testing of homogenisation methods: moving parameter experiments with ACMANT

P. Domonkos and D. Efthymiadis

Centre for Climate Change, University Rovira i Virgili, Campus Terres de l'Ebre, Tortosa, Spain

Correspondence to: P. Domonkos (peter.domonkos@urv.cat)

Abstract. During the European project COST ES0601 (HOME) a new homogenisation method, ACMANT has been developed for the automatic homogenisation of monthly temperatures. ACMANT turned out to be one of the best performing methods during the blind test experiments of HOME. The methodological development of ACMANT has been continued since then, and nowadays ACMANT is likely the best homogenisation method for large and spatially dense temperature datasets. Ensemble moving parameter experiments have been done to obtain more information about the performance of ACMANT. The HOME Benchmark was used as test dataset, thus the results of the latest experiments with ACMANT are comparable with the performance of the other homogenisation methods participated in HOME. The results indicate that the performance of ACMANT is generally not sensitive to its parameterisation, i.e. the change of the performance is generally small for quite a wide range of each parameter. The presented methodology of moving parameter experiments provides results in a fast and easy to evaluate form.

1 Introduction

In the time series of climatic observations, temporal biases from the true macroclimatic characteristics often occur due to technical, personal or environmental changes. With the homogenisation of time series, the frequency and magnitudes of such biases can be reduced. Time series homogenisation can be done with the use of documentary information about the changes in the settings of the observation or with statistical procedures or with the combination of both (Aguilar et al., 2003; Auer et al., 2005, etc.). When the spatial density of observing sites and the spatial correlation of the observed values are high, data quality can be significantly improved by statistical homogenisation. The conditions mentioned are generally met for European and North American temperature data from the last century (Menne and Williams, 2009; Domonkos and Štěpánek, 2009).

A large number of statistical homogenisation methods have been developed in the last decades. Between 2007 and 2011 action COST ES0601 (HOME, www.homogenisation.org) was dedicated to test the efficiency of the existing methods and foster further methodological developments. Under HOME, an international blind test experiment was organised, in which MASH (Szentimrey, 1999), PRODIGE (Caussinus and Mestre, 2004), USHCN (Menne and Williams, 2009), Craddock-test (Craddock, 1979) and ACMANT (Domonkos, 2011a, hereafter D2011) produced the best results in homogenising monthly temperature data (Venema et al., 2012). The development of ACMANT has been continued since the blind tests, and as ACMANT is a fully automatic method, its performance can be objectively tested also in non-blind mode. The aim of the present study is to examine the sensitivity of the performance of ACMANT to its parameterisation, because low sensitivity could confirm the leading position of ACMANT, while high sensitivity would indicate the uncertainty in the rank order of method efficiencies. The efficiency tests were made with ensemble moving parameter experiments, which is a useful tool for testing automatic and semi-automatic methods (McCarthy et al., 2008; Tichner et al., 2009; Williams et al., 2012). The methodology that we present allows the fast and easy evaluation of the results, therefore we recommend its use also to the examination of other automatic algorithms. The paper allows insight also into the theoretical aspects of the latest developments and the performance of the newest version of ACMANT.

The organisation of the paper is as follows. In Sect. 2, the latest development of ACMANT is presented. In Sect. 3, the methodology and the results of the moving parameter experiment are presented. Finally, in Sect. 4, the results and the further tasks ahead the method developers are discussed and the conclusions are summarised.

2 ACMANT

The full description of the first version of ACMANT (Applied Caussinus-Mestre Algorithm for homogenising Networks of Temperature series) has already been published (D2011). That version was referred as ACMANT late by Venema et al. (2012) and ACMANTv1 in web. However, the method has been developed since then. Here, the structure of ACMANT will be briefly described first, then the recent changes will be presented in details.

2.1 Structure of ACMANT

The ACMANT was developed on the basis of detection and correction algorithms in PRODIGE. PRODIGE was selected, because its inhomogeneity detection part and correction method (Caussinus and Mestre, 2004) have been turned out to be highly effective in comparison with many other methods (Domonkos, 2011b; Domonkos et al., 2013). However, ACMANT differs from PRODIGE in many details, first of all ACMANT uses reference series instead of pairwise comparisons and ACMANT is fully automatic.

In ACMANT reference series (i.e. series with which the so-called candidate series is compared to find the inhomogeneities in the series of spatial differences) are pre-homogenised before the main detection part. During pre-homogenisation, the use of the candidate series of the Main Detection is excluded from the calculation of adjustment-terms, thus double use of the same spatial connection is not allowed in ACMANT. Both in the pre-homogenisation and in the Main Detection, the optimal step function fitting with the Caussinus–Lyazrhi criterion is applied (Caussinus and Mestre, 2004). However, a novelty of ACMANT is that change-points are searched with bivariate tests, namely joint statistics of the annual means and summer-winter differences are examined (D2011). Correction-terms are always calculated with ANOVA (Caussinus and Mestre, 2004). After the main detection, further inhomogeneities are searched on a monthly scale. Outlier-filtering and filling of gaps of time series are applied three times in ACMANT: first, before pre-homogenisation, second, after pre-homogenisation and third, together with the final adjustments; see further details in D2011.

2.2 Latest development of ACMANT

Two kinds of development have been applied to ACMANT since its publication (D2011). One of these changes is that in the new ACMANT, ANOVA is applied also in the pre-homogenisation, while the other is the introduction of filtering of outlier periods. Together with these changes, the Secondary Detection is also modified.

2.2.1 Calculation of adjustment-terms

In the new ACMANT, ANOVA is always applied and thus unified relative time series, as well as all that were written in Sects. 3.3.4 and 3.6.2. of D2011 are no longer included in ACMANT.

ANOVA is applied in three modes in ACMANT: (a) In the pre-homogenisation it is applied on annual values at an outlier filtering step, (b) In the pre-homogenisation it is applied on annual values and with the exclusion of the further candidate series for preparing reference series to the Main Detection and (c) After the Main Detection and after the Secondary Detection it is applied on monthly data. See the description of ANOVA model in Caussinus and Mestre (2004).

2.2.2 Filtering of outlier-periods

Outlier-periods could be also referred to as short-term inhomogeneities, since their model is a short-term platform-like bias from the correct values. In such platforms the bias is constant for the outlier-period. In the real world the magnitude of the bias could vary within the period, so the platform is only a model. Note, however, that Domonkos (2011b) reported that observed temperature series of Hungary can be modelled well with the inclusion of a large number of short-term, platform-shaped inhomogeneities. Note also that short-term inhomogeneities can be detected only when the size of the bias is large, since with the shortening of the duration of biases the signal to noise ratio worsens. So that, the aim of the filtering of outlier-periods is to filter the short-term, large biases from time series. Both the detection and the correction of outlier-periods are more similar to the common outlier-filtering than to the inhomogeneity detection.

In ACMANT, filtering of outlier-periods is applied for 2–30 month long periods always after the common outlier filtering. The maximum length of such periods is controlled by a parameter ($c1$, Table 1), its value was between 18 and 30 in the latest examinations. The values of the detected outlier-periods are treated in the same way as data gaps or outliers, and the interpolation of Sect. 3.2 in D2011 is applied to them. However, for outlier-periods reaching a given threshold ($c2$), the starting and ending dates of outlier-periods are considered as change-points in the final calculations of correction-terms by ANOVA.

In searching outlier-periods, relative time series are used on monthly scale. First the values are standardised, i.e. they converted to have 0 mean and 1 standard deviation. Note that seasonal cycles had been filtered earlier (cf. Eq. 4 in D2011). The standardised relative time series are denoted with $\mathbf{E} = [e_1, e_2, \ldots e_{nm}]$, the length of series with nm and

Table 1. Moving parameters and their values in the experiments. Six values are allowed for each parameter. In column "basic" the values in ACMANTv1 (D2011) are shown.

		basic	1	2	3	4	5	6
$c1$	outlier-p.: max. effective window (month)	–	18	20	22	24	27	30
$c2$	distance between change-points (month)	–	3	4	5	6	8	10
$c3$	significance threshold in outlier p. filtering	–	16.0	20.0	25.0	32.0	42.0	60.0
$c4$	outlier period: power of duration	–	0.5	0.6	0.7	0.8	0.9	1.0
$c5$	outlier p.: length of outer periods (month)	–	24	24	24	36	36	36
$c6$	outlier period: correction for seasonality	–	0.10	0.15	0.20	0.30	0.40	0.50
$c7$	outlier period: precision of timing (month)	–	6	8	10	12	15	18
$c8$	summer–winter difference	0.50	0.25	0.33	0.42	0.50	0.59	0.67
$c9$	pre-homogenisation, penalty-coefficient	1.0	1.0	1.2	1.4	1.6	1.8	2.0
$c10$	Secondary Detection, penalty coefficient	1.0	1.0	1.25	1.5	1.8	2.3	3.0
$c11$	overlap of relative time series (year)	9	5	7	8	9	10	12
$c12$	window width in Secondary Det. (month)	60	36	42	50	60	70	80
$c13$	MA5 threshold	2.0	1.6	1.8	2.0	2.2	2.4	2.6
$c14$	MA10 threshold	1.4	1.2	1.3	1.4	1.5	1.65	1.8
$c15$	min. length for harmonic functions (month)	10	8	9	10	12	14	16
$c16$	Main Detection, monthly precision (month)	25	17	21	25	29	33	37
$c17$	r-threshold for 2 reference composites	0.5	0.7	0.8	0.8	0.85	1.0	1.0
$c18$	r-threshold for 3 reference composites	0.5	0.5	0.6	0.7	0.7	0.7	0.8
$c19$	r-threshold for 4 reference composites	0.5	0.4	0.48	0.6	0.6	0.5	0.6
$c20$	r-threshold for 5 reference composites	0.5	0.4	0.4	0.5	0.5	0.4	0.5

section means with upper stroke. Further denotations: l – length of outlier period, i and j – starting and ending months (respectively) of the outlier-period in the first estimation, m_1 – number of summer months (of June, July or August) in the outlier-period, m_2 – number of winter months (of January, February or December) in the outlier-period, int – integer part, sgn – sign of expression, mod – module 12, $c3$, $c4$,... are parameters.

Only one outlier-period is identified and selected in a particular step, i.e. the one with the most significant statistic ($c3$). Then its values are adjusted (temporal adjustments, which are valid only in the section of filtering the outlier-periods), and further outlier-periods are searched as long as at least one can be found with significant statistic. The identification of an outlier-period comprises two phases. In the first phase (A), the most significant outlier-period of the time series is selected and a first estimate is made for its position. In the second phase (B) the starting and ending months of the outlier-period are determined.

Phase (A): the outlier-period with the maximal $c3$ is searched for each i, j pairs ($1 \leq j - i < 30$) of standardised relative time series. If more than one relative time series are available for a given i, j pair, always the one with the highest sum of squared correlations of the reference composites is selected for the examination.

$$c3 = l'^{c4} d^2 \quad (1)$$

where d (magnitude-characteristic) and l' (duration-characteristic) are determined by Eqs. (2) and (3):

$$d = \overline{e_{i-c5}, e_{i-1}} + \overline{e_{j+1}, e_{j+c5}} - \overline{e_i, e_j} \quad (2)$$

$$l' = \mathrm{int}(\max\{l - c6\,|m_1 - m_2|, 1 + (1 - c6)(l - 1)\}) \quad (3)$$

Further conditions are that

$$\mathrm{sgn}(\overline{e_i, e_j} - \overline{e_{i-c5}, e_{i-1}}) = \mathrm{sgn}(\overline{e_i, e_j} - \overline{e_{j+1}, e_{j+c5}}) \quad (4)$$

$$c5 \geq 24 \quad (5)$$

$$\mathrm{mod}(c5) = 0 \quad (6)$$

Phase (B): the first and last months of the outlier-period are re-estimated with fitting optimal step-function in window $[e_{i-c5}, e_{j+c5}]$. This procedure is made in the same way as in the Secondary Detection (see Sect. 3.5.2. in D2011) with the exception that solutions with exactly two change-points are accepted only, and the first and second change-points are expected in the periods $[i - c7, i - 1]$ and $[j, j + c7 - 1]$, respectively. So that, the final duration of an outlier-period is equal or greater than the pre-estimated duration. If the resultant duration is longer than $c1$, then the temporary adjustment is applied for allowing the search of further outlier-periods, but otherwise such outlier periods are left out of consideration.

Notes:

a. Outlier-periods can be detected also in the ends of time series, and slightly different rules are applied for them. At the ends of the series Eq. (4) cannot be expected, $c5 \geq 36$, in phase B the window around the pre-estimated position is not symmetric and exactly 1 change-point is searched in that phase.

b. $c5$ has a minimum threshold for providing a relatively large sample-size of the surrounding values in the evaluation of the deviation of a pre-assumed outlier-period. Naturally, too great values for $c5$ are not advisable, either.

c. Equation (6), as well as the consideration of m_1 and m_2 are necessary due to the seasonal behaviour of inhomogeneity-caused biases. Otherwise a long-term inhomogeneity with large bias in the seasonal cycle could be detected as short-term outlier-period when a seasonal peak of the long-term bias and random noise accidentally add up.

d. The role of $c1$ is to separate which biases are considered short-term outlier-periods and which are considered long enough to be treated by the main detection part of ACMANT.

e. In Eq. (3) the coefficient of $(l-1)$ is $1-c6$. It is an arbitrary choice, the coefficient could be independent from $c6$.

2.3 Modification in the Secondary Detection

Parameter $c2$ (see Sect. 2.2) is used also in the Secondary Detection to separate outlier-periods from more persistent inhomogeneities. If the period between two adjacent change-points is shorter than $c2$, then that period is treated in the same way as outlier-periods detected by Eqs. (1)–(6).

3 Moving parameter experiments

Moving parameter experiments have been done to examine the sensitivity of the performance of ACMANT to changes in its parameterisation, as well as to find optimal parameter values if this sensitivity is significant. An objective was to vary each of the arbitrarily set parameters of ACMANT that might influence its performance.

3.1 Methods

3.1.1 Moving parameter experiments

Twenty parameters of ACMANT were varied. Seven from the twenty are defined in Sect. 2.2. of this study, while we refer to D2011 presenting the definition of the others: $c8$ is the coefficient of summer–winter difference (c_0^2 in Eq. 25). $c9$ and $c10$ are coefficients for the penalty term of the Caussinus–Lyazrhi criterion in the Pre-homogenisation and the Secondary Detection, respectively (Eq. 26). $c11$ is the length of overlap when more than one relative time series are used (Eq. 28). $c12$ is the window width in the Secondary Detection (Sect. 3.5). $c13$ ($c14$) is the threshold for accumulated anomaly MA5 (MA10) (Sect. 3.5.1). $c15$ is the minimum duration for fitting harmonic functions in the Secondary Detection (Sect. 3.5.2), as well as in the new version, also in determining the position of outlier period. $c16$ is the length of the period in which the timing of change-point is expected in the monthly precision section (Sect. 4, Part III, point 4). $c17 \ldots c20$ are minimum thresholds for the spatial correlation of increment-series (r). In the previous versions of ACMANT two reference components with $r \geq 0.5$ was satisfactory to derive relative time series. In the present experiments various thresholds of r are applied according to the number of reference composites. $c17, c18, c19$ and $c20$ are thresholds for the cases of 2, 3, 4 and 5 reference composites, respectively. $c17$, $c18$, $c19$ and $c20$ indicate four conditions for the reference composites, and reference series was built when at least 1 condition of them was met.

Two thousand experiments were made, each with the 20 networks of the surrogated temperature section of HOME benchmark. All the listed parameters were varied randomly, but only six values were allowed for each parameter (except for $c5$, for that parameter only 2 values seemed applicable). Hereafter we refer to the six allowed values as value "1", value "2", etc. The restriction to six values allows us to evaluate the impact of any of the twenty parameters with the comparison of six sub-samples, in which the examined parameter is constant.

The values of $c17$, $c18$, $c19$ and $c20$ were mutually dependent in the experiment in a way that if $c17$ had value "a", $c18, c19$ and $c20$ also had the values "a", so in reality only six sets of values were allowed for them. Due to this dependence, these parameters and their effects were jointly examined.

Table 1 presents the 20 parameters each together with a key-word referring to its role in ACMANT, with its value in ACMANTv1 (D2011) and with its values in the moving parameter experiments. It can be seen that each parameter is varied in a rather wide range.

3.1.2 Efficiency measures

We apply four efficiency measures: (i) root mean squared error (RMSE) of monthly values (°C), (ii) RMSE of annual values (°C), (iii) RMSE of individual trend-slope biases (°C/100 yr) and (iv) RMSE of network-mean trend-slope biases (°C/100 yr). All these measures characterise the residual errors in homogenised series, the first three are for the entire period of time series, while type (iv) is for period 1925–1999. For visualising the improvement in the data accuracy due to the use of ACMANT, raw data errors, and the residual errors after the blind homogenisation of HOME with various

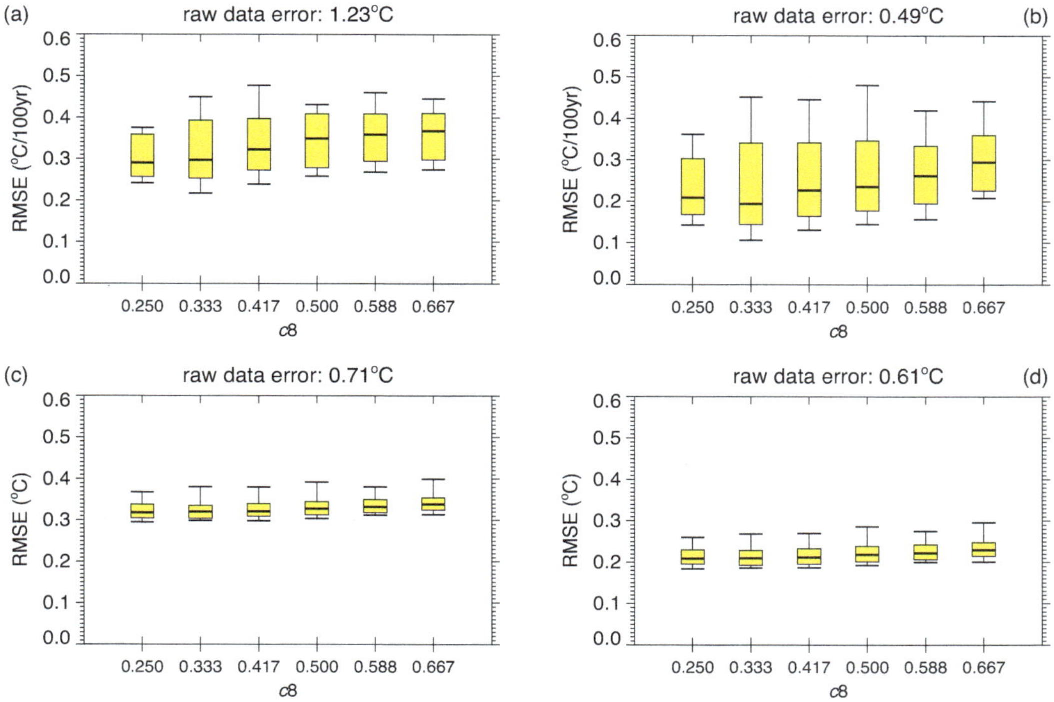

Figure 1. The sensitivity of residual errors to parameter $c8$. **(a)** Trend bias of station series, **(b)** trend bias of network-mean series, **(c)** RMSE of monthly values, **(d)** RMSE of annual values.

methods (Venema et al., 2012) are also presented in some figures. The involved homogenisation methods of the HOME experiments are PRODIGE, MASH, USHCN and C3-SNHT (Venema et al., 2012), since apart from ACMANT, they produced the highest efficiency for the entire surrogated dataset.

In HOME, only 15 networks from the 20 were evaluated formerly, and for monthly and annual residual errors Centred RMSE (CRMSE) were used (Venema et al., 2012). These differences did not result in visible deviations of the efficiencies presented in our study, relative to the HOME results.

3.2 Results

The moving parameter experiments proved the generally low sensitivity of ACMANT to its parameterisation. For most parameters, there are no significant differences of performance with changes of the value within the range examined. For the other parameters slight but significant declines of performance can be observed in one or both tails of the examined parameter ranges. The relatively highest sensitivity of performance is to $c8$ (Fig. 1), and for this parameter the basic value (ACMANTv1, Table 1) is clearly suboptimal. However, even in this case, the sensitivity is still moderate.

ACMANT has a generally high efficiency in homogenising temperature series and the moving parameter experiments described allows us to characterise the stability of the performance of ACMANT. Figures 2–5 present the results in comparison with the raw data errors and with the remaining errors after the HOME experiments with four homogenisation methods. Each of Figs. 2–5 shows the results of ACMANT in three columns. The first column shows the results of all the 2000 experiments, thus it includes also the experiments with suboptimal parameter values. For the second column, experiments with seven suboptimal parameter values are excluded, but each basic parameter value (Table 1) is remained in the accepted parameter ranges. For the last column, further four parameter values are excluded and two of them are basic parameter values in ACMANTv1. As at least one of the excluded parameter values frequently occurred, only 496 (197) experiments remained suitable for column 2 (column 3) of Figs. 2–5. The excluded parameter values from columns 2 and 3 (only column 3) are as follows. $c3$: value "1", $c4$: "6" ("5"), $c8$: "5", "6" ("4"), $c9$: "5", "6" ("1"), $c12$: "1", ($c17$–$c20$: "6"). Figures 2–5 show that ACMANT generally has the highest performance among the examined homogenisation methods. The advantage of ACMANT is even more enhanced when some suboptimal parameter values are excluded.

The advantage of ACMANT is not the same in the examined efficiency measures. In the residual errors of trends of station data (Fig. 2), PRODIGE and MASH have similar performance to that of ACMANT, although in the right column 95 % of the results with ACMANT are better than those with any other method. In estimating network-mean trends

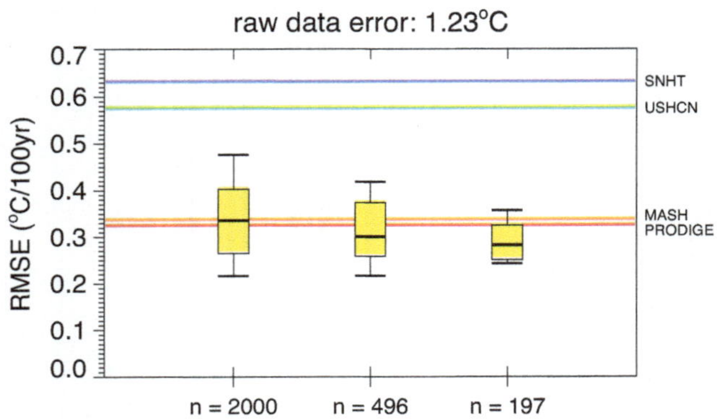

Figure 2. Boxplot of the residual errors in trend biases of station series after homogenising with ACMANT the HOME surrogated temperature data. Left column is for all the 2000 experiments, while for the middle and right columns some suboptimal parameters are excluded. Boxes include the values between 5 and 95 percentiles. Raw data errors and residual errors after blind homogenisation with various methods under HOME are also shown.

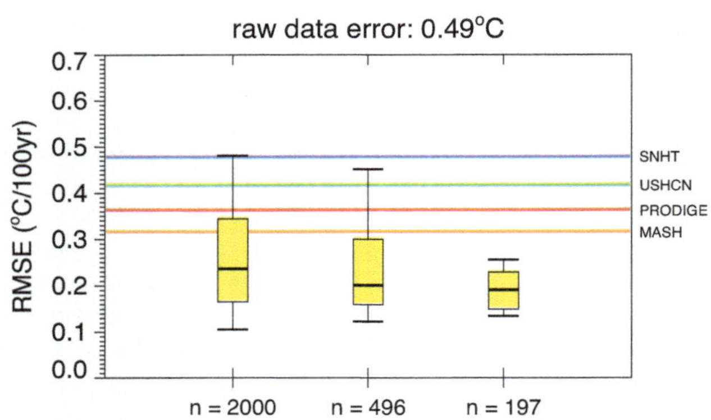

Figure 3. The same as Fig. 2, but for trend biases in network mean series.

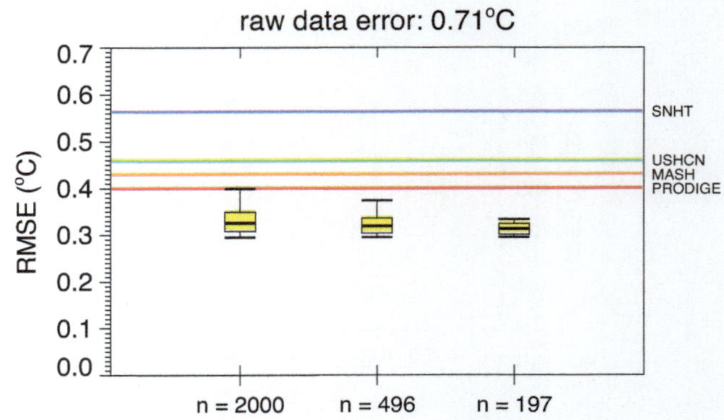

Figure 4. The same as Fig. 2, but for the RMSE of monthly values.

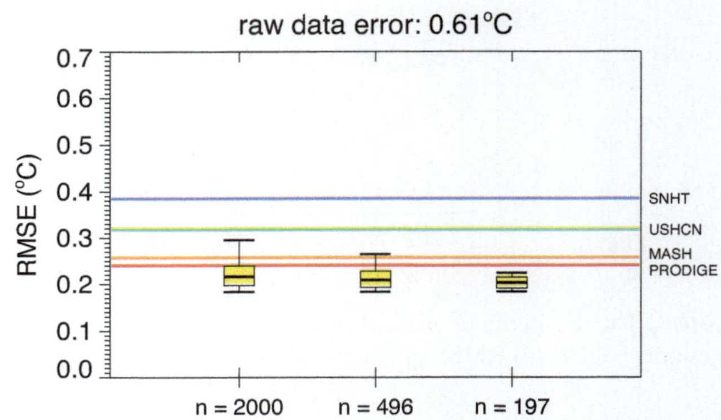

Figure 5. The same as Fig. 2, but for the RMSE of annual values.

(Fig. 3) the lead of ACMANT is even more dominant than in estimating trends of station data. On the other hand, the scattering of the results is great (except in the right column), and in the least successful experiments with ACMANT the residual error is as large as in the raw data.

The most important novelty of ACMANT is the harmonisation of the work between monthly and annual scales. As a consequence, the residual monthly RMSE (Fig. 4) is always the best with ACMANT, more precisely, the worst result among the 2000 experiments with ACMANT equals to the result of the second best method (PRODIGE). The residual annual RMSE (Fig. 5) has similar features than the previous efficiency measures, i.e. the results with ACMANT are the best except for a few experiments with suboptimal parameter values in ACMANT.

4 Discussion and conclusions

The results of the moving parameter experiments prove that the good performance of ACMANT late reported in Venema et al. (2012) is not for the overfitting of its parameters, since the performance has low sensitivity to the parameterisation. The high performance is a stable characteristic of ACMANT and it is a consequence of its good methodological properties.

The results indicate that after excluding some suboptimal parameter values, ACMANT always performs better than any other homogenisation methods in homogenising monthly temperatures. However, this result must be treated with some reservations, since (i) Twenty networks is a relatively small sample, because the efficiency of homogenising station series is strongly interdependent within networks; (ii) The HOME benchmark dataset captures well some characteristics of observed temperature data, but it does not mean that the frequency and magnitude distribution of inhomogeneities and some other characteristics are the same in each observed temperature dataset; (iii) Not all the known homogenisation methods are tested in HOME and even the tested methods could produce better results in repeated examinations either due to their new developments, or accidentally. Note here that relying on the experiences of HOME, the new homogenisation method HOMER has been developed (Mestre et al., 2013). Moreover, a new, automatic version of MASH has been reported, as well as Climatol and RHTest also have newly developed versions (www.climatol.eu/DARE).

For the reasons listed above, further tests are needed to confirm or deny the leading position of ACMANT. However, some other open questions seem to be even more important than the rank order among the best homogenisation methods. The efficiency of any homogenisation method strongly depends on the number and spatial correlations of time series in networks and on the frequency and other characteristics of inhomogeneities as well. When one or more of these characteristics are unfavourable, homogenisation could even worsen the data quality. For instance, it happened to the monthly precipitation data in the HOME experiments. The lesson of that failure is not that monthly precipitation data cannot be homogenised, but it is that the successful homogenisation has also some conditions others than applying good statistical methods. Our present knowledge is yet limited in this field, what means that at present we cannot quantify the necessary conditions. Similarly, we cannot separate confidentially the tasks for automatic homogenisation from the ones for manual homogenisation with intensive metadata use. We have the general knowledge that while huge datasets can hardly be homogenised manually, in case of small number of time series the human control is always advisable. However, this qualitative knowledge is not always sufficient to make optimal decisions in the selection of methods, therefore further tests are necessary to widen our knowledge in this field. Tests with automatic homogenisation methods can be executed relatively easily and their results are partly applicable also to non-automatic methods. Domonkos (2013) describes the objectives and the expected benefits of various kinds of efficiency tests, while Guijarro (2012) shows examples of comparative efficiency tests for various automatic homogenisation methods.

Our main conclusions are as follows:

- The performance of ACMANT has low sensitivity to the parameterisation of the method.
- ACMANT is one of the best methods for homogenising monthly temperatures, and it is likely the best method for homogenising large and spatially dense networks of temperature data.
- Although ACMANT is only for monthly temperature data, its development will help in the future to improve the efficiency of homogenisation also in other climatic variables and in other time resolutions.
- Further efficiency tests are needed to quantify the connections between dataset properties and method performance. First the automatic methods (such as e.g. ACMANT) must be tested, and their results will be partly applicable also for non-automatic methods.

Acknowledgements. The research was supported by the EURO4M FP7-SPACE-2009-1/242093.

Edited by: M. Brunet-India

References

Aguilar, E., Auer, I., Brunet, M., Peterson, T. C., and Wieringa, J.: WMO Guidelines on climate metadata and homogenization, WCDMP-No. 53, WMO-TD No 1186, WMO, Geneva, 2003.

Auer, I., Böhm, R., Jurković, A., Orlik, A., Potzmann, R., Schöner, W., Ungersböck, M., Brunetti, M., Nanni, T., Maugeri, M., Briffa, K., Jones, P., Efthymiadis, D., Mestre, O., Moisselin, J.-M., Begert, M., Brazdil, R., Bochnicek, O., Cegnar, T., Gajić-Čapka, M., Zaninović, K., Majstorović, Ž., Szalai, S., Szentimrey, T., and Mercalli, L.: A new instrumental precipitation dataset for the greater Alpine region for the period 1800–2002, Int. J. Climatol., 25, 139–166, 2005.

Caussinus, H. and Mestre, O.: Detection and correction of artificial shifts in climate series, J. Roy. Stat. Soc., C53, 405–425, doi:10.1007/s11200-009-0015-9, 2004.

Craddock, J. M.: Methods of comparing annual rainfall records for climatic purposes, Weather, 34, 332–346, 1979.

Domonkos, P.: Adapted Caussinus-Mestre Algorithm for Networks of Temperature series (ACMANT), Int. J. Geosci., 2, 293–309, doi:10.4236/ijg.2011.23032, 2011a.

Domonkos, P.: Efficiency evaluation for detecting inhomogeneities by objective homogenisation methods, Theor. Appl. Climatol., 105, 455–467, doi:10.1007/s00704-011-0399-7, 2011b.

Domonkos, P.: Measuring performances of homogenisation methods, Időjárás, 117, 91–112, in press, 2013.

Domonkos, P. and Štěpánek, P.: Statistical characteristics of detectable inhomogeneities in observed meteorological time series, Stud. Geophys. Geod., 53, 239–260, doi:10.1111/j.1467-9876.2004.05155.x, 2009.

Domonkos, P., Venema, V., Auer, I., Mestre, O., and Brunetti, M.: The historical pathway towards more accurate homogenisation, Adv. Sci. Res., 8, 45–52, doi:10.5194/asr-8-45-2012, 2012.

Guijarro, J. A.: After COST/ES0601: Automatic comparison of series homogenization methods, 8° Congreso de Associación Española de Climatológia, 25–28 September 2012, Salamanca, 2012.

McCarthy, M. P., Titchner, H. A., Thorne, P. W., Tett, S. F. B., Haimberger, L., and Parker, D. E.: Assessing bias and uncertainty in the HadAT adjusted radiosonde climate record, J. Climate, 21, 817–832, 2008.

Menne, M. J. and Williams Jr., C. N.: Homogenization of temperature series via pairwise comparisons, J. Climate, 22, 1700–1717, doi:10.1175/2008JCLI2263.1, 2009.

Mestre, O., Domonkos, P., Picard, F., Auer, I., Robin, S., Lebarbier, E., Böhm, R., Aguilar, E., Guijarro, J., Vertacnik, G., Klancar, M., Dubuisson, B., and Stepanek, P.: HOMER: HOMogenisation softwarE in R – methods and applications, Időjárás, 117, 47–67, in press, 2013.

Szentimrey, T.: Multiple Analysis of Series for Homogenization (MASH), in: Second Seminar for Homogenization of Surface Climatological Data, edited by: Szalai, S., Szentimrey, T., and

Szinell, Cs., WCDMP 41, WMO-TD 962, WMO, Geneva, 27–46, 1999.

Titchner, H. A., Thorne, P. W., McCarthy, M. P., Tett, S. F. B., Haimberger, L., and Parker, D. E.: Critically reassessing tropospheric temperature trends from radiosondes using realistic validation experiments, J. Climate, 22, 465–485, doi:10.1175/2008JCLI2419.1, 2009.

Venema, V. K. C., Mestre, O., Aguilar, E., Auer, I., Guijarro, J. A., Domonkos, P., Vertacnik, G., Szentimrey, T., Stepanek, P., Zahradnicek, P., Viarre, J., Müller-Westermeier, G., Lakatos, M., Williams, C. N., Menne, M. J., Lindau, R., Rasol, D., Rustemeier, E., Kolokythas, K., Marinova, T., Andresen, L., Acquaotta, F., Fratianni, S., Cheval, S., Klancar, M., Brunetti, M., Gruber, C., Prohom Duran, M., Likso, T., Esteban, P., and Brandsma, T.: Benchmarking homogenization algorithms for monthly data, Clim. Past, 8, 89–115, doi:10.5194/cp-8-89-2012, 2012.

Williams, C. N., Menne, M. J., and Thorne, P. W.: Benchmarking the performance of pairwise homogenization of surface temperatures in the United States, J. Geophys. Res.-Atmos., 117, D5, doi:10.1029/2011JD016761, 2012.

Comparison of Large Eddy Simulations of a convective boundary layer with wind LIDAR measurements

J. G. Pedersen[1], **M. Kelly**[1], **S.-E. Gryning**[1], **R. Floors**[1], **E. Batchvarova**[1,2], **and A. Peña**[1]

[1]Department of Wind Energy, Risø Campus, Technical University of Denmark, Roskilde, Denmark
[2]National Institute of Meteorology and Hydrology, Sofia, Bulgaria

Correspondence to: J. G. Pedersen (jegp@dtu.dk)

Abstract. Vertical profiles of the horizontal wind speed and of the standard deviation of vertical wind speed from Large Eddy Simulations of a convective atmospheric boundary layer are compared to wind LIDAR measurements up to 1400 m. Fair agreement regarding both types of profiles is observed only when the simulated flow is driven by a both time- and height-dependent geostrophic wind and a time-dependent surface heat flux. This underlines the importance of mesoscale effects when the flow above the atmospheric surface layer is simulated with a computational fluid dynamics model.

1 Introduction

Knowledge about the flow of air in the atmospheric boundary layer (ABL) is an important issue in many areas of society today. These include energy production, pollutant dispersion, transportation and weather forecasting.

Accurate high-resolution observations of the ABL are, however, difficult to obtain. Tower-based measurements usually cover only the atmospheric surface layer (ASL), and are expensive to establish. Radio soundings generally provide a nearly instantaneous profile of the entire ABL, but only infrequently. Modern remote sensing systems such as LIDARs can provide information about the ABL up to a few kilometers at high resolution under favorable conditions (Floors et al., 2011), but the technology is still maturing and the instruments suffer from periods of malfunction and accuracy issues.

These difficulties make computational fluid dynamics (CFD) modeling of the ABL a tempting alternative to field experiments. CFD models allow one to do numerical experiments in a controlled environment and to study the influence of individual physical parameters on the atmospheric flow. For this reason, CFD models have been used for decades in research to simulate surface-layer flows, generally without considering mesoscale meteorological phenomena. Above the ASL however, mesoscale effects become increasingly important and should not be neglected (Kumar et al., 2010; Conzemius and Fedorovich, 2010; Gryning et al., 2007; Brown et al., 2006). One important issue is therefore how well CFD models represent the real atmosphere.

In this paper we compare LIDAR measurements with Large Eddy Simulations (LESs) of a daytime/convective ABL. By looking at a specific set of LIDAR measurements from the Danish National Test Station for Wind Turbines, we illustrate how the agreement between observations and LES can depend on mesoscale effects.

The measurements cover a five-hour period on 6 May 2010, during which the ABL was characterized by turbulence generated by both surface stress and buoyancy. Details of the measurements and the numerical model are given in Sect. 2. Results are presented and discussed in Sects. 3 and 4, respectively. Section 5 provides a conclusion of our findings.

2 Method

2.1 Observations

During the period from the beginning of April 2010 to the end of March 2011, the ongoing tower measurements at the National Test Station for Wind Turbines at Høvsøre, Denmark, were supplemented by LIDAR measurements as part of the Tall Wind project. A pulsed long range wind LIDAR

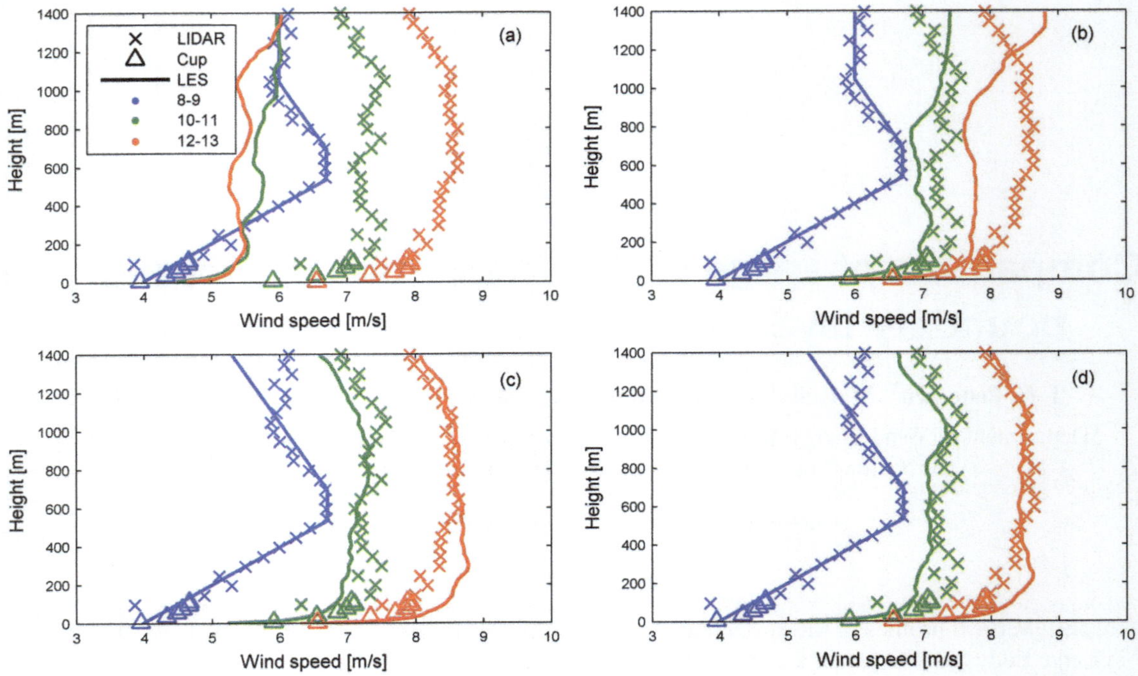

Figure 1. Wind speeds from simulations A, B, C and D compared to measurements.

Table 1. Applied forcing in simulations A, B, C and D.

Simulation	Geostrophic wind	Surface heat flux
A	Constant	Constant
B	Time dependent	Constant
C	Time and height dependent	Constant
D	Time and height dependent	Time dependent

(WindCube™ WLS70 by LEOSPHERE) was used to measure the wind speed from a height of 100 m up to around 2000 m, with a vertical resolution of 50 m; for details see Floors et al. (2011). Measurements with a carrier to noise ratio (CNR) below −30 dB are discarded. Sonic and cup anemometers mounted on a meteorological mast provide flux and wind speed measurements from 10 m above ground level up to 100 m and 116.5 m, respectively.

The test station is located approximately 2 km inland from the west coast of Jutland. In this comparison we focus on the period between 08:00 and 13:00 CET on 6 May 2010. During this period it was partly cloudy, and the wind was from the east, varying between 80 and 100 degrees at a height of 10 m. The terrain east of the test station is rural, flat and homogeneous.

2.2 Model

LES is one of the most powerful types of CFD modeling available today for studying turbulent flows. With this technique the Navier-Stokes equations are solved explicitly for all fluctuations at scales larger than the effective grid resolution. Smaller eddies and their effects on the flow are parameterized through a subgrid-scale (SGS) model.

We use the NCAR pseudo-spectral LES code descended from Moeng (1984) and Sullivan and Patton (2008), employing the SGS model of Sullivan et al. (1994). The code assumes a dry atmosphere and horizontal homogeneity. The simulated flow is driven by a pressure gradient across the computational domain, specified directly as the geostrophic wind components U_g and V_g, and a heat flux imposed at the lower boundary. The turbulent motions are initialized by adding small random perturbations to the initial fields of horizontal wind speed and potential temperature near the surface.

We conducted four different types of simulations (A, B, C and D), trying to reproduce the measured wind profiles. An overview of the simulations is given in Table 1.

All simulations were made in a domain of 6 km extent in the horizontal directions, and 2 km in the vertical direction. This corresponds approximately to 6 and 2 times the observed boundary layer height. The domain was covered by 256 grid points in all three directions, giving a resolution of 23 m in the horizontal directions and 8 m in the vertical direction. A sensitivity test was carried out by repeating simulation D with 64^3, 128^3, 256^3 and 512^3 grid points (D1, D2, D3 and D4). The mean wind profiles and the boundary layer growth of simulation D3 and D4 are very similar indicating that 256^3 grid points are sufficient to obtain

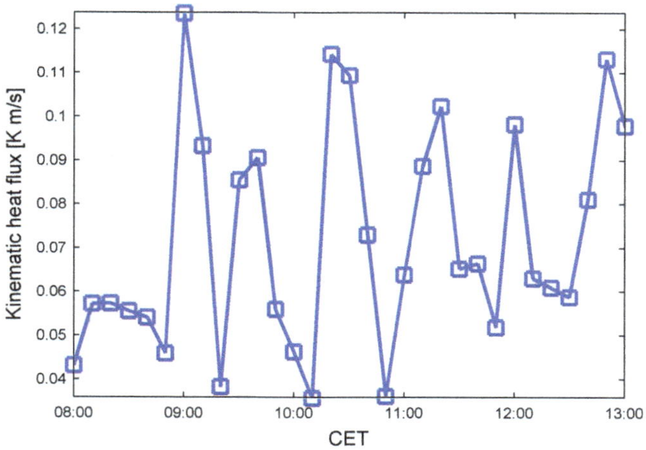

Figure 2. Measured kinematic heat flux at 10 m.

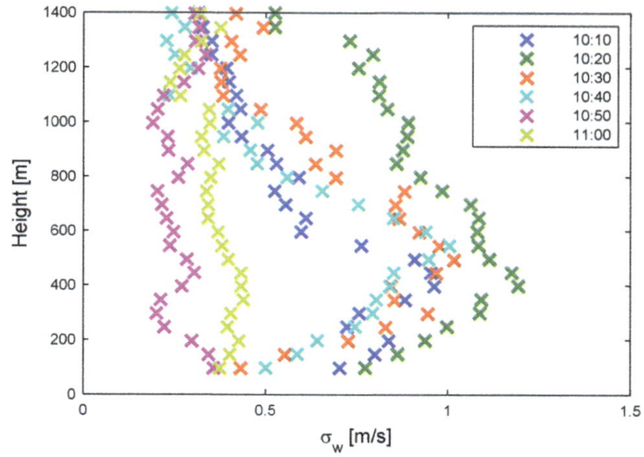

Figure 3. Profiles of σ_w measured by the LIDAR.

a resolution-independent solution. The number of surface-layer grid points in simulation D3 and the resulting values of the effective LES Reynolds number and the resolved-to subgrid-stress ratio were furthermore sufficiently high to place the simulation in or near the high accuracy zone as discussed by Brasseur and Wei (2010).

Each simulation ran for 4.5 h and was initialized at 08:30 CET with a wind speed profile fitted to the mean of the measured profiles between 08:00 and 09:00 CET up to 1000 m. Above this height the initial profile was set to follow the prescribed geostrophic wind. In simulations A and B the geostrophic wind was specified to be constant with height and was initially set to $6\,\mathrm{m\,s^{-1}}$. In simulations C and D the geostrophic wind was set to decrease linearly with height throughout the domain with a rate of $2\times10^{-3}\,\mathrm{s^{-1}}$, and in simulations B, C and D the geostrophic wind was furthermore set to increase linearly with time at a rate of $2\times10^{-4}\,\mathrm{m\,s^{-2}}$ at all heights. The geostrophic wind was updated at the beginning of each time step. The specified time and height dependencies were inferred from the measured wind speed profiles from approximately 1200 m and above.

The time-dependent surface heat flux in simulation D was updated every 10 min following the measured kinematic heat flux at 10 m. In A, B and C a constant value of $7.5\times10^{-2}\,\mathrm{K\,m\,s^{-1}}$ was used. In all the simulations, the initial potential temperature was constant at 280 K up to 600 m and then increasing with $3.4\times10^{-3}\,\mathrm{K\,m^{-1}}$ to the top of the domain.

3 Results

Wind speeds from simulations A, B, C and D (Table 1) are compared to measured wind speeds in Fig. 1 (a, b, c and d). The blue, green and red markers represent hourly averaged wind speeds from the periods 08:00 to 09:00, 10:00 to 11:00 and 12:00 to 13:00 CET. Crosses are LIDAR measurements and triangles are cup anemometer measurements. The solid lines are wind profiles from the simulations. The blue lines are the initial velocity profiles, while the green and the red are hourly averaged profiles between 1.5 and 2.5 and 3.5 and 4.5 h of simulation time, taken from a position in the middle of the computational domain.

Figure 2 shows the kinematic heat flux measured at 10 m between 08:00 and 13:00 CET. Figure 3 shows LIDAR profiles of the standard deviation of the vertical wind velocity (σ_w) over 10 min periods between 10:00 and 11:00 CET. Figure 4 (a and b) shows corresponding profiles from simulations C and D.

4 Discussion

The mean wind speed profiles of simulation A and B show poor agreement with the measured profiles. The profiles of simulation C and D are quite similar and show much better agreement with the measurements. This suggests that in this specific case, the decrease of the geostrophic wind with height and increase with time have a significant impact on the mean wind profiles. The fluctuations of the surface heat flux, on the other hand, seem only to have a small impact.

Regarding the σ_w-profiles, the varying heat flux has a considerable influence. In both simulations C and D the shape of the profiles are quite similar to the measured profiles, but the profiles of simulation D are spread over a wider range of values. This additional scatter gives results more closely resembling the measurements. At high altitudes however, the measured values are generally higher than the simulated values. The fluctuations of the observed heat flux are likely caused by varying cloud cover. In the model we only apply this through the varying surface heat flux in simulation D, but in reality it will also have other effects; for instance the entrainment of air from above the ABL may be influenced. It should be noted that variances measured by LIDAR differ

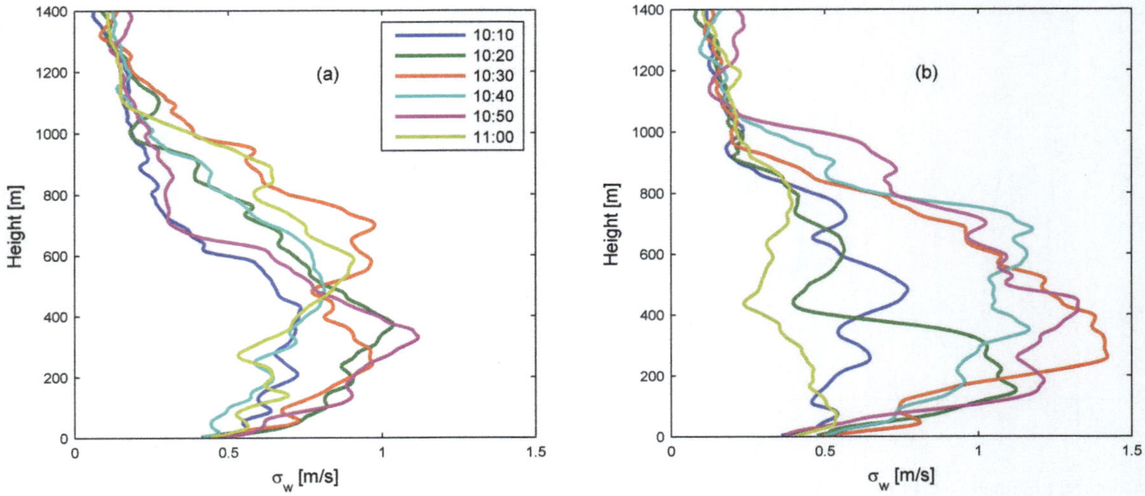

Figure 4. Profiles of σ_w from simulation C **(a)** and D **(b)**.

somewhat from the true variances, as discussed by Sathe et al. (2011). LIDAR measurements of mean wind speeds are generally quite accurate over homogeneous terrain.

5 Conclusions

By comparing four different LESs of a convective ABL to a specific set of LIDAR measurements, we have illustrated the importance of considering mesoscale phenomena when using a CFD model to simulate the flow above the ASL. In the specific case studied here, the simulation with a geostrophic wind varying with height and time gave much better agreement with the measured mean wind speed profiles than the simulations where geostrophic wind variations were neglected. Furthermore, we have shown that including fluctuations of the surface heat flux in the CFD model has little effect on the hourly mean wind speed profiles but it improves the prediction of the variability of the 10 min σ_w-profiles.

Acknowledgements. The work is supported by the Danish Research Agency Strategic Research Council (Sagsnr. 2104-08-0025) "Tall wind project" and is related to COST Action ES1002 (WIRE) and the Nordforsk CRAICC project.

Edited by: L. Wald

References

Brasseur, J. G. and Wei, T.: Designing large-eddy simulation of the turbulent boundary layer to capture law-of-the-wall scaling, Phys. Fluids., 22, 1–21, 2010.

Brown, A. R., Beljaars, A. C. M., and Hersbach, H.: Errors in parametrizations of convective boundary-layer turbulent momentum mixing, Q. J. Roy. Meteor. Soc., 132, 1859–1876, 2006.

Conzemius, R. and Fedorovich, E.: Large eddy simulation of realistic wind fields in daytime atmospheric boundary layer, in: Proceedings of the 5th International Symposium on Computational Wind Engineering, Chapel Hill, North Carolina, USA, 22–27 May 2010.

Floors, R., Batchvarova, E., Gryning, S.-E., Hahmann, A. N., Peña, A., and Mikkelsen, T.: Atmospheric boundary layer wind profile at a flat coastal site – wind speed lidar measurements and mesoscale modeling results, Adv. Sci. Res., 6, 155–159, doi:10.5194/asr-6-155-2011, 2011.

Gryning, S.-E., Batchvarova, E., Brümmer, B., Jørgensen, H., and Larsen, S.: On the extension of the wind profile over homogeneous terrain beyond the surface boundary layer, Bound.-Lay. Meteorol., 124, 251–268, 2007.

Kumar, V., Svensson, G., Holtslag, A. A. M., Meneveau, C., and Parlange, M. B.: Impact of surface flux formulations and geostrophic forcing on large-eddy simulations of diurnal atmospheric boundary layer flow, J. Appl. Meteorol. Clim., 49, 1496–1516, 2010.

Moeng, C.-H.: A large-eddy-simulation model for the study of planetary boundary-layer turbulence, J. Atmos. Sci., 41, 2052–2062, 1984.

Sathe, A., Mann, J., Gottschall, J., and Courtney, M. S.: Can wind lidars measure turbulence?, J. Atmos. Ocean. Tech., 28, 853–868, 2011

Sullivan, P. P. and Patton, E. G.: A highly parallel algorithm for turbulence simulations in planetary boundary layers: Results with meshes up to 1024^3, 18th Conference on Boundary Layer and Turbulence, Stockholm, Sweden, 2008.

Sullivan, P. P., McWilliams, J. C., and Moeng, C.-H.: A subgrid-scale model for large-eddy simulation of planetary boundary-layer flows, Bound.-Lay. Meteorol., 71, 247–276, 1994.

23

Synoptic-mesoscale analysis and numerical modeling of a tornado event on 12 February 2010 in northern Greece

I. T. Matsangouras[1], P. T. Nastos[1], and I. Pytharoulis[2]

[1]Laboratory of Climatology and Atmospheric Environment, Faculty of Geology and Geoenvironment, National and Kapodistrian University of Athens, Greece

[2]Department of Meteorology and Climatology, School of Geology, Aristotle University of Thessaloniki, Greece

Abstract. Tornadoes are furious convective weather phenomena, with the maximum frequency over Greece during the cold period (autumn, winter).This study analyzes the tornado event that occurred on 12 February 2010 near Vrastama village, at Chalkidiki's prefecture, a non urban area 45 km southeast of Thessaloniki in northern Greece. The tornado developed approximately between 17:10 and 17:35 UTC and was characterized as F2 (Fujita Scale). The tornado event caused several damages to an industrial building and at several olive-tree farms.

A synoptic survey is presented along with satellite images, radar products and vertical profile of the atmosphere. Additionally, the nonhydrostatic WRF-ARW atmospheric numerical model (version 3.2.0) was utilized in analysis and forecast mode using very high horizontal resolution (1.333 km × 1.333 km) in order to represent the ambient atmospheric conditions. A comparison of statistical errors between WRF-ARW forecasts and ECMWF analysis is presented, accompanied with LGTS 12:00 UTC soundings (Thessaloniki Airport) and forecast soundings in order to verify the WRF-ARW model. Additionally, a comparison between WRF-ARW and ECMWF thermodynamic indices is also presented. The WRF-ARW high spatial resolution model appeared to simulate with significant accuracy a severe convective event with a lead period of 18 h.

1 Introduction

Tornadoes are one of nature's most extreme and violent phenomena with high localized damage potential associated with severe convective storms. Extremely high winds, inside and around the tornado's funnel, are able to cause destruction even in strong buildings and in many cases loss of life. Despite the significant improvement in understanding and forecasting tornadoes (short term), owning to both theoretical insight progress and the use of Doppler Radars (Holum et al., 1997; Bluestein et al., 1997), their small spatial scale and short lifetime seem to be the determinant factors about the various natural mechanisms that produce these intensely swirling air columns (Church et al., 1993). Tornadoes occur in many parts of the world (Fujita, 1973) and several publications during the last two decades indicate their occurrence in many European countries (Dessens, 1984; Dessens and Snow, 1987; Gayá et al., 1997; Paul, 1999; Reynolds, 1999; Dotzek, 2001; Holzer, 2001; Bertato et al., 2003; Dotzek, 2003; Giaiotti et al., 2007; Giaiotti and Stel, 2007).

However, remarkable tornado events in Greece occurred between 1900 and 1999 have been reported only in limited academic research (Livadas, 1954; Kanellopoulou, 1977; Ogden, 1984; Flocas, 1992; Sioutas, 2002; Nastos and Matsangouras, 2010). Matsangouras and Nastos (2010) as well as Matsangouras et al. (2010) focus on exceptional Greek tornado events that took place after 2000, such as the tornado in Athens' international airport "Eleftherios Venizelos" and the one registered in Varda, at Elias prefecture western Greece. In due course, a systematic tornado recording research initiated by the Laboratory of Climatology and Atmospheric Environment (LACAE) of National and Kapodistrian University in Athens (NKUA) 4 yr ago. For academic purposes, an online tornado report system (http://tornado.geol.uoa.gr) was established developing an online tornado database for

(a) (b) (c) (d)

Figure 1. Tornado's damages near Vrastama Chalkidiki on 12 February 2010 (Photos: http://forum.snowreport.gr/forum_posts.asp?TID=23403&PN=1).

Greece that records general public data. In this study a thorough analysis of the catastrophic tornado event near Vrastama village of Chalkidiki (Northern Greece) is carried out. It is apparent that this tornado was not the first in history over Greece and definitely it will not be the last one. For the first time in Greece a high spatial WRF-ARW simulation was carried out in order to investigate the capabilities of this NWP model on severe convective events, such as this tornado event.

2 Tornado survey

The tornado occurred 2.5 km south of Chalkidiki's Vrastama village, a non urban area 45 km southeast of Thessaloniki in northern Greece (Lat: 40.36, Lon: 23.54; Figs. 2d, 4), on 12 February 2010. The system developed approximately between 17:10 and 17:35 UTC, caused significant damages at a green olive processing unit and on several olive oil farms. Based on the caused damages (Fig. 1), it could be characterized at least as an F2 (Fujita scale) or T4–T5 of TORRO scale (Meaden, 1976). The tornado carried away by 200-400 m several metallic parts of the oil processing building while scattered broken branches of olives oil trees were found along the tornado path.

3 Synoptic survey

The upper air analyses showed that during the day of the event a closed cyclonic circulation that prevailed at the isobaric level of 500 hPa over central Italy (Fig. 2a) was associated with cold air masses (−35 °C). The closed low induced a SW upper air stream over the area of interest at 12:00 UTC (ECMWF ERA-Interim data used in this study have been obtained from the ECMWF data server). The cyclonic circulation extended throughout the lower troposphere. At Mean Sea-Level (MSL), the low was located over central Adriatic Sea at 12:00 UTC (not shown) and was associated with a cold front (over western Greece) propagating eastwards. At 18:00 UTC the UKMO MSL pressure analysis shows that the cold front extended from northern Greece to Peloponnesus (south Greece) and northern Libya through the area of interest (Fig. 2b). MSG-2 satellite imagery at Ch9 (10.8 µm)

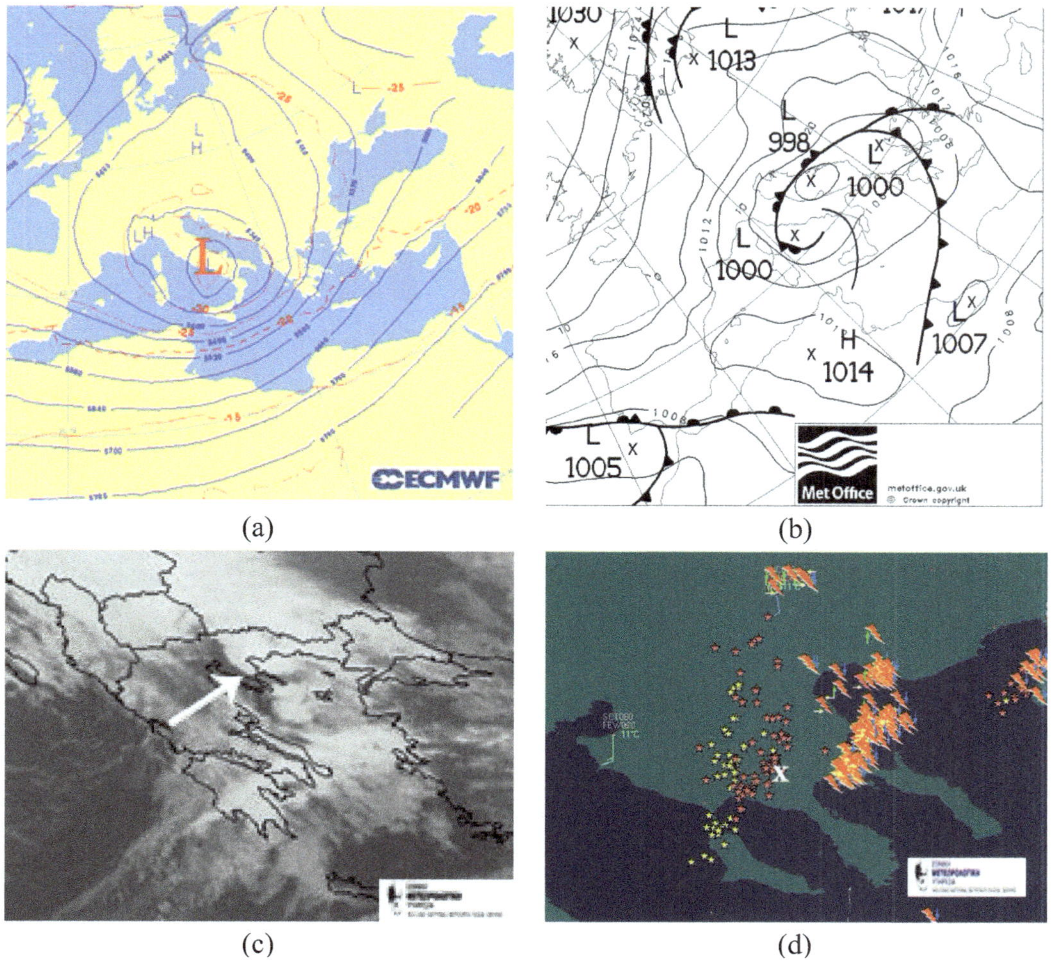

Figure 2. ECMWF Re-analysis at 12:00 UTC **(a)**, UKMO MSL analysis at 18:00 UTC **(b)**, IR satellite image (Ch9) at 17:30 UTC (white arrow indicates the associated storm) **(c)** and lightning activity at 17:00 UTC (yellow stars), 17:30 UTC (red stars) and 18:00 UTC (red spikes) from HNMS **(d)**, on 12 February 2010.

revealed convective clouds (Cb) associated with the cold front affecting the area of interest at 17:30 UTC (Fig. 2c). At the time of event, the nearest meteorological station of Thessaloniki-LGTS (~45 km NW of the event) reported significant weather phenomena such as Cb.

The thermodynamic environment about the closest upper air station (LGTS) at 12:00 UTC (~4 h prior to the event) did not reveal any significant storm index (K Index: 21.6 °C, Showalter Index (SI): 4.52 °C, Severe Weather Threat Index (SWEAT): 69.76, Convective Available Potential Energy (CAPE): 14.12 ($J\,kg^{-1}$) and Lifted Index (LIFT): 2.39 °C. Figure 2d illustrates the spatiotemporal variability of lightings from 17:00 to 18:00 UTC, which were acquired by Hellenic National Meteorological Service (HNMS) The white "X" denotes the location of the tornado while the yellow stars, the red stars and the red spikes show the lightning at 17:00, 17:30 and 18:00 UTC, respectively.

The images of "Aeolus 13" which is the closest radar of the HNMS Radar Network "Alkioni" depicted the convective cloud system propagating eastwards. Aeolus 13 that is located at LGTS, was established in 1980 and its type is WSR-74S (S Band wavelength 10 cm). During 2007 its receiver and software (IRIS) were both upgraded. In Fig. 3 the CAPPI (Constant Altitude Plan Position Indicator) at 1km height (Fig. 3a–c) and MAX (MAXimum Reflectivity) products (Fig. 3d–f) are presented in a 15 min time step. At 17:30 UTC maximum reflectivities of 44-50 dbz appeared in the area of interest.

4 WRF numerical model simulation

The WRF-ARW V3.2.0 nonhydrostatic numerical model (Skamarock et al., 2008; Wang et al., 2010) was used in order to simulate the weather conditions at 12 February 2010. Three 2-way interactive nested domains were

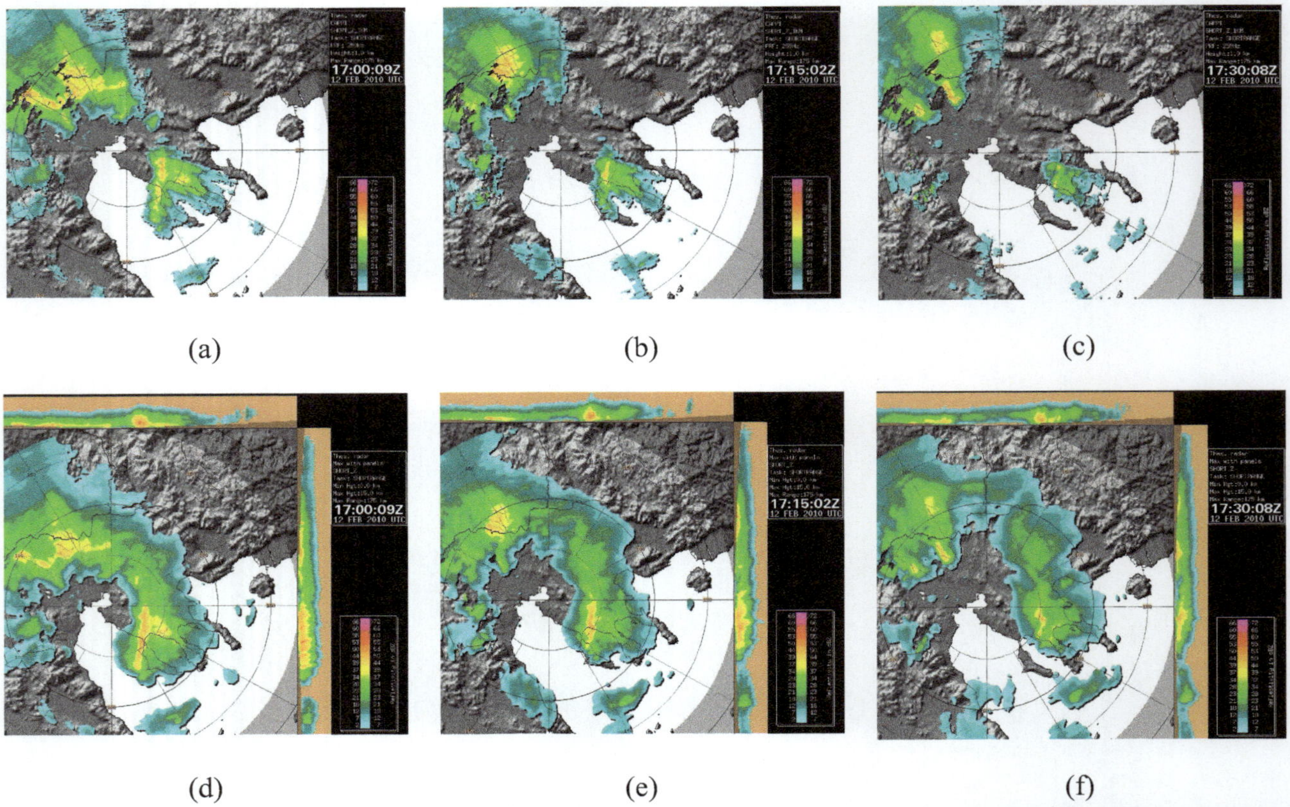

Figure 3. Radar images from LGTS Radar: CAPPI at 1 km height in 15 min time step **(a–c)** and MAX (MAXimum Reflectivity) in 15 min time step **(e–f)**.

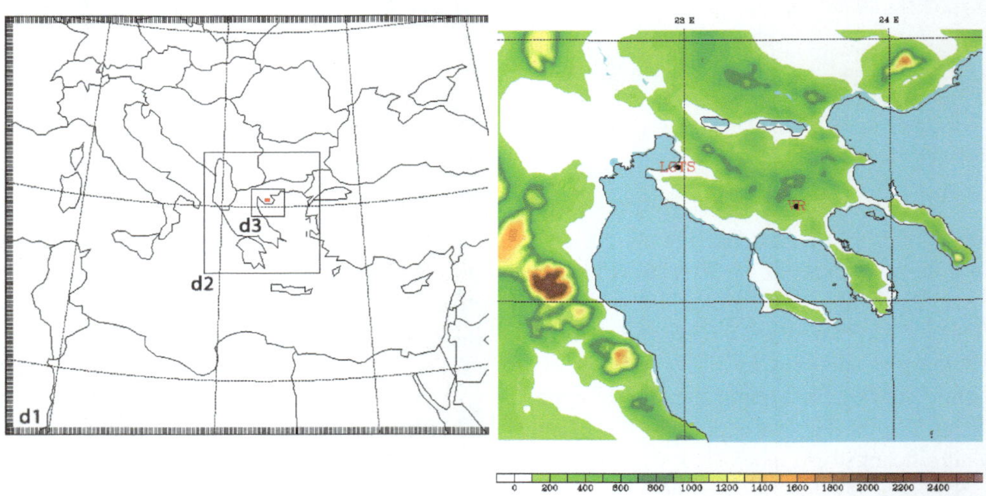

Figure 4. The three domains (D1-D2-D3) of WRF runs (the red spot indicates the tornado event area) **(a)** and the topography (m) of domain D3 **(b)**; LGTS is the airport of Thessaloniki and VR stands for Vrastama.

utilized (Fig. 4a). The spatial resolution of the model was 12km for D1 (over central and eastern Mediterranean Sea), 4 km for D2 (over Greece) and 1.333 km for D3 (over northern Greece-Chalkidiki; Fig. 4b). The runs started at 00Z on 12 February 2010 and the forecast horizon was 24 h.

The ECMWF analysis and 3-hourly forecasts of the 00:00 UTC cycle of 12 February 2010 with a spatial resolution of 0.125×0.125 degrees were used as initial and lateral boundary conditions, respectively, for the domain D1. The sea-surface temperatures were also derived from the ECMWF

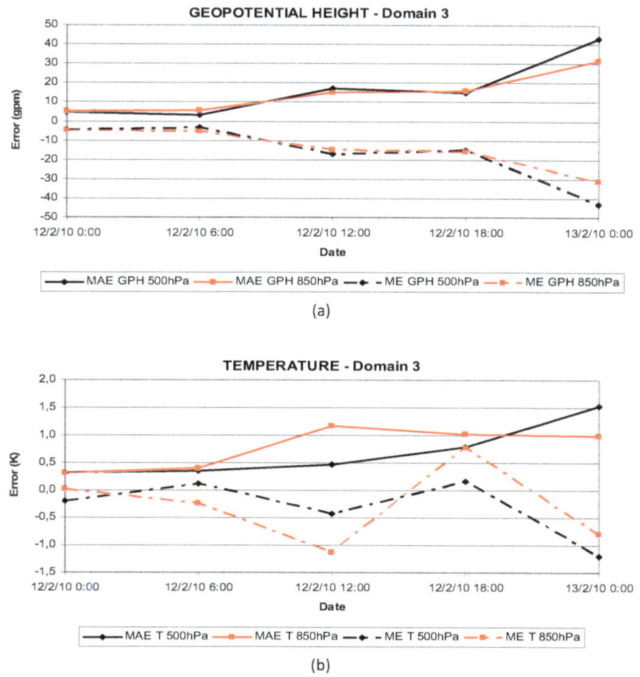

Figure 5. All statistics (Mean Absolute Error – MAE and Mean Error – ME) were calculated in the inner domain (D3) through the comparison of ECMWF gridded analyses and WRF forecasts for Geopotential Height at 500 and 850 hPa **(a)** and for temperature at 500 and 850 hPa **(b)**.

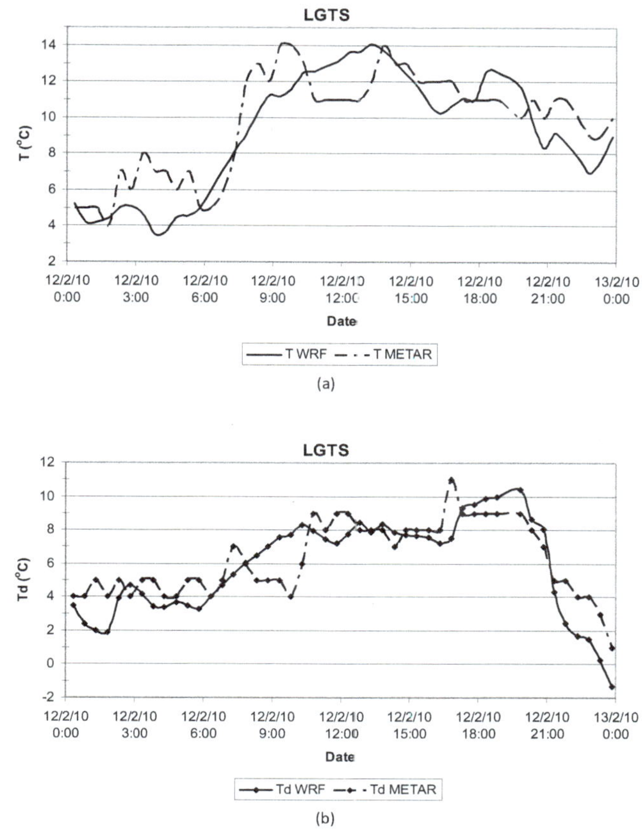

Figure 6. Time series of **(a)** Temperature and **(b)** Dew-point Temperature observations (METAR) and WRF forecasts at LGTS (Time in UTC).

analysis and were kept fixed to their initial values throughout the simulations. In the vertical, 39 sigma levels (up to 50 hPa) with increased resolution in the boundary layer were used by all nests. The GFDL scheme, the Monin-Obukhov (Eta), the Mellor-Yamada-Janjic level 2.5 and the NOAH Unified model were used in all nests to represent longwave/shortwave radiation, surface layer, boundary layer and soil physics, respectively. Cumulus convection was parameterized only in nests D1 and D2 by the Kain-Fritsch (new Eta) scheme. The Ferrier (new Eta) microphysical scheme is used by several operational NWP models as SKIRON, Eta, WRF-NMM and for this reason we stacked to this microphysical scheme to simulate our case study.

5 Results and discussion

With continuously improving computer technology, high-resolution model simulations can be conducted to study severe weather phenomena such as tornadoes in order to determine how far in advance these processes can be predicted. In our case study a tornado formed near Vrastama village, (Chalkidiki's prefecture) in northern Greece, between 17:10–17:30 UTC on 12 February 2010. The tornado intensity could be characterized as F2 (Fujita scale) based on the significant damages. The tornado formation was associated with the passage of a cold front propagating eastward-northeastward.

In order to evaluate the WRF forecasts we proceed through the comparison of ECMWF gridded analyses and WRF forecasts. All statistics (Mean Absolute Error – MAE and Mean Error – ME) were calculated in the inner domain (D3) for Geopotential Height of 500 hPa and 850 hPa, as well as the temperature for these levels (Fig. 5). The WRF forecasts were interpolated at the locations of ECMWF data (0.125 × 0.125 deg.) and all statistics were resulted from a total of 234 pairs of forecast/analysis at each time. The WRF underestimated (∼15 gpm) the Geopotential Height (GH) at 500 and 850 hPa after 06:00 UTC (thus predicting a deeper approaching through), but did very well for both temperatures (T) at 500 and 850 hPa at 18:00 UTC.

Additionally to GH-T comparison of middle and lower atmosphere, we calculated the statistics (Mean Error, Mean Absolute Error and Root Mean Square Error) of Temperature (T) and Dew-Point Temperature (T_d) at LGTS on 12 February 2010 (Fig. 6). The LGTS is the closest (∼45 km) meteorological station of HNMS Meteorological Station Network. The statistics were derived from a total of 47 pairs of Metar

Table 1. Statistics (Mean Error, Mean Absolute Error and Root Mean Square Error) of Temperature (T) and Dew-Point Temperature (T_d) at LGTS on 12 Feb 2010. Negative values of Mean Error indicate underestimation by WRF.

	Mean Error	Mean Absolute Error	Root Mean Square Error
T	−0.61	1.50	1.78
T_d	−0.41	1.28	1.59

Observations and WRF forecasts at LGTS. The WRF output Temperature (T) after 06:00 UTC overestimated the real temperature and 3 h later underestimated it until 12:00 UTC. After this time it followed too close the real temperature. On the other hand, the Dew-Point Temperature (T_d) after 12:00 UTC did pretty well with the real Dew-Point Temperature. In sum the Temperature (T) was underestimated by WRF with a mean error of 0.6181 and Dew-Point (T_d) was also underestimated with a mean error of 0.4188 (Table 1). The statistical scores of temperature and dew-point temperature at the airport of Thessaloniki (LGTS) are lower than those exhibited by the statistical evaluation of modern high resolution NWP models in Greece. Gofa et al. (2008) evaluated the operational SKIRON (0.06° ×0.06°) and COSMO-GR (0.0625° × 0.0625°) forecasts produced by HNMS at 30 Greek stations (including LGTS) in 2007. They found RMSE between 1.9–2.5 °C for temperature and 2.3–3.3 °C for dew-point temperature in the first 24 forecast hours. Certainly, one must keep in mind that the scores of our study (a) correspond to a single high-impact weather event at one station and not to a long period with different weather types at several stations and (b) are based on METAR observations in which the temperatures are rounded to the nearest integer (introducing a maximum error of ±0.5 °C).

Moreover, we carried out a verification of Vertical profiles of predicted Temperature (T WRF), observed Temperature (T sounding), predicted Dew-Point Temperature (T_d WRF) and observed Dew-Point Temperature (T_d sounding) at LGTS on 12:00 UTC 12/02/10 (not shown). The predicted vertical profile of T (T WRF) was identical with the observed Temperature (T Sounding). The predicted T_d vertical profile was identical with the observed T_d (T_d Sounding) up to 600 hPa. Futher higher a predicted T_d was ranged ±10 °C from the observed T_d.

Lift Index, K Index and others thermodynamic indices were calculated for tornado's spot at 18:00 UTC. The indices were based on ECMWF analyses and WRF 18 h forecasts and carried through the RIP software calculations. The 18 h predicted indices via WRF were too close with the ECMWF analysis indices (Table 2) revealing conditions for convective event. In addition to the above predicted thermodynamic indices, the WRF-ARW model revealed the Bulk Richardson Number Shear (BRNSHR) over the area of interest with maximum values of 300–400 $m^2 s^{-2}$ at 17:30 UTC (Fig. 7a). The Energy Helicity Index (EHI) that combines helicity and instability into one number was also was depicted at 17:30 UTC over the area of tornado formation (Fig. 7b) as the storm was approaching. Similar conclusions are drawn from the investigation of the following forecasts products: the Storm-Relative Environmental Helicity (SREH, Sfc – 3 km) in Fig. 7c and Vorticity Generation Potential (VGP) in Fig. 7d.

Table 2. Thermodynamic indices for Vrastama tornado's formation point at 18:00 UTC of 12/2/2010, calculated by RIP software. LI is lifted index, K is K-index, TT is Total Totals Index, SWI is Showalter Index, CAPE is Convective Available Potential Energy (J kg^{-1}), SWEAT Sweat Index.

Index	ECMWF	WRF
LI	−1.4	−2.7
K	30	34
TT	57	58
SWI	−0.5	−1.8
CAPE	254	545
SWEAT	416	537

It is worth to mention that the WRF forecasts products of maximum reflectivity predicted for 17:40 UTC and 17:50 UTC values equal to 49.5 and 52 dBz respectively. These values are in agreement with the RADAR products with an offset of 10 min (MAX reflectivity 44–50 dBz).

It is apparent that the analysis described in this paper is based on a single case study, therefore it may not be appropriate to generalize the results obtained by the WRF-ARW. An ambitious project is in progress by the authors in order to present a thermodynamic climatology of indices for tornado formation over specific sub regions of Greece. However, this case study demonstrates the usability and capability of high resolution WRF-ARW model in simulating severe convective events with a lead period of 18 h.

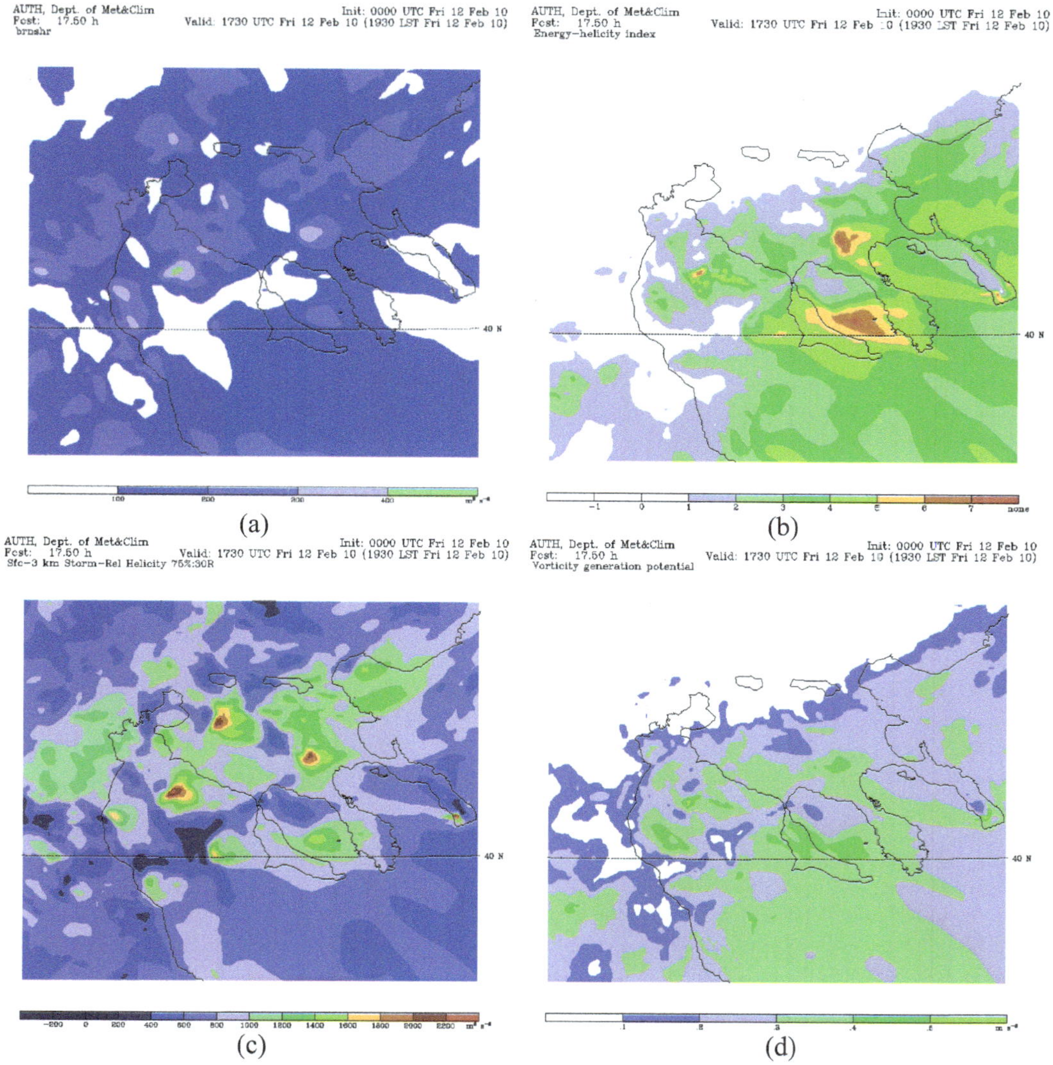

Figure 7. WRF-ARW outputs based on Ferrier microphysics scheme at 17:30 UTC on 12 February 2010. Bulk Richardson Number Shear ($m^2 s^{-2}$) **(a)**, Energy Helicity Index **(b)**, Storm-Relative Environmental Helicity ($m^2 s^{-2}$) **(c)** and Vorticity Generation Potential ($m s^{-2}$) **(d)**. Calculated using the RIP software.

Acknowledgements. The authors would like to acknowledge ECMWF, Eumetsat, MetOffice, HNMS for the processed data that were used in order to complete this study and NCAR for providing the WRF-ARW model. Pictures of the tornado provided by http://forum.snowreport.gr/ are thankfully acknowledged. Also authors need to express their thanks to the anonymous who reported the tornado event to our online reporting system http://tornado.geol.uoa.gr.

Edited by: F. Stel

References

Bertato, M., Giaiotti, D. B., Manzato, A., and Stel, F.: An interesting case of tornado in Friuli-Norteastern Italy, Atmos. Res., 67–68, 3–21, 2003.

Bluestein, H. B., Wesley, P. U., Dowell, D. C., Hutchinson, T. A., Crawford, T. M., and Wood, A. C.: Doppler radar analysis of the Northfield, Texas, tornado of 25 May 1994, J. Appl. Meteorol., 125, 212–230, 1997.

Church, C. R., Burgess, D., Doswell, C., and Davis-Jones, R. P. (Eds.): The Tornado: Its Structure, Dynamics, Prediction and Hazards, Geophys. Monogr., Vol. 79, American Geophysical Union, Washington DC, 637 pp., 1993.

Dessens, J.: Les Trombes en France. Climatologie et Caracteristiques Physiques, Lannemezan, France, 1984.

Dessens, J. and Snow, J. T .: Tornadoes in France, Weather Forecast., 4, 110–132, 1987.
Dotzek, N.: Tornadoes in Germany, Atmos. Res., 56, 233–251, 2001.
Dotzek, N.: An updated estimate of tornado occurrence in Europe, Atmos. Res., 67–68, 153–161, 2003.
Fujita, T. T.: Tornadoes around the world, Weatherwise, 26, 56–62, 1973.
Flocas, A. A.: Lessons of Meteorology and Climatology, 296–299, 1992 (in Greek).
Gayá, M., Ramis, C., Romero, R., and Doswell III, C. A.: Tornadoes in the Balearic Islands (Spain): Meteorological setting, Proceedings, INM/MO International Symposium on Cyclones and Hazardous Weather in the Mediterranean (Palma de Mallorca, Spain), Universitat de les Isles Baleares, 525–534, 1997.
Giaiotti, D. B. and Stel, F.: A multiscale observational case study of an isolated tornadic supercell, Atmos. Res., 83(2–4), 152–161, 2007.
Giaiotti, D. B., Giovannoni, M., Pucillo, A., and Stel, F.: The climatology of tornadoes and waterspouts in Italy, Atmos. Res., 83, 534–541, 2007.
Gofa, F., Pytharoulis, I., Andreadis, T., Papageorgiou, I., Fragkouli, P., Louka, P., Evgoustoglou, E., and Tyrli, V.: Evaluation of the operational NWP forecasts of the Hellenic National Meteorological Service, Proceedings of the 9th Conference on Meteorology, Climatology and Atmospheric Physics, 28–31 May, Thessaloniki, 51–58, 2008.
Holzer, A. M.: Tornado climatology of Austria, Atmos. Res., 56, 203–211, 2001.
Holum, D. K., Riordan, A. J., Monahan, J., and Keeter, K. K.: Severe thunderstorm and tornado warnings at Raleigh, North Carolina, B. Am. Meteorol. Soc., 78(11), 2559–2575, 1997.
Kanellopoulou, H.: The waterspout of 1st September 1973 in Corfu, Bull. Greek Met. Soc., 5(5), 41–50, 1977 (in Greek).
Livadas, G.: Fish rain in western Macedonia, Meteorologika, No. 2, Publ. Met. Inst. Univ. Thess., Thessaloniki, Greece, 1954.
Matsangouras, J. T. and Nastos, P. T.: The 27 July 2002 tornado event in Athens, Greece, Adv. Sci. Res., 4, 9–13, doi:10.5194/asr-4-9-2010, 2010.
Matsangouras, J. T., Nastos, P. T., and Nikolakis, D. J.: Study of meteorological conditions related to the tornado activity on 25-3-2009 over NW Peloponnesus, Greece, Proceedings of 10th COMECAP, 25–38 May, Patras, Greece, 417–425, 2010 (in Greek).
Meaden, G. T.: Tornadoes in Britain: their intensities and distribution in space and time, J. Meteorol., 1, 242–251, 1976.
Nastos, P. T. and Matsangouras, J. T.: Tornado activity in Greece within the 20th century, Adv. Geosci., 26, 49–51, doi:10.5194/adgeo-26-49-2010, 2010.
Ogden, D.: Account of the Corfu waterspout 16 June 1983, J. Meteorol., 9(91), 214–215, 1984.
Paul, J. F.: An inventory of tornadoes in France, Weather, 54, 217–219, 1999.
Reynolds, D. J.: European tornado climatology 1960–1989, J. Meteorol., 24, 376–403, 1999.
Sioutas, M. V.: TORRO Tornado Report for Greece, December 2001: a T5-6 tornado at Lassi on 16th, J. Meteorol., 27(270), 207–208, 2002.
Skamarock, W. C., Klemp, J. B., Dudhia, J., Gill, D. O., Barker, D. M., Duda, M. G., Huang, X.-Y., Wang, W., and Powers, J. G.: A description of the Advanced Research WRF Version 3. NCAR/TN-475, 2008.
Wang, W., Bruyère, C., Duda, M., Dudhia, J., Gill, D., Lin, H.-C., Michalakes, J., Rizvi, S., Zhang, X., Beezley, J., Coen, J., and Mandel, J.: ARW Version 3 Modeling System User's Guide, NCAR-MMM, 352 pp., 2010.

Assessment of climate vulnerability in the Norwegian built environment

H. O. Hygen[1], C. F. Øyen[2], and A. J. Almås[3]

[1]Norwegian Meteorological Institute, Postboks 43 Blindern, 0313 Oslo, Norway
[2]SINTEF Building and Infrastructure, Postboks 124 Blindern, 0314 Oslo, Norway
[3]Department of Civil and Transport Engineering, Norwegian University of Science and Technology (NTNU), 7491 Trondheim, Norway

Abstract. The main trends expected for the change of Norwegian climate for this century are increasing temperatures, precipitation and wind. This indicates a probable increase of climate-related risks to the Norwegian built environment. Through co-operation between the Norwegian Meteorological Institute and SINTEF Building and Infrastructure, building and climate information have been combined to estimate changes in strain to the built environment due to climate change. The results show that the risk of wood decay will increase for the whole country. Almost two million buildings will be subject to an increase in risk of wood decay from medium to high level. Similar analyses have been performed for other climate indices, demonstrating a clear increase in potential damages due to water and humidity, while frost damage probably will decrease.

1 Introduction

The Norwegian Meteorological Institute has an archive of daily historic grids of precipitation and temperature with a resolution of 1 km by 1 km. Downscaling of scenarios for future climate based on global climate models and local climate knowledge has matured to a stage where the same kind of grids are available for various projections of future climate (Engen-Skaugen et al., 2007).

SINTEF Building and Infrastructure and the Norwegian Meteorological Institute have a history ofco-operation. In the research project Climate 2000, one of the projects focused on developing robust building techniques. A series of climate-indices were developed or adapted to the Norwegian climate. Examples of such indices are wood decay (Lisø, 2006), frost decay (Lisø et al., 2007) and driving rain (Rydock et al., 2005).

In 2010 an official Norwegian report on climate-adaptation (NOU, 2010) was launched. As a part of this report the vulnerability of the Norwegian built environment was surveyed. This survey consisted of a multipart approach with a summary of the Norwegian built environment and recorded damages. Another part focussed on climate indices, both historic and future, and the influence on the built environment.

2 Method

To assess the climate vulnerability of the Norwegian built environment, knowledge of the climate and of the built environment is combined. This work is based on the geolocated daily 1 km by 1 km grids of temperature and precipitation, knowledge of geographic distribution of damages and the climate indices developed in the Climate 2000 programme.

Risk of wood decay was first established for the North American continent by Shäffer and adapted to Norway by Lisø et al. (2006). The wood decay index is described in Eq. (1).

$$\text{Wood Decay} = \frac{\sum_{\text{January}}^{\text{December}} (T_{\text{mean}} - 2)(D - 3)}{16.7} \tag{1}$$

T_{mean} is the monthly mean temperature and D is the number of days per month with more than 0.254 mm (i.e. 0.01 inch) precipitation.

The Norwegian Map Authority administers a database of all Norwegian buildings, called "The Matrix", where all buildings are represented by a set of coordinates. This gives two sets of geolocated data, climate indices and buildings. GIS-systems, in our case ArcGIS, facilitate the possibility of joining these datasets to analyse geographically distributed impacts. In detail, the method classifies all the indices in zones according to threat level, and counts the number of buildings in each zone both for historical climate data and future climate data. The threat zones were simply generated by splitting the range of the index in three equal parts. Three

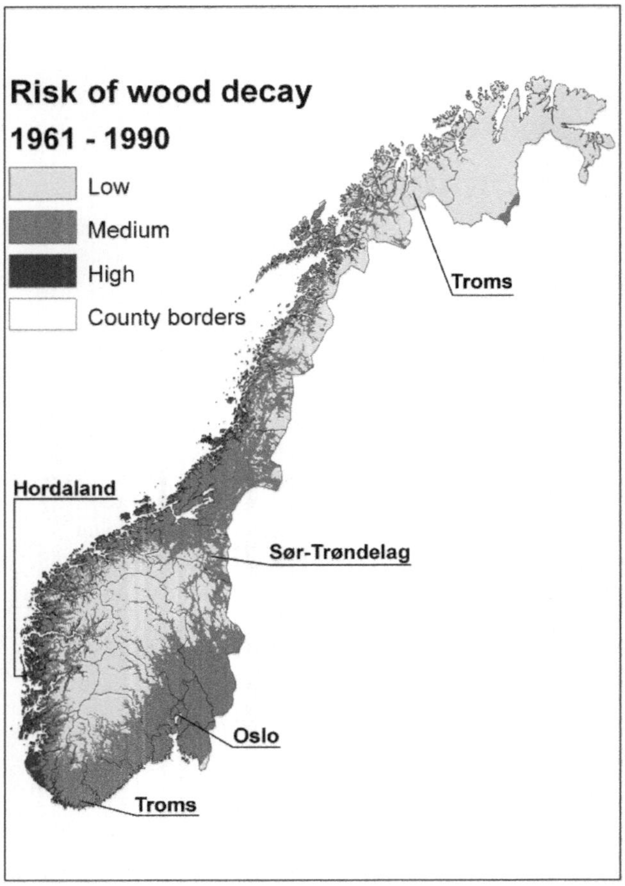

Figure 1. Map of Norway classified in risk of wood decay in a normal climate. Counties selected in Table 1 are marked.

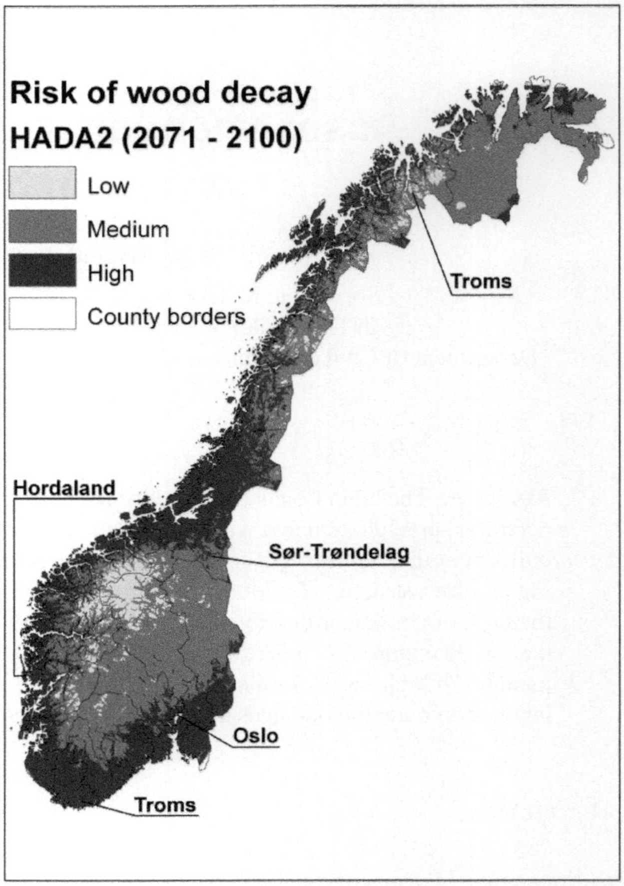

Figure 2. Map of Norway classified in risk of wood decay in future climate, based on projections by the Hadley Centre and scenario A2. Counties selected in Table 1 are marked.

zones are employed for practical reasons, as a balance between detailed results and computing time. To give a more detailed picture, the output results were sorted by counties.

There were some problems with the method. Some of the buildings were geolocated in the wrong UTM zone, thus locating some buildings in Sweden or the north Atlantic. Due to severe time restrictions in the project, there was no time to correct these kinds of error. Even greater uncertainty in the classification is related to how the shoreline was defined in the datasets. The climate grids were trimmed inside the shoreline so that tiles with both sea and land were defined as sea. This resulted in some buildings being defined as situated in the sea. Approximately 300 000 buildings (8 %) were unclassified due to these two errors, mostly due to the problem of missing the shoreline. A more thorough description of the method is described in Almås et al. (2011).

3 Results

Some of the indices used in this work are more robust than others. One of the strongest indices in this study is the risk of wood decay. Figure 1 shows a map of Norway classified in three zones of wood decay risk for the normal period 1961–1990. Figure 2 shows the risk of wood decay at the end of the century according to the projection by the Hadley Centre and scenario A2. Figure 3 shows a comparison of the number of houses in different risk classes based on historical (1961–1990) and future climate data (2071–2100) on a national level. The future climate is emission scenario A2 calculated with the model of the Hadley Centre, and downscaled at the Norwegian Meteorological Institute. Numbers of buildings on a national level in each zone are shown in Fig. 3, while numbers (in thousands) for selected counties are shown in Table 1.

Risk of frost decay is an index giving the risk for frost decay of mineral materials, developed by Lisø et al. (2006). A comparison of present and future risk of frost decay shows

Table 1. Number, in thousands, of buildings in wood decay risk zones for selected counties in Norway. Upper number is the normal climate (1961–1990), and lower number is projected climate (2071–2100).

County	Period	Un.	Low	Med.	High
Oslo	Normal	2	0	126	0
	HADA2	3	0	0	125
Vest Agder	Normal	11	2	78	30
	HADA2	12	0	0	108
Hordaland	Normal	34	3	35	192
	HADA2	35	0	10	220
Sør – Trøndelag	Normal	9	6	184	25
	HADA2	9	0	25	190
Troms	Normal	18	11	82	0
	HADA2	20	0	13	78

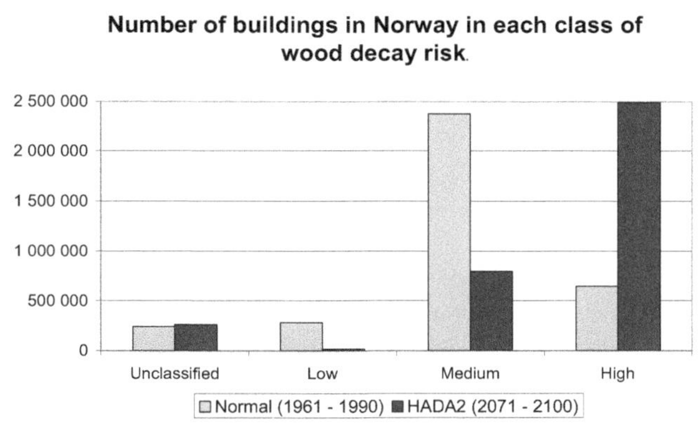

Figure 3. Number of buildings in Norway in normal climate and future climate according to scenario A2. Normal climate in grey and future climate in black.

a clear decrease. This decrease in vulnerability prompted the decision to not count the buildings for this index, due to the current building practices being robust to future climate.

A new index was also examined in this work: Wet Winter precipitation. This index indicates the risk of a snow covered roof being hit by rainfall. The potential impacts are a sudden increase in snow load and water penetrating cracks in the roofing. The index had one major weakness; information about previous snow cover on roofs was not available. This error had to be compensated for manually, resulting in an estimate of 600 000 buildings in Norway being affected by increased risk of Wet Winter precipitation.

4 Discussion

Multiple dimensions can be analysed to estimate the vulnerability of the Norwegian built environment. We have chosen to estimate the numbers of buildings affected by various climate parameters in combination with a SINTEF Building and Infrastructure survey of reported damage to the present Norwegian building stock (Ingvaldsen, 2001 and Lisø, 2006). The results clearly indicate that precipitation, in combination with other elements, is the main threat to the built environment, both in present and future climates. This knowledge gave a sound platform for advice on how to prepare the buildings for the future. The main advice is:

- carry out more frequent inspections and maintenance;
- use robust construction and detail solutions, components, and building materials with proper documentation of abilities to withstand the challenges of different climate zones of Norway (geographically differentiated building solutions).

Edited by: J. Prior

References

Almås, A., Øyen, C. F., and Hygen, H. O.: An approach to impact assessments of buildings in a changing climate, Build. Res. Inf., 39, 227–238, 2011.

Engen-Skaugen, T., Haugen, J. E., and Tveito, O. E.: Temperature scenarios for Norway: from regional to local scale, Clim. Dynam., 29, 441–453, doi:10.1007/s00382-007-0241-1, 2007.

Ingvaldsen, T.: Skader på bygg: Grunnlag for systematisk måling (Building defects: A basis for systematic evaluation, in Norwegian), NBI Project Report 308. Oslo: Norwegian Building Research Institute, 2001.

Lisø, K. R.: Building envelope performance assessments in harsh climates: Methods for geographically dependent design, Doctoral Theses, NTNU, 185 pp., 2006.

Lisø, K. R. , Hygen, H. O., Kvande, T., and Thue, J. V.: Decay potential in wood structures using climate data, Build. Res. Inf., 34, 546–551, 2006.

Lisø, K. R., Kvande, T., Hygen, H. O., Thue, J. V., and Harstveit, K.: Frost decay exposure index for porous, mineral building materials, Build. Environ., 42, 3547–3555, 2007.

NOU: Tilpassing til eit klima i endring. Servicenteret for departementa, Informasjonforvaltning, Oslo, 2010 (in Norwegian).

Rydock, J. P., Lisø, K. R., Førland, E. J., Nore, K., and Thue, J. V.: A driving rain exposure index for Norway, Build. Environ., 40, 1450–1458, 2005.

European wind variability over 140 yr

P. E. Bett, H. E. Thornton, and R. T. Clark
Met Office Hadley Centre, Exeter, UK

Correspondence to: P. E. Bett (philip.bett@metoffice.gov.uk)

Abstract. We present initial results of a study on the variability of wind speeds across Europe over the past 140 yr, making use of the recent *Twentieth Century Reanalysis* data set, which includes uncertainty estimates from an ensemble method of reanalysis. Maps of the means and standard deviations of daily wind speeds, and the Weibull-distribution parameters, show the expected features, such as the strong, highly-variable wind in the north-east Atlantic. We do not find any clear, strong long-term trends in wind speeds across Europe, and the variability between decades is large. We examine how different years and decades are related in the long-term context, by looking at the ranking of annual mean wind speeds. Picking a region covering eastern England as an example, our analyses show that the wind speeds there over the past ~ 20 yr are within the range expected from natural variability, but do not span the full range of variability of the 140-yr data set. The calendar-year 2010 is however found to have the lowest mean wind speed on record for this region.

1 Introduction

Knowing the form of the wind speed distribution is of critical importance when assessing the wind energy potential at a site. Typically, when wind farm developers or investors consider a site, they assess it using (at best) the past 20–30 yr, with data from direct observations, NWP models, and reanalyses. These recent decades reflect our personal experience of wind speeds, but they do not show the longer-term historical context. Understanding whether the most recent decades were more or less windy than normal, or if there are any significant long-term trends, is key to understanding the range of possible future windspeeds we might experience over the coming ~ 5 yr, or over the lifetime of a wind farm (~ 25 yr). This information is important not just for managing wind farms, but also for planning investment in future wind energy projects.

In this study, we show wind speed distributions for Europe over 140 yr (1871–2010), utilising the *Twentieth Century Reanalysis* data set (20CR, Compo et al., 2011). This reanalysis incorporates observations of sea-level pressure and surface pressure alone, with sea-surface temperature and sea-ice concentration data used as boundary conditions. The reanalysis used an experimental coupled atmosphere–land version of the NOAA NCEP Global Forecast Model, ran with 56 ensemble members. This allows an estimate of the uncertainties present due to episodes with less data (for example due to the reduction in Atlantic shipping during the World Wars, or the overall reduction in observations as one looks further back in time). We use the daily-mean wind speeds at the near-surface pressure level where $P/P_{\text{surface}} = 0.995$, on a regular lat–lon grid of resolution $2° \times 2°$. Other studies of European wind speed climatologies have used data sets that assimilate more observations or are at higher resolution. For example, the studies of Kiss and Jánosi (2008), Pryor et al. (2006), and Siegismund and Schrum (2001) use the ERA-40 (Uppala et al., 2005) and NCEP/NCAR reanalyses (Kalnay et al., 1996), which both span ~ 50 yr. The key feature of the use of the 20CR in our study is the unprecedented length of the time series, which allows us to compare decadal-scale fluctuations in the speed and variability of the wind.

In this short article, we present an initial sample of our results. We reserve a more full analysis for a later publication.

2 Results

2.1 Wind speed and variability

The simplest way of describing the long-term wind speed distribution over Europe is to map the long-term mean and standard deviation of the daily wind speeds. The long-term average is simply the mean of the daily wind speeds from all days from all ensemble members. For the standard deviation, we need to take account of the structure of the 20CR ensemble: the 56 members of the reanalysis were run as a series of independent "streams", each covering consecutive 5-yr periods[1] (see Table III in Compo et al., 2011). The result of this is that a time series from an ensemble member is only temporally continuous for 5 yr, and aggregating over any period longer than this risks mixing interannual variability with ensemble member-to-member variability. Therefore, for the long-term standard deviation, we first calculate standard deviations in 5-yr periods for each ensemble member. We then take ensemble means, giving us a single standard-deviation time series in 5-yr steps. These are then aggregated into a single long-term ensemble-mean value, using[2]

$$\sigma^2 = \frac{\sum_i N_i \left(\sigma_i^2 + \bar{U}_i^2\right)}{\sum_i N_i} - \bar{U}^2, \tag{1}$$

where N_i is the number of days in the 5-yr period i, the ensemble-means of the 5-yr mean and standard deviations of the daily wind speeds are $\bar{U}_i$ and σ_i respectively, and the long-term mean wind speed is $\bar{U} \equiv \left(\sum_i \bar{U}_i\right)/28$.

We show the resulting long-term maps in Fig. 1. The areas with the highest wind speeds, and the greatest day-to-day variability, are over the sea rather than the land. In particular, the highest-wind regions are in the north Atlantic, west of Ireland and north of Scotland. Relatively high winds also cover the rest of the British Isles and the coastal areas of north-western Europe, as well as the central Mediterranean.

The extremely low winds over the areas of more complex orography (in an arc from the Atlas Mountains, over Spain, the Alps, the Carpathians, and through to Turkey) are likely to be due to the drag scheme used to model the impact of unresolved orography. (Note that the horizontal scale of the mountains is much smaller than the grid size used here.) While such features are often seen in maps of model wind speeds (e.g. Howard and Clark, 2007), it is important to be particularly wary of the results in these areas, as it is not clear how to relate them to the real physical situation "on the ground".

We have tested for the presence of any long-term trends, by performing a simple linear regression on the ensemble-mean time series of 5-yr means and standard deviations of the wind speeds in each grid cell. The significance of any trend (i.e. whether the gradient of the linear fit was significantly different from zero) was assessed using a t test at the 0.1 % level. Figure 2 shows the long-term trends, shaded out when they are not statistically significant.

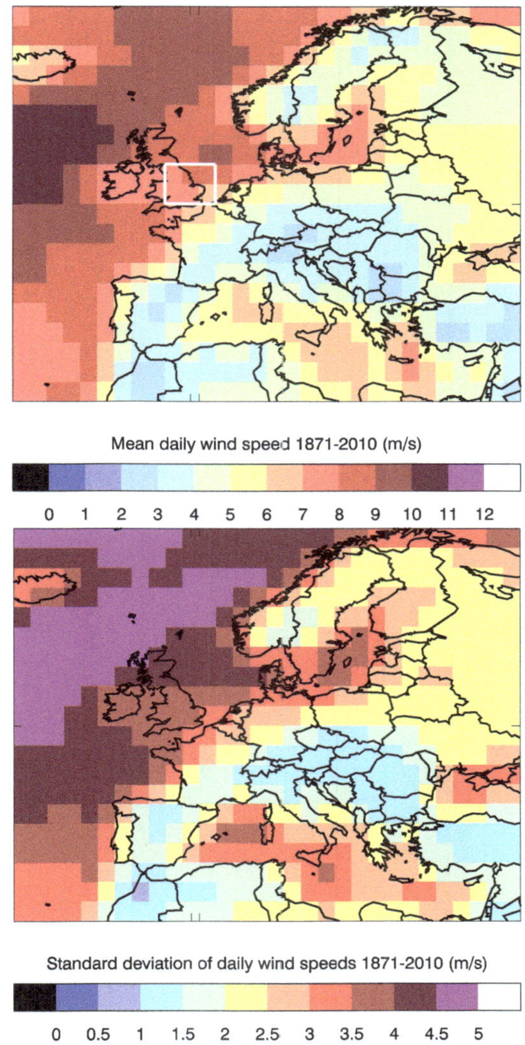

Figure 1. Maps of the 140-yr mean and standard deviation of the daily wind speeds from 20CR. The white box in the top plot marks the region we refer to as "eastern England".

While most areas over Europe show no significant trend, there are some areas of interest: the Atlantic off the north and west coasts of Ireland and Scotland shows a significant positive trend (with the mean wind speed increasing at a rate of $\sim 0.03\,\mathrm{m\,s^{-1}\,decade^{-1}}$), as does the eastern Mediterranean around Crete and Turkey ($\sim 0.07\,\mathrm{m\,s^{-1}\,decade^{-1}}$). There is a significant negative trend in the central Mediterranean around Sicily and Malta ($\sim -0.05\,\mathrm{m\,s^{-1}\,decade^{-1}}$). Note, however, that even where the trends are significant, they are still extremely small: a change of $\sim \pm 1\,\mathrm{m\,s^{-1}}$ per two centuries. The standard deviation trend map shows similar spatial patterns, but even fewer and smaller areas where those trends are significant.

[1]Streams 16 and 17 actually have lengths of 6 and 4 yr respectively, and stream 27 consists of 10 yr (1999–2010). We treat the whole series as 28 equal periods of 5-yr, for simplicity.

[2]Note that Eq. (1) is for the population standard deviation, which we use as we are aggregating data covering the complete time period, rather than estimating σ from a sample of years.

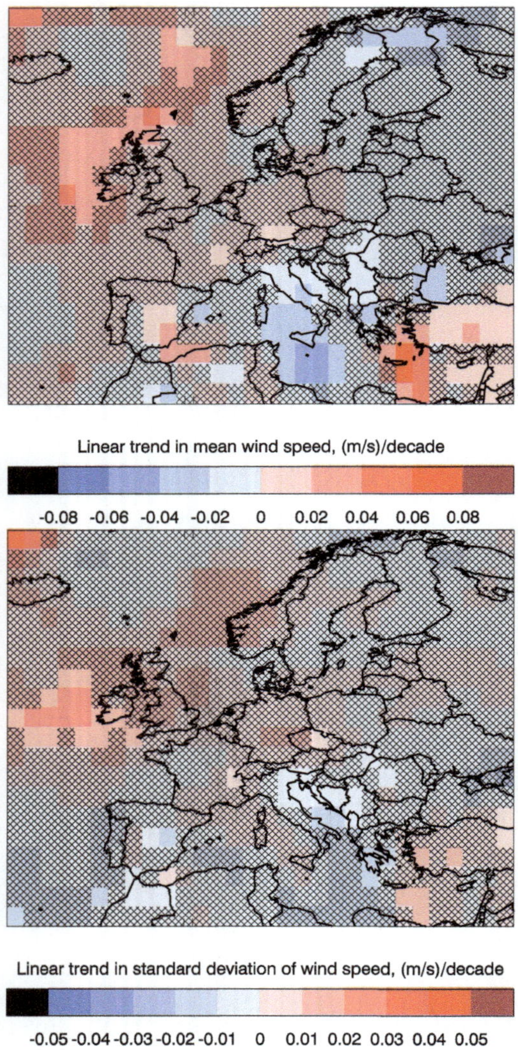

Figure 2. Gradients of the linear regressions to the 140-yr time series of 5-yearly ensemble-mean average wind speeds (top) and standard deviations of daily wind speeds (bottom). Shaded areas show where the trend is not statistically significant at the 0.1 % level.

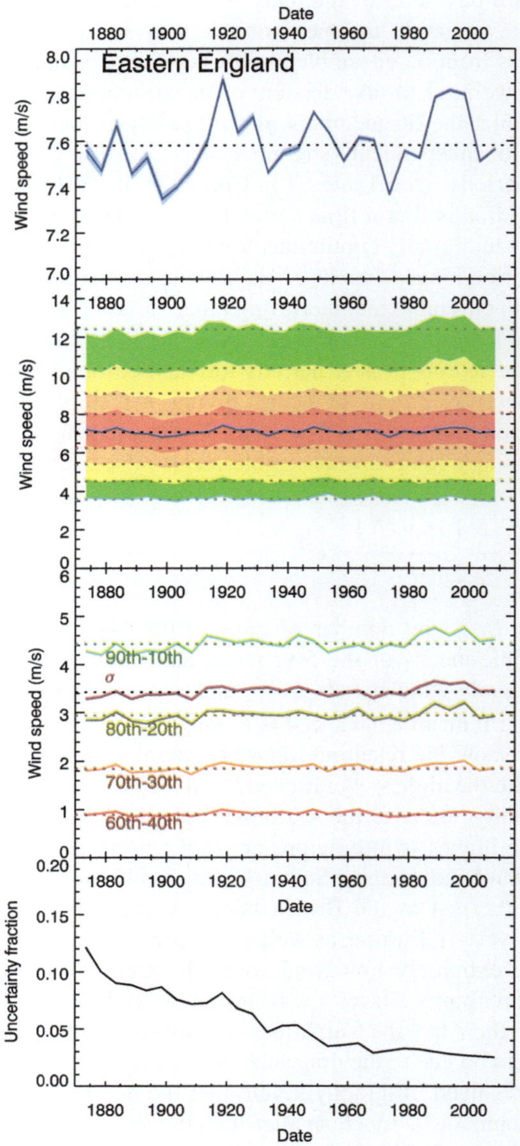

Figure 3. Time series of the wind speed distribution averaged over the eastern England region. Top: ensemble mean of the 5-yr mean wind speed, with lighter shading indicating the 10th and 90th percentiles of the ensemble member distribution (most visible at early times). Second from top: the ensemble mean of the 5-yr median wind speed (blue line), as well as the other deciles (coloured regions, from 10th percentile at the bottom to the 90th percentile at the top). The long-term means of the deciles are shown by dotted lines. Widths of the distribution are shown in the panel second from bottom: half the difference between ensemble-mean deciles (as labelled), and the ensemble mean of the standard deviation of ensemble members' wind speeds in 5-yr bins (labelled σ). Bottom: 5-yr means of the day-to-day ensemble spread (i.e. standard deviation of the ensemble members each day), as a fraction of the ensemble-mean 5-yr-mean wind speed.

We show an example time series in detail in Fig. 3, to illustrate some general features of the wind time series over Europe, and to show the benefits of being able to use such a long data set. It uses daily wind speeds averaged over a region covering eastern England (see box in Fig. 1). This region has been selected partly because it is an area of interest for future wind farm development[3], but also because it avoids some of the more exceptional features already discussed, such as the artificially low winds seen over complex terrain and the weak trends seen further to the north-west.

[3]See e.g. maps on the UK Government's RESTATS web site http://restats.decc.gov.uk/app/pub/map/map/ and the UK Crown Estate's web site http://www.thecrownestate.co.uk/energy-infrastructure/downloads/maps-and-gis-data/.

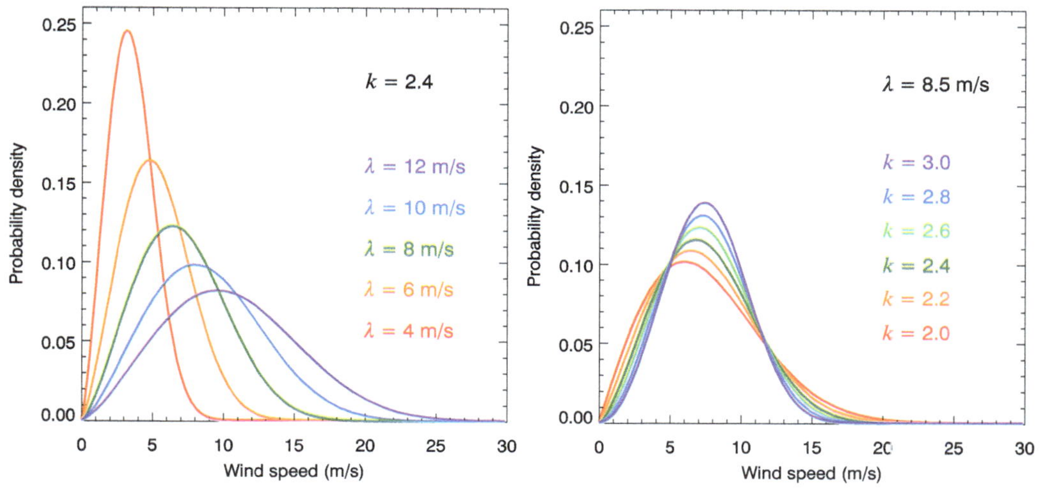

Figure 4. Demonstrations of Weibull distribution PDFs when varying the scale parameter (λ, left) and the shape parameter (k, right). Values chosen for display reflect those seen in our results.

We reserve a more detailed discussion of such features for subsequent papers.

The 5-yearly mean wind speeds (top panel of Fig. 3; note the magnified scale) clearly show the peak in the late-1980s up to around 2000, and the subsequent return to the long-term mean. We can also see that this excursion from the mean, while strong, is not exceptional in the long-term history of this region (see e.g. the peak in the 1915–1920 period, and the prolonged reduction over 1885–1910). There is also no clear trend.

By looking at the deciles of the distribution (second panel in Fig. 3), we can see how different features appear at different levels. For example, there is a clear trough in the early 1970s in the central region of the distribution; it does not appear in the 10th and 90th percentiles, and only forms part of a much broader reduction in the 80th percentile. The length and form of the reduction around the 1890–1910 period is also different in different deciles, with the upper deciles showing a longer reduction in wind speed. These features can be tracked through into the distribution widths (third panel of Fig. 3): for example, the dip in the widths over 1905–1910 is due to the lower deciles rising while the upper deciles are still low. The actual meteorological situations in these periods would need to be investigated in more detail to understand the underlying physical causes, but these features show the importance of analysing the whole distribution beyond just the mean.

The day-to-day ensemble spread as a fraction of the mean wind speed is shown in the bottom panel of Fig. 3. This will always be much greater than the uncertainty in the 5-yr mean shown in the top panel; nonetheless, it illustrates a general decrease in the uncertainty in the data, with an average daily spread between the ensemble members of about 2.5 % of the mean wind speed by the 2000s. This reduction in uncertainty is directly related to the dramatic increase in the amount of data assimilated over the course of the 20CR period, which constrains the ensemble members reducing their spread (see discussions in e.g. Compo et al., 2011; Ferguson and Villarini, 2012; Krueger et al., 2012; Wang et al., 2012). It is important to note that this panel shows the *uncertainty* in the reanalysis (through the daily ensemble spread), whereas the σ in the panel above refers to the *variability* of wind speeds over time in all ensemble members.

2.2 The Weibull distribution

Distributions of wind speeds at given locations are commonly modelled as being drawn from a Weibull distribution (e.g. Conradsen et al., 1984; Carta et al., 2009, and references therein), which has the probability density function (PDF)

$$P(U) = \frac{k}{\lambda}\left(\frac{U}{\lambda}\right)^{k-1} \exp\left[-\left(\frac{U}{\lambda}\right)^{k}\right], \quad (2)$$

where the random variable U is the wind speed, k is the (dimensionless) shape parameter, and λ is the scale parameter (in the same units as U). The form of the distribution for different values of k and λ is shown in Fig. 4. The shape parameter k determines the overall form of the distribution, including the skewness, and the width for a given λ; for wind speeds, k tends to lie between 2 and 3. For a given shape, the scale parameter λ determines the distribution's peak-location and width; higher values of λ result in a broader distribution with a peak at a higher wind speed. We can use the Weibull parameters as an alternative way of characterising the wind speed distribution. Examining changes in the Weibull parameters allows us to track the behaviour of the distribution as a whole, rather than e.g. just the mean value.

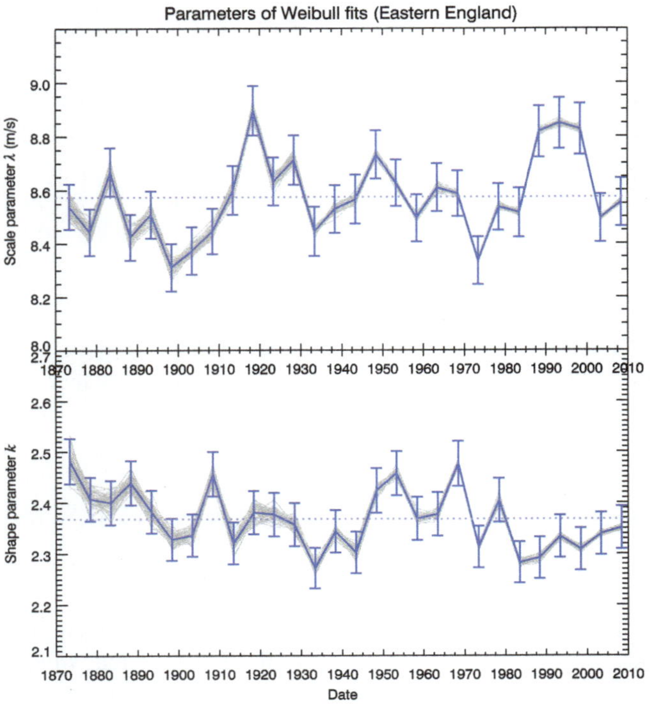

Figure 5. Time series (in 5-yr bins) of the Weibull scale and shape parameters for our sample region covering eastern England. The individual ensemble members are shown as grey lines (note that since ensemble members are discontinuous in these 5-yr steps, the lines do not reflect actual time series, but are plotted merely to guide the eye). The blue line shows the ensemble-means of the Weibull parameters in each step, with the error bars given by the ensemble-means of the estimated standard errors from the fits. The long-term means of the ensemble-mean time series are plotted as dotted lines.

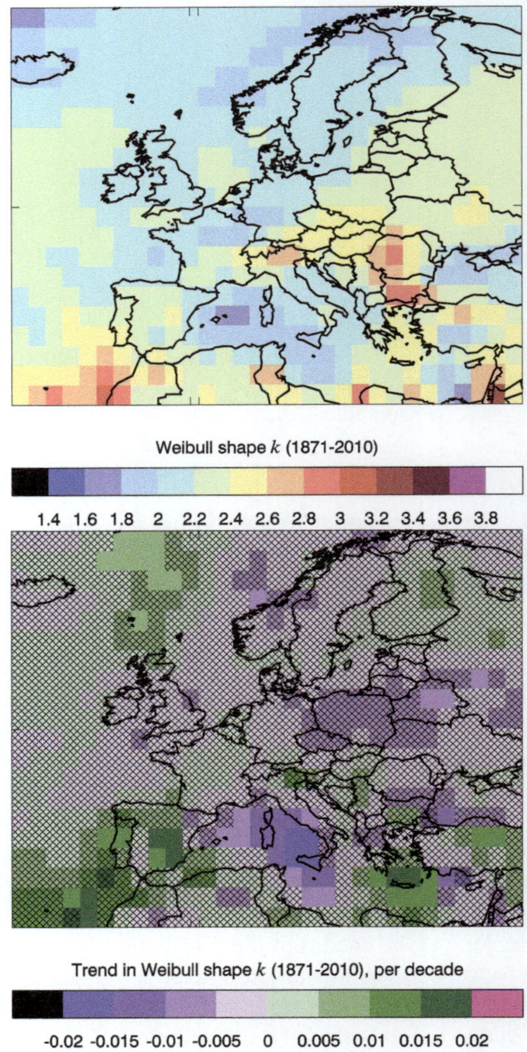

Figure 6. Maps of the long-term mean of the ensemble-mean time series of Weibull shape parameters (top), and the trend from linear regresson on those time series (bottom). As before, areas where the trend is not significantly different from zero are shaded.

In a given grid cell or region, we find maximum-likelihood estimates of the Weibull scale and shape for the distribution of daily wind speeds of each ensemble member in each 5-yr period. This yields ensemble estimates of the time series of Weibull parameters; an example from the same eastern England region as before is shown in Fig. 5. As expected, the behaviour of the scale parameter over time is similar to the mean and σ time series (Fig. 3), whereas this is not seen in the shape parameter. As the shape parameter governs the skewness of the distribution, the variation in k could be related to the presence of extreme high-wind events, as these could cause the form of the distribution to be fit better with a longer tail.

Once again, we can extract linear trends from these time series, test their significance, and map the results along with their long-term values. The long-term mean and trend of the shape parameter are shown in Fig. 6. The map for the average k shows very little spatial variability, staying in most areas between 2 and 3. Very few areas show any significant trend, although there is again an area of reduction over the central Mediterranean (i.e. tending to a more skewed distribution) and growth around Crete (towards a more symmetric distribution). The long-term mean and trend of the scale parameter is shown in Fig. 7. The mean-λ map is very similar to both the mean and standard deviation of the wind speeds (cf. Fig. 1), and the trend map is again familiar; the area of significant growth in the north-east Atlantic is in this case much larger.

The distribution width is related to *both* λ and k, but in opposite senses (e.g. narrower distributions can be obtained by reducing λ or increasing k, with corresponding impacts on the peak locations and skewness). Thus, it is not surprising that (for example) the negative trends in both k and λ over Italy and the Tyrrhenian Sea do not correspond with a significant trend in the wind speed standard deviation (cf. Fig. 2).

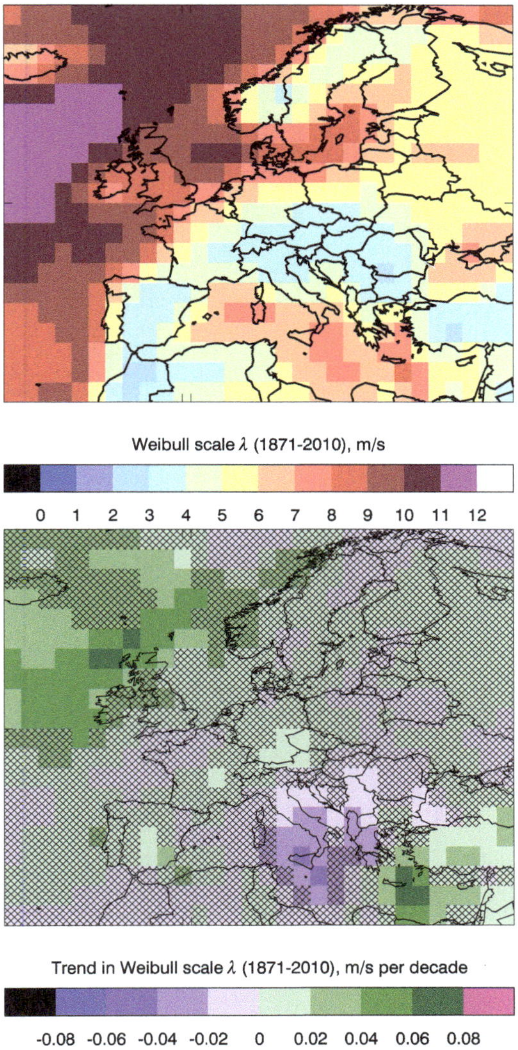

Figure 7. As Fig. 6, but for the Weibull scale parameters.

It is important to note again that, regardless of statistical significance, the trends present in both k and λ are still exceedingly small.

2.3 Year-ranking by wind speed

An additional way of relating the wind speeds in different years to the long-term context is to plot the annual mean wind speeds in rank order. This is shown qualitatively in Fig. 8, for the eastern England region. We can immediately see that the day-to-day variability in wind speed is much larger than the year-to-year variability (in annual-mean wind speed), and the mixture of colours shows the lack of a clear long-term trend.

We can get a more quantitative picture by highlighting individual decades. Figure 9 shows how the most recent three decades are positioned with respect to the 140-yr time series. Here we can see the differences in the spreads of the different decades: nine of the ten years of the 1990s (i.e. of the years 1990–1999 inclusive) are in the top half of the distribution, but only a couple are in the top 10. The 1980s includes some very high and very low wind years. The 2000s are also well-spread (6 yr in the top half, 4 yr in the bottom half; note 2008 is in the top 10), but it is 2010 that is the extreme low-wind year. This is because, since we are using calendar years, 2010 is hit by the low winds (or lack of high winds) associated with the extreme cold temperatures in January–February and November–December that year (e.g. Prior and Kendon, 2011a,b).

3 Summary and conclusions

Using the Twentieth Century Reanalysis, we have analysed the distribution of wind speeds over Europe, and how it has varied over 140 yr of daily data. The regions with the highest long-term average wind speeds, and highest day-to-day variabilities, are in the north-east Atlantic, and the data exhibits a strong land–sea contrast. Most areas show no clear trends over time in the mean wind speed or its variability; even in those areas where a trend is statistically significant, the magnitude is so small that it is still questionable whether it is physically meaningful or in any way relevant for applications such as wind energy. Long-term trends are difficult to distinguish from natural variability, and the variability is much greater than the uncertainties in the data.

However, the decadal-scale variations in the long time series help us understand the behaviour of recent decades in a long-term context. For the region that we have illustrated as an example (covering eastern England), we can see that the variations seen in wind speeds in the past 20–30 yr are typical of the natural variability of this region. Indeed, the past thirty years do not cover the full range of variability seen in the 140-yr time series, e.g. not capturing the the lower-wind episodes of the 1970s and 1900s in England. Note that, while the results do not depend qualitatively on the precise area in question, there is significant country-scale variability across Europe, both quantitatively and qualitatively.

Overall, we have shown that long-term data sets are essential in understanding wind speed distributions across Europe. In subsequent papers, we will describe our analysis and results in more detail, including a comparison with other data sets.

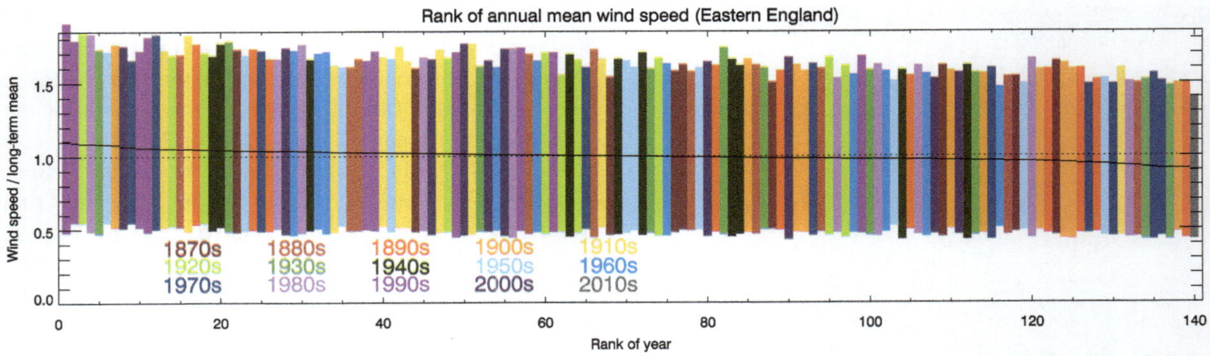

Figure 8. Annual mean wind speeds (black line) in the eastern England region, divided by the 140-yr mean, and plotted in rank order. The vertical bars (one for each year, coloured by decade) show the 10th and 90th percentile of the daily winds each year.

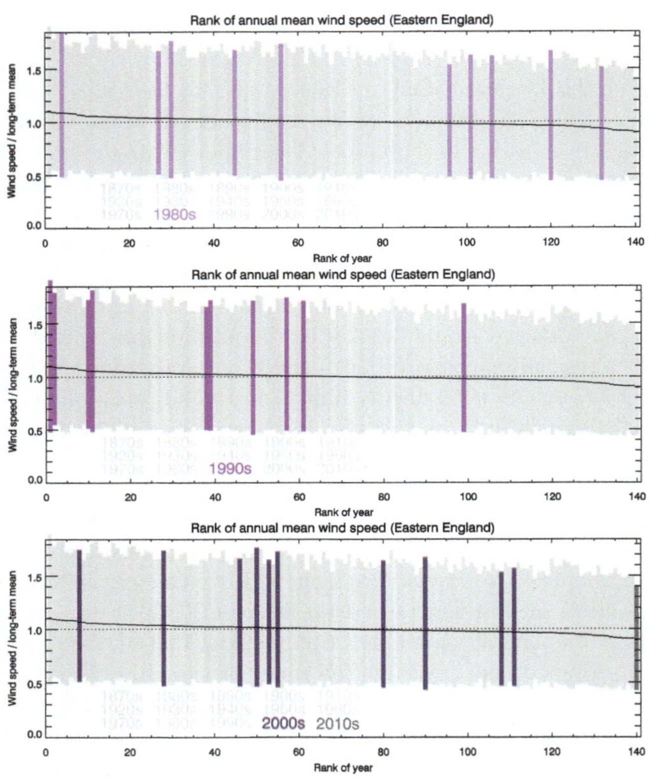

Figure 9. As Fig. 8, but highlighting the past three decades separately. We have included 2010 as an additional year in the 2000s' plot.

Acknowledgements. PB acknowledges the support of an EMS Young Scientist Travel Award. Support for the Twentieth Century Reanalysis Project dataset is provided by the US Department of Energy, Office of Science Innovative and Novel Computational Impact on Theory and Experiment (DOE INCITE) program, and Office of Biological and Environmental Research (BER), and by the National Oceanic and Atmospheric Administration Climate Program Office.

Edited by: S.-E. Gryning

References

Carta, J. A., Ramírez, P., and Velázquez, S.: A review of wind speed probability distributions used in wind energy analysis, Renew. Sust. Energ. Rev., 13, 933–955, doi:10.1016/j.rser.2008.05.005, 2009.

Compo, G. P., Whitaker, J. S., Sardeshmukh, P. D., Matsui, N., Allan, R. J., Yin, X., Gleason, B. E., Vose, R. S., Rutledge, G., Bessemoulin, P., Brönnimann, S., Brunet, M., Crouthamel, R. I., Grant, A. N., Groisman, P. Y., Jones, P. D., Kruk, M. C., Kruger, A. C., Marshall, G. J., Maugeri, M., Mok, H. Y., Nordli, Ø., Ross, T. F., Trigo, R. M., Wang, X. L., Woodruff, S. D., and Worley, S. J.: The Twentieth Century Reanalysis Project, Q. J. Roy. Meteor. Soc., 137, 1–28, doi:10.1002/qj.776, 2011.

Conradsen, K., Nielsen, L. B., and Prahm, L. P.: Review of Weibull Statistics for Estimation of Wind Speed Distributions, J. Clim. Appl. Meteorol., 23, 1173–1183, doi:10.1175/1520-0450(1984)023<1173:ROWSFE>2.0.CO;2, 1984.

Ferguson, C. R. and Villarini, G.: Detecting inhomogeneities in the Twentieth Century Reanalysis over the central United States, J. Geophys. Res. Atmos., 117, D05123, doi:10.1029/2011JD016988, 2012.

Howard, T. and Clark, P.: Correction and downscaling of NWP wind speed forecasts, Meteorol. Appl., 14, 105–116, doi:10.1002/met.12, 2007.

Kalnay, E., Kanamitsu, M., Kistler, R., Collins, W., Deaven, D., Gandin, L., Iredell, M., Saha, S., White, G., Woollen, J., Zhu, Y., Leetmaa, A., Reynolds, R., Chelliah, M., Ebisuzaki, W., Higgins, W., Janowiak, J., Mo, K. C., Ropelewski, C., Wang, J., Jenne, R., and Joseph, D.: The NCEP/NCAR 40-Year Reanalysis

Project, B. Am. Meteorol. Soc., 77, 437–471, doi:10.1175/1520-0477(1996)077<0437:tnyrp>2.0.co;2, 1996.

Kiss, P. and Jánosi, I. M.: Comprehensive empirical analysis of ERA-40 surface wind speed distribution over Europe, Energ. Convers. Manage., 49, 2142–2151, doi:10.1016/j.enconman.2008.02.003, 2008.

Krueger, O., Schenk, F., Feser, F., and Weisse, R.: Inconsistencies between Long-Term Trends in Storminess Derived from the 20CR Reanalysis and Observations, J. Climate, 26, 868–874, doi:10.1175/jcli-d-12-00309.1, 2012.

Prior, J. and Kendon, M.: The UK winter of 2009/2010 compared with severe winters of the last 100 years, Weather, 66, 4–10, doi:10.1002/wea.735, 2011a.

Prior, J. and Kendon, M.: The disruptive snowfalls and very low temperatures of late 2010, Weather, 66, 315–321, doi:10.1002/wea.874, 2011b.

Pryor, S. C., Barthelmie, R. J., and Schoof, J. T.: Inter-annual variability of wind indices across Europe, Wind Energy, 9, 27–38, doi:10.1002/we.178, 2006.

Siegismund, F. and Schrum, C.: Decadal changes in the wind forcing over the North Sea, Climate. Res., 18, 39–45, doi:10.3354/cr018039, 2001.

Uppala, S. M., Kållberg, P. W., Simmons, A. J., Andrae, U., Bechtold, Fiorino, M., Gibson, J. K., Haseler, J., Hernandez, A., Kelly, G. A., Li, X., Onogi, K., Saarinen, S., Sokka, N., Allan, R. P., Andersson, E., Arpe, K., Balmaseda, M. A., Beljaars, A. C. M., Berg, Bidlot, J., Bormann, N., Caires, S., Chevallier, F., Dethof, A., Dragosavac, M., Fisher, M., Fuentes, M., Hagemann, S., Hólm, E., Hoskins, B. J., Isaksen, L., Janssen, P. A. E. M., Jenne, R., Mcnally, A. P., Mahfouf, J. F., Morcrette, J. J., Rayner, N. A., Saunders, R. W., Simon, P., Sterl, A., Trenberth, K. E., Untch, A., Vasiljevic, D., Viterbo, P., and Woollen, J.: The ERA-40 re-analysis, Q. J. Roy. Meteor. Soc., 131, 2961–3012, doi:10.1256/qj.04.176, 2005.

Wang, X., Feng, Y., Compo, G. P., Swail, V. R., Zwiers, F. W., Allan, R. J., and Sardeshmukh, P. D.: Trends and low frequency variability of extra-tropical cyclone activity in the ensemble of twentieth century reanalysis, Clim. Dynam., 1–26, doi:10.1007/s00382-012-1450-9, 2012.

Finite element simulation of a local scale air quality model over complex terrain

A. Oliver[1], **G. Montero**[2], **R. Montenegro**[2], **E. Rodríguez**[2], **J. M. Escobar**[2], **and A. Perez-Foguet**[1]

[1]Laboratori de Càlcul Numèric (LaCàN), Departament de Matemàtica Aplicada III, Universitat Politècnica de Catalunya – BarcelonaTech, Jordi Girona 1-3, 08034 Barcelona, Spain

[2]University of Las Palmas de Gran Canaria, University Institute for Intelligent Systems and Numerical Applications in Engineering (SIANI), Spain

Correspondence to: A. Oliver (albert.oliver@upc.edu)

Abstract. In this paper we propose a finite element method approach for modelling the air quality in a local scale over complex terrain. The area of interest is up to tens of kilometres and it includes pollutant sources. The proposed methodology involves the generation of an adaptive tetrahedral mesh, the computation of an ambient wind field, the inclusion of the plume rise effect in the wind field, and the simulation of transport and reaction of pollutants. We apply our methodology to simulate a fictitious pollution episode in La Palma island (Canary Island, Spain).

1 Introduction

In this paper, we present a new methodology for local scale air quality simulations, summarized in Algorithm 1, by using a non-steady finite element method with unstructured and adaptive tetrahedral meshes. The aim of this proposal is to introduce an alternative to the standard implementation of current models, improving the computational cost of methods that use structured meshes (Hanjalić and Kenjereš, 2005).

Three remarkable uses of unstructured meshes in atmospheric pollution problems are the two-dimensional regional-global examples presented in Lagzi et al. (2004); Ahmad et al. (2006), the three-dimensional regional examples, including local refinement with element sizes of 2 km, presented in Tomlin et al. (2000), and the three-dimensional tetrahedral meshes for local wind field analysis with element sizes ranging from two meters up to two kilometres, see Montenegro et al. (2004); Montero et al. (2004). The ideas of this last approach are considered to include the plume rise effect in the wind field that is used in the air quality problem.

The wind field is crucial for the pollutant transport, specially in complex terrain areas. In order to simulate it, we have used a mass-consistent model. Several two-dimensional (Winter et al., 1995) and three-dimensional (Montero et al., 1998, 2005; Ferragut et al., 2010) adaptive finite element solutions have been developed by the authors.

The convection, diffusion and reaction problem is usually solved using splitting schemes (Martín et al., 2003; Chock et al., 2005) and specific numerical solvers for time integration of photochemical reaction terms (Saylor and Ford, 1995; Sandu et al., 1997; Ahmad et al., 2006). In this paper a stabilized finite element formulation with a Crank-Nicolson temporal integration is proposed to solve the problem (Donea and Huerta, 2003; Rodríguez-Ferran and Sandoval, 2007). The chemistry is simulated with the RIVAD chemical model (Scire et al., 2000). The transport and chemical terms are treated separately with Strang splitting operators (Ropp et al., 2004), and the non-linear chemical part is solved with a second order Rosenbrock method (Verwer et al., 1999).

The paper is organised as follows. In Sect. 2 we describe the main steps of the proposed methodology. Results are shown in Sect. 3, and finally the conclusions and future work are presented in Sect. 4.

2 Algorithm description

In this section a brief description of the different steps of Algorithm 1 is presented.

Algorithm 1 Wind and air pollution modeling algorithm

1: Construct an adaptive tetrahedral mesh of the 3-D domain
2: Wind field simulation from experimental data
3: Wind field modification including the plume rise effect
  3.1: Compute plume rise trajectory
  3.2: Mesh refinement along plume rise trajectory
  3.3: Apply step 2 on the refined mesh
  3.4: Modify the vertical wind field along the plume rise
4: Air pollution simulation from stack emission data

2.1 Adaptive tetrahedral mesh

The studied domain is limited at the bottom by the terrain and at the top by a horizontal plane. The lateral walls are formed by four vertical planes. A uniform distribution of nodes is defined on the upper boundary. A refinement/derefinement algorithm (Ferragut et al., 1994) is applied on this uniform mesh to construct a node distribution adapted to the terrain surface and stacks. Once the node distribution is defined both on the terrain and the upper boundary, we distribute the nodes located between both layers by using a vertical spacing function. Next, a three-dimensional mesh generator based on Delaunay triangulation (Escobar and Montenegro, 1996) is applied. Finally, the untangling and smoothing procedure described in (Escobar et al., 2003) is used to get a valid mesh and to improve its quality. A detailed description of the mesh generation procedure can be seen in Montenegro et al. (2002).

2.2 Wind field simulation

A mass-consistent model (Montero et al., 1998, 2005; Ferragut et al., 2010) is used to compute a wind field $\boldsymbol{u}$ in the three-dimensional domain Ω, with a boundary $\Gamma = \Gamma_a \cup \Gamma_b$, that satisfies the continuity equation and the impermeability condition on the terrain Γ_a,

$$\begin{aligned} \nabla \cdot \boldsymbol{u} &= 0 \quad \text{in } \Omega \\ \boldsymbol{n} \cdot \boldsymbol{u} &= 0 \quad \text{on } \Gamma_a \end{aligned} \tag{1}$$

where $\boldsymbol{n}$ is the outward-pointing normal unit vector.

The model formulates a Least-Squares problem in the domain Ω to find a wind field $\boldsymbol{u} = (u, v, w)$, such that it is adjusted as much as possible to an interpolated wind field $\boldsymbol{u}_0 = (u_0, v_0, w_0)$. The adjusting functional for a field $\boldsymbol{v} = (\widetilde{u}, \widetilde{v}, \widetilde{w})$ is defined as

$$E(\boldsymbol{v}) = \frac{1}{2} \int_\Omega (\boldsymbol{v} - \boldsymbol{u}_0)^t \, \mathbf{P} (\boldsymbol{v} - \boldsymbol{u}_0) \, \mathrm{d}\Omega \tag{2}$$

where $\mathbf{P}$ is a 3×3 diagonal matrix with $P_{1,1} = P_{2,2} = 2\alpha_1^2$ and $P_{3,3} = 2\alpha_2^2$. The Lagrange multiplier technique is used to minimise the functional (Eq. 2), with the restrictions (Eq. 1). Considering the Lagrange multiplier λ, the Lagrangian is defined as

$$L(\boldsymbol{v}, \lambda) = E(\boldsymbol{v}) + \int_\Omega \lambda \nabla \cdot \boldsymbol{v} \, \mathrm{d}\Omega \tag{3}$$

and the solution $\boldsymbol{u}$ is obtained by finding the saddle point $(\boldsymbol{u}, \phi)$ of the Lagrangian Eq. (3). This resulting wind field satisfies the Euler-Lagrange equation,

$$\boldsymbol{u} = \boldsymbol{u}_0 + \mathbf{P}^{-1} \nabla \phi \tag{4}$$

where ϕ is the Lagrange multiplier. As α_1 and α_2 are constant in Ω, the variational approach results in an elliptic problem in ϕ, by substituting Eq. (4) in Eq. (1), that is solved by using the finite element method.

$$-\nabla \cdot \left(\mathbf{P}^{-1} \nabla \phi \right) = \nabla \cdot \boldsymbol{u}_0 \qquad \text{in } \Omega \tag{5}$$

$$-\boldsymbol{n} \cdot \mathbf{P}^{-1} \nabla \phi = \boldsymbol{n} \cdot \boldsymbol{u}_0 \qquad \text{on } \Gamma_a \tag{6}$$

$$\phi = 0 \qquad \text{on } \Gamma_b \tag{7}$$

The interpolated wind field $\boldsymbol{u}_0$ can be constructed from experimental data or meteorological forecasting models. In this paper we consider the first case. Therefore, we consider an horizontal interpolation and a vertical extrapolation of the available measurements to construct $\boldsymbol{u}_0$ in the whole computational domain.

2.2.1 Horizontal interpolation

The most common technique of interpolation at a given point, placed at a height z_m over the terrain, is formulated as a function of the inverse of the squared distance between that point and the measurement stations, and the inverse of their height differences (Montero et al., 1998)

$$\boldsymbol{u}_0(z_m) = \xi \frac{\sum_{n=1}^{N} \frac{\boldsymbol{u}_n}{d_n^2}}{\sum_{n=1}^{N} \frac{1}{d_n^2}} + (1 - \xi) \frac{\sum_{n=1}^{N} \frac{\boldsymbol{u}_n}{|\Delta h_n|}}{\sum_{n=1}^{N} \frac{1}{|\Delta h_n|}} \tag{8}$$

where the value of $\boldsymbol{u}_n$ is the velocity measured at station n, N is the number of stations considered in the interpolation, d_n is the horizontal distance from station n to the point of the domain where we are computing the wind velocity, $|\Delta h_n|$ is the height difference between station n and the studied point, and ξ is a weighting parameter ($0 \le \xi \le 1$), that allows to give more importance to one of these interpolation criteria.

2.2.2 Vertical extrapolation

In this work, a log-linear wind profile is considered (Lalas and Ratto, 1996) in the surface layer, which takes into account the horizontal interpolation (Montero and Sanín, 2001) and the effect of roughness on the wind intensity and the direction. These values also depend on the air stability (neutral, stable or unstable atmosphere) according to the Pasquill stability class. Above the surface layer, a linear interpolation is carried out using the geostrophic wind. The logarithmic profile is given by,

$$\boldsymbol{u}_0(z) = \frac{\boldsymbol{u}^*}{k}(\log\frac{z}{z_0} - \Phi_m) \qquad z_0 < z \le z_{sl} \tag{9}$$

where $\boldsymbol{u}^*$ is the friction velocity, k is von Karman's constant, z_0 is the roughness length (McRae et al., 1982) and z_{sl} is the height of the surface layer. The values of Φ_m depend on the Pasquill stability class (Zannetti, 1990), and the friction velocity is obtained from Eq. (9) at any point (x, y) by using the horizontal interpolated velocity $\boldsymbol{u}_0(z_m)$.

The linear interpolation is given by,

$$\boldsymbol{u}_0(z) = \rho(z)\,\boldsymbol{u}_0(z_{sl}) + [1 - \rho(z)]\boldsymbol{u}_g \qquad z_{sl} < z \le z_{pbl} \tag{10}$$

where $\boldsymbol{u}_g$ is the geostrophic wind, z_{pbl} is the height of the planetary boundary layer, and $\rho(z)$ is defined as

$$\rho(z) = 1 - \left(\frac{z - z_{sl}}{z_{pbl} - z_{sl}}\right)^2\left(3 - 2\frac{z - z_{sl}}{z_{pbl} - z_{sl}}\right) \tag{11}$$

Finally, this model assumes $\boldsymbol{u}_0(z) = \boldsymbol{u}_g$ if $z > z_{pbl}$ and $\boldsymbol{u}_0(z) = 0$ if $z \le z_0$.

2.3 Plume rise

The plume rise phenomenon is mainly due to the difference of temperature between the released substance and the environment air, and the initial momentum. The trajectory of the plume rise has been widely studied in the past (Briggs, 1969a,b; Moore, 1974). These works differentiate between two kinds of cases: *predominant buoyancy rise* and *predominant momentum rise*. The characterization of these types essentially depends on the ratio between the intensities of the pollutant emission velocity and the wind velocity at the top of the stack.

Gaussian plume models (Olcese and Toselli, 2005) approximate the effective height of a plume z_H and the horizontal distance d_f from the stack to the point where the plume height reaches z_H, depending on the emission characteristics, the ambient wind and the atmospheric stability. The gas elevation mainly depends on the density difference between the emitted gas and the atmospheric air (buoyancy rise) and the emission velocity (momentum rise).

2.3.1 Predominant buoyancy rise

In all cases with d_f different from zero, the driving force is buoyancy, except for stable conditions and calm wind. In order to know the plume rise trajectory, we propose to combine an horizontal and a vertical motion, satisfying certain known conditions.

The vertical motion along the mean trajectory of the plume is defined by an acceleration $a_0(t)$, a velocity $w_0(t)$ and $z(t)$, from the initial time $t = 0$ to the final time $t = t_f$ when the plume reaches the effective height, satisfying the following conditions

$$z(0) = z_c' \qquad z(t_f) = z_H \tag{12}$$

$$w_0(0) = w_c \qquad w_0(t_f) = 0 \tag{13}$$

Since there are four conditions on the vertical motion, we propose a cubic approximation of $z(t)$, and therefore a quadratic approximation of $w_0(t)$, and a linear approximation of $a_0(t)$.

The horizontal motion is defined by a uniformly accelerated motion, with a constant positive acceleration vector $\boldsymbol{a}_d = (a_{dx}, a_{dy})$, a velocity $\boldsymbol{u}_d(t) = (u_d(t), v_d(t))$, and an horizontal relative position vector $\boldsymbol{d}(t) = (x(t) - x_c, y(t) - x_c)$ with respect to the centre of the stack, satisfying the following conditions

$$|\boldsymbol{d}(t_f)| = d_f \tag{14}$$

$$\boldsymbol{u}_d(0) = \boldsymbol{u}(x_c, y_c, z_c) \tag{15}$$

In order to define the mean trajectory of bent curved plumes considering the influence of complex terrains, we approximate it by a three-dimensional polygonal line taking into account the ambient wind directions, such that the longitude of its projection on the horizontal plane approximates the longitude d_f. In addition, the final height coincides with the effective height z_H. Therefore, this approximation tries to satisfy the main values of the end of the plume considering Briggs' equations.

2.3.2 Predominant momentum rise

In all cases where d_f is equal to zero, that is when the driving force is momentum or when the driving force is buoyancy with calm wind, the horizontal motion of the plume until reaching the effective height can be considered negligible. Thus the trajectory of the gases is nearly vertical.

In this case, we propose a vertical motion along the trajectory of the plume with a constant negative acceleration a_0, a linear velocity $w_0(t)$ and a quadratic trajectory $z(t)$. Imposing the conditions (12) and (13), this vertical motion is completely defined.

$$t_f = \frac{2(z_H - z_c')}{w_c} \tag{16}$$

$$a_0 = \frac{-w_c}{t_f} \tag{17}$$

$$w_0(z) = w_c\sqrt{1 - \frac{2(z - z_c')}{w_c t_f}} \tag{18}$$

In order to modify the ambient vertical wind velocity (w) along the region of the plume rise, we need to have a sufficient mesh resolution in this area. For this reason, we propose to refine locally the mesh (González-Yuste et al., 2004) along the Gaussian plume (Green et al., 1980) until all the tetrahedra inside that region fulfill a size criterion.

Finally, a new ambient wind field $\boldsymbol{u}$ is obtained on the refined mesh with the mass-consistent model described in Sect. 2.2. The effect of the gas emission is introduced in this field by modifying its vertical component along the plume.

2.4 Air pollution simulation

The air pollution simulation consists of solving the unsteady convection-diffusion-reaction formulation with an stabilized finite element method, specifically Least-Squares method, with a Crank-Nicolson temporal discretization. The equation system governing the problem can be expressed with the following vectorial equation:

$$\frac{\partial \mathbf{c}}{\partial t} + \mathbf{u} \cdot \nabla \mathbf{c} = \nabla \cdot (\mathbf{K} \nabla \mathbf{c}) + \mathbf{e} + \mathbf{s}(\mathbf{c}) \qquad (19)$$

for the spatial coordinates $\boldsymbol{x}$ and time t, $(\mathbf{x},t) \in \Omega \times (0, t^{\text{end}}]$, with initial condition $\mathbf{c}(\mathbf{x},0) = \mathbf{c}^{\text{ini}}(\mathbf{x})$ on $\mathbf{x} \in \Omega$, and the following boundary conditions:

$$\begin{cases} \mathbf{c}(\mathbf{x},t) = \mathbf{c}^{\text{emi}}(\mathbf{x}) & \text{in } \Gamma_S\text{: Top of the stack} \\ \mathbf{n} \cdot \mathbf{K} \nabla \mathbf{c} = -\mathbf{V}^{\text{d}} \mathbf{c} & \text{in } \Gamma_R\text{: Terrain} \\ \mathbf{n} \cdot \nabla \mathbf{c} = 0 & \text{in } \Gamma_{W_{\text{out}}}\text{: Outlet wind boundaries} \\ \mathbf{c}(\mathbf{x},t) = \mathbf{c}^{\text{out}}(\mathbf{x}) & \text{in } \Gamma_{W_{\text{in}}}\text{: Inlet wind boundaries} \end{cases} \qquad (20)$$

where ∇ is the gradient with respect to $\mathbf{x}$, and $\mathbf{c}$, $\mathbf{u}$, $\mathbf{e}$ and $\mathbf{s}(\mathbf{c})$ are respectively the concentration, the perturbed wind velocity, the emission and the chemical vectors with a dimension n_c (the number of pollutant species). $\mathbf{K}$ is the diffusion matrix of dimension $3 \times n_c$, $\mathbf{V}^{\text{d}}$ is the deposition diagonal matrix with dimension n_c, and $\mathbf{n}$ is the outward-pointing normal unit vector, $\mathbf{c}^{\text{emi}}$ is the concentration of the emission in the top of the stack, and $\mathbf{c}^{\text{out}}$ the outside concentration at the inlet wind boundaries. Scalar product "·" is applied n_c times: the first argument is multiplied by each one of the n_c components of the second argument.

The complete description of photochemical reaction of atmospheric species is highly complex (Finlayson-Pitts and Pitts, 1997; Kley, 1997; Andreae and Crutzen, 1997; Ravishankara, 1997). For instance, detailed Volatile Organic Components decomposition involves hundreds of thousand reactions (Atkinson and Arey, 2003; Szopa et al., 2005) that needs special methodologies to reduce the number of the modelled reactions and species. Reference models for gaseous phase reactions involve some tens of compounds (Jimenez et al., 2003; Kirchner, 2005). The most simplified models just involve about ten reactive species (Zlatev, 1995). On the other side, depending on the application, it can be necessary to take into account aqueous phase reactions, that involve several other reactions and species. The RIVAD model is one of the most simplified models that permit to simulate both processes, aqueous and gaseous, involving transport and reaction of four species (Scire et al., 2000). In this paper, we have considered the RIVAD model for the chemical term $\mathbf{s}(\mathbf{c})$.

This model is a pseudo first-order chemical scheme for acid rain simulation, specially calibrated for being used in non-urban areas. The concentration $\boldsymbol{c}$ involves four species, $c_1 = [SO_2]$, $c_2 = [SO_4]$, $c_3 = [NO_2]$ and $c_4 = [NO_3]$, and the components of the reaction vector $\boldsymbol{s}(\boldsymbol{c})$ are:

$$s_1(\mathbf{c}) = -s_2(\mathbf{c}) = -\alpha_1(\mathbf{c})c_1 \qquad (21)$$

$$s_3(\mathbf{c}) = -s_4(\mathbf{c}) = -\alpha_3(\mathbf{c})c_3 \qquad (22)$$

where $\alpha_1(\mathbf{c}) = \gamma_1/(c_1 + \delta_1 c_3)$ and $\alpha_3(\mathbf{c}) = \gamma_3/(c_3 + \delta_3 c_1)$. Note that for values close to zero of the concentration of the primary species c_1 and c_3, both $\alpha_1(\mathbf{c})$ and $\alpha_3(\mathbf{c})$ requires a proper numerical treatment in order to avoid excessively high reaction rates.

Next we will treat the linear and the non-linear problem separately. The development of the linear chemical problem focuses on the temporal and spatial discretization of the corresponding linear convection-diffusion-reaction Eq. (19). The non-linear case focuses on the development of an splitting method that combines the solution of Eq. (19) in a linear case, considering a null chemical term, and the solution of an ordinary differential equation system that approximates the evolution of the chemical reaction separately.

2.4.1 Linear chemical problem

In this case the chemical term is linear, that is, $\boldsymbol{s}(\boldsymbol{c}) = \mathbf{A}\boldsymbol{c}$ where $\mathbf{A}$ is constant matrix. The resulting Eq. (19) is solved with a Crank-Nicolson time integration scheme, and an spatial discretization with a stabilized finite element method, Least-Squares. Concentrations $\boldsymbol{c}^n$ and $\boldsymbol{c}^{n+1}$ at times t^n and $t^{n+1} = t^n + \Delta t$ are related using a Crank-Nicolson scheme as $\boldsymbol{c}^{n+1} = \boldsymbol{c}^n + \frac{\Delta t}{2}\left[\frac{\partial \boldsymbol{c}^{n+1}}{\partial t} + \frac{\partial \boldsymbol{c}^n}{\partial t}\right]$. We define the differential operator $\mathcal{L}$ as

$$\mathcal{L} = \boldsymbol{u} \cdot \nabla - \nabla \cdot (\boldsymbol{K} \nabla) - \mathbf{A} \qquad (23)$$

and a function $\mathcal{F}$ in Ω from the known $\boldsymbol{c}^n$

$$\mathcal{F} = \boldsymbol{c}^n - \frac{\Delta t}{2}\mathcal{L}\boldsymbol{c}^n + \frac{\Delta t}{2}\left[\boldsymbol{e}^{n+1} + \boldsymbol{e}^n\right] \qquad (24)$$

Applying the Crank-Nicolson scheme, we can rewrite Eq. ((19)) as

$$\left[\mathcal{I} + \frac{\Delta t}{2}\mathcal{L}\right]\boldsymbol{c}^{n+1} = \mathcal{F} \qquad (25)$$

where $\mathcal{I}$ is the identity operator.

Using Least-Squares we obtain a symmetric problem such that the weak form of the Eq. (25) is

$$\left(\left[\mathcal{I} + \frac{\Delta t}{2}\mathcal{L}\right]v, \left[\mathcal{I} + \frac{\Delta t}{2}\mathcal{L}\right]c^{n+1}\right) = \left(\left[\mathcal{I} + \frac{\Delta t}{2}\mathcal{L}\right]v, \mathcal{F}\right) \qquad (26)$$

where v is the test function, and $(\mu, v) = \int_\Omega \mu v \, d\Omega$ is the inner product

We define the next bilinear forms

$$a(v,\boldsymbol{c}) = (v, \boldsymbol{u}\cdot\nabla\boldsymbol{c} - \mathbf{A}\boldsymbol{c}) + (\nabla v, \boldsymbol{K}\nabla\boldsymbol{c}) + \int_{\Gamma_R} \mathbf{V}^{\mathrm{d}}\boldsymbol{c}\, v\, d\Gamma_R \tag{27}$$

$$b(v,\boldsymbol{c}) = \sum_{\mathrm{e}} (\boldsymbol{u}\cdots\nabla v - \mathbf{A}v, \boldsymbol{c})_{\mathrm{e}} \tag{28}$$

$$g(v,\boldsymbol{c}) = \sum_{e} (\boldsymbol{u}\cdot\nabla v - \mathbf{A}v, \boldsymbol{u}\cdot\nabla\boldsymbol{c} - \mathbf{A}\boldsymbol{c})_{\mathrm{e}} \tag{29}$$

where $(\mu, v)_{\mathrm{e}} = \int_{\Omega_{\mathrm{e}}} \mu v\, \mathrm{d}\Omega$, and $\sum_{\mathrm{e}}$ represents the sum over all the mesh elements. Applying the operators $\mathcal{L}$ and $\mathcal{F}$ in the weak form (Eq. 26), we obtain

$$(v, \boldsymbol{c}^{n+1}) + \frac{\Delta t}{2}\left[a(v,\boldsymbol{c}^{n+1}) + b(v,\boldsymbol{c}^{n+1}) + \frac{\Delta t}{2} g(v,\boldsymbol{c}^{n+1})\right] = (v,\boldsymbol{c}^{n}) - \frac{\Delta t}{2}\left[a(v,\boldsymbol{c}^{n}) - b(v,\boldsymbol{c}^{n}) + \frac{\Delta t}{2} g(v,\boldsymbol{c}^{n})\right] + \Delta t\left[(v, \boldsymbol{e}^{n+\frac{1}{2}}) + \frac{\Delta t}{2} b(v, \boldsymbol{e}^{n+\frac{1}{2}})\right] \tag{30}$$

where $\boldsymbol{e}^{n+\frac{1}{2}} = \frac{\boldsymbol{e}^{n+1}+\boldsymbol{e}^{n}}{2}$.

Equation (30) can be written as an equation system

$$\mathbf{B}\boldsymbol{c}^{n+1} = \boldsymbol{f} \tag{31}$$

where $\boldsymbol{c}^{n+1}$ is the concentration vector approximation at t^{n+1} in the degrees of freedom of the finite element discretization, $\boldsymbol{f} = [2\mathbf{F} - \mathbf{B}]\boldsymbol{c}^{n} + \frac{\Delta t}{2}\mathbf{F}[\boldsymbol{e}^{n+1} + \boldsymbol{e}^{n}]$, and $\mathbf{B}$ and $\mathbf{F}$ are square matrices with dimension $(n_c \times n_{\mathrm{dof}})$, being n_{dof} the number of degrees of freedom.

In order to solve this linear system it is necessary to find an efficient solver, using a sparse matrix storage. Since $\mathbf{B}$ is a symmetric positive definite matrix, we have considered a solver based on a conjugate gradient method preconditioned with an incomplete Cholesky factorisation density type (Lin and Moré, 1999). The left-preconditioning is used to improve the convergence of the conjugate gradient method. The original linear system is transformed into $\boldsymbol{T}^{-1}\mathbf{B}\boldsymbol{c}^{n+1} = \boldsymbol{T}^{-1}\boldsymbol{f}$, where $\boldsymbol{T}$ is the symmetric positive definite preconditioner obtained with the incomplete Cholesky factorisation. The large fill-in of the complete (i.e. standard) Cholesky factorisation is completely or partially avoided by discarding coefficients along the factorisation process. We have considered an incomplete Cholesky factorisation with no fill-in, such that the incomplete factor $\boldsymbol{L}$ has the same sparsity pattern as the lower triangle of matrix $\mathbf{B}$. The main advantage of the Cholesky method is that the incomplete factorisation of matrix $\mathbf{B}$ can be amortized over many time-steps. More details about the implementation of this system equation solver can be found in Donea and Huerta (2003) and Rodríguez-Ferran and Sandoval (2007).

2.4.2 Non-linear chemical problem

To deal with the non-linearity of the reactive term in the convection-diffusion-reaction Eq. (19), we have considered a splitting method that separates this equation into a convection-diffusion equation and a reaction equation. We will make use of the second order splitting operator (*Strang splitting*) proposed by Ropp et al. (2004):

$$\frac{\mathrm{d}\boldsymbol{c}^{*}}{\mathrm{d}t} = \boldsymbol{s}(\boldsymbol{c}^{*}) \quad \text{for } t \in \left[0, \frac{\Delta t}{2}\right] \text{ and } \boldsymbol{c}^{*}(\boldsymbol{x},0) = \boldsymbol{c}^{n}(\boldsymbol{x}) \tag{32}$$

$$\frac{\partial\boldsymbol{c}^{**}}{\partial t} + \boldsymbol{u}\cdot\nabla\boldsymbol{c}^{**} = \nabla\cdot(\boldsymbol{K}\nabla\boldsymbol{c}^{**}) + \boldsymbol{e} \quad \text{for } t \in [0,\Delta t] \text{ and } \boldsymbol{c}^{**}(\boldsymbol{x},0) = \boldsymbol{c}^{*}(\boldsymbol{x},\frac{\Delta t}{2}) \tag{33}$$

$$\frac{\mathrm{d}\boldsymbol{c}^{***}}{\mathrm{d}t} = \boldsymbol{s}(\boldsymbol{c}^{***}) \quad \text{for } t \in \left[\frac{\Delta t}{2}, \Delta t\right] \text{ and } \boldsymbol{c}^{***}(\boldsymbol{x},\frac{\Delta t}{2}) = \boldsymbol{c}^{**}(\boldsymbol{x},\Delta t) \tag{34}$$

Once we have split the Eq. (19), we solve three equations in different time steps; the reaction Eq. (32), the convection-diffusion Eq. (33), and the reaction Eq. (34), being finally $\boldsymbol{c}^{n+1}(\boldsymbol{x}) = \boldsymbol{c}^{***}(\boldsymbol{x},\Delta t)$.

The convection-diffusion Eq. (33) is solved using the same method proposed in the previous Section, with $\mathbf{A} = 0$. The non-linear chemical Eqs. (32) and (34) are solved node by node with a second order Rosenbrock method (ROS2) (Verwer et al., 1999). To use the ROS2 method, the Jacobian square matrix of $\boldsymbol{s}(\boldsymbol{c})$ of dimension n_c has to be computed.

3 Results

The approach has been used to simulate the transport and reaction of pollutants from a fictitious stack in La Palma island.

The studied domain taken under consideration is a rectangular area with dimensions 15 600 × 22 803 m. The topography of the terrain is highly complex ranging from the sea level up to a maximum height of 2279 m with several deep valleys. The upper boundary of the domain has been placed at $h = 9000$ m. The digital elevation model of the area is defined over a uniform grid with a spacing step of 200 m in directions x and y. We add the stack geometry to the topographical data, a stack with a height of 150 m over the terrain and the diameter at its top of 15 m.

The wind field is obtained from four meteorological stations placed in the studied region. Figure 1 represents the interpolated and resulting wind field at a height of 20 m above the terrain. While the interpolated wind field is almost uniform and crosses the terrain surface, the resulting wind field has higher velocities in the peaks of the mountains and follows the terrain and valleys, verifying the wind incompressibility and terrain impermeability conditions.

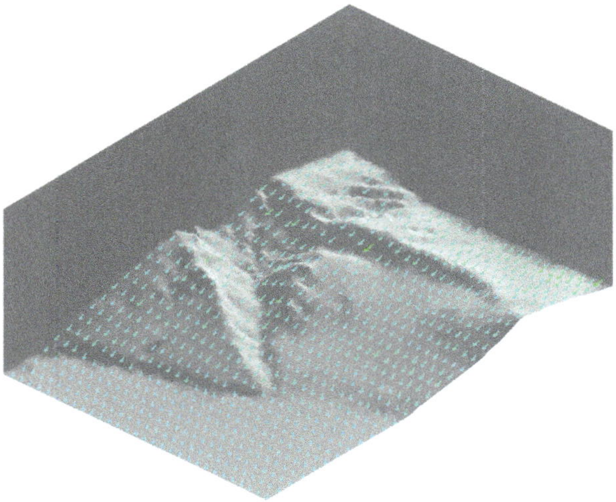

(a)

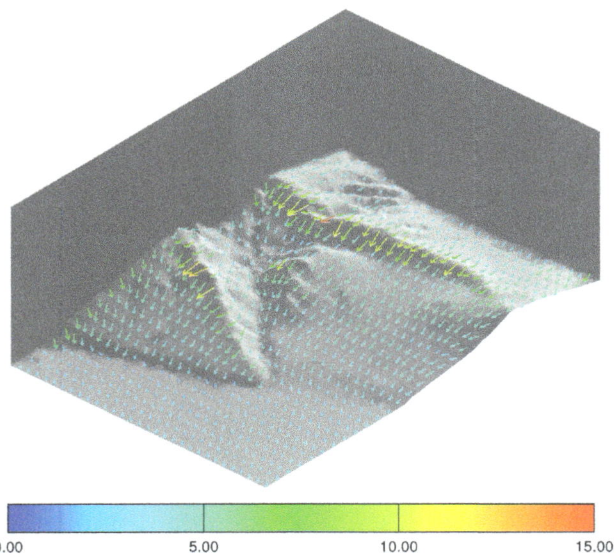

(b)

Figure 1: Interpolated **(a)** and resulting **(b)** wind field ($m\,s^{-1}$) at 20 m over the terrain.

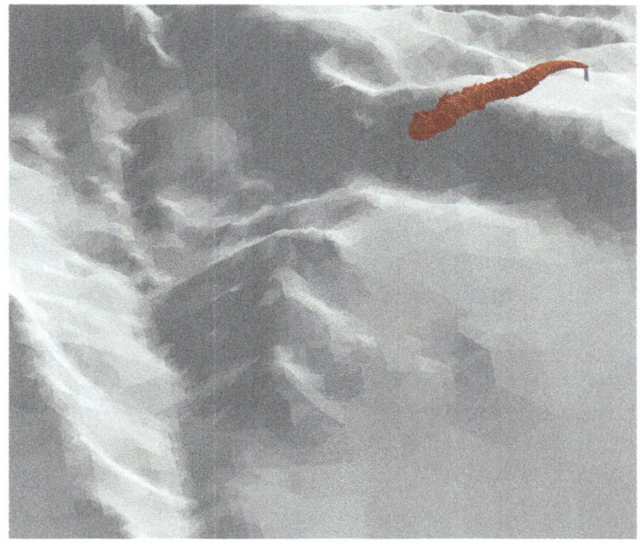

Figure 2: Refined region along the plume rise.

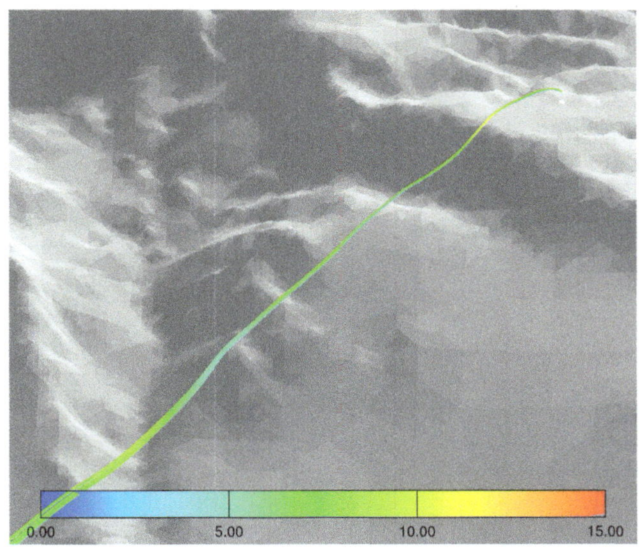

Figure 3: Modified wind field streamlines from the top of stack.

Figure 2 shows the plume rise region where a local mesh refinement has been performed. Figure 3 represents the streamlines of the modified wind field from the top of the stack. Note that streamlines follow the trajectory of the plume rise, so the vertical component of the wind field has been successfully modified.

The plume rise has been calculated using the following values. The stack exit velocity is $5\,m\,s^{-1}$ and the gas temperature is 573 K. The velocity of the wind field at the top of the stack is $V_o = 7.13\,m\,s^{-1}$. With these values, it results that the effective height of the plume is $z_H = 2347.52$ m and its horizontal distance is $d_f = 3700.04$ m.

Finally the transport and reaction of the pollutants has been simulated. The concentration of the primary pollutant at the top of the stack has been fixed to $6\,g\,m^{-3}$. We have considered a horizontal diffusion of $8\times10^{-6}\,m^2\,s^{-1}$, and a vertical diffusion of $4\times10^{-6}\,m^2\,s^{-1}$. The time-dependent problem has been simulated with a time step of 10 s.

Figure 4 shows the immission concentration distributions at 30 min for SO_2 and SO_4, respectively. These Figures are interesting since they give information about the pollutant concentration at the ground-level. Note that primary pollutant tends to have the highest concentrations near the

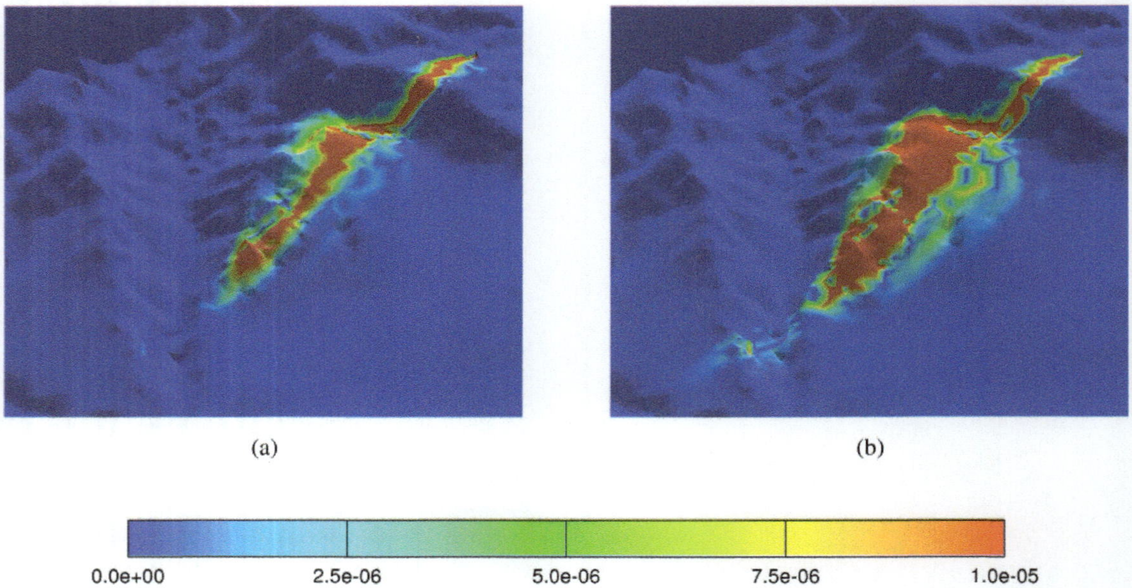

Figure 4: Immission concentration distribution (g m^{-3}) on the terrain of the primary **(a)** and secondary **(b)** pollutants at 30 min.

emission source, while the highest concentrations of the secondary pollutant are located further. This is consistent with the chemical reaction effect.

A final comment about the computational complexity of the evolution process should be done. For each time step we have to solve a finite element problem with a number of degrees of freedom about the number of nodes multiplied by the number of species, i.e. $455\,953 \times 4 = 1\,823\,812$. The number of time steps in the simulation period (about 30 min) is $30 \times 60/10 = 180$. Therefore, in the whole evolution process about 180 linear equation systems with 1 823 812 unknowns must be solved. The computational cost corresponding to the mesh generation, wind simulation, and the resolution of the ordinary differential equations in the splitting method are insignificant with respect to the resolution of the unsteady convection-diffusion equation. In a computer with 128 GB of RAM memory and 2.34 GHz, the total computing time is about 40 min. In a future work, the present computational complexity will be significantly reduced by using a refinement/derefinement strategy that follows the front of the pollutant plume, minimising the number of degrees of freedom in each time step.

4 Conclusions

The presented methodology promises to be useful to simulate air quality over complex terrains. The modified wind field has more reasonable trajectories and magnitudes than a simple interpolation of the wind data. The local mesh refinement along the Gaussian plume, allows to perturb the ambient wind field to introduce the effect of the pollutant emissions. The convection-diffusion-reaction equation obtains the values of concentration for all the pollutants in the whole three-dimensional domain. To really prove the usefulness of the proposed methodology, it will be validated against measured data in the near future.

Acknowledgements. This work has been supported by the Spanish Government, "Ministerio de Ciencia e Innovación", and FEDER, grant contracts: CGL2008-06003-C03, CGL2011-29396-C03-01, UNLP08-3E-010 and CSD2006-00032C.

Edited by: M. Piringer

References

Ahmad, N. N., Bacon, D. P., Hall, M. S., and Sarma, A.: Application of the multidimensional positive definite advection transport algorithm (MPDATA) to environmental modelling on adaptive unstructured grids, Int. J. Numer. Meth. Fl., 50, 1247–1268, doi:10.1002/fld.1113, 2006.

Andreae, M. O. and Crutzen, P. J.: Atmospheric Aerosols: Biogeochemical Sources and Role in Atmospheric Chemistry, Science, 276, 1052–1058, doi:10.1126/science.276.5315.1052, 1997.

Atkinson, R. and Arey, J.: Atmospheric degradation of volatile organic compounds, Chem. Rev., 103, 4605–4638, doi:10.1021/cr0206420, 2003.

Briggs, G. A.: Plume rise, Tech. rep., U.S. Atomic Energy Commission, Division of Technical Information, 1969a.

Briggs, G. A.: Optimum Formulas for Buoyant Plume Rise, Philos. T. R. Soc. S.-A, 265, 197–203, 1969b.

Chock, D. P., Whalen, M. J., Winkler, S. L., and Sun, P.: Implementing the trajectory-grid transport algorithm in an air quality model, Atmos. Environ., 39, 4015–4023, doi:10.1016/j.atmosenv.2005.03.037, 2005.

Donea, J. and Huerta, A.: Finite Element Methods for Flow Problems, John Wiley and Sons Ltd, West Sussex., 2003.

Escobar, J. M. and Montenegro, R.: Several aspects of three-dimensional Delaunay triangulation, Adv. Eng. Softw., 27, 27–39, doi:10.1016/0965-9978(96)00006-3, 1996.

Escobar, J. M., Rodríguez, E., Montenegro, R., Montero, G., and González-Yuste, J. M.: Simultaneous untangling and smoothing of tetrahedral meshes, Comput. Method. Appl. M., 192, 2775–2787, doi:10.1016/S0045-7825(03)00299-8, 2003.

Ferragut, L., Montenegro, R., and Plaza, A.: Efficient refinement/derefinement algorithm of nested meshes to solve evolution problems, Commun. Numer. Meth. En., 10, 403–412, doi:10.1002/cnm.1640100506, 1994.

Ferragut, L., Montenegro, R., Montero, G., Rodríguez, E., Asensio, M., and Escobar, J. M.: Comparison between 2.5-D and 3-D realistic models for wind field adjustment, J. Wind Eng. Ind. Aerod., 98, 548–558, doi:10.1016/j.jweia.2010.04.004, 2010.

Finlayson-Pitts, B. J. and Pitts, J. N.: Tropospheric Air Pollution: Ozone, Airborne Toxics, Polycyclic Aromatic Hydrocarbons, and Particles, Science, 276, 1045–1051, doi:10.1126/science.276.5315.1045, 1997.

González-Yuste, J. M., Montenegro, R., Escobar, J. M., Montero, G., and Rodríguez, E.: Local refinement of 3-D triangulations using object-oriented methods, Adv. Eng. Softw., 35, 693–702, doi:10.1016/j.advengsoft.2003.07.003, 2004.

Green, A. E. S., Singhal, R. P., and Venkateswar, R.: Analytic extensions of the Gaussian Plume Model, JAPCA J. Air Waste Ma., 30, 773–776, 1980.

Hanjalić, K. and Kenjereš, S.: Dynamic simulation of pollutant dispersion over complex urban terrains: A tool for sustainable development, control and management, Energy, 30, 1481–1497, doi:10.1016/j.energy.2004.05.001, 2005.

Jimenez, P., Baldasano, J. M., and Dabdub, D.: Comparison of photochemical mechanisms for air quality modeling, Atmos. Environ., 37, 4179–4194, doi:10.1016/S1352-2310(03)00567-3, 2003.

Kirchner, F.: The chemical mechanism generation programme CHEMATA – Part 1: The programme and first applications, Atmos. Environ., 39, 1143–1159, doi:10.1016/j.atmosenv.2004.09.086, 2005.

Kley, D.: Tropospheric Chemistry and Transport, Science, 276, 1043–1044, doi:10.1126/science.276.5315.1043, 1997.

Lagzi, I., Kármán, D., Turányi, T., Tomlin, A. S., and Haszpra, L.: Simulation of the dispersion of nuclear contamination using an adaptive Eulerian grid model, J. Environ. Radioactiv., 75, 59–82, doi:10.1016/j.jenvrad.2003.11.003, 2004.

Lalas, D. and Ratto, C.: Modelling of Atmospheric Flow Fields, World Scientific Publishing, Singapore, 1996.

Lin, C.-J. and Moré, J. J.: Incomplete Cholesky factorizations with limited memory, SIAM J. Sci. Comput., 21, 24–45, doi:10.1137/S1064827597327334, 1999.

Martín, M. J., Singh, D. E., Mouriño, J. C., Rivera, F. F., Doallo, R., and Bruguera, J. D.: High performance air pollution modeling for a power plant environment, Parallel Comput., 29, 1763–1790, doi:10.1016/j.parco.2003.05.018, 2003.

McRae, G. J., Goodin, W. R., and Seinfeld, J. H.: Development of a second-generation mathematical model for Urban air pollution – I. Model formulation, Atmos. Environ, 16, 679–696, doi:10.1016/0004-6981(82)90386-9, 1982.

Montenegro, R., Montero, G., Escobar, J. M., Rodríguez, E., and González-Yuste, J. M.: Tetrahedral Mesh Generation for Environmental Problems over Complex Terrains, in: Computational Science — ICCS 2002, edited by: Sloot, P., Hoekstra, A., Tan, C., and Dongarra, J., vol. 2329 of *Lecture Notes in Computer Science*, 335–344, Springer Berlin/Heidelberg, 2002.

Montenegro, R., Montero, G., Escobar, J. M., Rodríguez, E., and González-Yuste, J. M.: 3-D Adaptive Wind Field Simulation Including Effects of Chimney Emissions, in: Proceedings of WCCM VI/APCOM'04, Beijing, China, Tsinghua University Press and Springer-Verlag, 2004.

Montero, G. and Sanín, N.: 3-D modelling of wind field adjustment using finite differences in a terrain conformal coordinate system, J. Wind Eng. Ind. Aerod., 89, 471–488, doi:10.1016/S0167-6105(00)00075-1, 2001.

Montero, G., Montenegro, R., and Escobar, J. M.: A 3-D diagnostic model for wind field adjustment, J. Wind Eng. Ind. Aerod., 74–76, 249–261, doi:10.1016/S0167-6105(98)00022-1, 1998.

Montero, G., Montenegro, R., Escobar, J. M., Rodríguez, E., and González-Yuste, J. M.: Velocity Field Modelling for Pollutant Plume Using 3-D Adaptive Finite Element Method, in: Computational Science – ICCS 2004, edited by: Bubak, M., van Albada, G., Sloot, P., and Dongarra, J., vol. 3037 of *Lecture Notes in Computer Science*, 642–645, Springer Berlin/Heidelberg, 2004.

Montero, G., Rodríguez, E., Montenegro, R., Escobar, J. M., and González-Yuste, J. M.: Genetic algorithms for an improved parameter estimation with local refinement of tetrahedral meshes in a wind model, Adv. Eng. Softw., 36, 3–10, doi:10.1016/j.advengsoft.2004.03.011, 2005.

Moore, D. J.: A comparison of the trajectories of rising buoyant plumes with theoretical/empirical models, Atmos. Environ., 8, 441–457, doi:10.1016/0004-6981(74)90060-2, 1974.

Olcese, L. E. and Toselli, B. M.: Development of a model for reactive emissions from industrial stacks, Environ. Modell. Softw., 20, 1239–1250, doi:10.1016/j.envsoft.2004.08.008, 2005.

Ravishankara, A. R.: Heterogeneous and Multiphase Chemistry in the Troposphere, Science, 276, 1058–1065, doi:10.1126/science.276.5315.1058, 1997.

Rodríguez-Ferran, A. and Sandoval, M. L.: Numerical performance of incomplete factorizations for 3D transient convection-diffusion problems, Adv. Eng. Softw., 38, 439–450, doi:10.1016/j.advengsoft.2006.09.003, 2007.

Ropp, D. L., Shadid, J. N., and Ober, C. C.: Studies of the accuracy of time integration methods for reaction–diffusion equations, J. Comput. Phys., 194, 544–574, doi:10.1016/j.jcp.2003.08.033,

2004.
Sandu, A., Verwer, J., Loon, M. V., Carmichael, G., Potra, F., Dabdub, D., and Seinfeld, J.: Benchmarking stiff ODE solvers for atmospheric chemistry problems – I. Implicit vs explicit, Atmos. Environ., 31, 3151–3166, doi:10.1016/S1352-2310(97)00059-9, 1997.
Saylor, R. D. and Ford, G. D.: On the comparison of numerical methods for the integration of kinetic equations in atmospheric chemistry and transport models, Atmos. Environ., 29, 2585–2593, doi:10.1016/1352-2310(95)00187-4, 1995.
Scire, J. S., Strimaitis, D. G., and Yamartino, R. J.: A User's Guide for the Calpuff Dispersion Model (version 5), Earth Tech., Inc, Concord, MA., 2000.
Szopa, S., Aumont, B., and Madronich, S.: Assessment of the reduction methods used to develop chemical schemes: building of a new chemical scheme for VOC oxidation suited to three-dimensional multiscale HO_x-NO_x-VOC chemistry simulations, Atmos. Chem. Phys., 5, 2519–2538, doi:10.5194/acp-5-2519-2005, 2005.
Tomlin, A., Ghorai, S., Hart, G., and Berzins, M.: 3-D Multiscale air pollution modelling using adaptive unstructured meshes, Environ. Modell. Softw., 15, 681–692, doi:10.1016/S1364-8152(00)00038-4, 2000.
Verwer, J. G., Spee, E. J., Blom, J. G., and Hundsdorfer, W.: A Second-Order Rosenbrock Method Applied to Photochemical Dispersion Problems, SIAM J. Sci. Comput., 20, 1456–1480, doi:10.1137/S1064827597326651, 1999.
Winter, G., Montero, G., Ferragut, L., and Montenegro, R.: Adaptive strategies using standard and mixed finite elements for wind field adjustment, Sol. Energy, 54, 49–56, doi:10.1016/0038-092X(94)00100-R, 1995.
Zannetti, P.: Air Pollution Modeling, Computational Mechanics Publications, Boston, 1990.
Zlatev, Z.: Computer Treatment of Large Air Pollution Models, Kluver Academic Publishers, Dordrecht, 1995.

Effects of the climate change on regional ozone dry deposition

E. Kolozsi-Komjáthy, R. Mészáros, and I. Lagzi
Department of Meteorology, Eötvös Loránd University, Budapest, Hungary

Abstract. This impact study investigates connections between the regional climate change and the tropospheric ozone deposition over different vegetations in Hungary due to the possible changes of atmospheric and environmental properties. The spatial and temporal variability of the dry deposition velocity of ozone was estimated for different time periods (1961–1990 for reference period and two future scenarios: 2021–2050 and 2071–2100). Simulations were performed with a sophisticated deposition model using the RegCM regional climate model results as an input. We found a significant reduction of the ozone deposition velocities during summer months, which predicts less ozone damage to the vegetation in the future. However elevated ozone concentration and changed plant physiology can compensate the effect of this reduction.

1 Introduction

Tropospheric ozone plays an important role in the formation of the photochemical air pollution and affects both vegetation and human health. Recently it has also been shown that the indirect radiative forcing of climate change through ozone effecting on the land carbon sink could be an important factor and can induce a positive feedback for global warming (Sitch et al., 2007; Klingberg et al., 2010).

Harmful effects of near surface ozone on different surfaces can be quantified by different flux-based metrics (e.g., Ashmore et al., 2004; Gerosa et al., 2009) derived from the ozone concentration and deposition velocity. Both terms, and therefore the ozone load, can be modified due to the effects of global warming. The deposition velocity of ozone depends on both atmospheric and other environmental (soil, vegetation, surface) properties. Spatial and temporal distributions of the deposition velocity can also be predicted based on the regional climate model results (e.g., Bartholy et al., 2008). According to changed climatic conditions and emission patterns of ozone precursors, the concentration of ozone will also change (Meleux et al., 2007; Langner et al., 2005). The main aim of this research is to explore the spatial and temporal variability of the dry deposition velocity of ozone over Hungary regarding to possible changes in the future based on regional climate model results.

2 Model description

2.1 Input database

Regional Climate Model (RegCM) results are used for the input data of the ozone deposition model. RegCM has originally been developed at the National Center for Atmospheric Research (NCAR) and has been mostly applied to studies of the regional climate and the seasonal predictability around the world. The 3.1 model version has been adapted and used at the Eötvös Loránd University (Torma et al., 2008). The horizontal resolution of the model database is 10 km and the model domain is from 45.73° N to 48.70° N latitude and from 16.02° E to 23.04° E longitude. From the results of RegCM the following meteorological data are used in this study: average daily values of air temperature at 2 m, specific humidity at 2 m, u and v component of the horizontal wind at 10 m, root zone soil water content, total snow amount, incident solar energy flux, net absorbed solar energy flux, net infrared energy flux, surface air pressure. The Biosphere-Atmosphere Transfer Scheme's (BATS) land use categories are used (Fig. 1) in our study. Calculations for Hungary are performed for three time periods: two future scenarios for 2021–2050 and 2071–2100; and reference period (1961–1990). For the reference period calculations are made with the initial conditions obtained by the ECHAM global circulation model. These initial meteorological fields are used to compare future results to the reference period. Results with initial conditions from the ERA40 database are compared with the CRU climate database to verify the regional

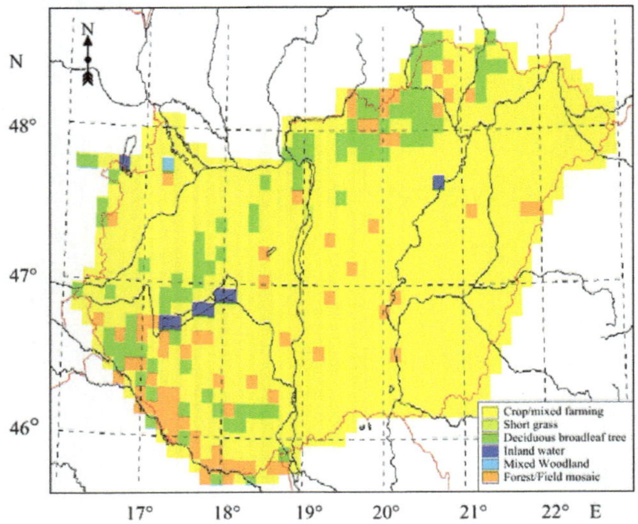

Figure 1. Land use categories over Hungary in RegCM-3 model.

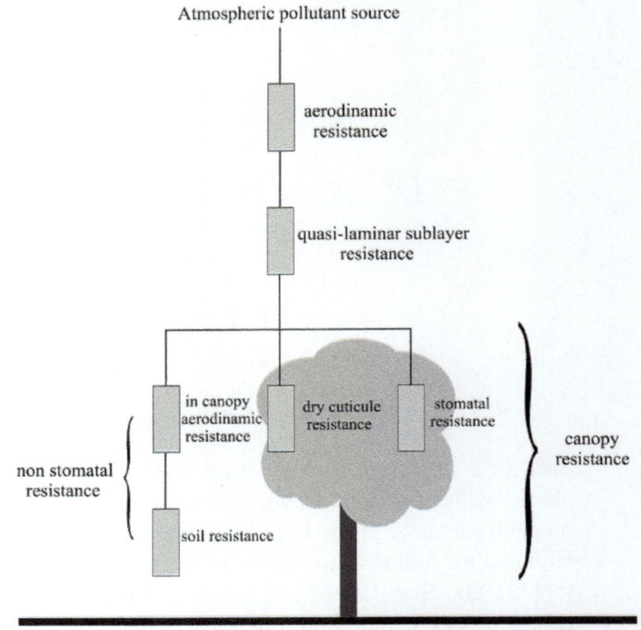

Figure 2. Calculation scheme of the resistance submodel.

climate model (Torma et al., 2008). In our study the comparison in changes of the ozone deposition velocity for the future are obtained by similar manner. This comparison is based on the fact that if the errors of a climate model are systematic, the right conclusions can be drown for the future using the results derived from ECHAM for the control period.

2.2 Deposition model

A sophisticated deposition model (Mészáros et al., 2009) is used to calculate the deposition velocity of ozone. The model has been developed at the Eötvös Loránd University, and it has been adapted and reconstructed for this specific task. The model uses a resistance analogy method in calculation of the deposition velocity. The scheme of the resistance model is depicted in Fig. 2. The parameterizations of the resistances in this model version are based on Zhang et al. (2003).

3 Results and conclusion

Using the results of the regional climate model as an input database of the deposition model based on daily values, a 30-yr averaged dry deposition velocity of ozone with its spatial and temporal variability is estimated over a Central European region including Hungary.

The spatial distribution of an average daily dry deposition velocity of ozone for each season in the control period (1961–1990) can be seen in Fig. 3. The highest values (up to 0.4 $cm\,s^{-1}$) occur in the summer period, the lowest ones in winter (around 0.1 $cm\,s^{-1}$). In spring and autumn, the deposition velocity varies around 0.2 $cm\,s^{-1}$. Yearly courses of the deposition velocity are similar for the three periods, but in the future a larger decrease is expected from May to August, while no considerable changes are found in other months (Fig. 4). The average values for the whole area are presented in this figure. According to the regional climate model results a major warming is expected for the far future in summer (+3.1 °C in average). For the near future this warming is 1.1 °C. In a warmer climate, the stomata of the plants are more closed to reduce water loss. Therefore, a decrease in the deposition velocity is expected for the future. In addition, the soil moisture and air temperature can also effect stomatal closure. The RegCM results show a decrease in precipitation over Hungary for the future. This drying and warming can cause decreased deposition velocities in our estimation.

The spatial variability of the changes in each season is presented in Fig. 5, where the relative differences of deposition velocities between the predicted (2021–2050 and 2071–2100, respectively) and the control periods are illustrated. The expected decrease in summer is higher in the Southern region than in the North. This decrease is due to the increasing stress factors on leaf stomata (higher temperature, relative humidity and soil moisture stresses). In the other seasons the changes are rarely greater than ±10%. In all seasons, a South-West – North-East axis can be found with a small decrease or increase of the deposition velocity in the North Eastern part of Hungary, and in general, larger decrease in the South, South-Western regions.

Figure 6 presents the average changes of the deposition velocity for each season. The magnitudes of the changes are larger for the period 2071–2100 except for spring. Larger changes are expected in summer, when a decrease of around

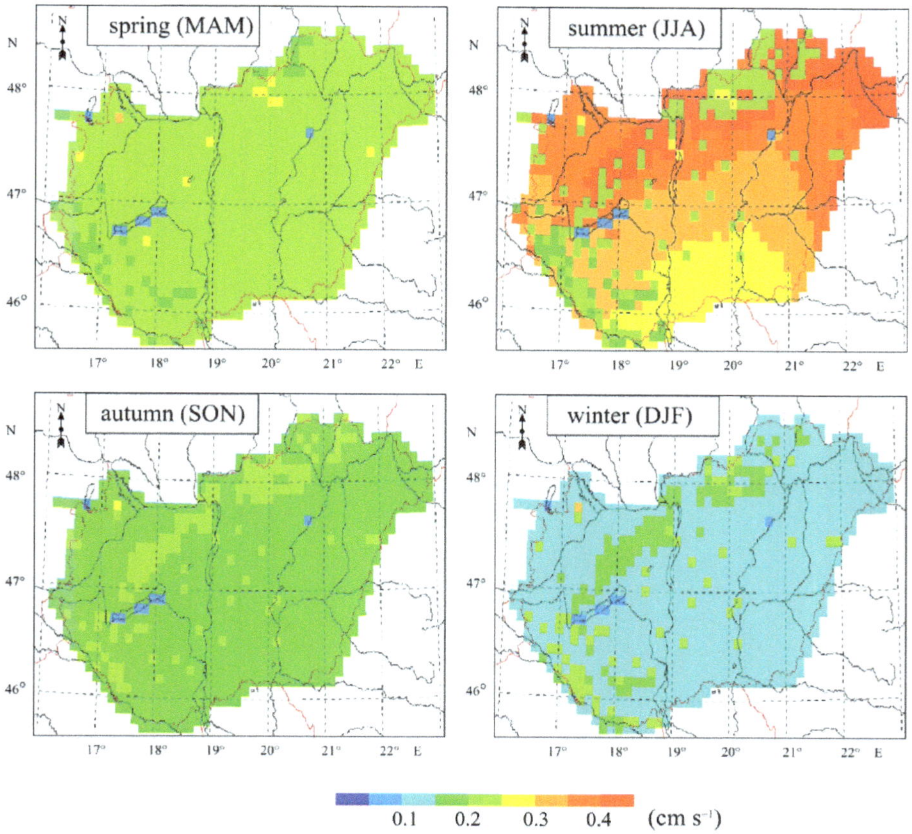

Figure 3. Average daily dry deposition velocity of ozone for different seasons in the control period (1961–1990) over Hungary.

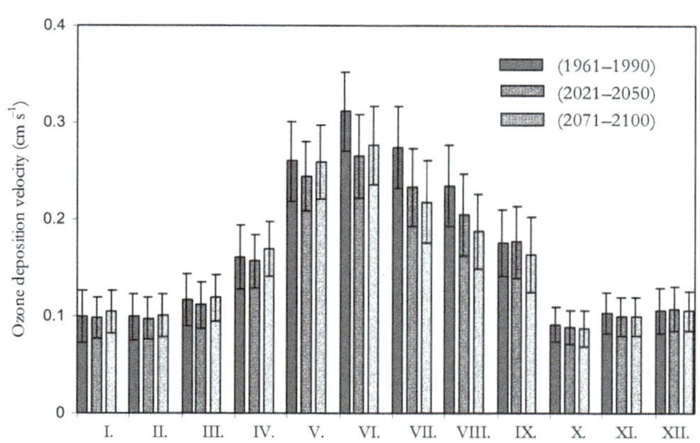

Figure 4. Average daily dry deposition velocity of ozone in each month for the following 30-yr periods: 1961–1990, 2021–2050, 2071–2100, (averages for all grid points; error bars are standard deviations).

12% and 14% is predicted in the values of the deposition velocity for the 2021–2050 and the 2071–2100 periods, respectively. A small increase of the deposition velocity is expected in winter for both future periods and in spring for the 2071–2100 period.

Based on these results, a lower deposition velocity is expected in the future in the vegetation period, which means a lower environmental load to vegetation assuming the same near surface ozone concentrations. However, the possible changes in ozone concentrations could modify these results. An expected increase of the ozone concentration may compensate or exceed the effect of the decreased deposition velocity. It must also be noted that these results correspond to Hungary (Central European region), and for other parts of Europe other changes in direction and magnitude might be expected, which require further investigations.

Acknowledgements. The authors would like to thank Judit Bartholy, Rita Pongrácz, Csaba Torma for RegCM data. This research is supported by Hungarian Research Found (OTKA K68253, K81933 and K81975) and the European Union and co-financed by the European Social Fund (grant agreement no. TÁMOP 4.2.1./B-09/1/KMR-2010-0003).

Edited by: M. Piringer

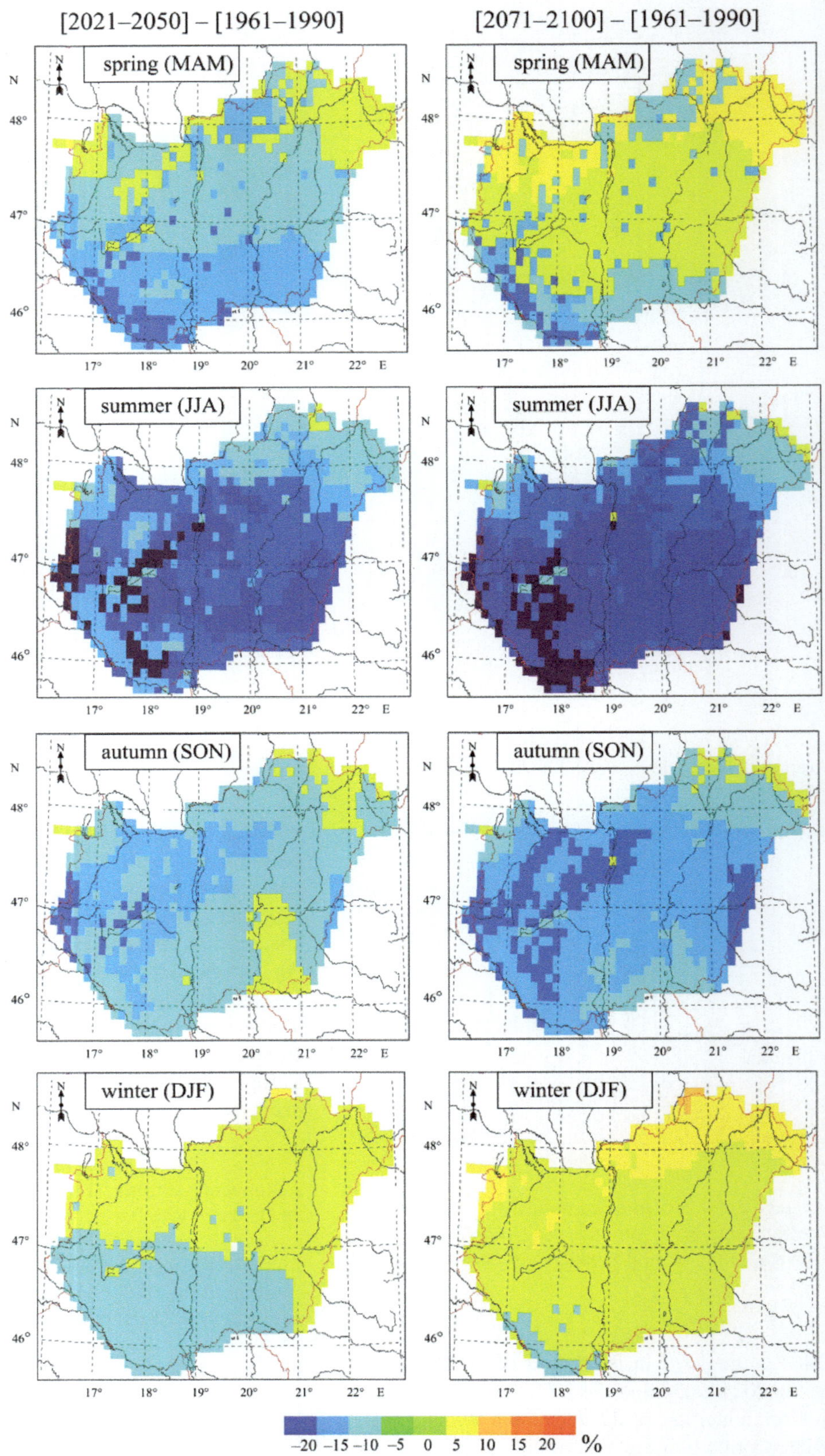

Figure 5. Spatial distribution of relative differences between ozone deposition velocities in each period for four seasons.

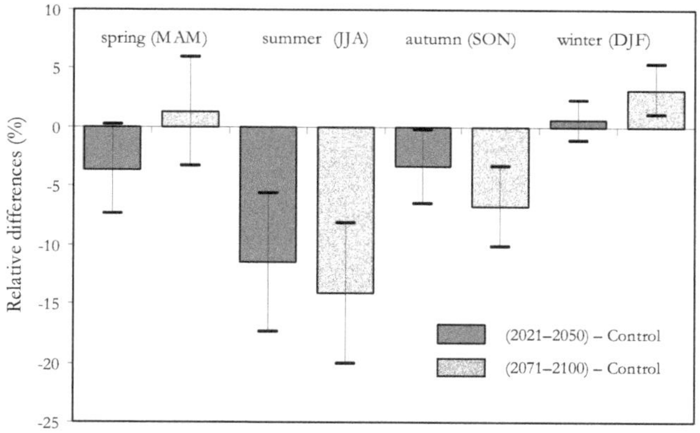

Figure 6. Averages and standard deviations of relative differences between ozone deposition velocities in each period.

References

Ashmore, M., Emberson, L., Karlsson, P. E., and Pleijel, H.: New directions: A new generation of ozone critical levels for the projection of vegetation in Europe, Atmos. Environ., 38, 2213–2214, 2004.

Bartholy, J., Pongrácz, R., Gelybó, Gy., and Szabó, P.: Analysis of expected climate change in the Carpathian Basin using the PRUDENCE results, Időjárás, 112, 249–264, 2008.

Gerosa, G., Marzuoli, R., Desotgiu, R., Bussotti, F., and Ballarin-Denti, A.: Validation of the stomatal flux approach for the assessment of ozone visible injury in young forest trees, Results from the TOP (transboundary ozone pollution) experiment at Curno, Italy, Environ. Pollut., 157, 1497–1505, 2009.

Klingberg, J., Engardt, M., Uddling, J., Karlsson, P. E., and Pleijel, H.: Ozone risk for vegetation in the future climate of Europe based on stomatal ozone uptake calculations, Tellus A, 63, 174–187, 2010.

Langner J., Bergström, R., and Foltescu, V.: Impact of climate change on surface ozone and deposition of sulphur and nitrogen in Europe, Atmos. Environ., 39, 1129–1141, 2005.

Meleux, F., Solmon, F., and Giorgi, F.: Increase in summer European ozone amounts due to climate change, Atmos. Environ., 41, 7577–7587, 2007.

Mészáros, R., Zsély, I. Gy., Szinyei, D., Vincze, Cs., and Lagzi, I.: Sensitivity analysis of an ozone deposition model, Atmos. Environ., 43, 663–672, 2009.

Torma, Cs., Bartholy, J., Pongrácz, R., Barcza, Z., Coppola, E., and Giorgi, F.: Adaptation of the RegCM3 climate model for the Carpathian Basin, Időjárás, 112, 233–247, 2008.

Sitch, S., Cox, P. M., Collins, W. J., and Huntingford, C.: Indirect radiative forcing of climate change through ozone effects on the land-carbon sink, Nature, 448, 791–794, 2007.

Zhang, L., Brook, J. R., and Vet, R.: A revised parameterization for gaseous dry deposition in air-quality models, Atmos. Chem. Phys., 3, 2067–2082, doi:10.5194/acp-3-2067-2003, 2003.

Wind variability in a coastal area (Alfacs Bay, Ebro River delta)

P. Cerralbo[1], M. Grifoll[1,2], J. Moré[3], M. Bravo[3], A. Sairouní Afif[3], and M. Espino[1,2]

[1]Maritime Engineering Laboratory, Polytechnic University of Catalonia (LIM/UPC), c./Escar, 6, 08039 Barcelona, Spain

[2]International Centre of Coastal Resources Research, Polytechnic University of Catalonia (CIIRC/UPC), c./Jordi Girona, 1–3, 08034 Barcelona, Spain

[3]Servei Meteorològic de Catalunya (SMC), c./Berlín, 38–46, 08029 Barcelona, Spain

Correspondence to: P. Cerralbo (pablo.cerralbo@upc.edu; pablocerralbo@gmail.com)

Abstract. Wind spatial heterogeneity in a coastal area (Alfacs Bay, northwestern Mediterranean Sea) is described using a set of observations and modelling results. Observations in three meteorological stations (during 2012–2013) along the coastline reveal that wind from the N–NW (strongest winds in the region) appears to be affected by the local orography promoting high wind variability on relatively short spatial scales (of the order of few kilometres). On the other hand, sea breezes in late spring and summer also show noticeable differences in both spatial distribution and duration. The importance of wind models' spatial resolution is also assessed, revealing that high resolution (= 3 km) substantially improves the results in comparison to coarse resolution (9 km). The highest-resolution model tested (400 m) also presents noticeable improvements during some events, showing spatial variability not revealed by coarser models. All these models are used to describe and understand the spatial variability of the typical wind events in the region. The results presented in this contribution should be considered on hydrodynamic, ecological and risk management investigations in coastal areas with complex orography.

1 Introduction

In the open sea and ocean, wind variability responds mostly to mesoscale structures like cyclones and anticyclones, as well as more permanent structures such as easterly (polar) and westerly (middle latitudes) winds. But when orographic constraints appear, such as oceanic islands (Chavanne et al., 2002) or mountains in coastal areas (Jiang et al., 2009), the wind presents high spatial variability, showing important curl gradients and becoming less predictable. In coastal areas, several examples of high spatial variability due to topographic constraints have been described (e.g. Herrera et al., 2005; Boldrin et al., 2009). In recent years, the application of numerical models in both the atmosphere and oceans has contributed to improving the understanding and description of this variability (Schaeffer et al., 2011). Moreover, modelling studies have revealed that the model resolution is a key factor for the correct representation of wind patterns, which could be essential in a correct prediction of flood episodes (Brecht and Frank, 2014) and could allow for the correct application of hydrodynamic modelling (Signell et al., 2005; Bignami et al., 2007). Several authors have described wind variability in lakes (e.g. Venäläinen et al., 2003), whilst only few studies focused on the wind description in a small-scale domain such as small estuaries or coastal areas. Considering the importance of wind on hydrodynamics, water mixing, waves and air quality, this contribution seeks to fulfill this gap, presenting an example of wind variability in a small-scale coastal area through observations and modelling results. A small bay, Alfacs Bay, in the northwestern Mediterranean Sea was selected.

This contribution is organized as follows. First, a short description of the study area (Sect. 2) and the set of observations is presented (Sect. 3). In this section the spatial and temporal variability observed is described. Then, the numeri-

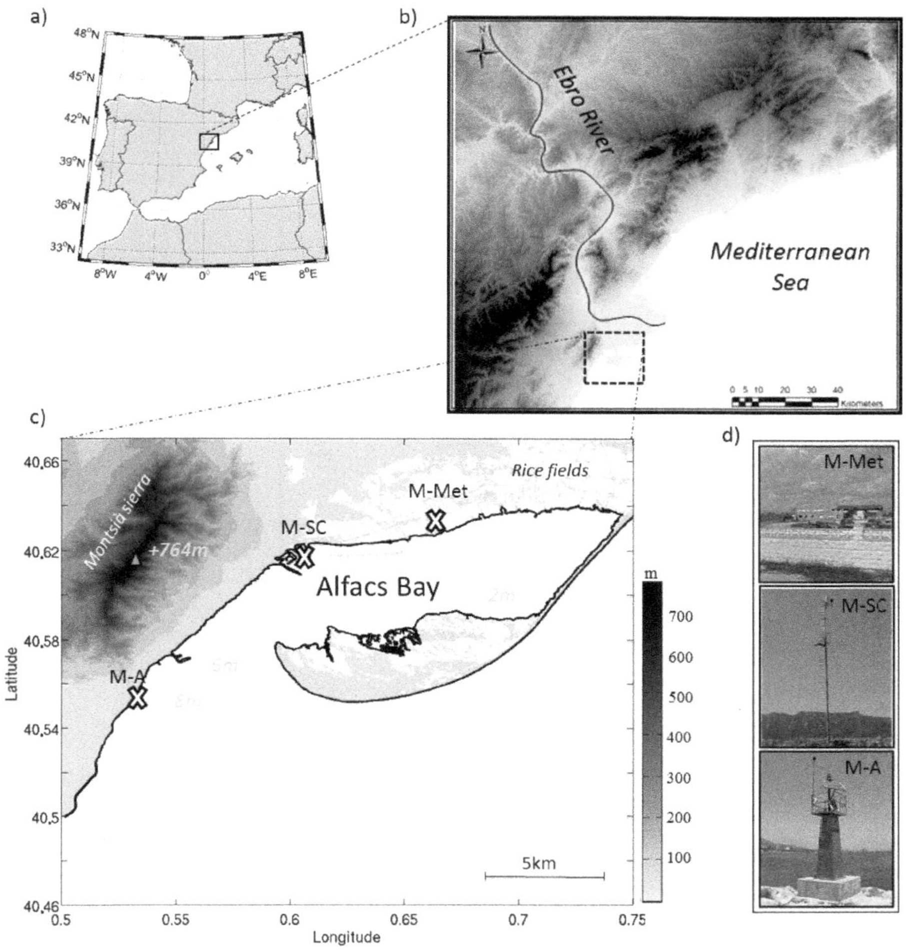

Figure 1. **(a)** Location map. **(b)** Ebro River delta region, with Alfacs Bay delimited by dashed square. **(c)** Study area showing meteorological stations (white crosses): M-A for Les cases d'Alcanar station, M-SC for Sant Carles de la Ràpita harbour station and M-Met for Meteocat station. Colour bar indicates altimetry above mean sea level. The bathymetry is also shown (isobaths, each 2 m). **(d)** Meteorological stations pictures.

cal weather prediction model implementation and outputs are shown, compared to the observed winds (Sect. 4). Finally, results for selected wind events and patterns observed are discussed.

2 Study area

The Ebro Delta (NE coast of Spain) forms two semi-enclosed bays, Fangar and Alfacs (to the north and south, respectively). The dimensions of Alfacs Bay are 16 km from head to mouth (Fig. 1c), 4 km wide and a mouth connection to the open sea of about 2.5 km. The bay is surrounded by rice fields to the north – which spill around 10 $m^3 s^{-1}$ of freshwater 10 months per year to the bay (Serra et al., 2007) – and a sand beach enclosing it on the eastern side, which can suffer breaching processes under severe storm conditions (Gracia et al., 2013). Serra de Montsià, with maximum altitudes around 700 m, closes the bay on the west side (Fig. 1).

The synoptic winds on the Catalan coast are affected by orographic constraints, such as the blocking winds of the Pyrenees that promote tramuntana (N) and mistral (NW) winds over some areas, and the wind channelling due to river valleys (Sánchez-Arcilla et al., 2008). Northerly winds in the region are mainly produced by high pressures over the Azores and lows over the British Isles and Italy; other synoptic situations could also lead to strong winds from the NW in the Ebro Delta (Martín Vide, 2005). Winds in the bay have been characterized as having a northwestern and southwestern predominance, with the strongest ones coming from the NW (channelized by the Ebro River valley; see Fig. 1b), being also the most common strongest winds on the Catalan coast during autumn and winter (Bolaños et al., 2009). On the other hand, some authors have reported the high spatial

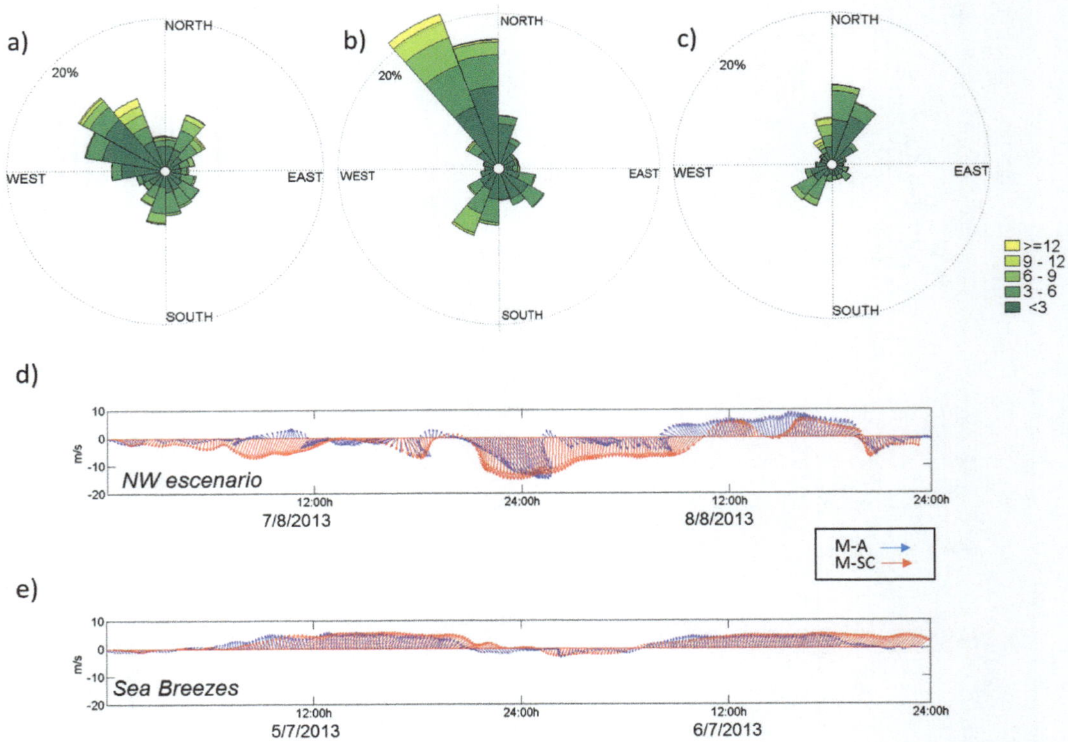

Figure 2. Wind roses for M-A **(a)**, M-Met **(b)** and M-SC **(c)** during the period 2012–2013. Velocities – colour bar in **(c)** – are in $\mathrm{m\,s^{-1}}$. Dotted line indicates frequency of 20 %. Wind time series for northwesterlies and sea breeze events in **(d)** and **(e)**, respectively. For **(d)** and **(e)**, a Lanczos filter of 2 h has been applied to the 10 min wind data.

heterogeneity of the wind fields inside the bay (Camp, 1994), in agreement with observed events during field campaigns by the authors of this manuscript, in which winds from the northwest were blowing inside the bay whilst in the mouth of the bay the wind was almost calm.

3 Observations

Atmospheric data – wind, atmospheric pressure, solar radiation and humidity – were obtained from three fixed land stations: Alcanar (M-A); Sant Carles (M-SC) from Xarxa d'Intruments Meteorològics de Catalunya (XIOM) described in Bolaños et al. (2009); and Alfacs-Meteocat (M-Met), which belongs to the automatic weather stations network of the Meteorological Service of Catalunya (http://www.meteocat.cat). Pictures of the stations are found in Fig. 1d. Both M-A and M-SC are at 10 m above the ground, while M-Met measures at 2 m. In order to compare wind intensities from all stations, we have adapted the measurements at 2m to the standard height of 10 m. The method adapted in Herrera et al. (2005) from Oke (1987) is used to compute the velocities at 10 m (w_{10}) from the observed values (w_h), following $w_{10} = w_h \frac{\log\left(\frac{10}{z_s}\right)}{\log\left(\frac{h}{z_s}\right)}$, where h represents the measurement height (2 m). Following Agterberg and Wieringa (1989), we have considered a typical roughness (z_s) for plains with low vegetation (rice fields) of ≈ 0.03 m. The roughness variability as a function of the wind direction is not considered.

Observations from June 2012 to June 2013 at three meteorological stations show noticeable differences (Fig. 2), confirming high variability among them. This period has been chosen for this analysis because it is the only one with data from all three stations. In M-SC (Fig. 2c) a bimodal behaviour with the most common winds from the southwest and north–northeast are shown. These winds are in agreement with data acquired at the same station for about 16 years (not shown), indicating high representation for the period selected. The highest intensities correspond to N–NW winds ($> 9\,\mathrm{m\,s^{-1}}$). For M-A the directions are more scattered, the most common winds being from W-NW to NW. The highest-intensity winds still come from the NW–NE, but the purely northern winds are less common. Finally, winds from M-Met show also a clear bimodal behaviour, with winds from the NW and SW–SE being the most common. However, M-Met is altered by the human buildings on the south side, which would alter on the wind measurements. Among the three stations most of the differences are clearly seen in land winds due to possible effects of the Serra de Montsià mountain range, showing high heterogeneity in wind fields in short distances. To better understand the wind variability, directional

Table 1. Summary of the main characteristics of the three different model configurations used in this study.

Model	Domain	Nominal resolution	Lead time	Outputs
WRF9	Iberian Peninsula	9 km	72 h	3 h
WRF3	NE Spain	3 km	48 h	1 h
CALMET	SW Catalonia	0.4 km	48 h	1 h

scatter plots for stations with data at 10m height are shown in Fig. 3a. The corresponding mean velocity for the two stations (M-A and M-SC) is also shown in colours. This figure shows that winds from the SW–SE in M-A are rotated 20–30° clockwise in M-SC, as well as that winds from the NW–N in M-A seem to be rotated a bit clockwise in M-SC (also seen in Fig. 2). On the other hand, winds from W to NW in M-SC are not observed as in M-A; M-SC concentrates more on the N–NE winds. All these data illustrate that Serra de Montsià (Fig. 1) could act as a physical barrier to some type of northerly winds, redirecting them. These effects are probably most clear on mistral winds (NW), which are not completely reproduced by M-SC (Fig. 2d) but oriented to the north. Another meteorological station was planned to be operating between these two stations, but in the end it was no possible to deploy it. The winds from the SW–SE correspond mainly to see breezes in spring and summer (Bolaños et al., 2009). The time and spatial evolution of sea breeze differences between stations are also observable (Fig. 2e). Weak nocturnal offshore winds (NW) rotate and increase in intensity until the maximum of around 5 m s^{-1} and S–SW direction in M-A is reached. Similar behaviour is observed in M-SC, but the nocturnal winds come from the NE and the rotation is very noticeable, as breezes arise in the afternoon from the SW. In M-Met the behaviour is almost the same as M-A. All these data reflect that even during sea breezes the orography and probably variation of land uses (rice fields in the delta's plain versus brush forest in Serra de Montsià) affect the direction, intensity and durability of winds. Winds from the S–SE (not related to sea breezes) are also probably affected by topography (Fig. 3), but there are not enough observations to investigate the spatial variability in this case. The lack of wind data just in front of Serra de Montsià does not allow us to know the exact behaviour of NW (mistral) winds and sea breezes in the mouth of the bay, but the use of numerical models would allow us to approximate the corresponding theoretical wind patterns. The wind module reveals good agreement between both stations (not shown). Thus indicating most of the variability related to direction. All the observations are based on 1-year-long data, so no climatic conclusion is expected from our analysis.

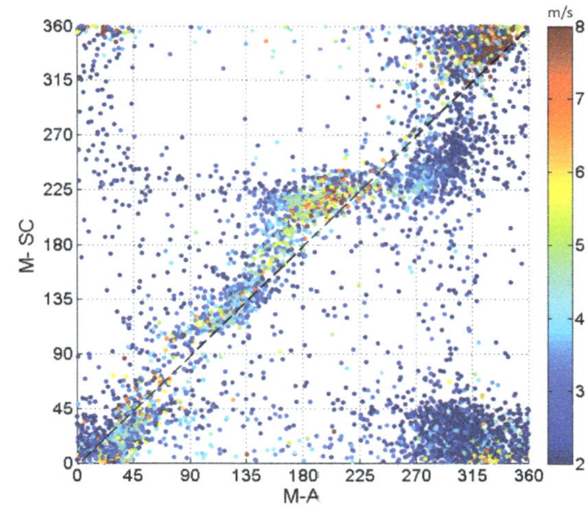

Figure 3. Wind direction comparison between M-A and M-SC for 1 year. The coloured data indicate mean wind velocity values (m s^{-1}) from both stations.

4 Numerical modelling

4.1 Description

Two different atmospheric models, provided by Meteocat, are used in this study to assess the role of spatial heterogeneity in the bay. For this purpose, the Weather Research and Forecasting (WRF) model has been selected (see http://www.wrf-model.org/), in particular the dynamical solver called Advanced Research WRF (WRF-ARW). The WRF-ARW is a model that has a worldwide community of users. It is a fully compressible, non-hydrostatic mesoscale model (Skamarock et al., 2008) and is based on solving the primitive equations of the atmosphere with different physical parameterizations available. Version 3.1.1 is used, with some minor code changes that consider an increase of the surface stress parameter (u-star) by a factor of 20 % in order to address the surface wind overestimation in the WRF model that has been reported by several authors (Mass and Ovens, 2011). By introducing this modification, a better agreement with meteorological observations is obtained. In later versions of the WRF model, more sophisticated corrections have been incorporated in the same direction, such as a surface drag parameterization scheme which allows enhancing the u-star depending on the sub-grid terrain variance (Mass and Ovens, 2011). The main parameterizations used are the Thompson scheme (Thompson et al., 2004) for microphysics, the YSU scheme (Hong et al., 2006) for planetary boundary layer and the unified Noah land surface model (Tewari et al., 2004) for surface physics. More configuration details are available at http://www.meteo.cat. This model is run using Global Forecast System (GFS) data as initial and boundary conditions (with 0.5° and 6 h spatial and temporal resolution). Two different configurations depending on grid resolution have been analysed: 9 and 3 km (named hereafter WRF9 and WRF3).

The main model parameters are summarized in Table 1. Not only spatial resolution is different between them; temporal resolution is also different. These configurations correspond to the available products of the Meteocat meteorological operational forecast system.

On the other hand, an additional simulation has been considered to derive atmospheric data at a very high resolution (400 m). In particular, the WRF-ARW outputs at 3 km are downscaled by a diagnostic meteorological model called CALMET. CALMET, a component of the CALPUFF Modeling System designed for the simulation of atmospheric pollution dispersion (Scire et al., 1999), is a diagnostic three-dimensional meteorological model which includes parameterized treatments of slope flows, kinematic terrain effects and terrain blocking effects, among others. These particular aspects help to better represent regional flows with an efficient computational cost.

The operational numerical model resolutions (spatial and temporal) were selected considering both reliability and computational costs. The computational cost for a 3-day forecast for CALMET (at 400 m resolution) is about 2 h using 10 CPUs. On the other hand, a similar domain and forecast horizon using WRF3 (3 km of resolution) take around 1 h (30 CPUs). In this sense, the application of WRF at higher resolution is not currently considered as an operational product due to computational limitations. For all the verification processes the daily first 24 h of prediction from the operational system are used.

4.2 Model verification

In this section, we present the results of verification studies to assess the performance of wind velocity and direction prediction from the models against the observation. The verification of both WRFs (WRF3 and WRF9) and CALMET is shown in Fig. 4 for summer 2013 (Fig. 4a) and winter 2014 (Fig. 4b). Different oceanographic campaigns were developed at Alfacs Bay during the same periods. We chose that period in order to coincide the meteorological observations and model results with oceanographic data. Verification of all the models and resolutions is done against hourly means from observational data. The wind module velocities measured in both M-A and M-SC are graphically compared to modelling results for both systems (model data is interpolated through bi-lineal interpolation to corresponding points) using a Taylor diagram (Taylor, 2001). In this diagram we can observe the comparison between observations and model through the correlation, the centred root-mean-square difference (CRMSD) and the standard deviations (SDs; see Appendix). In order to compare in the same figure different models against different observations, the standard deviations and CRMSD are normalized over the standard deviation of the corresponding observations (Grifoll et al., 2013). The model skill improves as the points get closer to the observation reference point in the diagram. In summer 2013, better correlation (around 0.6) is observed in M-SC, while in M-A the values decrease to ≈ 0.5. The standard deviation shows that the model presents lower-amplitude variations than the observed values. In spring 2014 only values from M-A were available. In this period, the correlations are higher for all models, and the SDs are more similar to the observations. Differences through different models are only observable in M-SC locations, revealing the highest correlations and lower errors in higher-resolution models (best results corresponding to CALMET). In M-A, the models do not show noticeable differences.

The wind module velocities measured in both M-A and M-SC are compared in Fig. 5a and b, respectively, for summer 2013 with modelling results using a Weibull distribution, which is defined as a two-parameter function commonly used to fit the wind speed frequency distribution. In both stations, the best fit between the model and observational Weibull distributions is observed for CALMET. In M-A, CALMET and the observational distribution show almost equally shaped coefficient; even observational data present stronger winds. WRF3 also has similar shape, with even more energy distributed at medium wind intensities ($= 3\,\mathrm{m\,s^{-1}}$). WRF9 seems to overestimate the mean winds. In M-SC, winds from CALMET are the most similar to observations, even overestimating the frequency of mean winds, and then not reproducing the maximum intensities (6–10 $\mathrm{m\,s^{-1}}$). On the other hand, WRF3 and WRF9 overestimate both mean wind intensities and corresponding frequencies. In the winter season, observations show higher wind intensities, and both CALMET and WRF3 present Weibull shapes (not shown) similar to observations.

Some characteristic events representing the most usual winds in the area have also been analysed in order to understand the behaviour of each model under different conditions (resumed in Table 2). A period of 3 days is considered to have enough data to compare. Results show that winds from CALMET and WRF3 have higher correlations (except in northwest 2) than WRF9. The worst results are observed during northwesterly winds in summer 2013. This is clearly affected by the topography, and the observed wind in M-A is not reproduced by any of the models. Slightly better results are obtained in M-SC, especially by CALMET (but still with poor correlation). This event was characterized by its shortness (less than 6–8 h) and unsteadiness. On the other hand, in the winter period, another NW wind event (lasting for more than 1 day) was reproduced with noticeable agreement in M-A. In this case, the simulation WRF9 seems to reproduce quite well the wind time series in M-A, but no data for M-SC are available to compare. Southern winds – southeast and diurnal regime of sea breeze – are better reproduced by the finest models, being the highest improvement between WRF3 and WRF9. During sea breezes and considering the complete diurnal–nocturnal cycle, the CALMET model seems to reproduce winds better than the coarser ones. Considering the daily duration of the sea breeze – between 6 and 8 h – all the

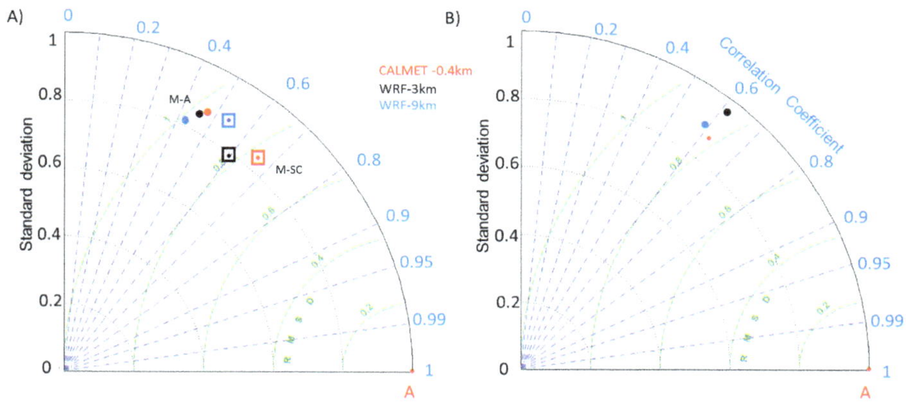

Figure 4. Taylor diagram for summer 2013 **(a)** and winter 2014 **(b)** for both M-A (coloured dot) and M-SC (coloured square filled with dot) model configurations (red for CALMET, black WRF3 and blue for WRF9) compared to corresponding meteorological stations. Both modelled standard deviations and RMSD are normalized over observational standard deviation.

Table 2. Correlation among the three different atmospheric models and observational data for 3-day-long events during summer 2013 and winter 2014. No correlation for winter 2014 (M-SC dismantled on September 2013).

	Day	M-A			M-SC		
		WRF9	WRF3	CALMET	WRF9	WRF3	CALMET
Northwest	8 Aug 2013	.02	.01	.12	.21	.40	.46
Northwest 2	4 Apr 2014	.80	.72	.75	–	–	–
Southeast	13 Aug 2013	.43	.64	.64	.58	.71	.75
Sea breezes	6 Jul 2013	.64	.67	.74	.64	.66	.76
Northeast	28 Mar 2014	.75	.86	.83	–	–	–

models would be able to reproduce it (Table 2). However, the temporal variability of such processes could only be reproduced using high-temporal-resolution models (≈ 1 h).

4.3 Spatial patterns and wind variability

Model wind snapshots for the three different resolutions are used to understand the spatial structures associated with most common winds in the area (Fig. 6). Three events have been chosen, representing a case with higher variability (Fig. 6a, d and g) to one with an almost homogeneous wind field (Fig. 6c, f and i). For northwesterly winds (left column panels in Fig. 6) it is clear that Serra de Montsià exerts a physical barrier on wind fields, thus revealing areas in the inner bay with high wind intensities and areas down the mountain with almost calm winds or with different direction – shadow effect, described in other environments such as the Hawaiian Islands in Chavanne et al. (2002). These effects were also observed for winds coming from the north (not shown). Atmospheric pressure at surface on 4 April 2014 shows low pressure over the North Atlantic and a high-pressure area over north Africa. This synoptic situation promotes winds from the north–northwest (triggered by the Ebro River valley) in the study area. The modelled winds corresponding to observations in M-A and M-SC locations are similar, not showing all the direction variability measured in observations (Figs. 2 and 3). However, the wind patterns in both WRF3 and CALMET are similar and show spatial wind variability inside the bay, thus indicating that the medium-resolution model is able to reproduce topographic constraints under these circumstances. On the other hand, the coarser model (Fig. 6a) does not reveal such a variability – expected for the dimensions of the bay and model resolution, with pixels almost half of the bay size. In summary, CALMET reproduces with higher accuracy these kind of winds in M-SC, while in M-A the errors are similar to coarser models. Both stations are located near the maximum transition zone between high- and lower-intensity winds (Fig. 6g), corresponding to the areas where modelling errors would be more sensitive to topographic effects.

An intensification of sea breezes (Fig. 6b, e and h) at midday in inner areas of the bay is clearly represented as well as a clockwise gyre of wind in M-SC related to M-A. The modelled highest intensities in the inner bay are not able to be validated, due to the lack of more observational data in this area (M-Met has been defined as a bad indicator of wind field). On the other hand, differences from coarser to the finest model configurations are noticeable. Both WRF3 and CALMET show some spatial structures in daily regimes

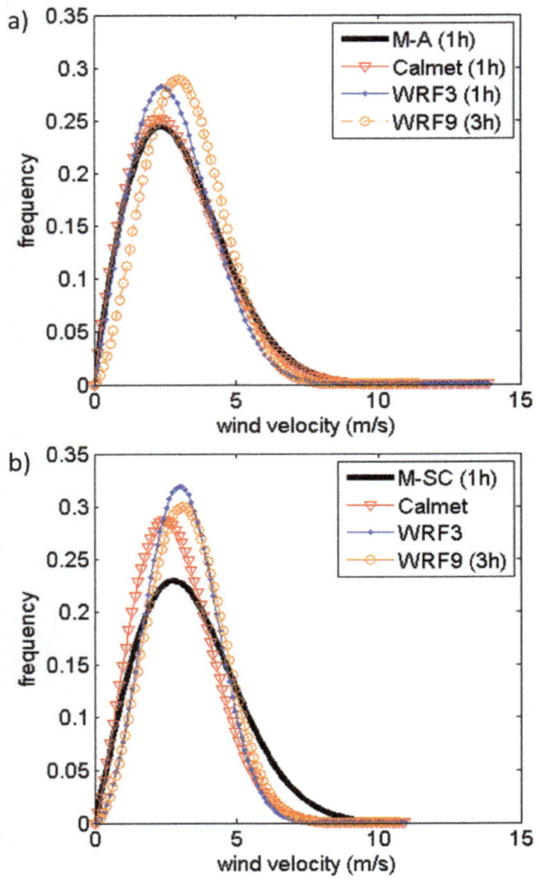

Figure 5. Weibull distributions for summer 2013 in M-A **(a)** and M-SC **(b)**. Black line for observational data (hourly) and coloured lines for each model configurations (red for CALMET, blue for WRF3 and orange for WRF9).

not solved by WRF9. In the time series, sea breezes time-lag between M-A and M-SC observations is not reproduced by any of the models.

On the other hand, spatially homogeneous wind fields have also been observed during several events. In this case, winds from the northeast are shown in Fig. 6c, f and i. The wind fields reproduced by observations and atmospherical models indicate homogeneous spatial winds, not affected by topography in the Ebro Delta (winds coming along the coast). For these winds, the coarser model does reproduce the wind pattern similar to the finest model. In Fig. 3a there is an area where directions around 0–45° shows high correlation between M-SC and M-A. However, in Table 2 the NE case shows that CALMET and WRF3 have better correlation than WRF9, indicating that in some cases the temporal resolution would also play an important role in wind prediction.

5 Final remarks

This contribution presents an example of high wind variability in a coastal area (Alfacs Bay, NW Mediterranean Sea) during the period 2012–2013. Observational data demonstrate that wind direction seems to be affected by the surrounding mountains. These effects are maximized during northwesterly winds, in which the local mountains exert noticeable shadowing effects over the mouth of the bay. These results are in agreement with Camp (1994), showing high wind spatial variability at Alfacs Bay, as well as other similar studies which show the wind-channelling effects in some rias of Galicia (Herrera et al., 2005). Other winds, like sea breezes, also show noticeable variability, not only in space but also in time, and probably related to orography, land uses and different sea-water temperature in the bay and in the open sea. Winds from the S–SE not related to sea breezes are likely affected by local orography, but not enough events were recorded to confirm the observed pattern. Due to the short length of observational period (around 1 year), the results have no climatic significance.

The spatial heterogeneity also plays an important role in wind modelling results in this coastal area. The coarse model (9 km) does not reproduce the spatial variability associated with most topographically influenced winds (northwesterlies and sea breezes). The medium-resolution model (3 km) has proven to represent the wind spatial fields with enough accuracy according to the observations. This indicates that the effects of the main topographic structures on the area are recognized by this resolution model, contrary to other places where similar model resolution was not able to reproduce all the wind variability due to orography (Cerralbo et al., 2012). The predicted wind from the CALMET model at the highest resolution (400 m) also improves the spatial variability and shows the highest correlation with observational data under some circumstances. In this sense, the CALMET model could be an interesting and useful product for ocean and wave modelling, minimizing the information losses due to the downscaling processes. In summary, all the systems analysed reproduce with enough accuracy some of the characteristic winds observed at Alfacs Bay. The highest-resolution model shows better responses, since it reproduces more realistically wind fields and discriminates topographic structures such as mountains and gaps between them. However, correlation in some cases is higher in coarse models (WRF9), agreeing with Signell et al. (2005), who demonstrate that sometimes the higher-resolution models would present lower correlation due to higher "noise" (more variability) compared to the coarser models. Other authors (e.g. Miglietta et al., 2012; De Biasio et al., 2014) also argue that local models could show more details (more detailed flow patterns) although worse statistics due to errors in timing and location, whilst global models would produce smoother results and probably much skillful forecasts. At this point, the computational cost

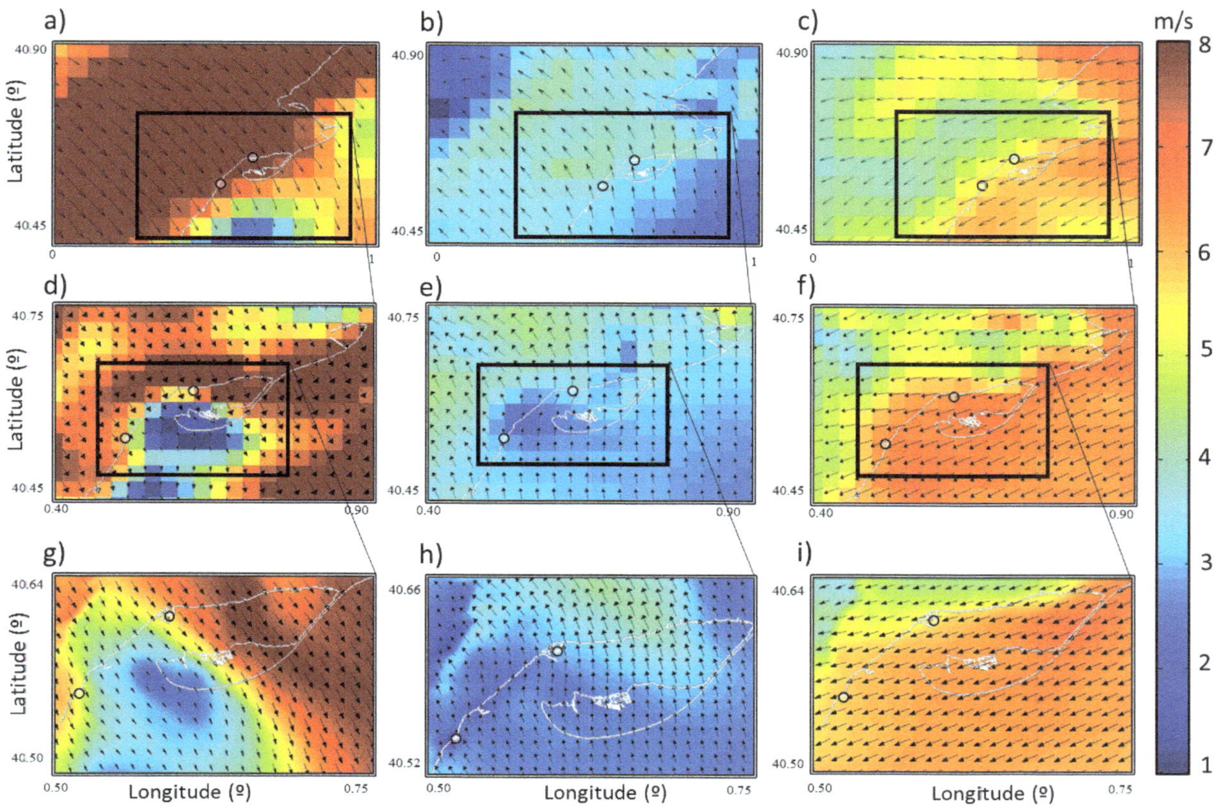

Figure 6. The three different models configurations are plotted for three snapshots of typical wind events at Alfacs Bay. **(a–c)** for WRF9 km, **(d–f)** for WRF3 and **(g–i)** for CALMET 400 m. Events represent winds from the northwest (left column panels) on 4 April 2014, sea breezes (central panels) on 6 July 2013 and northeasterly winds (right panels) on 12 March 2014. White circles indicate meteorological station locations (M-A and M-SC).

would indicate which should be the atmospheric model to be considered depending on the skill assessment requirements.

The effects of wind spatial variability on relatively short length scales would be an important factor to be considered in studies dealing with biology and ecology hazards, e.g. harmful algal blooms as described in Quijano-Scheggia et al. (2008), hydrodynamics (Cucco and Umgiesser, 2006) and water quality parameters (Grifoll et al., 2011) in coastal waters.

Appendix A: Statistics

The statistics used in the normalized Taylor diagram are defined as follows, where "obs" corresponds to observations, m corresponds to model results and the over bar (–) denotes all data length mean values:

$$\mathrm{SD} = \frac{\left(\sqrt{\frac{\sum_{i=1}^{n}(m_i - \overline{m})}{n}}\right)}{\mathrm{SD_{obs}}}, \tag{A1}$$

$$\mathrm{CRMSE}(\mathrm{obs}, m) = \frac{\sqrt{\frac{\sum_{i=1}^{n}\left[(\mathrm{obs}_i - \overline{obs}) - (m_i - \overline{m})\right]^2}{n}}}{\mathrm{SD_{obs}}}, \tag{A2}$$

$$\mathrm{Correlation}(\mathrm{obs}, m) = \frac{\sum_{i=1}^{n}(m_i - \overline{m})\cdot\left(\mathrm{obs}_i - \overline{\mathrm{obs}}\right)}{n \cdot \mathrm{SD_{obs}} \cdot \mathrm{SD}_m}. \tag{A3}$$

Acknowledgements. This work was supported by a FPI-UPC pre-doctoral fellowship from the European project FIELD_AC (FP7-SPACE-2009-1-242284 FIELD_AC), the Spanish project PLAN_WAVE (CTM2013-45141-R) and the Secretariat d'Universitats i Recerca del Dpt. d'Economia i Coneixement de la Generalitat de Catalunya (Ref 2014SGR1253). The campaigns were carried out thanks to the MESTRAL project (CTM2011-30489-C02-01). We would like to thank Joan Puigdefàbregas, Jordi Cateura, Joaquim Sospedra and Elena Pallarés from Laboratori d'Enginyeria Marítima for all their help with campaigns and data analysis, as well as the Ebro Irrigation Community (Comunitat de Regants de la dreta de l'Ebre, http://www.comunitatregants.org) and the XIOM network (Xarxa d'Instruments Oceanogràfics de Catalunya; http://www.xiom.cat) for the information and their commitment to the study. Finally, we thank the two anonymous reviewers for their in-depth criticisms and suggestions.

Edited by: M. M. Miglietta

References

Agterberg, R. and Wieringa, J.: Mesoscale terrain roughness mapping of the Netherlands, Technical Report TR-115, Royal Netherlands Meteorological Institute, Ministerie van Verkeer en Waterstaat, 1989.

Bignami, F., Sciarra, R., Carniel, S., and Santoleri, R.: Variability of Adriatic Sea coastal turbid waters from SeaWiFS imagery, J. Geophys. Res., 112, C03S10, doi:10.1029/2006JC003518, 2007

Bolaños, R., Jorda, G., Cateura, J., Lopez, J., Puigdefabregas, J., Gomez, J., and Espino, M. The XIOM: 20 years of a regional coastal observatory in the Spanish Catalan coast, J. Mar. Syst., 77, 237–260, doi:10.1016/j.jmarsys.2007.12.018, 2009.

Boldrin, A., Carniel, S., Giani, M., Marini, M., Bernardi Aubry, F., Campanelli, A., Grilli, F., and Russo, A.: Effects of bora wind on physical and biogeochemical properties of stratified waters in the northern Adriatic, J. Geophys. Res., 114, C08S92, doi:10.1029/2008JC004837, 2009.

Brecht, B. and Frank, H.: High resolution modelling of wind fields for optimization of empirical storm flood predictions, Adv. Sci. Res., 11, 1–6, doi:10.5194/asr-11-1-2014, 2014.

Camp, J.: Aproximaciones a la dinamica estuarica de una bahia micromareal Mediterranea, Thesis, Universitat de Barcelona, Barcelona, 1994.

Cerralbo, P., Grifoll, M., Espino, M., and López, J.: Predictability of currents on a mesotidal estuary (Ria de Vigo, NW Iberia), Ocean Dynam., 63, 131–141, doi:10.1007/s10236-012-0586-9, 2012.

Chavanne, C., Flament, P., Lumpkin, R., Dousset, B., and Bentamy, A.: Scatterometer observations of wind variations induced by oceanic islands: Implications for wind-driven ocean circulation, Can. J. Remote Sens., 28, 466–474, doi:10.5589/m02-047, 2002.

Cucco, A. and Umgiesser, G.: Modeling the Venice Lagoon residence time, Ecol. Model., 193, 34–51, doi:10.1016/j.ecolmodel.2005.07.043, 2006.

De Biasio, F., Miglietta, M. M., Zecchetto, S., and della Valle, A.: Numerical models sea surface wind compared to scatterometer observations for a single Bora event in the Adriatic Sea, Adv. Sci. Res., 11, 41–48, doi:10.5194/asr-11-41-2014, 2014.

Gracia, V., García, M., Grifoll, M., and Sánchez Arcilla, A.: Breaching of a barrier under extreme events, The role of morphodynamic simulations, J. Coast. Res., 951–956, doi:10.2112/SI65-161.1, 2013.

Grifoll, M., Del Campo, A., Espino, M., Mader, J., González, M., and Borja, A.: Water renewal and risk assessment of water pollution in semi-enclosed domains: Application to Bilbao Harbour (Bay of Biscay), J. Mar. Syst., 109–110, S241–S251, doi:10.1016/j.jmarsys.2011.07.010, 2011.

Grifoll, M., Aretxabaleta, A. L., Pelegrí, J. L., Espino, M., Warner, J. C., and Sánchez-Arcilla, A.: Seasonal circulation over the Catalan inner-shelf (northwest Mediterranean Sea), J. Geophys. Res.-Oceans, 118, 5844–5857, doi:10.1002/jgrc.20403, 2013.

Herrera, J. L., Piedracoba, S., Varela, R. A., and Rosón, G.: Spatial analysis of the wind field on the western coast of Galicia (NW Spain) from in situ measurements, Cont. Shelf Res., 25, 1728–1748, doi:10.1016/j.csr.2005.06.001, 2005.

Hong, S.-Y., Noh, Y., and Dudhia, J.: A new vertical diffusion package with an explicit treatment of entrainment processes, Mon. Weather Rev., 134, 2318–2341, doi:10.1175/MWR3199.1, 2006.

Jiang, H., Farrar, J. T., Beardsley, R. C., Chen, R., and Chen, C.: Zonal surface wind jets across the Red Sea due to mountain gap forcing along both sides of the Red Sea, Geophys. Res. Lett., 36, L19605, doi:10.1029/2009GL040008, 2009.

Martín Vide, J.: Los mapas del tiempo, Volumen 1 de Colección Geoambiente XXI. Davinci Continental, Editorial Davinci, Mataró, Barcelona, 219 pp., 2005.

Mass, C. F. and Ovens, D.: Fixing WRF's high speed wind bias: a new subgrid scale drag parameterization and the role of detailed verification, in: 24th Conference on Weather and Forecasting and 20th Conference on Numerical Weather Prediction, Preprints, 91st American Meteorological Society Annual Meeting, 23–27 January 2011, Seattle, WA, 2011.

Miglietta, M. M., Thunis, P., Pederzoli, A., Georgieva, E., Bessagnet, B., Terrenoire, E., and Colette, A.: Evaluation ofWRFmodel performances in different European regions with the DELTA-FAIRMODE evaluation tool, Int. J. Environ. Pollut., 50, 83–97, 2012.

Oke, T. R.: Boundary Layer Climates, 2nd Edn., Routledge, London, 1987.

Quijano-Scheggia, S., Garcés, E., Flo, E., Fernandez-Tejedor, M., Diogene, J., and Camp, J.: Bloom dynamics of the genus Pseudo-nitzschia (Bacillariophyceae) in two coastal bays (NW Mediterranean Sea), Scientia Marina, 72, 577–590, 2008.

Sánchez-Arcilla, A., González-Marco, D., and Bolaños, R.: A review of wave climate and prediction along the Spanish Mediterranean coast, Nat. Hazards Earth Syst. Sci., 8, 1217–1228, doi:10.5194/nhess-8-1217-2008, 2008.

Schaeffer, A., Garreau, P., Molcard, A., Fraunié, P., and Seity, Y.: Influence of high-resolution wind forcing on hydrodynamic modeling of the Gulf of Lions, Ocean Dynam., 61, 1823–1844, doi:10.1007/s10236-011-0442-3, 2008.

Scire, J. S., Robe, F. R., Fernau, M. E., and Yamartino, J.: A User's Guide for the CALMET Meteorological Model (Version 5.0), Earth Tech, Concord, MA, 1999.

Serra, P., More, G., and Pons, X.: Monitoring winter flooding of rice fields on the coastal wetland of Ebre delta with multitemporal remote sensing images, IEEE International Geoscience and Remote Sensing Symposium, Barcelona, 2495–2498, doi:10.1109/IGARSS.2007.4423350, 2007.

Signell, R. P., Carniel, S., Cavaleri, L., Chiggiato, J., Doyle, J. D., Pullen, J., and Sclavo, M.: Assessment of wind quality for oceanographic modelling in semi-enclosed basins, J. Mar. Syst., 53, 217–233, doi:10.1016/j.jmarsys.2004.03.006, 2005.

Skamarock, W. C., Klemp, J. B., Dudhia, J., Gill, D. O., Barker, D. M., Duda, M., Huang, X.-Y., Wang, W., and Powers, J. G.: A description of the advanced research WRF version 3, NCAR Tech. Note NCAR/TN-475+STR, Nat. Cent. for Atmos. Res., Boulder, Colorado, 2008.

Taylor, K. E.: Summarizing multiple aspects of model performance in a single diagram, J. Geophys. Res., 106, 7183–7192, 2001.

Tewari, M., Chen, F., Wang, W., Dudhia, J., LeMone, M. A., Mitchell, K., Ek, M., Gayno, G., Wegiel, J., and Cuenca, R. H.: Implementation and verification of the unified NOAH land surface model in the WRF model, in: 20th conference on weather analysis and forecasting/16th conference on numerical weather prediction, American Meteorological Society, Seattle, WA, 11–15, 2004.

Thompson, G., Rasmussen, R. M., and Manning, K.: Explicit forecasts of winter precipitation using an improved bulk microphysics scheme, Part I: Description and sensitivity analysis, Mon. Weather Rev., 132, 519–542, 2004.

Venäläinen, A., Sahlgren, V., Podsechin, V., and Huttula, T.: Small-scale variability of the wind field over a typical Scandinavian lake, Boreal Environ. Res., 8, 71–81, 2003.

Numerical models sea surface wind compared to scatterometer observations for a single Bora event in the Adriatic Sea

F. De Biasio, M. M. Miglietta, S. Zecchetto, and A. della Valle

Institute of Atmospheric Sciences and Climate, Italian National Research Council, Corso Stati Uniti 4, 35127 Padova, Italy

Correspondence to: F. De Biasio (f.debiasio@isac.cnr.it)

Abstract. We compare the sea surface wind fields forecasted by a Global Circulation Model (GCM) and three Limited Area Models (LAMs) in an operational-like set-up, with the wind remotely sensed by the NASA QuikSCAT scatterometer. The comparison is performed for a single case of Bora wind in the Adriatic Sea, with the purpose to understand the ability of the model forecasts in reproducing the mesoscale features captured by the scatterometer, and to investigate on the suitability of LAM and GCM forecasts as possible forcing in storm surge models (SSMs). The performance is evaluated by means of statistical parameters regarding wind speed and direction showing that, at least in terms of classical statistical parameters, the GCM offer the most advantageous choice in terms of cost/benefit.

1 Introduction

An open question for Numerical Weather Prediction (NWP) modelling is how atmospheric fields simulated by Global Circulation Models (GCMs) and Limited Area Models (LAMs) compare to observations, and how the wind field in presence of complex orography are represented. On the marine environment, satellite-borne scatterometers offer a huge number of observations of the so called "ocean wind vector" (OWV), which expresses the wind blowing at the sea surface. OWV observations are essential to describe the state of the bottom layer of the atmosphere and to forecast the state of the atmosphere and the underlying sea surface. The sea is, thus, an important test-site for model performance, particularly when enclosed and semi-enclosed basins surrounded by complex orography are considered, because the model performance is affected by the interaction of the flow with both the sea surface and the orography. This is of particular interest for the Adriatic Sea, where specific synoptic configurations, combined with the local orography (Zecchetto et al., 2012; Zecchetto and Accadia, 2014), often cause atmospheric conditions favourable to high water levels at the sea boundaries, particularly in the Venice Lagoon, whose location in the Adriatic Sea is shown in the left panel of Fig. 1. NWP models are used to simulate the atmospheric fields, and their output variables are used as forcing to storm surge models in order to forecast the sea level therein.

The objective of this work is to compare the 10 m wind fields simulated by four NWP models with the OWV provided by the NASA QuikSCAT scatterometer, in view of a possible use of the models as forcing in storm surge modelling systems. Here we start from a single case study of interest for the Adriatic Sea. The four NWP models are the Integrated Forecast System (IFS), the operational GCM of the European Centre for Medium-Range Weather Forecasts (ECMWF), and three different LAMs: ALADIN, COSMO-LAMI and the Weather Research and Forecasting (WRF).

Previous works comparing scatterometer observations and NWP model forecasts were based on the analysis of a high number of observed and simulated collocated data, providing monthly-seasonal statistics. Accadia et al. (2007) analysed the seasonal fields of several statistics parameters, using two years of QuikSCAT L2B 25 km observations and the wind field forecasts produced by the Quadrics Bologna Limited-Area Model (QBOLAM) over the Mediterranean Sea on a $0.1^\circ \times 0.1^\circ$ regular grid. They found areas of lower

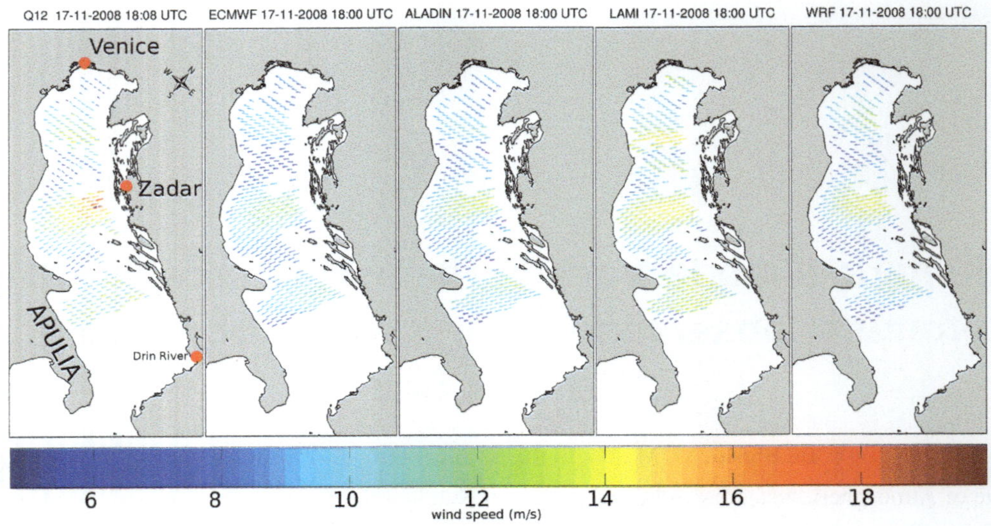

Figure 1. Sea surface wind fields on 17 November. From left to right: the scatterometer pass at 18:08 UTC, and the models wind fields (ECMWF, ALADIN, COSMO-LAMI, WRF) at 18:00 UTC. The maps are rotated 40° clockwise. The colourbar corresponds to the wind intensity, in m s^{-1}, of the vectors plotted in the maps. The first map on the left shows also the geographic positions of Venice (Italy), Zadar (Croatia), the Drin River (Albania) and Apulia region (Italy). The title of the fourth map from left should be read "COSMO-LAMI".

than average forecast skill, identified as semi-enclosed basins surrounded by rough orography. The discrepancies found between model forecasts and scatterometer observations were ascribed mainly to three factors; the representation of the orography in the model; the numerical damping required for computational stability (resulting in a model spatial resolution coarser than the nominal grid size); other forecast errors, which are associated with the spatial and temporal misplacement of the meteorological phenomena (fronts, depressions, etc.) and the misrepresentation of data needed by the model as initial or boundary conditions (e.g. imposing constant sea surface temperature fields during the model run).

Kara et al. (2009) compared, on a monthly basis, the QuikSCAT 25 km data with "analysis quality" NWP winds (A. Wallcraft, personal communication, 2014) coming from the NOGAPS dataset at 0.5°-resolution, in the Mediterranean Sea. They found major differences near coastal boundaries due to contamination of model winds over sea by land data, as a consequence of the coarse resolution of the model.

In our study we focused on the comparison of model data and scatterometer winds on a per-swath basis, in a single case of bora wind (17–18 November 2008) in the Adriatic Sea (Fig. 1), with the objective of operational SSM applications, keeping the scatterometer wind as the reference field.

The paper is organised as follows. The instruments and the methods are described in Sect. 2. Section 3 presents the case study, followed by the discussion in Sect. 4. The conclusions are drawn in Sect. 5.

2 Instruments and methods

2.1 Scatterometer data and preparation of the dataset

Scatterometers are active microwave instruments providing measurements of the wind field over the ocean. In this work we use the OWV fields at 10 m over the sea with a grid spacing of 12.5 km, provided by the SeaWinds scatterometer onboard the NASA QuikSCAT satellite (Dunbar et al., 2006), available from 1999 to 2009 and processed by the SeaWinds Processing and Analysis Center (SeaPAC) at the Jet Propulsion Laboratory (JPL).

Operating in Ku-band, the data retrieved by SeaWinds could be contaminated by rain. However, the rainfall rate records at several ground stations along the coast of the Adriatic Sea in the present case were from zero to extremely low, so that the scatterometer data were completely retained. The swath of QuikSCAT is 1800 km wide: data corresponding to wind cells at nadir and at the outer sides of the swath have been discarded, since in these regions have lower quality and more noise. Only winds in the range 3–25 m s^{-1} have been used for comparison.

The accuracy of the QuikSCAT winds at 25 km grid spacing has been investigated, among several authors, by Ebuchi et al. (2002), which reported uncertainties of ±1.01 ms^{-1} and of ±23° respectively for wind speed and direction in open ocean. Tang et al. (2004) evaluated the high-resolution L2B QuikSCAT OWV at 12.5 km grid spacing data also in coastal regions, and reported a slight deterioration of the accuracies in the proximity of the coastline: ±1.70 ms^{-1} in speed and ±26° in direction for wind speed > 3 m s^{-1}.

Table 1. LAMs set-up.

	Nominal grid (km)	Time step (hours)	Initial/boundary conditions	Microphysics scheme	Radiation scheme	Closure scheme	Orography	Convection scheme
ALADIN	8	1	ARPEGE analysis/ forecast	ALARO0	Geleyn-Hollingworth radiation	1.5 order closure turbulence	land-surface ISBA scheme	modified Bougeault (1985)
COSMO-LAMI	7	1	ECMWF analysis/ forecast	bulk	Ritter-Geleyn radiation	TKE Mellor-Yamada 2.5 order	Terra land surface	Tiedtke
WRF	10	1	ECMWF analysis/ forecast	Thompson et al. (2004)	Kain cumulus	Yonsei University (YSU) boundary layer	Unified Noah land-surface model	Rapid Radiative Transfer Model

Winds fields provided by scatterometers are given for neutral stability conditions. While in the open ocean the stability conditions are close to neutral (Chelton and Freilich, 2005), in the Mediterranean Sea, on average, they range from neutral to unstable (Zecchetto and De Biasio, 2007). The correction of scatterometer winds from neutral to real stability conditions affects mainly low winds, implying a reduction of 20–30 % for unstable conditions and an increase of +20 % for stable conditions in the Mediterranean Sea (Zecchetto and De Biasio, 2007). To compare scatterometer data with model winds, the adjustment can be done in two ways: modifying the scatterometer values to real stability conditions, or vice-versa, adjusting model winds to neutral stability. As we have in mind real applications, we consider the former approach. The adjustment is calculated with the model of Liu et al. (1979), which requires, as input, the air and dew point temperatures, the sea-surface temperature and the atmospheric pressure at sea level. All these parameters have been derived from the ECMWF analysis fields.

2.2 Atmospheric model data

In this section we present the NWP models used for the comparison. They are briefly described below. More details on the LAMs set-up and characteristics are reported in Table 1.

2.2.1 The ECMWF operational global circulation model

The ECMWF operational forecast model, known as Integrated Forecast System (Simmons et al., 1989), is a global spectral model that produces deterministic forecasts up to 15 days in advance.

For the case considered here, the model implementation includes 91 vertical levels and horizontal truncation level 799 (TL799, approximately 25 km). The model output fields are available every 3 h. For the sake of comparison, in this study we use the forecast produced at 12:00 UTC the day before each temporal match-up. The parameters used are the 10 m U and V wind components. One may object that the comparison with QuikSCAT is not fair, since global models assimilate QuikSCAT data. Anyway, the effective horizontal resolution of the scatterometer data used for the assimilation in the ECMWF model was about 100 km (Stoffelen and Anderson, 1997), which is too coarse for an effective modification of the mesoscale fields in the analysis over the Adriatic sea; also, the fields used for the comparison are the 24 h forecasts, which are only marginally affected by the assimilation of wind data over sea surface. Moreover, the three LAMs use ARPEGE and ECMWF IFSs as initial and boundary conditions, so that the supposed advantage given by QuikSCAT assimilation into GCMs is shared with LAMs too.

2.2.2 The ALADIN model

ALADIN (Aire Limitée Adaptation dynamique Développement InterNational; Bubnova et al., 1995) is a limited-area spectral high-resolution model, nested into the ARPEGE (Action de Recherche Petite Echelle Grande Echelle) IFS GCM. It has been mainly developed at Météo-France in cooperation with ECMWF. The sea surface wind fields used in this work are the results of operational simulations at DHZ (Državni Hidrometeorološki Zavod), the meteo-hydrological service of Croatia.

2.2.3 The COSMO-LAMI model

The COSMO model (Steppeler et al., 2003) is a limited area model developed and maintained by national and regional European meteorological services members of the Consortium for Small-scale Modeling (COSMO). A particular set up, called COSMO-LAMI (Limited Area Model Italy), is run operationally by the Agenzia Regionale per la Prevenzione e Protezione Ambientale-Servizio Idro Meteo Clima (ARPA-SIMC) of the Emilia Romagna Region.

2.2.4 The WRF model

The WRF model (see http://www.wrf-model.org; Wang et al., 2010) is a numerical weather prediction system designed to serve both operational forecasting and atmospheric

Table 2. Statistical parameters of the five wind datasets for the two passes, on 17 November (left) and 18 November (right) 2008. The statistical parameters are the mean wind direction and speed with their standard deviations (σ), and root mean square error (RMSE), bias, centred root mean square difference (cRMSD) and correlation coefficient (R) of the model wind speed with respect to the scatterometer observations. In bold the best scores.

	17 November								18 November							
	$\langle wdir \rangle$ (°)	σ_{wdir} (°)	$\langle ws \rangle$ ($m\,s^{-1}$)	σ_{ws} ($m\,s^{-1}$)	$RMSE_{ws}$ ($m\,s^{-1}$)	$bias_{ws}$ ($m\,s^{-1}$)	$cRMSD_{ws}$ ($m\,s^{-1}$)	R_{ws}	$\langle wdir \rangle$ (°)	σ_{wdir} (°)	$\langle ws \rangle$ ($m\,s^{-1}$)	σ_{ws} ($m\,s^{-1}$)	$RMSE_{ws}$ ($m\,s^{-1}$)	$bias_{ws}$ ($m\,s^{-1}$)	$cRMSD_{ws}$ ($m\,s^{-1}$)	R_{ws}
QUIK SCAT	48.3	19.4	10.65	2.00	–	–	–	–	45.4	14.1	8.15	2.95	–	–	–	–
ECMWF	39.0	16.9	9.27	1.36	1.91	1.38	**1.32**	**0.75**	31.2	22.5	6.79	2.05	2.31	1.36	1.87	0.78
ALADIN	46.6	**18.5**	9.49	1.53	1.96	1.16	1.57	0.63	33.8	**17.5**	7.74	**2.63**	2.15	0.41	2.11	0.72
COSMO-LAMI	43.5	15.7	**11.03**	**1.93**	**1.87**	**−0.37**	1.83	0.57	**49.7**	27.4	9.17	3.62	2.83	−1.02	2.64	0.69
WRF	**47.7**	17.4	9.51	1.79	2.01	1.15	1.65	0.63	39.9	19.6	**7.93**	2.19	**1.67**	**0.22**	**1.66**	**0.83**

research needs, mainly developed at the National Center for Atmospheric Research (NCAR). The version 3.0 of the model, already implemented at the ISAC-CNR facilities, has been used for this study in a single grid configuration. The current configuration uses 40 vertical levels, more closely spaced in the boundary layer.

2.3 Temporal and spatial collocation

The scatterometer pass time has been identified as the reference time. Then the wind fields of the four models at the synoptic hour closer to the pass time have been used for the match-up, thus allowing a maximum time difference of ±1.5 h.

Scatterometer data have been interpolated on a regular grid of 0.125° by 0.125°, to compare with the ECMWF wind. The interpolation method is pure Laplacian, to avoid excessive data smoothing: it has been demonstrated not to change the wind speed and direction statistics (Accadia et al., 2007). LAMs winds have been mapped to the same grid with the nearest neighbour algorithm. After interpolation, only data points farther than 15 km from the coast, the nominal distance for the QuikSCAT 12.5 km dataset, have been retained, using for this purpose the Global 30 Arc-Second Elevation (GTOPO30) digital elevation model of the U.S. Geological Survey, which has an accuracy of about 1 km.

3 The Bora case

We focused the comparison on a case of Bora wind in the Adriatic Sea during 17 to 18 November 2008. Bora (Yoshino, 1976; Jurčec, 1981; Smith, 1987; Pandžić and Likso, 2005), is a north-easterly strong and cold wind affecting the entire Adriatic Sea. Atkinson (1989) classified this wind as orographic or downslope. It is generally due to a polar high over the plateau behind the Dinaric Alps and a pressure low farther south over the warmer Adriatic Sea. Such a meteorological situation favours the funnelling of the stratified cold air behind the Dinaric Alps through the mountain chain gaps towards the Adriatic basin.

The scatterometer passes over the area considered for comparison with the NWP model fields are at 18:08 UTC on 17 November, and 03:50 UTC on 18 November. Consequently, the model outputs have been extracted at 18:00 UTC on 17 November, and 03:00 UTC on 18 November.

At the first pass (Fig. 1, left), the scatterometer data shows an intense north-easterly flow across the Adriatic sea, with a maximum of about 20 $m\,s^{-1}$ to the west of the Croatian coast near Zadar (Zadar is shown with a red mark in Fig. 1). Two additional but weaker low level jets are identified in the northern and southern part of the Adriatic sea. All the models are able to capture these features pretty well, being the WRF model wind the closest to that of the scatterometer; a slight underestimation of the wind speed occurs for all the models, apart from COSMO-LAMI, which slightly overestimates the northern and the southern jets. Wind roses (not shown) confirm the good performance of the models in reproducing the wind intensity and direction. The statistical parameters in Table 2 show that LAMs generally perform better than ECMWF model in terms of wind direction and intensity (for ECMWF IFS, the wind is weaker and slightly rotated anticlockwise).

The relative frequency of wind speed bins (not shown) has similar distribution for ECMWF, ALADIN and WRF, which slightly underestimate the scatterometer data, while COSMO-LAMI is shifted towards higher values of wind speed bins.

Similar considerations can be drawn at the later pass, when the Bora structures shift southward (Fig. 2). At that time, the most intense jet blows across the southern Adriatic, from the valley of Drin river towards Apulia (south-eastern Italy); the jet across Zadar is still present, but slightly weaker and less extensive than eight hours earlier; a third, weaker jet can be identified in between these two features.

Again, COSMO-LAMI overestimates the wind speed, especially the two southernmost jets, and the wind speed frequency is shifted towards higher bins (not shown), but the distribution in terms of average wind direction is the closest to that of the scatterometer (see Table 2), although standard deviation is the highest; ECMWF significantly underestimates the wind speed and distributes the wind over a narrower range of speed bins.

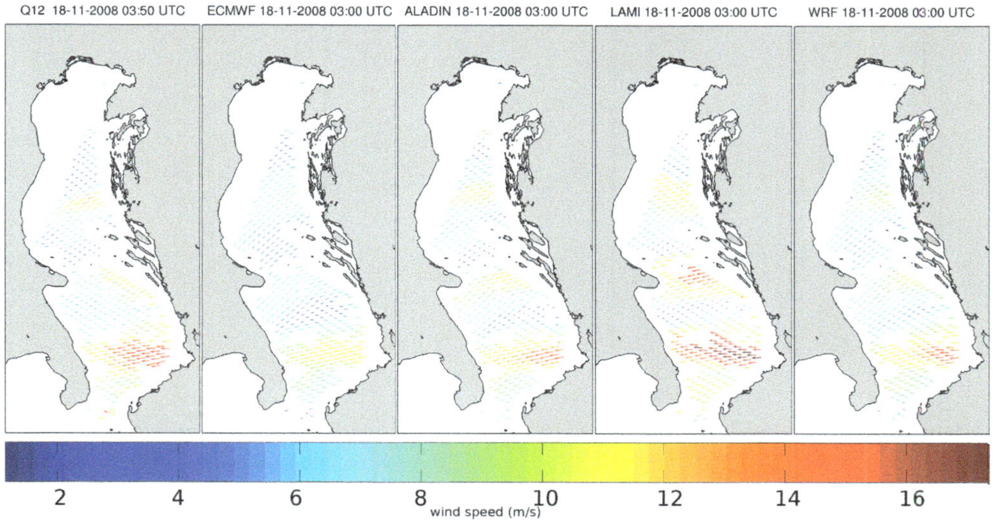

Figure 2. Sea surface wind fields on 18 November. From left to right: the scatterometer pass at 03:50 UTC, and the models wind fields (ECMWF, ALADIN, COSMO-LAMI, WRF) at 03:00 UTC. The maps are rotated 40° clockwise. The colorbar corresponds to the wind intensity, in m s^{-1}, of the vector plots in the maps.

4 Discussion

The spatial scales typical of storm surges in the Adriatic Sea are rather large, as a consequence of the large temporal scales (hours) of the phenomenon, due to the dependence of the local storm surge level on the wind stress over the entire Adriatic Sea. In the perspective of SSM operational applications, scatterometer data are appropriate tools: SeaWinds, as well as ASCAT and OSCAT, the nowadays operational scatterometer sensors, provide data in the Mediterranean Sea twice a day, ensuring the possibility to follow the evolution of meteorological phenomena with time scales of some hours, and their distribution is relatively fast and free of charge.

From a statistical point of view, however, the spatial and temporal sampling of the Adriatic Sea surface wind by scatterometer observations accounts for more than 90 % of the wind energy of the statistically significant mean wind field (Zecchetto and Cappa, 2001), assuring that scatterometer observations supply an adequate description of the structure of the sea surface wind up to spatial scales in the meso-beta range (Orlanski, 1975).

On the other hand, satellite-borne Synthetic Aperture Radars (SAR), have a much finer spatial resolution (from metres to hundreds of metres, depending on the sensor); however, their images are not frequent; their acquisition has often to be agreed in advance with the space agencies; the use of their full spatial resolution would be unpractical for operational use; finally, there is no standard SAR wind product operationally provided by the agencies: the algorithms to extract the wind direction from them are still in a development phase and depend on external sources of information (Zecchetto and De Biasio, 2008).

As explained before, scatterometer data have been corrected for stability making use of ECMWF analyses. As Bora wind transports cold air over the Adriatic Sea, the stability conditions were unstable during the case study: the ECMWF mean air-sea temperature difference were respectively −3.2 °C and −5.3 °C during the two scatterometer passes. Consequently, the bias between stability-corrected winds and neutral winds was −0.19 m s^{-1} during the first pass and −0.34 m s^{-1} during the second. These biases are not particularly large, but are comparable to the model biases (see Table 2), and thus have to be taken into account. This is even more important considering that the storm surge depends linearly on the wind stress, which depends from the squared wind speed.

A possible source of error could be the ability of the scatterometer to accurately record high and low wind speeds. The scatterometer data used in this work are in the wind speed range suitable to assure the nominal accuracy. This is evident in Fig. 3, which reports the scatterplots of scatterometer and model wind speeds, for the pass of 17 November (left) and 18 November (right). During the first pass the scatterometer wind speeds are concentrated in the range 6–17 m s^{-1}, with some outliers at 20 m s^{-1}. Model data are in the range 4–14 m s^{-1}. During the second pass the wind speed ranges are more broadened, with the scatterometer data between 3 and 16 m s^{-1}, and the model data between 1 and 17 m s^{-1}. In both scatterplots all the best-fits are far from the perfect agreement, apart from that of the COSMO-LAMI model in the second passage. This should be ascribed to the particular basin chosen for the comparison and its rough orography, and for the proximity of all the collocation points to the coastline.

Table 2 shows the modelled statistical parameters for the two passes in comparison with the scatterometer data. At the

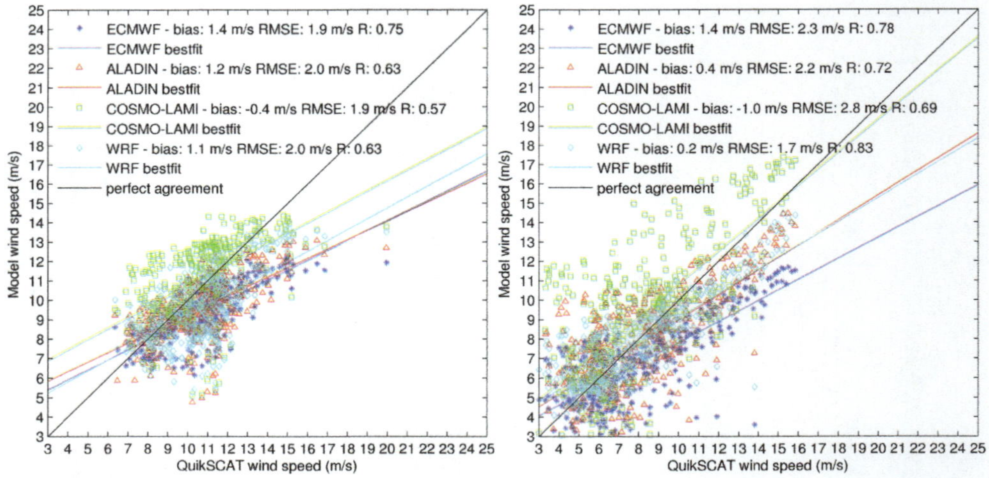

Figure 3. Scatterplot of the wind speed of the QuikSCAT scatterometer and the four NWP models for the satellite passes of 17 November (left) and 18 November (right). The legends report the bias, the RMSE and the correlation coefficient of the models with respect to the scatterometer observations. In black solid the perfect agreement line.

first pass, ECMWF has the largest correlation coefficient (R), but the worst bias, and the standard deviation (σ) is strongly underestimated. COSMO-LAMI, which is the only model overestimating the wind speed, has at this time the smallest bias, the best σ and the smallest root mean square error (RMSE); WRF and ALADIN behave similarly, with bias and centred root mean square difference (cRMSD: it is the RMSE calculated on the unbiased datasets) very close to those of ECMWF model. At later time (right panel), WRF outperforms the other models in terms of R, cRMSD, RMSE and bias. ALADIN shows a slightly worse R (but still larger than 0.7), and has the σ most similar to the scatterometer data. As a consequence of the overestimation (bias $= -1.02\ \mathrm{m\,s^{-1}}$) of the low level jets, COSMO-LAMI has the largest σ and the smallest R. Finally, although ECMWF strongly underestimates the wind speed and σ, it has still one of the largest R.

Really, the traditional statistical indices should be taken into consideration with caution. For example, the forecast of a jet predicted with the correct intensity but with a slightly wrong timing and location, is worse in terms of correlation than the forecast of a global model, which predicts a smoother jet in the correct location, but is definitely much more skilful for a forecaster (Miglietta et al., 2012). Thus, a smooth forecast can lead to better verification statistics than a forecast with a correct amplitude in structural features (Warner, 2011). This aspect is also highlighted by Signell et al. (2005), who conducted extensive comparison of NWP forecast wind fields from a GCM and three LAMs of different spatial resolution with in-situ data. They found that the temporal correlation between in-situ data and GCM was better than with LAMs, and ascribed this result to the fact that large-scale features can be predicted in a deterministic way (the GCM nominal grid size was 40 km), while small-scale features are stochastic, and their prediction can thus be misrepresented in time and space, especially by LAMs.

About the significance of traditional statistical indices, a special attention should be paid to the wind speed σ. Skamarock (2004), analysing the deviation of NWP model spectra from the theoretical behaviour, estimated that the NWP models effective resolution resulted to be several times larger than their nominal resolution. Therefore, given a modelled σ comparable to that of the observations, it is fair to expect that part of the signal in σ is due to model noise, and thus the effective σ, due to the modelled geophysical signal, should be lower. With this observation in mind it is clear that the comparison of the model σ with that of observations has to be considered cautiously.

5 Conclusions

This study compares the wind speed and direction retrieved from the scatterometer in a Bora event across the Adriatic sea with the wind simulated with four different models, i.e. a global circulation model and three limited area models. All the models are able to reproduce the different wind patterns emerging in the scatterometer wind field at two different passes, but with some differences. The global model (ECMWF IFS) is the best in terms of correlation, but it suffers from a larger bias (underestimation) and a reduced variability (smaller standard deviation). COSMO-LAMI is the only model among the four to overestimate, on average, the wind speed and produces the largest standard deviation; it is the worst model in terms of correlation. Generally, ALADIN and WRF behave in an intermediate way between the former two models. Being comparable with the other model statistics, the ECMWF model results as the best candidate for operational SSM applications in terms of cost/benefit, based on this statistical analysis. Such results should be tested in a longer time period to verify if they are statistically robust. A work is in progress to compare the scatterometer data with two different operational limited area models and one global circulation model over a three-year period.

Acknowledgements. The QuikSCAT SeaWinds ocean wind vector data were obtained from the Physical Oceanography Distributed Active Archive Center (PO.DAAC) at the NASA Jet Propulsion Laboratory, Pasadena, CA (http://podaac.jpl.nasa.gov).

The ECMWF GCM fields were obtained from ECMWF by mean of the Meteorological Service of Aeronautica Militare.

The ALADIN and COSMO-LAMI wind fields have been supplied by the Istituzione Centro Previsioni e Segnalazioni Maree (ICPSM) of the Venice Municipality.

Edited by: A. M. Sempreviva

References

Accadia, C., Zecchetto, S., Lavagnini, A., and Speranza, A.: Comparison of 10-m wind forecasts from a regional area model and QuikSCAT scatterometer wind observations over the Mediterranean Sea, Mon. Weather Rev., 135, 1946–1960, 2007.

Atkinson, B. W.: Meso-scale atmospheric circulation. Academic Press, Harcourt Brace Jovanovich Publishers, London, UK, 1989.

Bougeault, P.: A simple parameterization of the large-scale effects of cumulus convection, Mon. Weather Rev., 113, 2108–2121, 1985.

Bubnova, R. , Hello, G., Benard, P., and Geleyn, J. F., Integration of the fully elastic equations cast in the hydrostatic pressure terrain-following coordinate in the framework of the ARPEGE/Aladin NWP System, Mon. Weather Rev., 123, 515–535, 1995.

Chelton, D. B. and Freilich, M. H.: Scatterometer-based assessment of 10-m wind analyses from the operational ECMWF and NCEP numerical weather prediction models, Mon. Weather Rev., 133, 409–429, 2005.

Dunbar, R. S., Lungu, T., Weiss, B., Stiles, B., Huddleston, J., Callahan, P. S., Shirtliffe, G., Perry, K. L., Hsu, C., Mears, C., Wentz, F., and Smith, D.: QuikSCAT Science Data Product Users Manual, Version 3.0, Jet Propulsion LaBoratory, Pasadena, CA, JPL Document D-18053-Rev A, 2006.

Ebuchi, N., Graber, H. C., and Caruso, M. J.: Evaluation of Wind Vectors Observed by QuikSCAT/SeaWinds Using Ocean Buoy Data, J. Atmos. Ocean. Tech., 19, 2049–2062, doi:10.1175/1520-0426(2002)019<2049:EOWVOB>2.0.CO;2, 2002.

Jurčec, V.: On mesoscale characteristics of Bora conditions in Yugoslavia, Pure Appl. Geophys., 199, 640–657, 1981.

Kara, A. B., Wallcraft, A. J., Martin, P. J., and Pauley, R. L.: Optimizing surface winds using QuikSCAT measurements in the Mediterranean Sea during 2000–2006, J. Mar. Syst., 78, 119–131, doi:10.1016/j.jmarsys.2007.11.010, 2009.

Liu, W. T., Katsaros, K. B., and Businger, J. A.: Bulk parametrizations of air-sea exchanges of heat and water vapor including the molecular constraints at the interface, J. Atmos. Sci., 36, 1722–1735, 1979.

Miglietta, M. M., Thunis, P., Pederzoli, A., Georgieva, E., Bessagnet, B., Terrenoire, E., and Colette, A.: Evaluation of WRF model performances in different European regions with the DELTA-FAIRMODE evaluation tool, Int. J. Environ. Pollut., 50, 83–97, 2012.

Orlanski, I.: A rational subdivision of scales for atmospheric processes, B. Am. Meteorol. Soc., 56, 527–530, 1975.

Pandžić, K. and Likso, T.: Eastern Adriatic typical wind field patterns and large-scale atmospheric conditions, Int. J. Climatol., 25, 81–98, 2005.

Signell, R., Carniel, S., Cavaleri, L., Chiggiato, J., Doyle, J., Pullen, J., and Sclavo, M.: Assessment of wind quality for oceanographic modeling in semi-enclosed basins, J. Mar. Syst., 53, 217–233, 2005.

Simmons, A. J., Burridge, D. M., Jarraud, M., Girardm, M., and Wergen, W.: The ECMWF medium-range prediction model's development of the numerical formulations and the impact of increased resolution, Meteorol. Atmos. Phys., 40, 28–60, 1989.

Skamarock, W. C.: Evaluating Mesoscale NWP Models Using Kinetic Energy Spectra, Mon. Weather Rev., 132, 3019–3032, 2004.

Smith, R. B.: Aerial characteristics of Yugoslavian Bora, J. Atmos. Sci., 44, 2934–2939, 1987.

Steppeler, J., Doms, G., Schättler, U., Bitzer, H. W., Gassmann, A., Damrath, U., and Gregoric, G.: Meso-gamma scale forecasts using the nonhydrostatic model LM, Meteorol. Atmos. Phys., 82, 75–96, 2003.

Stoffelen, A. and Anderson, D.: Ambiguity removal and assimilation of scatterometer data, Q. J. Roy. Meteorol. Soc., 123, 491–518, 1997.

Tang, W. Q., Liu, W. T., and Stiles, B. W.: Evaluation of high-resolution ocean surface vector winds measured by QuikSCAT scatterometer in coastal regions, IEEE T. Geosci. Remote, 42, 1762–1769, 2004.

Thompson, G., Rasmussen, R. M., and Manning, K.: Explicit forecasts of winter precipitation using an improved bulk microphysics scheme. Part I: Description and sensitivity analysis, Mon. Weather Rev., 132, 519–544, 2004.

Wang, W., Bruyère, C., Duda, M., Dudhia, J., Gill, D., Lin, H.-C., Michalakes, J., Rizvi, S., and Zhang, X.: ARW Version 3 Modelling System User's Guide, 312 pp., NCAR, Boulder, Colorado, 2010.

Warner, T. T.: Numerical Weather and Climate Prediction, Cambridge University Press, 548 pp., 2011.

Yoshino, M.: Local wind Bora, University of Tokyo Press, Tokyo, Japan, 1976.

Zecchetto, S. and Cappa, C., The spatial structure of the Mediterranean Sea winds revealed by ERS-1 scatterometer, Int. J. Remote Sens., 22, 45–70, doi:10.1080/01431160175038848, 2001.

Zecchetto, S. and De Biasio, F.: Sea Surface Winds over the Mediterranean Basin from Satellite Data (2000–04): Meso- and Local-Scale Features on Annual and Seasonal Time Scales, J. Appl. Meteor. Climatol., 46, 814–827, doi:10.1175/JAM2498.1, 2007.

Zecchetto, S. and De Biasio, F.: A Wavelet-Based Technique for Sea Wind Extraction from SAR Images, IEEE T. Geosci. Remote, 46, 2983–2989, doi:10.1109/TGRS.2008.920967, 2008.

Zecchetto, S. and Accadia, C.: Diagnostics of T1279 ECMWF analysis winds in the Mediterranean basin by comparison with ASCAT 12.5 km winds, Q. J. Roy. Meteorol. Soc., published online in Wiley Online Library, doi:10.1002/qj.2315, 2014.

Zecchetto, S., De Biasio, F., and Accadia, C.: Scatterometer and ECMWF derived wind vorticity over the Mediterranean Basin, Q. J. Roy. Meteorol. Soc., 139, 674–684, doi:10.1002/qj.2001, 2012.

A vector auto-regressive model for onshore and offshore wind synthesis incorporating meteorological model information

D. Hill[1], K. R. W. Bell[1], D. McMillan[1], and D. Infield[2]

[1]Dept. of Electronic and Electrical Engineering, University of Strathclyde, Glasgow, UK
[2]Centre for Doctoral Training in Wind Energy Systems, University of Strathclyde, Glasgow, UK

Correspondence to: D. Hill (david.c.hill@strath.ac.uk)

Abstract. The growth of wind power production in the electricity portfolio is striving to meet ambitious targets set, for example by the EU, to reduce greenhouse gas emissions by 20 % by 2020. Huge investments are now being made in new offshore wind farms around UK coastal waters that will have a major impact on the GB electrical supply. Representations of the UK wind field in syntheses which capture the inherent structure and correlations between different locations including offshore sites are required. Here, Vector Auto-Regressive (VAR) models are presented and extended in a novel way to incorporate offshore time series from a pan-European meteorological model called COSMO, with onshore wind speeds from the MIDAS dataset provided by the British Atmospheric Data Centre. Forecasting ability onshore is shown to be improved with the inclusion of the offshore sites with improvements of up to 25 % in RMS error at 6 h ahead. In addition, the VAR model is used to synthesise time series of wind at each offshore site, which are then used to estimate wind farm capacity factors at the sites in question. These are then compared with estimates of capacity factors derived from the work of Hawkins et al. (2011). A good degree of agreement is established indicating that this synthesis tool should be useful in power system impact studies.

1 Introduction

The potential of wind power to contribute to demand for electric power needs to be fully assessed if effects of climate change on food production, water resources, insurance costs and so on are to be avoided. In order to facilitate such assessments, a Vector Auto-Regressive (VAR) model-based synthesis has been developed previously by the authors using just onshore data (Hill et al., 2012). These syntheses contain information about the annual and diurnal trends as well as the stochastic component. Many studies have addressed the potential impacts of wind power on a power system (e.g. Strbac, 2002; Gross et al., 2006; Cox, 2009). Different approaches have been taken to the representation of wind speed structure – numerical weather prediction models (e.g. Aigner and Gjengedal, 2010), an approach incorporating reanalysis and mast datasets (Kubik et al., 2013) or purely statistical approaches. A literature review of the latter appears in Hill et al. (2012).

The spatial correlation of wind speeds (and hence wind power) over scales over several hundred km is well known (Hill et al., 2012). This was discussed by Oswald et al. (2008) who looked at an extreme event and evaluated quantitatively by Sinden (2007) and Gibescu et al. (2006). Failure to model such correlations will result in excessive aggregate smoothing of wind power output, and an underestimate of the need for reserve and unrealistic inter-area power flows important in network constraint cost analysis.

A de-trending process is briefly described in the Methodology. The Results section depicts forecasting skill and also shows model to model comparisons of capacity factors with work carried out by Hawkins et al. (2011), which describes the meteorological model used in their work and incorporating a calibration process.

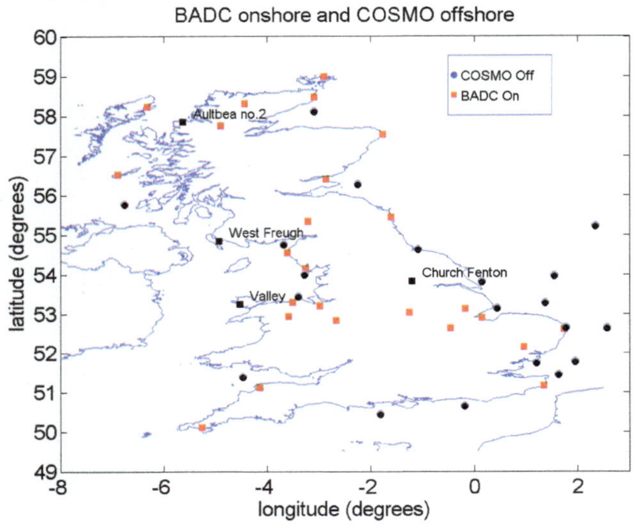

Figure 1. Map to show locations of sites onshore and offshore where time series input to VAR model, results of labelled sites appear in Fig. 3.

2 Methodology

A key innovation in the approach taken here is a de-trending process to represent the cyclical varying components of wind speed and then a stochastic VAR model is applied to the de-trended component. The MIDAS set of onshore weather data produced by the British Atmospheric Data Centre (UK Meteorological Office, 2012) and representing observations at 10 m above ground level was used as the initial source of onshore wind speeds. Modelled sites were chosen onshore that have > 90 % coverage for the period January 1996 to December 2005, taken as the trend formation period onshore. These sites also had > 90 % coverage from January 2006 to December 2007, for use in the VAR modelling process. Observations from multiple offshore locations could not be obtained. Instead, offshore data for the latter period was obtained from the Consortium for Small-scale Modelling (COSMO, 2010), meteorological model (Doms and Schättler, 2002), which does not suffer from data gaps and represent wind speeds at 10 m above sea level. See Fig. 1 for site locations, whose offshore locations are the same as those used in the Scottish Electricity Dispatch Model (SEDM) which made use of the weather model developed by Hawkins et al. (2011).

Some form of de-trending of the data is necessary to ensure a reasonable degree of stationarity in the de-trended series, an assumption of the VAR modelling. In Hill et al. (2012) the wind speed data were first de-trended site by site by the suitable application of harmonic analysis to whatever period of data existed for those sites. A concurrent period common to all sites was then taken for which to calculate the de-trended series. However, in previous work, for some sites, a longer period than that was used to calculate the

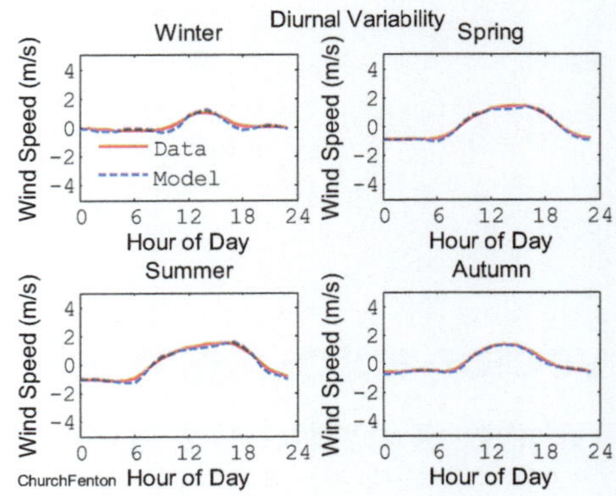

Figure 2. Diurnal trends at Church Fenton, particularly pronounced in Spring and Summer.

trends. In the current work, the trends for all the onshore sites are calculated based on data from the same period, i.e. January 1996 to December 2005, while the trends for all the offshore sites are based on data from January 2006 to December 2006, referred to henceforth simply as the year 2006. These sets of site-by-site trends are subtracted from observations at each site in 2006 to form the training set for the VAR model.

The harmonic analysis described in Hill et al. (2012) remains the same in this work, with an annual component modelled with Fourier terms Ω, 2Ω and 3Ω (where Ω is the annual angular frequency) and a seasonally varying diurnal component with ω and 2ω terms (where ω is the 24 h angular frequency) – see Fig. 2 for an example of the diurnal trends.

These deterministic trends, represented by $\boldsymbol{y}_{\mathrm{a}}$ (annual) and $\boldsymbol{y}_{\mathrm{ds}}$ (seasonally varying diurnal), are then subtracted from the data, $\boldsymbol{y}$, as in Eq. (1), to obtain the de-trended series $\boldsymbol{y}_{\mathrm{dt}}$, before commencing the process of Vector Auto-Regression.

$$\boldsymbol{y}_{\mathrm{dt}}(t) = \boldsymbol{y}(t) - \boldsymbol{y}_{\mathrm{a}}(t) - \boldsymbol{y}_{\mathrm{ds}}(t) \quad (1)$$

The VAR model is represented by Eq. (2) where $\boldsymbol{\Phi}_1$ and $\boldsymbol{\Phi}_2$ are 50×50 matrices (known simply as the VAR coefficients) reflecting the influences between the 50 sites under consideration at lags t-1 and t-2 h respectively. There is no simple physical interpretation of $\boldsymbol{\Phi}_1$ and $\boldsymbol{\Phi}_2$ but an equation in the multivariate case can be derived which is similar, but not identical, to the Yule-Walker equations of the univariate case, which in turn relates the covariances of the observations to the auto-regressive coefficients (see for example Sect. 3 in Lütkepohl, 2005).

$$\boldsymbol{y}_{\mathrm{dt},t} = \boldsymbol{\Phi}_1 \boldsymbol{y}_{\mathrm{dt},t-1} + \boldsymbol{\Phi}_2 \boldsymbol{y}_{\mathrm{dt},t-2} + \boldsymbol{e}_t \quad (2)$$

The vector $\boldsymbol{y}_{\mathrm{dt},t}$ is a column vector containing the de-trended series, one site per row and $\boldsymbol{e}_t$ is a Gaussian (normally distributed) noise term. Equation (2) is first applied to the training period of year 2006 to define the VAR model by finding the VAR coefficients $\boldsymbol{\Phi}_1$ and $\boldsymbol{\Phi}_2$ which minimise the sum of squares of the residuals, $\boldsymbol{e}_t\, \boldsymbol{e}_t^{\mathrm{T}}$ in a least squares approach. The year 2007 is reserved for assessing the forecasting accuracy and the determination of the standard deviation of the noise term. The VAR coefficient matrices are used to drive a synthesis of wind speeds by utilising Eq. (2) as a recursive equation using observations at each hour t-1 and t-2 with the output at time t given simply as Eq. (2) with no noise term. This output is compared with observed value at time t to determine the one step ahead forward prediction error. The process is stepped forward in time and repeated so that a succession of error values is found enabling its standard deviation σ to be calculated. Along with a mean of 0, this is used in a Normally distributed ($N(0,\sigma)$) noise term $\boldsymbol{e}_t$ thus completing the VAR model.

The number of VAR matrix terms in the model was chosen so that this one-step ahead error was minimised. The addition of the 2nd term provides a small improvement of 2.4 %, whilst a possible t-3 term did not provide further improvements.

Application of the model to synthesise a wind speed time series starts from some initial condition at times $\boldsymbol{y}_{\mathrm{dt},0}$ and $\boldsymbol{y}_{\mathrm{dt},1}$ together with the one-step ahead forward prediction error. Finally, the value of the trend for the synthesised hour is added to the output of the VAR model.

The aim of this modelling process is to allow the underlying local meteorological characteristics to be captured so that the developed model is able to describe all sites and their interactions, in particular the differences between actual conditions, hour by hour, and the underlying trends. The model also provides a prediction (forecast) of the next hour, given observations in the current and previous hours. These observations depend on the time of day and day of the year and are sensitive to year to year variations. That is, a particular year might be "windier" or "less windy" than the year or years on which the trends were calculated; the de-trended data which is then used to determine the parameters of the auto-regressive model would, as a consequence, not necessarily reflect the current year and the differences would be handled as best it can by the VAR modelling. This highlights the particular challenge of accurately modelling inter-year variability. Particular sites may have observations in some years that differ markedly from the "current" year to which the model is applied for modelling or forecasting purposes.

It should also be noted that diurnal trends are not significant far offshore due to the absence of thermal effects, and that coastal and inland sites may also have distinct patterns associated with them.

The synthesized wind speed series can then be converted to hub height (of 80 m) using a power law (Wieringa, 1993) and then passed through a wind power curve such as that used in the TradeWind report (2009), to produce a wind power series. These power series are obtained at all the modelled offshore sites (chosen to represent the main existing or proposed wind farm locations) and then calculations of capacity factors are made (The capacity factor is the ratio of energy produced by a wind farm in a given period to the total amount of energy it would have generated had it been generating at full power). Comparisons are then made with the results achieved in the development of a prototype Scottish Electricity Dispatch Model (SEDM) for analysis of the GB power market and which used a large set of data derived from a meteorological model without any statistical model, based on (Hawkins et al., 2011), referred to below as "the Edinburgh model".

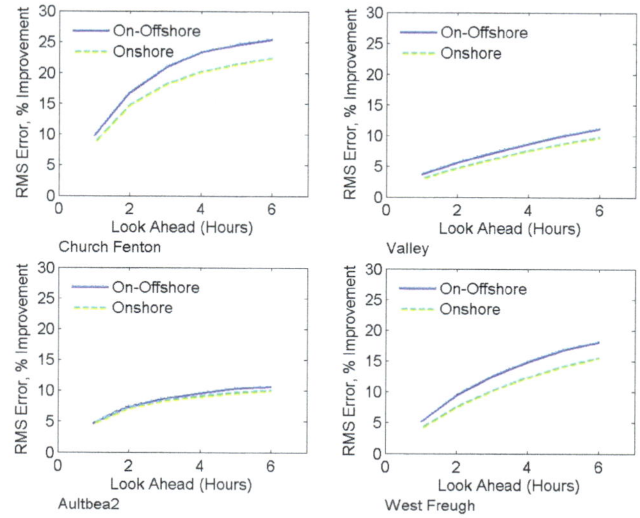

Figure 3. Forecasting skill of VAR model with improvements compared with persistence forecasting, at 4 sample sites for purely onshore model and mixed onshore/offshore model.

3 Results

The work here utilises a training window for model estimation and a forecasting window for its subsequent validation (it does not use "in-sample" data). As stated above, the model was trained on January 2006 to December 2006 and then the year 2007 was used to evaluate the model's predictive ability. Figure 3 shows the improvements gained over persistence forecasting (where the wind speed at time "now" is held fixed for all look aheads) in the RMS errors at 4 candidate sites. The results from the purely onshore model are seen to be inferior to the mixed onshore/offshore model at nearly all sites by a few per cent, except around the boundaries of the model, such as those at station Aultbea2 and Valley. For sites like these it is hard to see the level of improvement as the lines in Fig. 3 are so close, so further calculations have been made as described below. Improvements over persistence are easier to identify in Fig. 3 and can be seen to be up to 25 % and more at 6 h ahead, such as those at Church Fenton, although at sites such as Valley and Aultbea2, located around the edges of the model's geographical limits, percentages are down to 10 % at 6 h ahead.

Further calculations to explore the benefits of extending the data used to include offshore sites (albeit in the form of meteorological model output) have been undertaken for all sites for look ahead periods from 1 to 6 h. We concentrate here on the results for just 3 and 6 h ahead forecasts. The improvement (in percentage terms) achieved by the addition of offshore met. model data to the onshore data for the 6 h ahead forecasts, compared with the purely onshore analysis, varied across the sites studied, ranging from −1.8 % (i.e. a slight degradation) for the Western Isles, up to a 22.1 % improvement for Walney Island, with an overall mean forecasting improvement across all sites of 10.4 %. Interestingly, at 3 h ahead, improvements rose by up to 59.6 % (for Tiree) with an overall mean improvement of 16.3 %. The main reason for the better performance of the onshore/offshore model would appear to be simply due to the presence of additional relevant (i.e. correlated) data, that happens in this case to cover a wider region. Additional onshore data of good quality would be expected to also improve the modelling in a similar manner. However if the offshore data was not of good quality, or there was no significant relationship between the offshore and onshore data, then no improvement would be anticipated. Further research is required to explore the strength of the relationship of any new data that might be added to an existing model, and the improvements (or otherwise) in model performance. It would be attractive then to derive mathematical constraints based on correlation that would determine when an additional measurement site would improve the performance of an existing model.

How well the VAR model presented here compares with other "state-of-the-art" techniques is difficult to say as model results are highly data dependent. A proper comparison would require an exchange of data sets, or the implementation in full of the methods used by others. This is beyond the scope of the present research. See the reviews by Giebel et al. (2011) and Ma et al. (2009) for more information on other modelling approaches.

Figure 4 shows the two sets of wind turbine capacity factors calculated for the Edinburgh model and the VAR model at all the offshore sites used in the VAR model and, whilst a few significant discrepancies do exist, the overall pattern of results is similar, even though the two models were based on completely independent data. It should be noted that the Edinburgh model was based on longer term information covering nearly 10 years (from April 2001 to December 2010). The capacity factors for the VAR model synthesis used here are based on just two years, 2006 and 2007; this period was seen from a GL Garrad Hassan paper (Hodgetts, 2011) to be fairly "typical" in that it was close to the long term mean.

A brief investigation of possible regional variations in the agreement in capacity factor estimates from the two methods was made. It was tentatively concluded that the South Eastern offshore sites on the whole showed better agreement than those on the West of the country, although the sites of Lynn and Inner Dowsing, and Dudgeon showed poor agreement

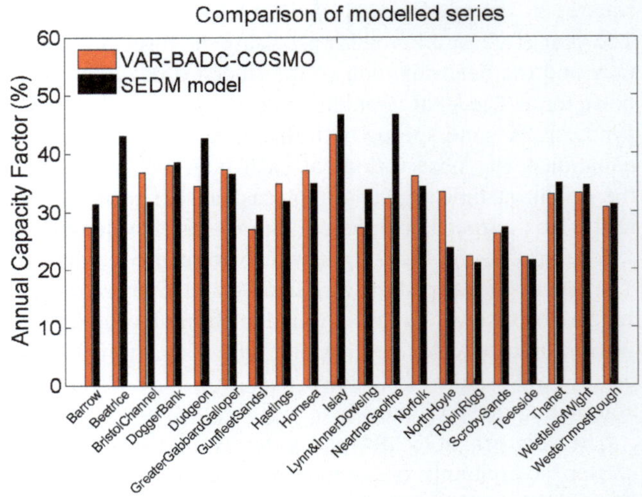

Figure 4. Comparison of capacity factors derived from VAR model and Edinburgh model.

with differences up around 24 %. It is thought that these results may be due to the greater concentration of input data to the VAR model in the regions showing better agreement.

A judgment of which of a number of candidate models is most "accurate" depends on the purpose to which the models are being put. For example, the priority might be to adequately reproduce monthly average wind speeds, or alternatively the hour-by-hour variability, and a given model might not achieve both of these equally well in comparison with other models. Moreover, such an evaluation depends on the availability of actual observations to which the models can be compared. Although many models are based on wind speed, to be of value in the energy sector, they must in the end quantify wind power, and this depends on an accurate wind speed to power conversion which, in itself, is far from trivial. For example, in practice it depends on wind direction, the size of a wind farm, the terrain in which it is located and the design of wind turbines at the farm and their spacing (mainly due to wake effects). Furthermore, the validation of a wind power time series depends on hour-by-hour observations of wind power. In Britain, only monthly total energy outputs are publicly available which limits the extent of the validation. The Edinburgh model was compared with these monthly outputs and here, a comparison is made with the results from that model.

4 Conclusions

A method capable of providing both wind forecasts and a synthesis of wind speeds (and wind power) across the GB power network system, useful for planning and operation purposes, has been presented. Further improvements to published work have been made, including a more rigorous approach to separate training from application. In addition, the

VAR model has been extended to include offshore meteorological model data. The performance of this extended VAR model forecasting has been assessed, and improvements over persistence by as much as 25 % have been demonstrated. Up to 3.7 % of this improvement can be attributed to the addition of the offshore model data at 6 h ahead.

Capacity factors at offshore wind farm sites (both existing and planned) have been compared with an approach based purely on a meteorological model and, with a few exceptions, it has been shown that the two approaches are in good agreement at most sites. Future work will involve the development of interpolation techniques (e.g. Kriging) to facilitate the estimation of capacity factors at onshore wind farm sites away from meteorological measurement sites, with which it will be possible to validate against measured wind power data from the Renewables Obligation Certificate register.

Acknowledgements. The authors kindly thank the EPSRC for providing funding under the "Top and Tail" grant (Project number EP/I031707/1, http://www.topandtail.org.uk/). The provision of data from the BADC and COSMO is also gratefully acknowledged and a special note of thanks goes to Tobias Aigner for his help in unpacking the COSMO model data.

Edited by: D. Brayshaw

References

Aigner, T. and Gjengedal, T.: Detailed wind power production in northern Europe, Renewable Energy Conference, 27 June–2 July, Yokohama, Japan, 2010.

COSMO (Consortium for Small-scale Modelling): available at: www.cosmo-model.org (last access: January 2013), 2010.

Cox, J.: Impact of intermittency: how wind variability could change the shape of the British and Irish electricity markets, Pöyry Energy (Oxford) Ltd, Oxford, 2009.

Doms, G. and Schättler, U.: A description of the nonhydrostatic regional model, LM, Part 1: Dynamics and Numerics, Deutscher Wetterdienst, 2002.

Gibescu, M., Ummels, B. C., and Kling, W. L.: Statistical wind speed interpolation for simulating aggregated wind energy production under system studies, in: Proc. 9th Int. Conf. Probabilistic Methods Applied to Power Systems KTH, Stockholm, Sweden, June, 1–7, 2006.

Giebel, G., Brownsword, R., Kariniotakis, G., Denhard, M., and Draxl, C.: The state-of-the-art in short-term prediction of wind power – A literature review, 2nd Edn., Technical Report, Technical University of Denmark, 2011.

Gross, R., Heptonstall, P., Anderson, D., Green, T., Leach, M., and Skea, J.: The costs and impacts of intermittency: an assessment of the evidence on the costs and impacts of intermittent generation on the British electricity network, UK Energy Research Centre, available at: http://www.ukerc.ac.uk/support/tiki-index.php?page=Intermittency, March, 2006.

Hawkins, S., Eager, D., and Harrison, G. P.: Characterising the reliability of production from future British offshore wind fleets, IET Renewable Power Generation Conference, September, 2011.

Hill, D., McMillan, D., Bell, K. R. W., and Infield, D.: Application of auto-regressive models to UK wind speed data for power system impact studies, IEEE Trans. Sustain. Energ., 3, 134–141, January, 2012.

Hodgetts, B.: A statistical review of recent wind speed trends in the UK, RenewableUK conference, Manchester, available at: http://www.gl-garradhassan.com/assets/downloads/A_Statistical_Review_of_Recent_Wind_Speed_Trends_in_the_UK.pdf, October, 2011.

Kubik, M. L., Brayshaw, D. J., Coker, P. J., and Barlow, J. F.: Exploring the role of reanalysis data in simulating regional wind generation variability over Northern Ireland", Renew. Energ., 57, 558–561, 2013.

Lütkepohl, H.: New introduction to multiple time series analysis, Springer-Verlag, New York, 2005.

Ma, L., Luan, S., Jiang, C., Liu, H., and Zhang, Y.: A review on the forecasting of wind speed and generated power", Renew. Sustain. Energ. Rev., 13, 915–920, 2009.

Oswald, J. O., Raine, M., and Ashraf-Ball, H.: Will British weather provide reliable electricity?," Energ. Pol., 36, 3212–3225, June, 2008.

Sinden, G.: Characteristics of the uk wind resource: Long term patterns and relationship to electricity demand, Energ. Pol., 35, 112–127, January, 2007.

Strbac, G.: Quantifying the system costs of additional renewables in 2020, report to UK DTI, available at: http://www.dti.gov.uk/files/file21352.pdf, 2002.

TradeWind Project Final Report: available at: http://www.trade-wind.eu/fileadmin/documents/publications/Final_Report.pdf, May 2009.

UK Meteorological Office: Met. Office Integrated Data Archive System (MIDAS) Land and Marine Surface Stations Data (1853–current), NCAS British Atmospheric Data Centre, available at: http://badc.nerc.ac.uk/view/badc.nerc.ac.uk__ATOM__dataent__ukmo-midas, 2012.

Wieringa, J.: Representative roughness parameters for homogeneous terrain, Bound. Layer Meteorol., 63, 323–363, 1993.

Comparison of the economic impact of different wind power forecast systems for producers

S. Alessandrini[1,2], F. Davò[1], S. Sperati[1], M. Benini[1], and L. Delle Monache[2]
[1]RSE, Ricerca Sistema Energetico, Milano, Italy
[2]National Centre for Atmospheric Research, Boulder, CO, USA

Correspondence to: S. Alessandrini (alessand@ucar.edu)

Abstract. Deterministic forecasts of wind production for the next 72 h at a single wind farm or at the regional level are among the main end-users requirement. However, for an optimal management of wind power production and distribution it is important to provide, together with a deterministic prediction, a probabilistic one. A deterministic forecast consists of a single value for each time in the future for the variable to be predicted, while probabilistic forecasting informs on probabilities for potential future events. This means providing information about uncertainty (i.e. a forecast of the PDF of power) in addition to the commonly provided single-valued power prediction. A significant probabilistic application is related to the trading of energy in day-ahead electricity markets. It has been shown that, when trading future wind energy production, using probabilistic wind power predictions can lead to higher benefits than those obtained by using deterministic forecasts alone. In fact, by using probabilistic forecasting it is possible to solve economic model equations trying to optimize the revenue for the producer depending, for example, on the specific penalties for forecast errors valid in that market. In this work we have applied a probabilistic wind power forecast systems based on the "analog ensemble" method for bidding wind energy during the day-ahead market in the case of a wind farm located in Italy. The actual hourly income for the plant is computed considering the actual selling energy prices and penalties proportional to the unbalancing, defined as the difference between the day-ahead offered energy and the actual production. The economic benefit of using a probabilistic approach for the day-ahead energy bidding are evaluated, resulting in an increase of 23 % of the annual income for a wind farm owner in the case of knowing "a priori" the future energy prices. The uncertainty on price forecasting partly reduces the economic benefit gained by using a probabilistic energy forecast system.

1 Introduction

Among the limiting factors to the penetration of wind energy is the variable nature and limits to predictability of wind speed. Electricity is distributed over a region via a grid, and in a deregulated market it is crucial for the utility companies that provide electricity to be able to deliver the committed energy to the grid. In this sense wind energy is at a disadvantage when compared to other types of energy sources because of its intrinsic limited predictability. For this reason, in the past wind energy producers have often been exempted from penalties for failing to deliver the day-ahead planned energy. However, with the increasing penetration of the wind energy in the energy market, in many European countries (e.g., Germany, Denmark and Italy) the requirement of an accurate energy forecast has been introduced, along with penalties proportional to the forecast errors. Wind power predictions can be categorized in two main groups: deterministic and probabilistic. A deterministic forecast consists in a single value for each time in the future for wind power. Probabilistic forecasting provides probability density functions (PDF) from which probabilities of future outcomes of events can be estimated. The value of probabilistic power predictions and the usefulness of uncertainty quantification that results from it have been explored (Roulston et al., 2003; Zugno et al., 2013). A significant application is when renewable energy is traded in day-ahead electricity markets. Trading future wind energy production using probabilistic wind power predictions can lead to higher economic benefits than those obtained by using deterministic forecasts alone. In fact, the maximum income for a producer is obtained by offering in the day-ahead market an amount of energy that can be different from the mean value of the forecasted PDF (see Roulston et al., 2003; Zugno et al., 2013). Indeed, it can be analytically shown that under the conditions imposed by the day-ahead energy market, the optimal day-ahead market bid for a wind energy producer is a certain quantile of the distribution of forecasted wind power (Bremnes, 2004; Linnet, 2005; Pinson et al., 2007). This optimal quantile is a function of the day-ahead and the imbalance prices, which are not known apriori and change on hourly basis. The optimal bids might significantly differ from the point forecasts of wind power production. A theoretical discussion about quantile forecasts being optimal bids in electricity markets can be found in Gneiting (2011).

In this work we prove, for the first time on a real case application of an Italian wind farm, that it is possible to limit the economic loss (caused by penalties regulation) by using a probabilistic forecast that choses the optimal quantile for bidding wind energy in the day-ahead market.

2 Site and power data description

The case study is based in one wind farm located in a complex-terrain mountainous area of Southern Italy, whose total installed power is about 8 MW. An hourly time series power data is available for the period November 2010–May 2012. The power data have been filtered and quality controlled following Alessandrini et al. (2013).

3 The probabilistic power forecast

The probabilistic wind power forecast has been obtained by using a deterministic meteorological model's simulation, performed by applying the non-hydrostatic model RAMS (Pielke et al., 1992) and the analog ensemble technique (AnEn) (Delle Monache et al., 2012, 2013).

RAMS has been run using the ECMWF (European Centre for Medium range Weather Forecast) analysis as initial conditions and 6-hourly ECMWF deterministic forecast fields as boundary conditions. The simulation set up consists of 2 nested grids with a horizontal grid spacing of 12 and 4 km and 30 vertical levels. A set of 59 h long forecast simulations have been run, starting at 12:00 UTC of every day of the two-year period 2011–2012.

The AnEn technique provides a set of likely power predictions (i.e., an ensemble that is a Monte Carlo approximation of the PDF associated with future power production) using a historical dataset of observations and deterministic model predictions. For each forecast lead time the ensemble set of forecasts of a certain variable is constituted by a set of its measurements from the past. These measurements (in this case wind power data) are those concurrent to past forecasts at the same lead time, chosen across the past runs most similar to the current forecast. The metric used to rank past forecasts' similarity to the current forecast is defined as follows (Delle Monache et al., 2012, 2013).

$$\|F_t, A_{t'}\| = \sum_{i=1}^{N_v} \frac{w_i}{\sigma_{f_i}} \sqrt{\sum_{j=-\tilde{t}}^{\tilde{t}} \left(F_{i,t-j} - A_{i,t'+j}\right)^2} \quad (1)$$

where F_t is the current forecast for the lead time t at a certain location, $A_{t'}$ is an analog forecast for the same lead time t' before F_t was issued and at the same location, N_v and w_i are the number of physical variables and their weights, respectively, σ_{f_i} is the standard deviation of the time series of the past forecast of a given variable at the same location, $\tilde{t}$ is an integer equal to half width of the time window over which the metric is computed (e.g., if $\tilde{t} = 1$ h, the distance will be computed over the 3 forecast lead times corresponding to $t - 1$ h, t, $t + 1$ h), and $A_{i,t'+j}$ and $F_{i,t-j}$ are the values of the analog and the forecast in the time window for a given variable. The goal is to find past forecasts of the meteorological variables (chosen among the ones with the highest correlation with the quantity to be predicted, wind power in this case) that were predicting similar values and temporal trend (a similar behaviour as a function of time along the time interval defined by $\tilde{t}$) compared to the current forecast. In the current application the number of the ensemble member has been kept constant and equal to 20.

In this study, wind speed and wind direction have been extracted at the wind farm location at the hub height, equal to 50 m above the ground, and are used as predictors for the AnEn's algorithm. Year 2011 has been chosen as the training period, while probabilistic wind power forecast are issued along year 2012. The hourly probabilistic forecasts in the interval between lead times equal to 36 and 59 h are retained for the economic computation, assuming they are used for the day-ahead market bids.

4 Economic model

In this section the economic impact of a probabilistic wind power forecast is evaluated introducing a very a simple model of an electricity market. This model mimic the real situation of some nations where penalties must be paid by the producers of the energy market on the unbalancing, which is computed as the difference between the hourly promised (a day-ahead) energy and the delivered one. The penalties are usually related to the day-ahead and spot market energy prices (the day-ahead market occurs once a day usually ending around 09:00 a.m. local time and defines the hourly energy production for the day after, while the spot market occurs every hour to balance the energy production with the real demand). In this kind of mechanism, Roulston et al. (2003) showed that the maximum income can be achieved by promising an hourly energy production that is not necessarily the mean value of the forecasted energy PDF (depending on the day-ahead and spot market prices). While the market model used here is a simplified version of real conditions and may not replicate all current situations of different nations, the wind power data are related to a real wind farm located in Italy. The market model simulates the mechanism by which producers promise a target amount of electricity, E_c, for a particular hour of the day-ahead. The producer is paid a fixed unit price, P_c, for the promised electricity. If at that particular hour the wind production is not sufficient to satisfy the contract, then the generator must satisfy the contract by purchasing the shortfall on the spot market at a cost of P_s per unit usually greater than P_c. If, in the opposite case, the real production exceeds the promised energy the excess of production is paid at a price P_{s1} usually lower than P_c. In this latter case there will be some other energy producers (usually from conventional sources) who must be reimbursed by the amount of energy they are not able to deliver in to the grid. Indicating E_a as the real generated energy, the hourly income I for the producer can be computed as follows:

$$I = \begin{cases} E_c P_c + (E_a - E_c) P_{s1} & \text{if } E_a \geq E_c \\ E_c P_c - (E_c - E_a) P_s & \text{if } E_a < E_c \end{cases} \quad (2)$$

In this model the penalties faced by the producers are asymmetric. It can be shown (see Roulston et al., 2003; Zugno et al., 2013) that under the conditions described by Eq. (2) the optimal day-ahead market bid for a wind energy producer is a certain quantile of the distribution of forecasted wind power. The optimal value of E_c is such that the marginal income from an additional unit of promised electricity is exactly balanced by the expected penalty from failing to meet E_c for that unit. This balance occurs when:

$$P_c - P_{s1} = P(E_c)(P_s - P_{s1}) \quad (3)$$

or,

$$P(E_c) = \frac{P_c - P_{s1}}{P_s - P_{s1}} \quad (4)$$

where the function $P(E)$ is the probability that the actual produced energy E_a is lower than any critical value E (i.e., the cumulative function of the distribution of the expected produced energy). Since P_s is greater than P_c, and P_{s1} is generally lower than P_c, the ratio at the right hand side in Eq. (4) is always between 0 and 1. The optimal quantile obtained by solving Eq. (4) is dynamic since P_s, P_c, and P_{s1} are not known a priori at the time of the bid and change every hour, while the cumulative function $P(E)$ can be obtained as an output of a probabilistic power forecast system and depends on the time too. The optimal bid E_c might significantly differ from the point forecasts of wind power production, which often is considered as the value of the energy with the expected highest probability of occurrence.

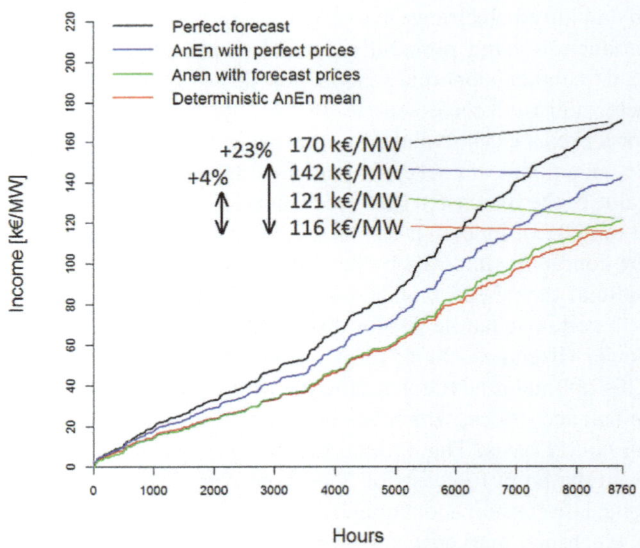

Figure 1. Cumulative income obtainable per 1 MW of installed capacity by a wind farm located in the south of Italy as a function of the hours of the year 2011. The different lines refer to the different approaches in bidding the wind energy in the day-market.

5 Results

In Fig. 1 the cumulative income of the studied wind farm is plotted as a function of the hours of one year. The hourly income is computed following Eq. (2) with P_s, P_{s1} and P_c being respectively the real occurred buy and selling prices of spot market and the spot prices of the day-ahead market. The black line refers to the income obtainable per 1 MW of installed capacity and a perfect forecast (the promised energy in the day-ahead market is exactly equal to the produced energy). The red line refers to the income achievable by selling during the day-ahead market an amount of energy obtained by a deterministic (or spot) wind power forecast (the mean of the 20 members used for the probabilistic forecast) that doesn't consider any economic optimization model. The blue line indicates the case when the amount of energy sold in the day-ahead market is set by solving Eq. (4). Since P_s, P_c, and P_{s1} are not known a priori at the time of the bid they would need to be forecasted as the cumulate function $P(E)$. In this case, the real prices are assumed known in advance. The next simulation (green line) has been carried out under the realistic condition of not knowing the market prices at the time of the bidding. A simple method for forecasting the market prices has been adopted. For each day of the test period the forecasted price for a given lead time is the average of the 7 prices occurred during the previous 7-day long period at the same lead time. The analog ensemble method is used to forecast the cumulative function $P(E)$ of Eqs. (3) and (4). The economic benefit of a probabilistic approach compared to a deterministic one corresponds to a potential increment of 23 % on the total annual income (blue line compared to

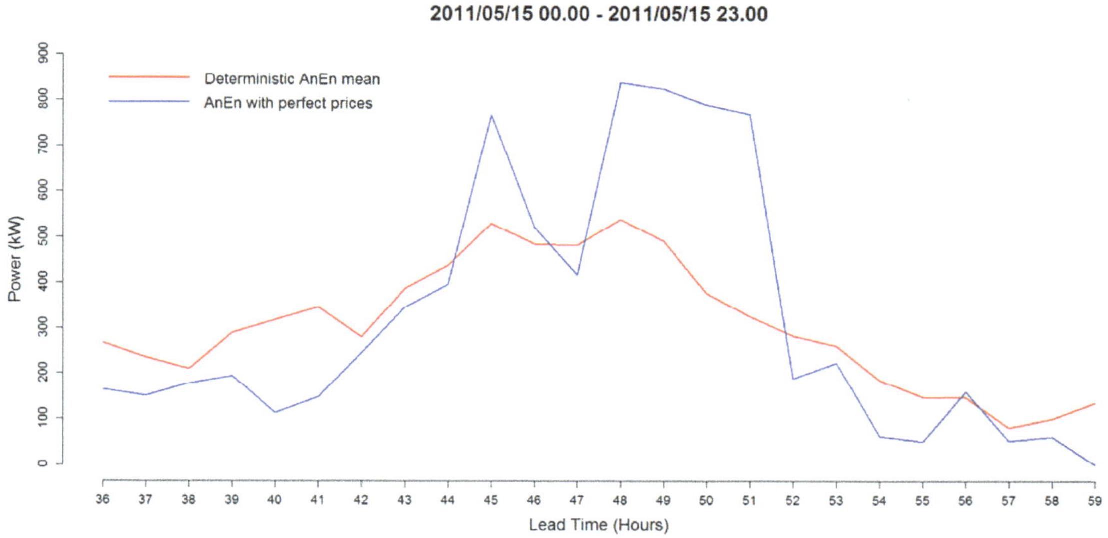

Figure 2. Hourly time-series as a function of forecast lead time of the deterministic forecast provided by AnEn mean (red), and optimal quantile forecast (blue) chosen, in the case of perfect prices forecast, by using the PDF of AnEn in the Eq. (4).

red line). The uncertainty of price forecasting significantly reduces this increment even if a greater annual income, compared to the deterministic approach, is still achieved.

In Fig. 2 an example of two bidding approaches are shown. A strategic bidding based on the choice of the optimal quantile (blue line) can lead to significantly different values from those of a deterministic approach (red line).

6 Conclusions

In this work, a practical application of probabilistic wind power forecast for selling wind energy in the day-ahead energy market has been shown. The energy corresponding to an optimal quantile of the forecasted distribution has been used for bidding in the day-ahead market. The economic benefit compared to a deterministic approach has been examined. This benefit corresponds to a potential increment of 23 % on the total annual income, under the hypothesis of knowing a priori the actual market prices that will occur during the day-ahead. The uncertainty in forecasting the market prices reduces the efficacy of the model in increasing the annual income. In fact, it has been shown that if the market prices are also predicted, the economic benefit achieved by choosing the "optimal quantile" approach is considerably reduced. Further investigations will consider the use of more accurate price forecasting techniques and different economic rules.

Acknowledgements. This work has been partly financed by Xcel Energy and the Research Fund for the Italian Electrical System under the Contract Agreement between RSE (formerly known as ERSE) and the Ministry of Economic Development - General Directorate for Nuclear Energy, Renewable Energy and Energy Efficiency stipulated on 29 July 2009 in compliance with the Decree of 19 March 2009.

The authors also acknowledge F. Fioretti and U. De Angelis of Enel Spa for providing meteorological and power data for the test case.

Edited by: D. Brayshaw

References

Alessandrini, S., Sperati, S., Pinson, and P. A: comparison between the ECMWF and COSMO Ensemble Prediction Systems applied to short-term wind power forecasting on real data, Appl. Energ., 107, 271–280, 2013.

Bremnes, J. B.: Probabilistic wind power forecasts using local quantile regression, Wind Energy, 7, 47–54, 2004.

Delle Monache, L., Nipen, T., Liu, Y., Roux, G., and Stull, R.: Kalman filter and analog schemes to post-process numerical weather predictions, Mon. Weather Rev., 139, 3554–3570, 2011.

Delle Monache, L., Eckel, A., Rife, D., and Searight, K.: Probabilistic weather predictions with an analog ensemble, Mon. Weather Rev., 141, 3498–3516, 2013.

Gneiting, T.: Quantiles as optimal point forecasts, Int. J. Forecasting, 27, 197–207, 2011.

Linnet, U.: Tools supporting wind energy trade in deregulated markets, Master's Thesis, Technical University of Denmark, 2005.

Pielke, R. A., Cotton, W. R., Walko, R. L., Tremback, C. J., Lyons, W. A., Grasso, L. D., Nicholls, M. E., Moran, M. D., Wesley, D. A., Lee, T. J., and Copeland, J. H.: A Comprehensive Meteorological Modeling System – RAMS, Meteorol. Atmos. Phys., 49, 69–91, 1992.

Pinson, P., Chevalier, C., and Kariniotakis, G.: Trading wind generation from short-term probabilistic forecasts of wind power, IEEE T. Power Syst., 22, 1148–1156, 2007.

Roulston, M. S., Kaplan, D. T., Hardenberg, J., and Smith, L. A.: Using medium-range weather forecasts to improve the value of wind power production, Renew. Energ., 28, 585–602, 2003.

Zugno, M., Pinson, P., and Jonsson, T.: Trading wind energy based on probabilistic forecasts of wind generation and market quantities, Wind Energy, 16, 909–926, 2013.

Large-Eddy Simulation on turbulent flow and plume dispersion over a 2-dimensional hill

H. Nakayama and H. Nagai
Japan Atomic Energy Agency, Ibaraki, Japan

Abstract. The dispersion analysis of airborne contaminants including radioactive substances from industrial or nuclear facilities is an important issue for air quality maintenance and safety assessment. In Japan, many nuclear power plants are located at complex coastal terrains. In these cases, terrain effects on the turbulent flow and plume dispersion should be investigated. In this study, we perform Large-Eddy Simulation (LES) of turbulent flow and plume dispersion over a 2-dimensional hill flow and investigate the characteristics of mean and fluctuating concentrations.

1 Introduction

The dispersion analysis of airborne contaminants including radioactive substances from industrial or nuclear facilities is an important issue in air quality maintenance and safety assessment. In a flat terrain, the mean concentration of a plume can be easily predicted by using a Gaussian plume model. However, many nuclear power plants in Japan are located in complex coastal terrains. In this case, the effects of the terrain should be taken into consideration when predicting plume dispersion. Terrain effects on turbulent flow and/or plume dispersion have been investigated in many studies. For example, Jackson and Hunt (1975) proposed a theoretical model for the analysis of turbulent structures over a gentle hill. In one of the earliest wind tunnel experimental studies, Khurshudyan et al. (1981) conducted wind tunnel experiments of turbulent flow and plume dispersion over 2-dimensional hills with different slopes and investigated the influence of hill slope and source location on the flow and concentration fields. Castro and Snyder (1982) conducted wind tunnel experiments on plume dispersion over 3-dimensional hills with various ratios of spanwise hill breadth to height and investigated the effect of hill shape on mean concentration distributions. Arya and Gadiyaram (1986) also conducted wind tunnel experiments on flow and dispersion in the wakes of 3-dimensional low hills and investigated the difference between the ground-level mean concentration of a plume released from a point source with and without the hill. Sada (1991) investigated the streamwise variation of vertical profiles of mean wind velocity, turbulence kinetic energy, and mean concentration over a 2-dimensional gentle hill using wind tunnel experiments. In one of the earliest numerical studies, Hino (1968) conducted numerical experiments for plume dispersion over a complex terrain and showed the contaminant distribution patterns. Recently, Castro and Apsley (1997) performed numerical simulations of plume dispersion over 2-dimensional hills using the standard k-ε turbulence model and compared this with the experimental data.

These previous studies have focused on the terrain effects on the characteristic of turbulent flow and/or mean concentration. Another issue related to plume dispersion is the potential problem of the accidental release of hazardous and flammable materials. During the assessment of the hazard to human health from toxic substances, the existence of high concentration peaks in a plume should be considered. For safety analysis of flammable gases, certain critical threshold levels should be evaluated. In such a situation, not only mean but also fluctuating concentrations should be estimated. In this study, we perform a numerical simulation of unsteady behaviors of turbulent flow and plume dispersion over a 2-dimensional hill by using an LES that can give more detailed information on the flow and concentration fields in comparison with the wind tunnel experiments of Sada (1991) and investigate the characteristics of mean and fluctuating concentrations.

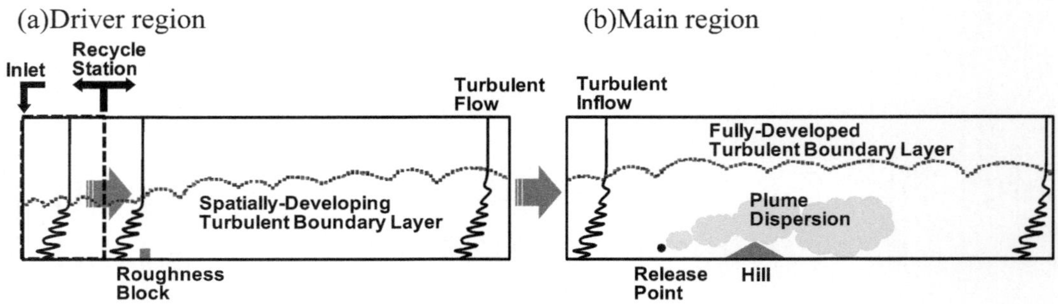

Figure 1. Schematic illustration of computational model.
(**a**) Driver region for generation of a spatially-developing TBL flow and (**b**) main region for plume dispersion over hill.

2 Numerical model and computational settings

The basic equations for the LES model are the spatially filtered continuity equation, Navier-Stokes equation and the transport equation for concentration. The subgrid-scale (SGS) Reynolds stress is parameterized by using the standard Smagorinsky model (Smagorinsky, 1963) with a Van Driest damping function (Van Driest, 1956), where the Smagorinsky constant is set to 0.12 (Iizuka and Kondo, 2004) for estimating the eddy viscosity. The subgrid-scale scalar flux is also parameterized by an eddy viscosity model and the turbulent Schmidt number is set to 0.5. Various SGS models for LES have been proposed besides the standard Smagorinsky model. For example, dynamic Smagorinsky models have been proposed by Germano et al. (1991), Lilly (1992), and Meneveau et al. (1996). However, Iizuka and Kondo (2004) examined the influence of various SGS models on the prediction accuracy of LESs of turbulent flow over a hilly terrain and showed that the prediction accuracy of LESs with the standard Smagorinsky model is better than that of the LESs with the dynamic Smagorinsky type models. This indicates that the dynamic Smagorinsky type models are not always effective for determining model constant. As a static type SGS model, Nicoud and Ducros (1999) proposed the wall-adapting local eddy-viscosity (WALE) model. This model can capture the effects of both the strain and the rotation rate of the small-scale turbulent motions without a damping function from the wall. According to Temmerman et al. (2003), the WALE model shows better performance when compared with the dynamic/standard Smagorinsky model. However, the conventional Smagorinsky model that has the advantage of simplicity and low computational costs is adopted in our LES model because the focus of our research is not the small order effects of turbulent flow.

The coupling algorithm of the velocity and pressure fields is based on the Simplified Marker and Cell (SMAC) method with the second-order Adams-Bashforth scheme for time integration. The SMAC method is the algorithm for solving numerically the Navier-Stokes equation and was proposed by Amsden and Harlow (1970). The Poisson equation is solved by the Successive Over-Relaxation (SOR) method which is an iterative method for solving a Poisson equation for pressure. For the spatial discretization in the governing equation of the flow field, a second-order accurate central difference is used. For the dispersion field, Cubic Interpolated Pseudo-particle (CIP) method proposed by Takewaki et al. (1985) and a second-order accurate central difference method are used for the advection and diffusion terms, respectively. The time step interval $\Delta t U_\infty / H$ is 0.005 (Δt: time step).

To perform LES of plume dispersion, a thick turbulent boundary layer (TBL) flow with strong velocity fluctuations should be simulated in order to mimic a plume released from an elevated point source. Various kinds of methods for producing a realistic TBL flow have been proposed. For example, Mochida et al. (1992) performed a preliminary LES of a channel flow for the inlet boundary condition. This method can be easily applied, but the flow is driven by a pressure gradient under the periodic boundary condition in the streamwise direction. Because it is known that the influence of a pressure gradient on TBL characteristics is large (Kline, 1967), it is desirable to reproduce an approaching flow without a pressure gradient.

Figure 1 shows a schematic illustration of the computational regions for plume dispersion over a hill immersed in a fully-developed TBL flow. First, a spatially-developing TBL flow with strong velocity fluctuations in the driver region is generated by incorporating the inflow turbulence generation method proposed by Kataoka and Mizuno (2002) into an upstream small fraction of the driver region with a 2-dimensional roughness bar placed at the ground surface. Next, the inflow turbulence data obtained near the exit of the driver region is imposed at the inlet of the main region at each time step, and calculation of the turbulent flow and plume dispersion over a hill is performed. The total size and number of grid points for the computational regions are $81.5H \times 12.5H \times 25.0H$ (H: hill height) with a Cartesian grid system and $810 \times 120 \times 100$ in x-, y-, and z-directions, respectively. At the exit of the driver and main regions, a Sommerfeld radiation condition (Gresho, 1992) is imposed. At the top, a free-slip condition for streamwise and spanwise velocity components is imposed and the vertical velocity

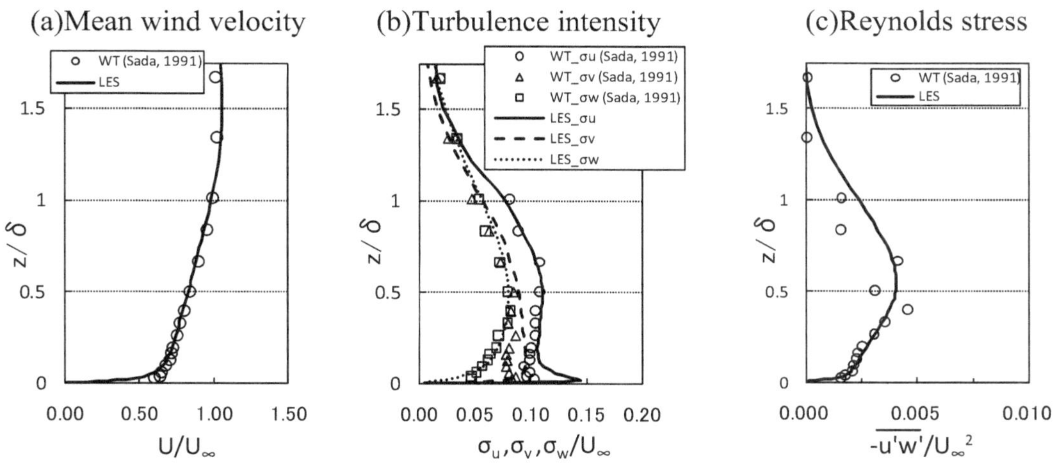

Figure 2. Turbulence characteristics of approaching flow.

component is 0. At the side, a periodic condition is imposed and at the flat ground surface, a non-slip condition for each velocity component is imposed. Here, in the flow field, Immersed Boundary Method proposed by Fadlun et al. (2000) is used in order to consider the terrain effects. In the concentration field, a zero gradient is imposed at all the boundaries and a fractional volume term is introduced into the scalar conservation equation to account for the space occupied by terrain. A 2-dimensional hill with a mean slope angle of 20° is placed in the main region. The position of the center of the hill is located at 12.5 H downstream of the inlet in the main region. The origin of the coordinates is the ground surface at the center of the hill. A release point of a tracer gas is located at a distance of 9.1 H from the center of the hill at an elevation of 0.45 H. The Reynolds number based on hill height and free-stream velocity (U_∞) at the hill height is almost 5000.

3 Previous wind tunnel experiments for evaluation of the model performance

Many wind tunnel experimental studies of the dispersion characteristics of a plume over a hilly terrain have been conducted. For example, Sada et al. (1991) investigated turbulence structures and dispersion characteristics of a plume over a 2-dimensional hill. The experiments were conducted under conditions of a neutrally stratified TBL flow in the wind tunnel of Central Research Institute of Electric Power Industry. The test section of the wind tunnel is 20-m long, 3-m wide and 1.5-m high. A TBL flow with strong velocity fluctuations from the ground surface to the upper level was generated using roughness elements with an L-shaped cross sections placed on the floor at the entrance of the wind tunnel section. The 2-dimensional hill was 0.11-m high, 0.6-m long, with a hill aspect ratio (ratio of hill length to height) of about 5.5. A release point was located at a distance of 1.0 m upstream of the center of a hill at a height of 0.05 m. The thickness of the TBL at the downstream position of the release point is 0.3 m. Using those parameters, the vertical profiles of mean wind velocity, turbulence kinetic energy and mean concentration over a 2-dimensional hill were obtained.

In this study, in order to evaluate the model performance, we compare the LES data of turbulent flow and plume dispersion over a 2-dimensional hill with the wind tunnel experimental data by Sada et al. (1991), described above.

4 LES results

4.1 Turbulence characteristics of approaching flow

Figure 2 shows a comparison of LES results with the experimental data obtained by Sada (1991) for vertical profiles of mean wind velocity, turbulence intensities and Reynolds stress in the driver region. Each turbulence statistics obtained by an LES is found to be almost consistent with the experimental data. Therefore, it is considered that the TBL flow is successfully simulated.

4.2 Turbulence characteristics over hill

Figures 3 and 4 show a comparison of LES results with the experimental data (Sada, 1991) for vertical profiles of mean wind velocity (U/U_∞) and turbulence kinetic energy (k/U_∞^2) at the positions of $x/H = -2.7$, 0.0, 1.8, 3.6, 8.2, 11.8, 14.5 and 18.2. There are slight discrepancies of mean wind velocity between the LES results and the experimental data at $x/H = 1.8$ and 3.6. The values for the reattachment length behind a hill for the previous experiment (Sada, 1991) and the LES are $L/H = 5.9$ and 6.9 (L: reattachment length), respectively. Iizuka and Kondo (2006) also performed an LES with the standard Smagorinsky model for a 2-dimensional hill with a hill aspect ratio of 5.0 and compared it with the reattachment length of the experimental data of Ishihara et al. (2001). The results show that the computed reattachment

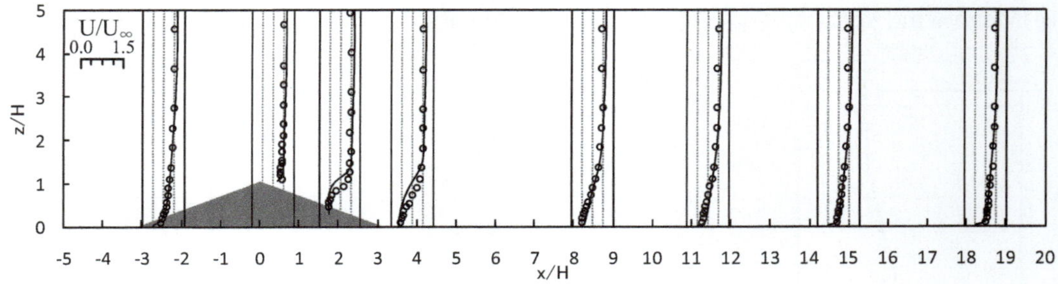

Figure 3. Streamwise variation of vertical profiles of mean wind velocity.

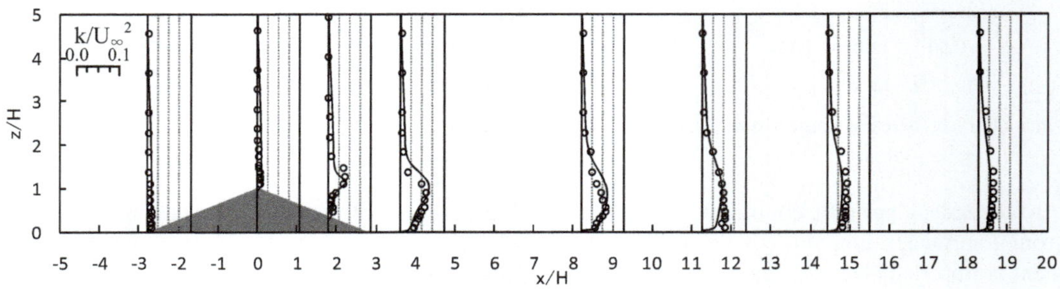

Figure 4. Streamwise variation of vertical profiles of turbulence kinetic energy.

Figure 5. Instantaneous plume dispersion field. The yellow areas on the isosurface indicate 0.01% of initial concentration.

length ($6.4\,H$) is consistent with the experimental data. The reattachment length obtained by our LES model shows an overprediction of $1.0\,H$ despite the fact that we employed the same conventional Smagronsky model as Iizuka's for our LES model. Our LES model needs to be further investigated and improved. For the turbulence kinetic energy, LES results are smaller than those of the experimental data at heights larger than $z/H = 1.0$ at the positions of $x/H = 1.8$, 3.6 and 8.2. However, basic characteristics of wind field such as the increase of wind speed at the crest, the decrease of wind speed behind a hill and the recovery of wind speed with downwind distance are almost similar to the experimental data. The streamwise variations of the turbulence field such as the occurrence of a sharp peak behind a hill due to the strong instability of separated shear layers and the formation of uniform turbulence kinetic energy profiles with downwind distance due to the active turbulent mixing motions are also similar to the experimental data.

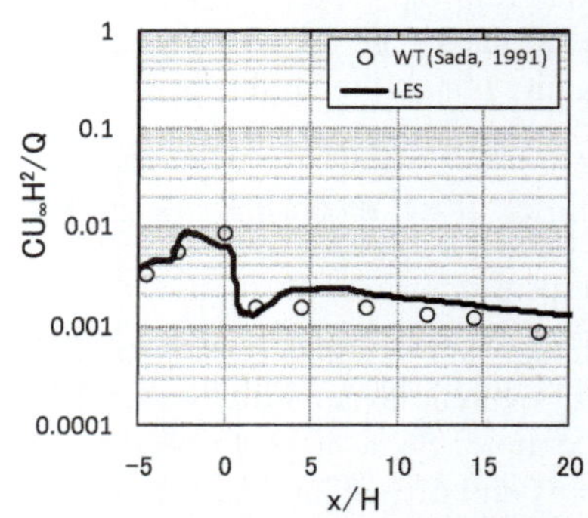

Figure 6. Streamwise variations of mean concentration near the ground level.

4.3 Characteristics of mean and fluctuating concentrations over a hill

Figure 5 the shows instantaneous plume dispersion field at times $t*(= tU_\infty/H) = 10.8$, 21.6 and 43.2 after the plume release. It shows that the plume moves upward above the hill at first, then a portion of the plume is entrained into the wake region. At $t* = 43.2$, the plume is entirely dispersed in

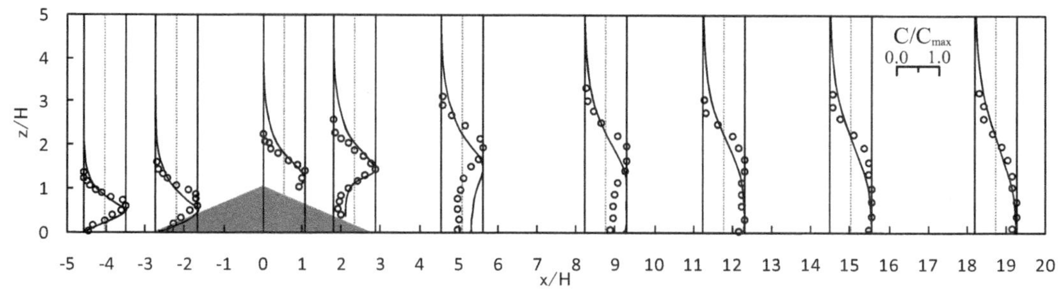

Figure 7. Streamwise variation of vertical profiles of mean concentration.

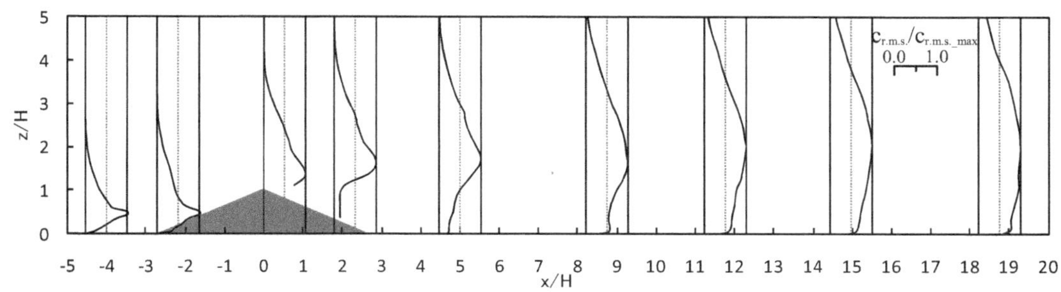

Figure 8. Streamwise variation of vertical profiles of r.m.s. concentration.

the wake region. Figure 6 shows a comparison of LES results with the experimental data (Sada, 1991) for streamwise variations of the mean concentration near ground level. The mean concentration (C) is normalized by the free-stream velocity, hill height and source strength (Q). It is found that the basic characteristics such as the increase of mean concentration towards the crest, the rapid decrease behind a hill and the gradual decrease with downwind distance are almost similar to the experimental data. Figure 7 shows a comparison of the LES results with the previous experimental data for mean concentration at $x/H = -4.5$, -2.7, 0.0, 1.8, 4.5, 8.2, 11.8, 14.5 and 18.2. The mean concentration is normalized by the maximum of mean concentration ($C_{\max}$) at each downstream position. The peak locations of the mean concentration obtained from the LES are lower than those obtained from the experiment, and the mean concentration values are overestimated in the wake region. However, the basic characteristics of plume dispersion such as the rapid vertical spread of a plume behind a hill and the formation of uniform mean concentration profiles with downwind distance are similar to the experimental data. Figure 8 shows streamwise variation of vertical profiles of r.m.s. concentration ($c_{\text{r.m.s.}}$). The r.m.s. concentration is normalized by the maximum of the r.m.s. concentration ($c_{\text{r.m.s.\_max}}$) at each downstream position. Here, we show the r.m.s. concentration profiles obtained by the LES, since the fluctuating characteristic of concentration was not discussed in the previous experiment (Sada, 1991). We found that its values become small inside the wake region due to the smoothing effects of recirculating flows, compared with the mean concentration values.

5 Conclusions

In this study, we performed a numerical simulation of unsteady behaviors of turbulent flow and plume dispersion over a 2-dimensional hill immersed in a fully-developed TBL by using an LES and investigated the characteristics of mean and fluctuating concentrations. First, a spatially-developing TBL flow with strong velocity fluctuations in the driver region was generated by incorporating Kataoka's method (2002) into an upstream small fraction of the driver region with a 2-dimensional roughness bar placed at the ground surface. Then, we obtained the inflow turbulence data at the inlet of the main region, and calculated the flow and dispersion over a hill. As compared to the experimental data (Sada, 1991), the main characteristics such as complex behaviors of turbulent flow and plume dispersion behind a hill were successfully simulated. Furthermore, the r.m.s. concentration is found to be smaller than the mean concentration inside the wake region due to the smoothing effects of recirculating flows.

Edited by: A. Baklanov

References

Amsden, A. A. and Harlow, F. H.: A simplified MAC technique for incompressible fluid flow calculations, Journal of Computer Physics, 6, 322–325, 1970.

Arya, S. P. S. and Gadiyaram, P. S.: An experimental study of flow and dispersion in the wakes of three-dimensional low hills, Atmos. Environ., 20, 729–740, 1986.

Castro, I. P. and Snyder, W. H.: A wind tunnel study of dispersion from sources downwind of three-dimensional hills, Atmos. Environ., 16, 8, 1869–1887, 1982.

Castro, I. P. and Apsley, D. D.: Flow and dispersion over topography: A comparison between numerical and laboratory data for two-dimensional flows, Atmos. Environ., 31, 6, 839–850, 1997.

Fadlun, E. A., Verzicco, R., Orlandi, P., and Mohd-Yusof, J.: Combined immersed-boundary finite-difference methods for three-dimensional complex flow simulations, J. Comput. Phys., 161, 35–60, 2000.

Germano, M., Piomelli, U., Moin, P., and Cabot, W.H.: A dynamic subgrid-scale eddy viscosity model, Phys. Fluids, A3, 7, 1760–1765, 1991.

Gresho, P. M.: Some interesting issues in incompressible fluid dynamics, both in the continuum and in numerical simulation, Adv. Appl. Mech., 28, 45–140, 1992.

Hino, M.: Computer experiment on smoke diffusion over a complicated topography, Atmos. Environ., 2, 541–558, 1968.

Iizuka, S. and Kondo, H.: Performance of various sub-grid scale models in large-eddy simulations of turbulent flow over complex terrain, Atmos. Environ., 38, 7083–7091, 2004.

Iizuka, S. and Kondo, H.: Large-eddy simulations of turbulent flow over complex terrain using modified static eddy viscosity models, Atmos. Environ., 40, 925–935, 2006.

Ishihara, T., Fujino, Y., and Hibi, K.: A wind tunnel study of separated flow over a two-dimensional ridge and a circular hill, Journal of Wind Engineering, 89, Proceedings of the Fifth Asia-Pacific Conference on Wind Engineering, Kyoto, Japan, 573–576, 2001.

Jackson, P. S. and Hunt, J. C. R.: Turbulent wind over a low hill, Q. J. Roy. Meteor. Soc., 101, 929–955, 1975.

Kataoka, H. and Mizuno, M.: Numerical flow computation around aeroelastic 3D square cylinder using inflow turbulence, Wind Struct., 5, 2–4, 379–392, 2002.

Khurshudyan, L. H., Snyder, W. H., and Nekrasov, I. V.: Flow and dispersion of pollutants over two-dimensional hills, United States Environmental Protection Agency Report EPA-600/4-81-067, 1981.

Kline, S. J., Reynolds, W. C., Schraub, F. A., and Runstadler, P. W.: The structure of turbulent boundary layers, J. Fluid Mech., 30, 741–773, 1967.

Lilly, D.: A proposed modification of the Germano subgrid-scale closure method, Phys. Fluids, A4, 3, 633–635, 1992.

Meneveau, C., Lund, T. S., and Cabot, W. H.: A Lagrangian dynamic subgrid-scale model for turbulence, J. Fluid Mech., 319, 353–385, 1996.

Mochida, A., Murakami, S., Shoji, M., and Ishida, Y.: Numerical simulation of flowfield around Texas Tech Building by Large Eddy Simulation, First International Symposium on Computational Wind Engineering, Tokyo, 42–47, 1992.

Nicoud, F. and Ducros, F.: Subgrid-scale stress modelling based on the square of the velocity gradient tensor, Flow Turbul. Combust., 62, 183–200, 1999.

Sada, K.: Development of atmospheric diffusion evaluation method incorporating thermal and topographical effects (5) – atomospheric diffusion model under topographical conditions, CRIEPI Report, T90040, 1991.

Smagorinsky, J.: General circulation experiments with the primitive equations, Mon. Weather Rev., 91, 3, 99–164, 1963.

Temmerman, L., Leschziner, M. A., Mellen, C. P., and Frohlich, J.: Investigation of wall–function approximations and subgrid–scale models in large eddy simulation of separated flow in a channel with streamwise periodic constrictions, Int. J. Heat Fluid Fl., 24, 157–180, 2003.

Takewaki, H., Nishiguchi, A., and Yabe, T.: Cubic Interpolated Pseudo-particle method (CIP) for solving hyperbolic-type equations, J. Comput. Phys., 61, 261–268, 1985.

Van Driest, E. R.: On turbulent flow near a wall, J. Aerosp. Sci., 23, 1007–1011, 1956.

33

Describing the possible climate changes in France and some examples of their effects on main crops used in livestock systems

F. Ruget[1], **J.-C. Moreau**[2], **M. Ferrand**[2], **S. Poisson**[2], **P. Gate**[3], **B. Lacroix**[3], **J. Lorgeou**[3], **E. Cloppet**[4], **and F. Souverain**[4]

[1]INRA, UMR 1114 EMMAH, 84914 Avignon cedex 9, France
[2]Institut de l'Élevage, Auzeville 31326 Castanet-Tolosan et 149 Rue de Bercy 75011, Paris, France
[3]Arvalis Institut du végétal, La Minière 78210 Guyancourt/31450 Baziège/91170 Boigneville, France
[4]Météo-France, 42 Av G Coriolis 31000 Toulouse, France

Abstract. The effects of climate change on forage and crop production are an important question for the farmers and more largely for the food security in the world. Estimating the effect of climate change on agricultural production needs the use of two types of tools: a model to estimate changes in national or local climates and an other model using climatic data to estimate the effects on vegetation. In this paper, we will mainly present the effects of climate change on climatic features, the variability of criteria influencing crop production in various regions of France and some possible effects on crops.

1 Introduction

The effects of greenhouse effect increase on future climate are assessed through GCM (General circulation model). We have used the outputs from one of them, the ARPEGE model developed by the CNRM (Météo-France) with the input of two scenarios of economic, technical and socio-economic development. These are the A2 scenario, with weak attention to GHG emissions leading to a high CO_2 concentration in the atmosphere at the end of the century (800 ppm) and the B1 scenario, a moderate scenario where the CO_2 concentration would be better controlled, allowing to reach only 550 ppm at the end of the century.

As crop indicator, it is possible to use either a full crop model using sophisticated focused entries for elaborated outputs, like yields either specific inputs for agroclimatic indicators, referring to some specific, especially sensitive, periods or processes. We will use both types of tools to obtain either information on crop chances to get over some difficult phases, either range of achievable yields.

The aim of this study is to estimate projected changes for main agricultural areas at the regional scale.

2 Material and methods

We have performed three types of analyses using the present and future climatic data. First, we analyzed the climatic data, with means, maps and multiple factor analysis. Second, we built and tested several agrometeorological criteria, in order to analyze the evolution of these criteria, and know the risks on practices for the main crops in livestock farms. Third, we used meteorological data to provide inputs for a crop model for grass, alfalfa (grass and legumes forages) and maize (representative summer arable crop in France), and we estimated the projected evolution of yields and water needs, without or with simple adaptations of agricultural practices.

2.1 Data, tools

2.1.1 GCM (general circulation model)

As GCM, we chose ARPEGE, from Météo-France for many reasons: it is one of the GCM used in the IPCC AR4, it has similar results with other GCM (Pérarnaud et al., 2005; Planton et al., 2008) and it has a variable resolution over the world, higher in our study zone, reaching 50 to 60 km over France (Déqué, 2007). Moreover, Déqué (2007) proposed regionalization methods to fit better with local conditions.

Our study has focused on two time periods, the near (2020–2049) and the distant (2070–2099) future. For each of them, we have only considered mean effect, without any attempt to catch the temporal evolution inside each period. Since the climatic results are similar in the near future for both scenarios, and they have the same tendencies in the distant future, which induces the higher impacts, we chose to only present here the results of the distant future for the scenario A2.

2.1.2 Use of meteorological data: RCM (regional climate model)

According to the aim of the parts of the study, we used different downscaling methods. For the two first parts (climate description and agrometeorological indices), we used direct outputs of the high-resolution GCM because they are convenient for studies on the frequency of events (for instance extreme events, like very high temperatures) or important thresholds for crop management (number of days without rain in spring for entering in the field or in summer for preventing drought effects. The study was performed with 235 points in the metropolitan France ($50 \times 50\,km^2$ grid),

For the third part (use of a crop model), we used variations between the present and the future estimated climatic values, so-called anomalies. It is the delta method proposed by Déqué (2007) and used in the same case by Bergot et al. (2004) or Gonzalez-Camacho et al. (2008). This choice was justified by the existence of significant differences between computed yield (mean and standard error) when directly using observed and estimated series in the present (Gonzalez-Camacho et al., 2008).

According to Déqué (2007), "this method needs observed daily data at the location of the study. It also assumes that the climate variability is unchanged in the scenario projection, i.e. is inherited from observed variability". In order to obtain pairs of actual and fictive stations, we chose 34 meteorological stations in the main agricultural areas, associating actual stations and centre of grids, as near as possible in distance and above all on altitude. The anomalies (ratio for rain, differences for other variables) were computed as 30 days moving averages for all the available years for each period of the future.

2.1.3 Crop model

The crop model STICS was designed at the end of the nineties to estimate yield and needs (water, nitrogen) of main arable crops and pastures (Brisson et al., 1998; Ruget et al., 2006). It has been validated in a large range of situations in France (Brisson et al., 2002; Ruget et al., 2006; Ruget, 2009) : in these studies, yields and water consumption have been correctly estimated in a wide range of climates (especially temperature) and with various levels of irrigation (arable crops) and fertilisation (grassland).

2.2 Methods

2.2.1 Climatic data, univariate and multivariate analysis

After mapping several climatic variables, we performed a multiple factor analysis (MFA) applied on spatial data in order to reach a global description of the climate and not independently on each variable (Lebart et al., 1995; Escoffier and Pagès, 1998; Ferrand et al., 2008). We analyzed 7 climatic variables (minimum and maximum temperature, rain, solar radiation, PET, wind, humidity) and an extra bloc containing the spatial coordinates. We used 7 rank statistics, that is, for each series of 30 years (except 37 years for the recent past period, hereafter named present) of each variable, the 7 values showing the distribution (median, minimum and maximum, 5th and 95th, 25th and 75th percentiles) for each series of climatic data corresponding to a time period (present or future), This allows the delimitation of zones having similar climates. In this way, MFA was used to define climatic zones in each period by using the future as reference. That means that we used criteria of zones defined for the distant future and we applied the criteria defining the zones for distant future to the present in order to see where the zones of the future are present now, and where new types of climates will appear.

2.2.2 Agroclimatic indicators

The second method used is the definition of agroclimatic indices and the comparison of values in the present and in a future period. The aim of using indicators is to give indications on specific periods of the year when crops might be affected by harsh climatic conditions: for instance, the occurrence of very low temperatures in spring, or very high temperatures in some consecutive days. We used pre-established and own-built indicators. Although some indicators are not specific for a period of the plant cycle, the effects on crop are often connected to the sensitive phases of the crops.

The frequency of sequences of 5 days with more than 60 mm of rain in spring preventing the beginning of grazing is one example combining field conditions (high rainfall estimating low soil portability) and period of production (heat sums after the beginning of growth).

For wheat, we used MFA to study the risks of wheat grain shrivelling. MFA allows the combination of several variables: the percentiles of number of days (or consecutive days) higher than thresholds of minimum and maximum temperatures and percentiles of the number of days of water deficit, known as factors stopping partially or totally grain filling. Among these criteria, we have used the number of days with maximum temperatures higher than 25 °C where begins the decrease of grain filling and the number of days with maximum temperature higher than 30 °C where grain filling is stopped. A high number of consecutive days with temperature higher than the threshold may stop definitely grain growth. MFA identifies zones with the same risk of grain shrivelling. They are defined as 7 classes with the

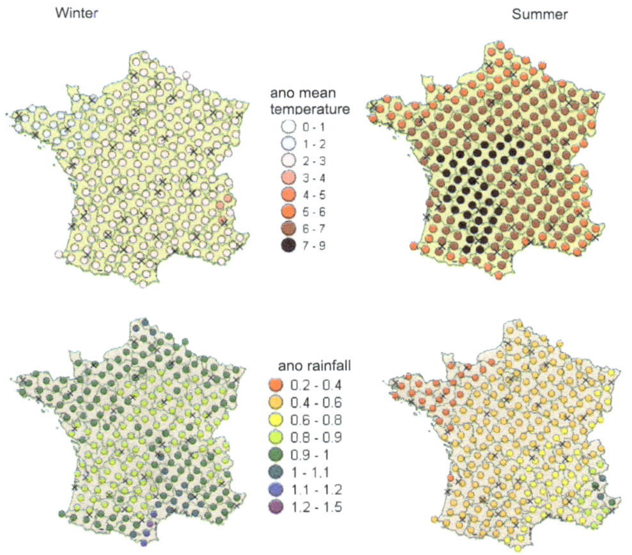

Figure 1. Seasonal anomalies for temperature and rainfall in the distant future (A2 scenario) in winter and summer: mean values of differences (future-present) for temperature(in °C), ratio (future/present) for rainfall (no change = 1) for the season.

present climatic data and the same criteria are applied on the future climatic data.

For maize, the risk of stopping grain filing was only studied through the frequency of one variable 1 day with maximum temperature higher than 38 °C.

2.2.3 Use of the crop model

The third method is the estimation of achievable yields, taking into account future climate. We present only one example on pastures. We assumed slight adaptations of practices, like the dates of mowing (or grazing), which are automatically adapted using lengths between uses expressed in heat sums.

3 Results

3.1 Climatic data

3.1.1 Univariate analysis

An important result of this study is the confirmation of higher temperatures and evaporation conditions, together with lower rainfall for the whole French territory. In Fig. 1a (temperature) and b (rainfall), it appears that climate is more affected in summer than in the other seasons, not only for temperature but also for rainfall.

For temperatures, the highest increase is predicted in summer to the western surround of the Massif Central (more than 7 °C for mean temperature in distant A2). The lower increases are in Brittany and Normandy both in summer and winter.

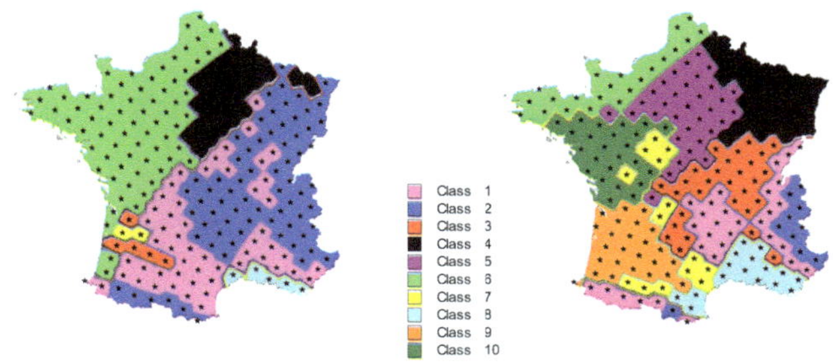

Figure 2. Climatic zones as defined by Multiple factor analysis (MFA) on spatial data: left, classes defined in the distant future applied on the present data, right classes as defined in the distant future.

For rainfall, the differences of the climate evolution between regions are high: some regions are slightly affected like Alps and South-East, while others are highly affected and they might lose 60% of their present summer rain. The zone with the highest decrease of rain in summer is the North-West part of France, mainly Normandy and Brittany.

3.1.2 Multivariate analysis

In the Fig. 2, we show the present situations of 10 zones defined by using the MFA of spatial data in the distant future climate. It can be seen that the zone of low mountain climates will go up, the extension of high mountain climate will be reduced, while the area of the Mediterranean climate will be larger. The Atlantic front will be dryer: 3 new types of climate will appear in the western part of France (except Brittany, Normandy and North on one hand and Pyrenees mountains on the other hand). The mean climatic characteristics of these zones are not far from those of the Mediterranean zone means (higher maximal temperatures and high evapotranspiration), but with higher rainfall and lower solar radiation.

3.2 Agro-climatic indicators

3.2.1 Grassland

In Fig. 3, showing the difference of the number of days with unfavourable conditions for grazing when production is sufficient (500 °C.day after the possible beginning of growth), conditions will benefit from an improvement in springtime in most cases, and especially in the South-East zone of France. Only a small area in the North appears to be faced to worsened conditions.

3.2.2 Winter wheat

The Table 1 shows the number of days with maximum temperature higher than 25 °C during a period of 650 °C after 5 June for each defined class, that is for each geographical zone in the present and the future: assuming that the

Table 1. Number of days where $T_{max} > 25\,°C$ for the 7 defined classes (classes defined on the present period), applied on present and future data, ear emergence at 5 June (like in the present conditions).

	1970–2006 number of days where $T_{max} > 25\,°C$				2070–2099 number of days where $T_{max} > 25\,°C$			
	mean	standard error	minimum	maximum	mean	standard error	minimum	maximum
Class 1	5.4	2	1.2	9.9	9.9	4.2	4.4	15.5
Class 2	2.1	1.7	0	5.4	2.7	2.5	0.1	7.4
Class 3	5.9	2.3	1.5	9.7	14.6	1.9	11.8	16.1
Class 4	11.2	3.6	3.1	17.4	18	3.8	8	25.4
Class 5	10.3	4.3	3.7	19.5	16.9	2.7	13.2	22.6
Class 6	13	3.6	5.3	15.9	24.8	5.4	7.6	36.2
Class 7	24.8	5.4	13.4	31.5	27.8	5.7	14.5	34.1

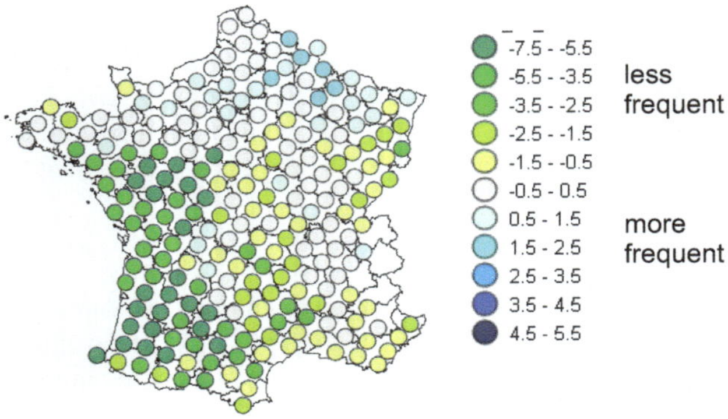

Figure 3. Agro-climatic indicator for the decrease of soil trafficability for early grazing, defined as 5 days with rainfall higher than 60 mm at the beginning of spring (500 °C.day after 1 February). Number of years among 10 when more (positive values) or less (negative values) unfavourable conditions will occur in the distant A2 scenario.

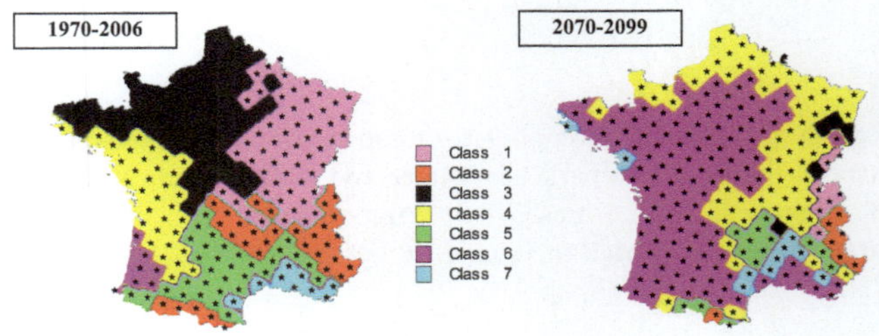

Figure 4. Risks of wheat shrivelling, using the MFA method to define zones. Left, definition of zones with present climatic data. Right, application to the future of the criteria defined in the present (distant A2 scenario).

ear emerges at 5 June, it illustrates the number of days with shrivelling risks in the classes defined with present climatic data.

Class 2 has very low risk now, and the mean value of this criterion is higher in the future. In the same way, classes 1 and 3 with low risk now will have higher risk later. Moreover (Fig. 4), these zones will become very small in the future. Class 6 will have the greatest increase – from extreme South-West to almost the half western part of France – and it is a class with high shrivelling risks. That means that without any management adaptation, the risks of poor grain filling will become high: choosing varieties with earlier ear emergence may allow to avoid high temperatures during grain filling.

3.2.3 Maize

In Fig. 5, the number of days with maximum temperature higher than 38 °C will greatly increase between the present period (left) and the distant future period (right, distant A2). Now, the risk of stopping grain filling of maize is mainly present in the South-western part of France, but this event will occur in all the regions of France in the distant future.

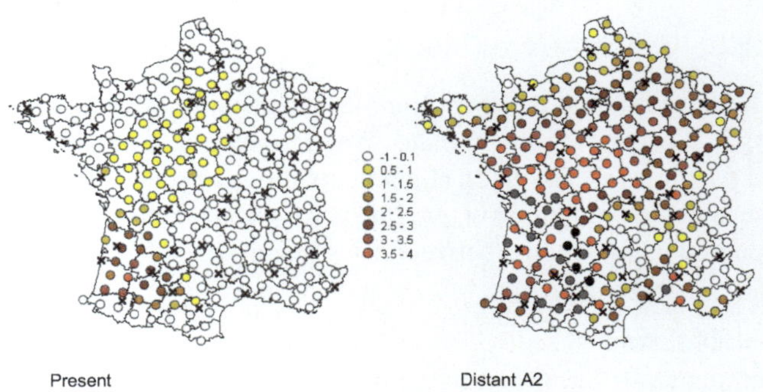

Figure 5. Indicator of risks for maize grain filling: difference of number of years among 10 with at least once a maximum temperature higher than 38 °C in the first half August, between the future (distant A2) and the present period, that is increase in the risk when positive values.

For wheat and maize, heat risks during grain filling will increase in the future, because of the increase of temperatures.

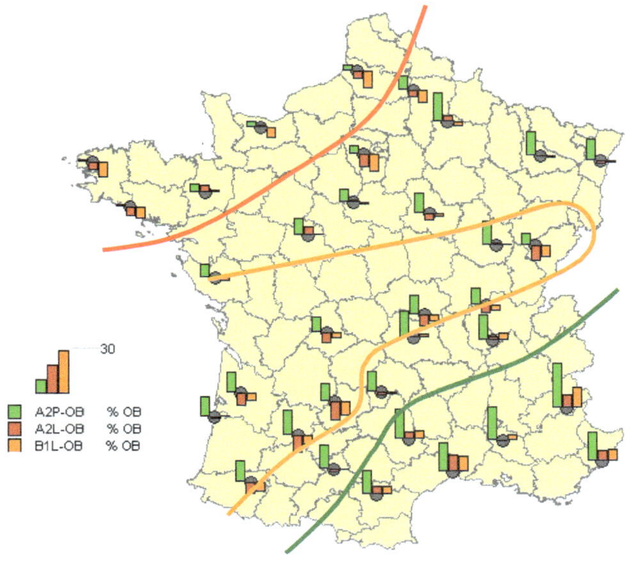

Figure 6. Grassland relative differences of yields between 3 future scenarios (A2P: near A2, green; A2L: distant A2, red; B1L: distant B1, orange) and present estimations (OB: observed) for 34 coupled actual and fictive stations.

That means that it will be essential to change practices (variety or cultivation dates) to avoid the risks of high fall of yield for temperature sensitive species.

3.3 Yields computed by the crop model

We show only one example of projected variation of yield, for grassland with frequent uses, cultivated on a soil with low available water (normal for grasslands) for all the studied scenarios (Fig. 6). Simulations take into account the positive CO_2 effects (increase of production, decrease of transpiration (Ruget et al., 1996; Bethenod et al., 2001; Long et al., 2006)). The evolution is mostly positive in the near future (green bar) and generally slowly negative in the distant future, except in the South-East of France where the increase of production will likely continue in the second half of the century.

The increase in yields comes mainly from the higher temperatures (development acceleration, leaf growth increase), occurring without high decrease of water availability and taking advantage from the possible reduction of transpiration by CO_2 increase. The decrease in the distant future comes from the lack of water, which seems not higher in the future than now in the South-East.

The comparison of grass and alfalfa in the same conditions shows an advantage of alfalfa, mainly due to the higher rooting depth avoiding water stress even in the distant future (Ruget et al., 2010).

4 Conclusions

In the climate analysis, apart from the confirmation of already known effects (higher temperature, lower precipitation), the most interesting result for agriculture comes from the spatial distribution of changes: all the regions of France are not affected in the same way. One interest of this study is also to show complementary and corroborating information from several analyses (univariate analyses, MFA on spatial data, agro-climatic indicators, and crop models).

The reliability of the chain through the use of meteorological data with a degree of uncertainty for elaborating criteria also affected by an other degree of uncertainty (thermal thresholds in agroclimatic indicators) or the filter of crop models is an open question.

Further work is needed to confirm and improve these conclusions, in order to be able to propose new adapted agricultural practices, especially when considering the system and farm levels.

Edited by: T. Cegnar

References

Bergot, M., Cloppet, E., Pérarnaud, V., Déqué, M., Marçais, B., and Desprez-Loustau, M. L.: Simulation of potential range expansion of oak disease caused by *Phytophthora cinnamomi* under climate change, Glob. Change Biol., 10, 1539–1552, 2004.

Bethenod, O., Ruget, F., Katerji, N., Combe, L., and Renard, D.: Impact of atmospheric CO_2 concentration on water use efficiency of maize, Maydica, 46, 2, 75–80, 2001.

Brisson, N., Mary, B., Ripoche, D., Jeuffroy, M. H., Ruget, F., Nicoullaud, B., Gate, P., Devienne-Barret, F., Antonioletti, R., Dürr, C., Richard, G., Beaudoin, N., Recous, S., Tayot, X., Plénet, D., Cellier, P., Machet, J.-M., Meynard, J. M., and Delécolle, R.: STICS : a generic model for the simulation of crops and their water and nitrogen balances. I. Theory and parametrization applied to wheat and corn, Agronomie, 18, 311–346, 1998.

Brisson, N., Ruget, F., Gate, P., Lorgeou, J., Nicoullaud, B., Tayot, X., Plénet, D., Jeuffroy, M.-H., Bouthier, A., Ripoche, D., Mary, B., and Justes, E.: STICS: a generic model for the simulation of crops and their water and nitrogen balances. II. Model validation for maize and wheat, Agronomie, 22, 69–92, 2002.

Déqué, M.: Frequency of precipitation and temperature extremes over France in a anthropogenic scenario: Model results and statistical correction according to observed values, Global Planet. Change, 57, 16–26, 2007.

Escoffier, B. and Pagès, J.: Analyses factorielles simples et multiples, DUNOD, Paris, 284 pp., 1998.

Ferrand, M., Lequenne, D., Manneville, V., Jannot, P., and Lopez C.: Apport de la spatialisation des données en analyse multidimensionnelle pour évaluer l'impact des activités agricoles sur la teneur en nitrates des eaux, Revue MODULAD, 39, 81–94, 2008.

Gonzalez-Camacho, M., Mailhol, J.-C., and Ruget, F.: Impacts locaux attendus de l'augmentation du CO_2 dans l'atmosphère sur la productivité de l'eau du maïs dans la Drôme, France, Irrigation and Drainage, 57, 2, 229–243, Published Online: 27 July 2007, online ISSN: 1531-0361, doi:10.1002/ird.332, 2008.

IPCC: Climate Change: The Physical Science Basis. Contribution of Working Group I to the Fourth Assessment, Report of the Intergovernmental Panel on Climate Change, Cambridge University Press, Cambridge, United Kingdom and New York, NY, USA, 2007, 996 pp., 2007.

Lebart, L., Morineau, A., and Piron, M.: Statistique Exploratoire Multidimensionnelle, DUNOD, Paris, 439 pp., 1995.

Long, S., Ainsworth, E., Leakey, A., Nösberger, J., and Ort, D.: Food for Thought: Lower-Than-Expected Crop Yield Stimulation with Rising CO_2 Concentrations, Science, 312, 1918–1921, 2006.

Pérarnaud, V., Seguin, B., Malezieux, E., Déqué, M., and Loustau, D.: Agrometeorological research and applications needed to prepare agriculture and forestry to 21st century Climate Change, Climatic Change, 70, 319–340, 2005.

Planton, S., Déqué, M., Chauvin, F., and Terray, L.: Expected impacts of climate change on extreme climate events, C. R. Geoscience, 340, 564–574, 2008.

Ruget, F., Bethenod, O., and Combe, L.: Repercussions of increased atmospheric CO_2 on maize morphogenesis and growth for various temperature and radiation levels, Maydica, 41, 3, 181–191, 1996.

Ruget, F., Novak, S., and Granger, S.: Du modèle STICS au système ISOP pour estimer la production fourragère, Adaptation à la prairie, application spatialisée, Fourrages, 186, 241–256, 2006.

Ruget, F.: Impact du changement climatique sur la luzerne et la prairie, disparités régionales, in: Changement climatique, Conséquences et enseignements pour les grandes cultures et l'élevage herbivore, 22 octobre 2009, Paris, Arvalis-Institut du Végétal Ed, 2009.

Ruget, F., Moreau, J.-C., Cloppet, E., and Souverain, F.: Effect of climate change on grassland production for herbivorous livestock systems in France, in: Grassland in a changing world, edited by: Schnyder, H., Isselstein, J., Taube, F., Schellberg, J., Wachendorf, M., Herrmann, A., Gierus, M., Auerswald, K., Wrage, N., and Hopkins, A., Grassland Science in Europe, Vol. 15, EGF 2010, Kiel, in press, 2010.

A robust method to identify cyclone tracks from gridded data

F. Kreienkamp, A. Spekat, and W. Enke
Climate & Environment Consulting Potsdam GmbH, David-Gilly-Str. 1, 14469 Potsdam, Germany

Abstract. A system to derive tracks of barometric minima is presented. It is deliberately using coarse input data in space (order of $2° \times 2°$) and time (6-hourly to daily) as well as information from just one geopotential level. It is argued that the results are, for one robust in the sense of an assumption of the IMILAST Project that *the use of as simple as possible metrics should be strived for* and for two tailored to the input from reanalyses and GCMs. The methodology presented is a necessary first step towards an automated storm track recognition scheme which will be employed in a second paper to study the future development of atmospheric dynamics in a changing climate. The process towards obtaining storm tracks is two-fold. In its first step cyclone centers are being identified. The performance of this step requires the existence of closed isolines, i.e., a topology in which a grid-point is surrounded by neighbours which all exhibit higher geopotential. The usage of this topology requirement as well as the constraint of coarse data may lead, though, to limitations in identifying centers in geopotential fields with shallow gradients that may occur in the summer months; moreover, some centers may potentially be missed in case of a configuration in which a small scale storm is located at the perimeter of a deep and very large low (a kind of "dent in a crater wall"). The second step of the process strings the identified cyclone centers together in a meaningful way to form tracks. By way of several examples the capability to identify known storm tracks is shown.

1 Introduction

This study focuses on the identification and tracking of moving low pressure systems. It is the result of a project initialized by the Regional Environment Agency of Saxonia and thus has an emphasis on tracks with relevance to Central Europe. The method applied has large scale as well as regional scale elements – the data source for the the identification consists of NCAR reanalysis data (Kalnay et al., 1996) from 1951–2006. Developing such an algorithm on a rather coarse spatially and temporally resolved basis leads to a robust tool to address issues such as the classification of dynamic circulation aspects and the study of cyclone tracks, e.g., in global climate model data.

Similar to the classification of circulation patterns into so-called Großwetterlagen, first attempts towards a classification system for storm tracks in the European sector were carried out at the end of the 19th century. The meteorologist Wilhelm Jacob van Bebber published his system – which has been in use ever since – in 1891 in Meteorologische Zeitschrift (van Bebber, 1891). The formative influence to his approach was maritime meteorology, in particular the identification of "heavy weather". His requirement was that there are preferred pathways for the movement of storms into central and eastern Europe. The van Bebber classification, as shown in Fig. 1, however, was based on observations from only a few years and station network of rather low density.

The system includes tracks with relevance for the British Isles, the North-Eastern Atlantic, the North Sea and the Baltic in its groups I through IV as well as one group, V, with a more southerly placement leading across Western Europe into the Mediterranean Sea, branching off towards Northern Africa, the Balkans or the Eastern Mediterranean area. To this date, track Vb of the latter group has remained in common use, unlike the large majority of van Bebber's tracks. Frequently, a connection is established between the advance of intense cyclones from the Mediterranean area on the one hand and particularly intense precipitation events on the other. So-called *Vb-situations* are, e.g., linked to flooding events in the summer of 1997 (Odra river) and 2002 (Elbe river). It should be added that the North-Western Mediterranean Sea and the Gulf of Genova in particular are not

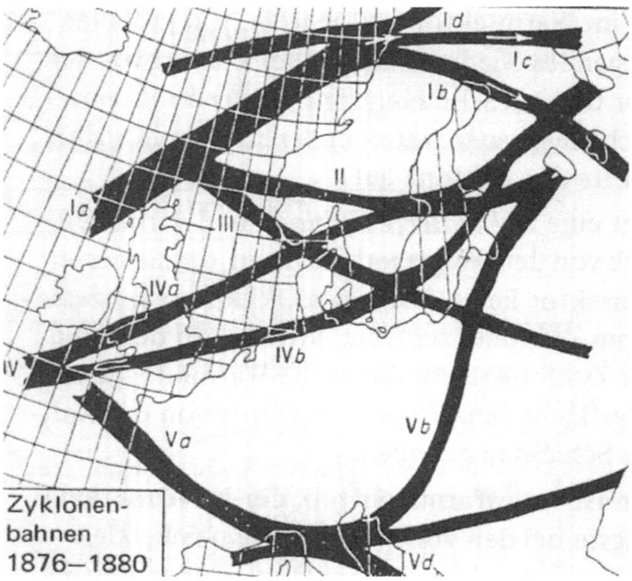

Figure 1. Storm tracks according to van Bebber (1891).

mere transition areas for passing cyclones but areas of frequent cyclogenesis. In Petterssen (1956) chapter 13, maps are given that show the occurrence and the genesis of cyclones in the entire Northern Hemisphere which highlight the cyclone activity in the Mediterranean Basin, particularly in winter. Recent publications, e.g., Lionello et al. (2006), Anagnostopoulou et al. (2006), Homar et al. (2007) or Lionello and Giorgi (2007) focus on present and future cyclone climatologies for this area.

Already in the first half of the 20th century synoptic studies of the connection between the strength of the atmospheric flow and the displacement of cyclones/anticyclones were being carried out (e.g., Rossby, 1939). In the 1980s and 1990s the discussion of existence, frequency and future modification of storm tracks was reviewed (e.g., Lau, 1988). The progress in the development of atmosphere models led to particular involvement of moving intense cyclones as identifiers of the modelled climate's dynamics, as e.g., Bengtsson (1991) summarized. Moreover, cyclones of that kind have a major impact, e.g. on the extremes of wind and precipitation. An interesting point of the atmospheric dynamics and storm tracks is discussed in Hoskins and Valdes (1990). The authors note a puzzling contradiction: A cyclone center moves as a consequence of the atmosphere system's tendency to achieve an energy balance. The baroclinic instability in a region should be reduced by a moving system. Why, the authors are asking, are there frequent similar tracks, often used in rapid succession? There ought to be a mechanism which sustains baroclinic instability despite this flagging process. Their conclusion, based on comparably coarse assumptions of the climate system's components and low resolution climate models was that the heating by ocean currents, such as the Gulf Stream triggers the formation of a corridor in which cyclones propagate.

The concept of a corridor has been picked up by other authors, who consider van Bebber's cyclone tracks as bundles in a genesis and propagation area, such as the entire North Atlantic. Studies in conjunction with future emission scenarios have been published, e.g. by Lambert (1995), Carnell et al. (1996) or Schubert et al. (1998). Bengtsson et al. (2006) conclude in a comprehensive study that the Northern Atlantic and Scandinavia in particular will be subjected to a future increase in cyclone activity while the track branches into the Mediterranean area (the V-group of van Bebber), rated as a secondary track system, will decrease in importance. A drop in Mediterranean cyclone activity as a consequence of CO_2 doubling had been found by Lionello et al. (2002). More recent studies (Lionello et al., 2008) based on climate models with a higher resolution confirm these findings. Pinto et al. (2006) show evidence of a decline in Mediterranen cyclone activity including cyclogenesis in the Gulf of Genova. They add that Central Europe is an area jointly influenced by cyclones moving in from the Atlantic as well as the Mediterranean Sea and that there is an indication in the scenarios that this will remain the case in the future.

The Fourth IPCC Assessment Report (IPCC, 2007) includes summaries concerning cyclone tracks in several chapters:

- *Observations* (Trenberth et al., 2007): There is a poleward displacement of the storm tracks; the frequency of cyclones is reduced but their intensity increases. Climate variability with respect to extratropical lows is large and the problem of track identification becomes larger because of constant changes in the observation system.

- *Climate Models* (Randall et al., 2007): Simulated cyclone tracks have a stronger zonal orientation as the observed ones.

- *Understanding and Attributing Climate Change* (Hegerl et al., 2007): The poleward displacement of the main tracks, as observed in the last 50 years will continue into the future with high probability.

- *Global Climate Projections* (Meehl et al., 2007): The poleward displacement of the tracks can be attributed in part to the expansion of the subtropical Hadley circulation – future changes in precipitation would thus occur even if the cyclone intensity remains unchanged.

- *Regional Climate Projections* (Christensen et al., 2007): There is a predominant probability that the wind will decrease in the Mediterranean area and increase over Northern Europe. The modifications of the cyclone tracks have a clear influence on winter precipitation.

As stated above, the identification of cyclone tracks is a research field that dates back at least to the late 19th century. A review of the methods employed in detecting and tracking cyclones, including the methods' evolution over time, can be found in Raible et al. (2008). In that paper it is also pointed out that cyclone frequency and track climatologies exhibit major differences depending on the methodologies employed. This sparked the establishment of the IMILAST (**I**ntercomparison of **MI**d-**LA**titude **ST**orm diagnostics) project in 2009[1] which is to be finished in 2011. The IMILAST tasks range from a survey of methods (more than 20 methods are documented!) to assess their individual uncertainties which may lead to a set of recommendations concerning purpose-specific tracking strategies. On the *Background* Web Page of IMILAST it is emphasized that *The use of as simple as possible metrics should be strived for*. Techniques that aim for simplicity and easy reproducibility are the focus of several studies, e.g., Hanson et al. (2004).

Common to most methods is a two step procedure: First a cyclone detection scheme is devised, followed by a cyclone tracking procedure. The methodologies employed by different working groups cover a broad range. In exemplary form, three strategies will be briefly described. The approach of Blender and Schubert (2008) identifies minima of the 1000 hPa geopotential field (z1000) within a neighbourhood of 8 grid points of a reference point. In order to be able to include very weakly developed centers, a minimum gradient (20 gpm per 1000 km) in a given area (radius 1000 km) is required. The centers which were so identified are lined up to form cyclone tracks by a next-neighbour search. An other strategy is suggested by Wernli and Schweiz (2006). They are also identifying the cyclones as local minima in the z1000 field. A local minimum is required to be surrounded by a closed contour which is 15 gpm higher than this local minimum. The cyclone tracks are sought by calculating a potential position for the next time step as a first guess. Then the candidate in greatest proximity of that first guess position within a 1000 km perimeter is chosen. Sinclair (1997) has focused particulary on the cyclone identification aspect. In a first step a preproceesing is applied to remove scale dependence in the needed atmospheric fields followed by a smoothing according to Cressman (1959). According to synoptic experience, it is rather difficult to identify atmospheric minima in the Southern Hemisphere; there, cyclones are identified as local maxima of the cyclonic gradient wind vorticity which is required to surpass a threshold. Then, the first guess strategy is applied when constructing the tracks: A predicted position is compared with the cyclone position as it occurs in the subsequent time step.

The authors have developed their own approach to produce a tracking algorithm that is as straightforward and simple as possible – similar to what IMILAST is proposing. Only the geopotential in a coarse grid of about $2° \times 2°$ is evaluated. This makes the method applicable to long-term climatologies such as NCEP reanalyses, as well as to GCM scenarios of the IPCC-AR4 generation. The term *robust* is to be interpreted in the context of the method's ability to identify the cyclone centers and produce cyclone tracks from coarse input.

[1] http://www.proclim.ch/imilast/index.html

2 Data

850 hPa geopotential height data from NCEP reanalyses (Kalnay et al., 1996) were used in the study to develop the identification methods for cyclone centers and tracks. They establish a three-dimensional climatology of the atmosphere for the time frame 1950 until now. The "alternative product", which has been generated by ECMWF (ERA40, Uppala et al., 2005) is available just until 2002 and cannot be used to study the immediate past. There are other products, such as ERA interim, which extend into the immediate past and have a higher resolution. Yet, they do not extend far enough back to establish a useful climatology – for that, at least 30 years would be required. It is noteworthy that due to a higher model internal resolution (ERA40 (T106) to NCEP (T62)) ERA40 is able to produce a greater number of cyclones (Raible et al., 2008). This may be one reason why not all observed cyclones can be found by the used method.

In conjunction with the aforementioned focus on Central Europe, the study has been carried out in an area enclosed by a rectangle of 38 zonal grid-points (28.125 W to 43.125 E) and 36 meridional grid-points (21.45 N to 88.57 N). Some cyclone centers in subsequent figures might not have been labelled, since they are outside this area.

3 Method and discussion

A straightforward and pragmatic system was developed which derives the positions of the minima in the gridded 6-hourly geopotential fields (cf. Sect. 3.1). Using a set of displacement criteria those minima are stringed together to form tracks. This produces a catalogue of storm tracks as well as a wealth of information concerning size, intensity and displacement of the moving lows.

3.1 Identification of cyclone centers

A *center* is defined as an individual, contained minimum which is surrounded by larger values in all directions. Small, shallow lows which are a mere dent in a much larger cyclone can only be detected if there are appropriate neighbouring grid points within the spatial resolution of the grid point data set. Finding the deepest cyclone in a geopotential field is a trivial exercise – it is just necessary to determine the field's minimum.

A thought experiment may be carried out to illustrate the next steps of the approach: Imagine the reservoir of a dam which has flooded a settlement. If we assume the water level

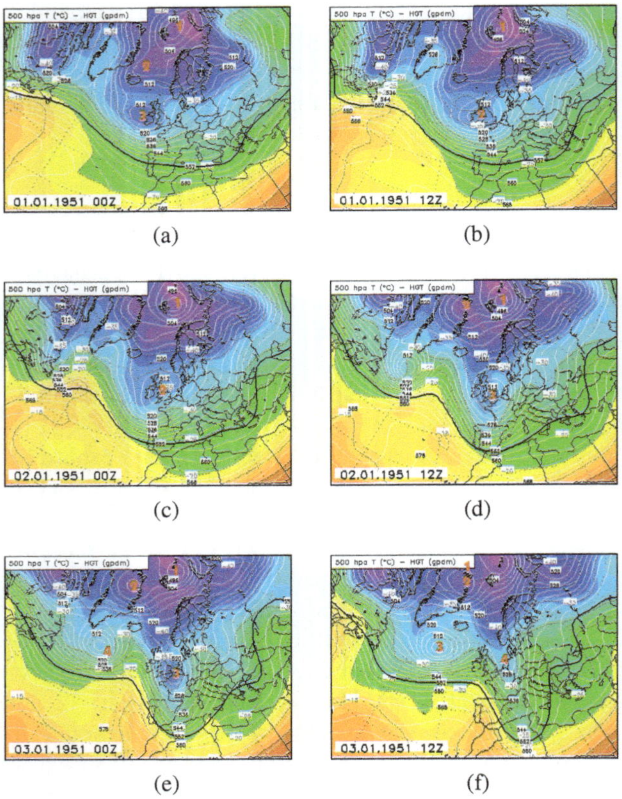

Figure 2. Series of 500 hPa charts from 1 to 3 January 1951 (Source: www.wetterzentrale.de). The centers found by the identification algorithm are plotted in red; they are sorted by the intensity of the cyclone.

to drop, individual structures and building re-appear[2]. Perhaps the bell tower of a church (center of the strongest cyclone). Then the top of a smaller belfry (center of a secondary cyclone with the second-lowest core pressure). Then maybe the top of a high building (center of a tertiary cyclone). With further retreat of the water, another roof may appear, and so on. The search algorithm has to be capable of finding connected "buildings" as well as "building parts". This is achieved by stepwise lowering of the water level (raising of the reference level) until the first structure shows. A new structure is recognized as soon as subsequent water level lowering (reference pressure rise) yields only points which have lower pressure than the (successively raised) reference pressure. Figures 2 and 3 show examples for identified centers.

The 500 hPa level has been chosen just for illustration purposes, since centers are particularly clear in this level of low divergence – the identification itself uses the 850 hPa geopo-

[2]There is a twist to the thought experiment, since we are dealing with low pressure structures and thus the picture has to be imagined upside down.

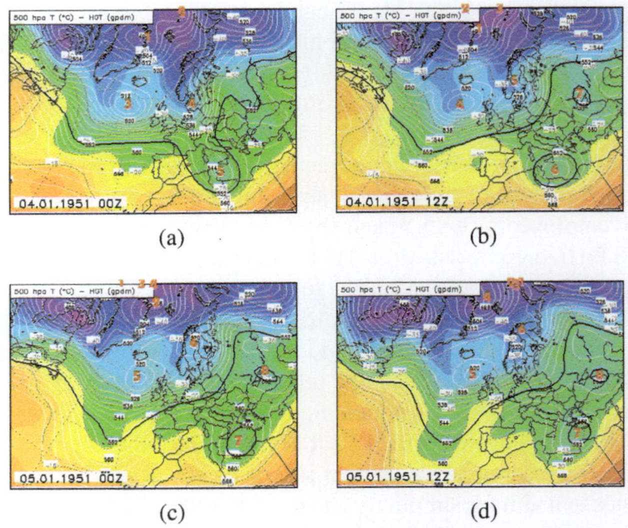

Figure 3. As in Fig. 2, but for 4 and 5 January 1951.

tential level. Surface pressure data would be an option but they pose several problems: (i) It would be problematic to track cyclones over elevated terrain[3] (ii) in comparison with upper air data, the structures which are to be identified exhibit a higher degree of detail which reduces the efficiency of the center-finding algorithm, and (iii) particularly in winter, the surface level is frequently decoupled from cyclone activity by inversions which also reduces the efficiency of the algorithm. The identification of centers is carried out in steps of 6 h – the chart archive used for illustration, on the other hand, contains just 00z and 12z information. A few remarks concerning the charts, Figs. 2 and 3:

- 1 Jan 00z: There were three cyclones identified, which are numbered according to their depth. The trough near the northern Ural was not identified as an individual cyclone because no closed isoline could be constructed within the grid resolution (cf. Sect. 2). However, low #2 fulfilled this criterium. The strong cyclone over the Hudson Bay is outside the identification window used for this study.

- 1 Jan 12z: Only two centers were identified, since the trough to the north-east of Iceland was not sufficiently large to fulfil the criteria – this is again the case on 2 Jan 00z (Fig. 2c), while on 2 Jan 12z (Fig. 2d) a size is re-gained that enables identification.

- The charts of 3 Jan 12z and subsequent dates (Figs. 2f and 3) frequently contain numbered centers on the upper edge. The related cyclones, identified by the algorithm

[3]In $2° \times 2°$ resolution, the topography of continental Europe is generally below 1500 m, making 850 hPa a more suitable level for the study.

are outside the range of the map. Moreover, it is possible that centers have no label; in such a case the cyclone would be outside the investigation window mentioned in Sect. 2.

In addition to the position and the geopotential height of the center, the "number of pixels" a measure of the area covered by the cyclone is registered - each pixel being a grid point. It is the number of grid points in the $38° \times 36°$ field which are topologically assigned to an individual cyclone.

3.2 Identification of tracks

As illustrated in the previous section each chart contains "candidates" for moving cyclones but in nature there are also formation and dissolution processes as well as phases in which the identification algorithm is unable to find a closed center. For each date, i.e., every six hours, all cyclones are entered into the calculation in ascending order of their core geopotential. The track-forming algorithm is considering all cyclones that are entirely within the $38° \times 36°$ field. In order to relate the individual positions P of the n centers at time step T to those at the subsequent time step $T+1$ a similarity matrix for each center $P_n(T) \Rightarrow P_m(T+1)$ is formed. T and $T+1$ denote the time steps which are 6 h apart. The maximum allowed D for two consecutive positions is 10. The similarity measure evaluates three components with individual weights:

1. Distance D between the centers (weight: 4) which is calculated as $D = |i_T - i_{T+1}| + |j_T - j_{T+1}|$ with i being the meridional component (number of the grid point in the $38° \times 36°$ field) and j being its zonal component.

2. Difference Z in geopotential between the centers at time steps T and $T+1$ (weight: 2).

3. Difference A in the number of pixels (see above) between the cyclones (weight: 1)

An aspect of item 1 above should be pointed out: The grid used is not equidistant and there is a potential for bias when assessing cyclone displacement in a kilometric sense to the effect that some tracks may not be captured if cyclones move (i) extremely fast, (ii) on a zonally oriented track and (iii) in very high latitudes. In an example shown later in this paper (cf. Table 2) a displacement of up to ≈ 400 km occurs for a very fast moving cyclone in 60° N, whereas the maximum D would be ≈ 1100 km in that latitude which compares to the 1000 km assumed in Wernli and Schweiz (2006). Since our study has a target area in Central Europe this bias, although it may occur, is of less relevance.

The similarity matrix is often incomplete because on a given date T there might be, e.g., cyclones 1, 4 and 7 eligible for track construction whereas on time step $T+1$ cyclones 2, 4, 5 and 6 might be candidates.

Table 1. Storm track of a cyclone near the British Isles between 1 and 4 January, 1951. Level: 500 hPa.

Day	Date	Storm No.	Latitude	Longitude
1 Jan 51	00	3	55.0N	11.3W
	06	2	53.2N	11.3W
	12	2	53.2N	5.6W
	18	2	55.0N	1.9W
2 Jan 51	00	2	55.0N	1.9W
	06	2	53.2N	1.9W
	12	3	51.3N	0.0W
	18	3	51.3N	1.9E
3 Jan 51	00	3	51.3N	3.8E
	06	4	51.3N	5.6E
	12	4	53.2N	5.6E
	18	4	55.0N	7.5E
4 Jan 51	00	4	55.0N	9.4E

Table 2. Track of the storm "Anatol" from Iceland across the German Bight into the area of the Baltic States, between 2 and 5 December 1999. Level: 500 hPa.

Day	Date	Storm No.	Latitude	Longitude
2 Dec 99	06	2	64.4N	15.0W
	12	2	64.4N	11.3W
	18	2	64.4N	11.3W
3 Dec 99	00	2	64.4N	11.3W
	06	2	58.8N	3.8W
	12	2	56.9N	3.8E
	18	1	58.8N	11.3E
4 Dec 99	00	1	58.8N	16.9E
	06	1	58.8N	22.5E
	12	1	58.8N	28.1E
	18	1	58.8N	31.9E
5 Dec 99	00	1	58.8N	35.6E
	06	3	60.6N	39.4E

The algorithm identifies the minima of D, Z and A between all candidate centers at two time steps. In that phase, tentative track segments are generated. If they are shorter than eight consecutive time steps, i.e., 48 h, they are discarded. Tracks would end at the time step with a displacements of $D > 10$ or if they are not completely in the investigation window any more or if they leave the window altogether. Examples are given in Tables 1 and 2.

The displacement of the cyclone can be traced in Figs. 2 and 3, too[4]. The longitude and latitude entries refer to the positions of the grid points that were evaluated. Column *Storm No* makes reference to the centers which are identified at that particular date and ordered by magnitude. On 1 January 1951 (Fig. 2a), the low, labelled "3", can be found

[4]Note that the figures just show the 00z and 12z conditions, whilst the Table lists positions every 6 h.

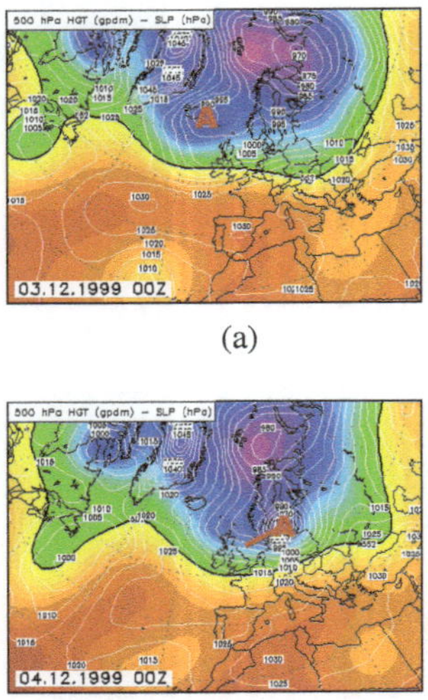

(a) (b) (c) (d)

Figure 4. Section of the track of storm "Anatol" (the center is indicated by the letter A) in the sea level pressure field on 3 and 4 December 1999 (white isolines). The 12-h displacement of the center is marked by a red line in subfigure (b)–(d). The center's positions found on 2 December 06z until 5 December 06Z are given in Table 2.

east of the British Isles. An intermediate stadium is included in the 06z row of that day in Table 1. It indicates a westward motion. The center can thus be found at 12z over Northern Ireland (Fig. 2a). Between 18z, 1 January and 12z, 2 January only very little motion is visible. Then the center begins its eastbound track along the 51st parallel. This motion receives an additional northward component (cf. center "4" in Fig. 2f) between 3 January 12z and 4 January 00z. Simultaneously, a bubble-like motion towards the Mediterranean Sea occurs. It develops into a cutting-off low which, on 4 January 00z (Fig. 3a) appears so markedly that an individual center is recognized. The new Mediterranean cyclone is moving only very little, nevertheless it can be identified until 6 January (not fully covered in the displayed sequence of charts). The previous track of the center that remained in Central Europe can be traced across the Baltic Sea into Northern Scandinavia. The cyclone arrives at the Kola Peninsula on 5 January 12z (cf. Fig. 3d).

Table 2 shows the example of a track identification for a very fast moving storm which occurred in the first December days of 1999 and received the name "Anatol" by the Meteorologists from the Free University of Berlin[5]. It developed from a trough in the North Atlantic between Greenland and Iceland and moved only little until 3 January 00z. After that it deepened quickly and moved swiftly across the northern Central Europa towards the Baltic States.

[5]http://www.met.fu-berlin.de/adopt-a-vortex/

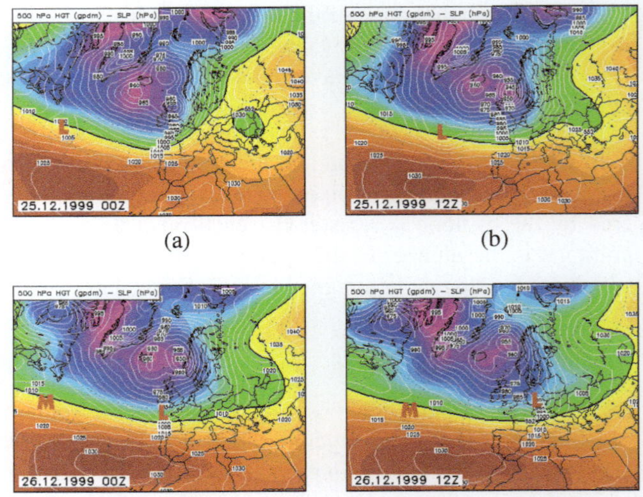

(a) (b) (c) (d)

Figure 5. Section of the track of storm "Lothar" (the center is marked by the letter L) in the sea level pressure fields on 25 and 26 December 1999 (white isolines. The track of storm "Martin" which followed right after "Lothar" has been marked with an M in subfigures (c) and (d).

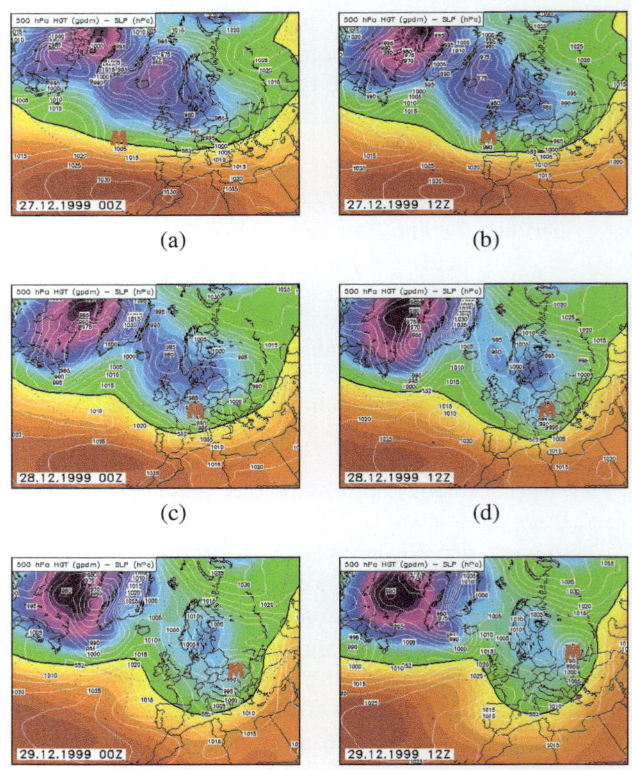

(a) (b) (c) (d) (e) (f)

Figure 6. Section of the track of storm "Martin" (the center is denoted "M") in the sea level pressure field between 27 and 29 December 1999 (white lines).

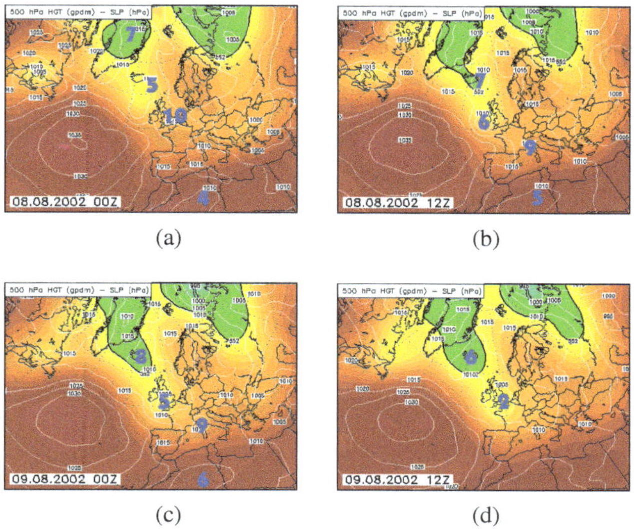

Figure 7. Example of a sequence of summer days: Sea level pressure field on 8 and 9 August 2002 (white lines). A selection of identified cyclone centers are numbered in blue.

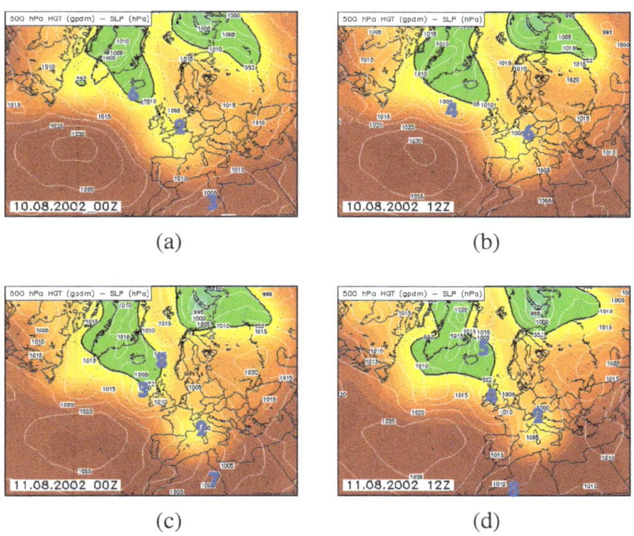

Figure 8. As in Fig. 7, but for 10 and 11 August 2002.

The episode which is relevant for Central Europe has been displayed in four surface pressure charts (Fig. 4). As shown in Table 2 the low was the second largest in the map area up to 3 December 12z – the strongest could be found near the hemispheric cold center over the Arctic Sea near Spitsbergen. Thereafter and until 5 December, "Anatol" had the lowest core pressure in the entire map sector.

The displacement speed exceeded, in particular in the phase over the Eastern Atlantic and the North Sea, 7 degrees of longitude (on the order of 400 km in that region) in 6 h!

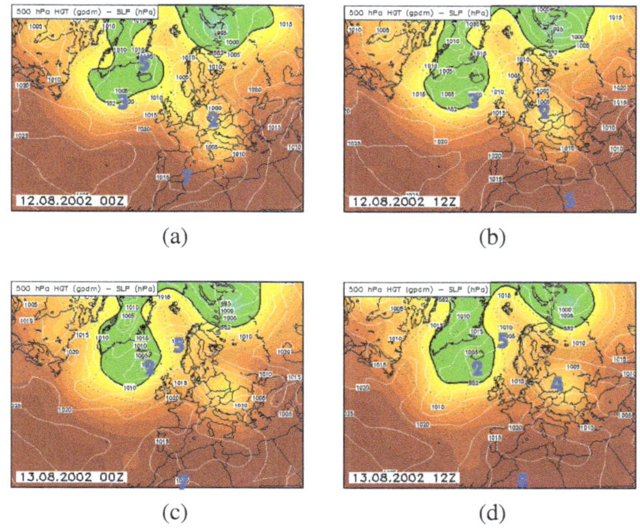

Figure 9. As in Fig. 7, but for 12 and 13 August 2002.

The tracking criteria of the identification algorithm are, as just shown, capable of a dependable recognition of a continuous storm track. The path of storm "Lothar", however, could not be identified in full. For one, the geographical resolution of the data set (2° × 2°) is too coarse, for two "Lothar" was moving towards Central Europe at the edge of a very large and intense Arctic Sea low – consisting of several centers – and for three, even in its most intense stadium, "Lothar" had a very small closed isobar (cf. Fig. 1b and d in Ulbrich et al. (2001)).

As can be seen in Fig. 5, a cyclone with a closed isobar around the center is located East of Newfoundland on 25 December 00z. Its position, west of 30 W, is outside the map area of this study. Subsequently, "Lothar" is a mere marginal dent in the outer reaches of an extended Arctic Sea cyclone with a strong pressure gradient.

The paradox stated in Hoskins and Valdes (1990), where it is pointed out that it appears unusual that storm tracks are used several times in succession has an interesting reflection in the events of the days following 2 December. Already on 26 December, a further cyclogenesis occurs near Newfoundland. The storm, known under the name "Martin" (denoted "M" in Fig. 5c and d), develops. The evolution during the following days is shown in Fig. 6.

In the case of "Martin" the algorithm which identifies storm centers was able to find closed isobars and generate a storm track. This is due to the changed synoptic situation, because the strong Artic Sea low which was dominant during the track of "Lothar" had weakened after 27 December – at a great distance near the hemispheric cold center, located over the Canadian Arctic at that time, another very strong low was forming. Thus a comparably weak gradient over the North Sea and Central Europe caused "Martin" to stand out much more clearly and a successful track identification was possible.

The flow regime of the transition seasons and the winter is characterized by strong gradients which simplifies the centers' identification and the tracking in comparison to summer. A final example from 2002 may help illustrating this fact.

As an important result the analysis of summer data yielded that the center identification algorithm frequently found 6 to 10 centers in a chart – in winter conditions this number is are clearly smaller. Yet the temporal sequence of positions only leads to a few tracks. Figures 7–9 show the genesis and propagation of a cyclone which caused large amounts of rain in Central Europe. As an extreme example, between 12 August 2002 07z and 13 August 2002 the rainfall amount at the Saxon station Zinnwald-Georgenfeld was 312 mm (Fritzschner and Lux, 2002). It caused the Elbe flooding, inflicting major damage to infrastructure across the area, including the historic center of Dresden.

The weather process in the first half of August 2002 was characterized by repeated cyclone advances from the Atlantic and the Mediterranean. The problematic track identification process is evident in the chart sequence Fig. 7. The cyclones have a weather impact rather because of their thermal and hydrometeorological properties than because of their dynamics (weak gradients and coherence)[6]. Charts of temperature and humidity of the lower troposphere (not shown here) indicate that there was a high potential for intense precipitation at that time.

Fritzschner and Lux (2002) attribute the precipitation event that led to the Elbe flooding to a van Bebber Vb situation. In fact, the charts indicate, starting on 10 August (Fig. 8a, intermittently the second strongest center in the investigation area) the development of a shallow extended low which expanded westward into Central and Southern Europe (Fig. 8c). The cyclogenesis was most pronounced to the south of the Alps (at that and the following dates indicated by the number "2"). The center moved slowly to the north-east, in phase with the upper air current (Fig. 9, denoted "2" in (a) and (b) and "4" in (c) and (d)), triggering sustained heavy rain. The criteria for a *Vb* cyclones were met almost in a textbook-like way. Nevertheless, Fritzschner and Lux (2002) point out that cyclone developments and tracks of that type seldom are the single cause for heavy rain.

4 Summary, conclusions and outlook

The aim of the study was the development of an algorithm that can identify storm tracks in an objective manner. An important application would be the analysis of GCM data in order to determine future developments in climate dynamics as they would emerge from the changing placement of storm tracks and intensity of storms.

[6]Please note that the coloured bands in the figures do not represent the temperature but the 500 hPa geopotential height.

A method has been presented that is capable of determining storm tracks in gridded geopotential/surface pressure data sets of rather coarse spatial and temporal resolution, as they are provided by reanalyses and GCMs. It combines a geometric identification scheme for local minima with a set of rules to determine highly probable tracks from the arrays of identified centers in subsequent fields.

The center recognition performance depends on the existence of closed isolines, i.e., a topology in which a gridpoint is surrounded by neighbours which all exhibit higher geopotential. For cyclones that are lined up to form storm tracks these requirements are well met. A few exceptions are the relatively shallow fields that may occur in the summer months and a configuration in which a small scale storm is located at the perimeter of a deep and very large low (a kind of "dent in a crater wall").

The actual assembly of the tracks from individual centers takes into account that the motion along the track should stick to several criteria: For one, a track is accepted if it lasts for at least 48 h (eight consecutive time steps); furthermore there should be no displacement larger than 10 grid points (of the $\approx 2° \times 2°$ input data) from time step to time step. By way of several examples the capability to identify knowns storm tracks was shown.

This paper focuses on the description of the tracking algorithm. In a subsequent paper the *classification* of storm tracks will be addressed. Comparisons with the historic classification of van Bebber will need to be made.

Acknowledgements. This work was made possible by funds from the State of Saxony, grant AZ 13-8802.3529/47.

Thanks are due to the two anonymous referees for their constructive comments that helped to improve the manuscript substantially.

Edited by: M. Piringer

References

Anagnostopoulou, Chr., Tolika, K., Flocas, H., and Maheras, P.: Cyclones in the Mediterranean region: present and future climate scenarios derived from a general circulation model (HadAM3P), Adv. Geosci., 7, 9–14, doi:10.5194/adgeo-7-9-2006, 2006.

Bengtsson, L.: Advances and prospects in numerical weather prediction, Q. J. Roy. Meteor. Soc., 117, 855–902, 1991.

Bengtsson, L., Hodges, K., and Roeckner, E.: Storm Tracks and Climate Change, J. Climate, 19, 3518–3543, 2006.

Blender, R. and Schubert, M.: Cyclone tracking in different spatial and temporal resolutions, Mon. Weather Rev., 128, 377–384, 2008.

Carnell, R., Senior, C., and Mitchell, J.: An assessment of measures of storminess: Simulated changes in northern hemisphere winter due to increasing CO_2., Clim. Dynam., 12, 467–476, 1996.

Christensen, J., Hewitson, B., Busuioc, A., Chen, A., Gao, X., Held, I., Jones, R., Kolli, R., Kwon, W.-T., Laprise, R., Rueda, V. M., Mearns, L., Menéndez, C., Räisänen, J., Rinke, A., Sarr, A., and Whetton, P.: Regional Climate Projections, in: Climate Change

2007: The Physical Science Basis. Contribution of Working Group I to the Fourth Assessment Report of the Intergovernmental Panel on Climate Change , edited by: Solomon, S., Qin, D., Manning, M., Chen, Z., Marquis, M., Averyt, K. B., Tignor, M., and Miller, H. L., Cambridge Unversity Press, Cambridge, UK and New York, USA, Chap. 11, 847–940, 2007.

Cressman, P.: An operational objective analysis system, Mon. Weather Rev., 87, 548–553, 1959.

Fritzschner, U. and Lux, G.: Starkniederschläge in Sachsen im August 2002, Tech. Rep., Deutscher Wetterdienst, 2002.

Hanson, C., Palutikof, J., and Davies, T.: Objective cyclone climatologies of the North Atlantic – a comparison between the ECMWF and NCEP Reanalyses, Clim. Dynam., 22, 757–769, 2004.

Hegerl, G., Zwiers, F., Braconnot, P., Gillett, N., Luo, Y., Orsini, J. M., Nicholls, N., Penner, J., and Stott, P.: Understanding and Attributing Climate Change, in: Climate Change 2007: The Physical Science Basis. Contribution of Working Group I to the Fourth Assessment Report of the Intergovernmental Panel on Climate Change, edited by: Solomon, S., Qin, D., Manning, M., Chen, Z., Marquis, M., Averyt, K. B., Tignor, M., and Miller, H. L., Cambridge University Press, Cambridge, UK and New York, USA, Chap. 9, 663–746, 2007.

Homar, V., Jansà, A., Campins, J., Genovés, A., and Ramis, C.: Towards a systematic climatology of sensitivities of Mediterranean high impact weather: a contribution based on intense cyclones, Nat. Hazards Earth Syst. Sci., 7, 445–454, doi:10.5194/nhess-7-445-2007, 2007.

Hoskins, B. and Valdes, P.: On the existence of storm-tracks, J. Atmos. Sci., 47, 1854–1864, 1990.

IPCC: Climate Change 2007: The Physical Science Basis. Contribution of Working Group I to the Fourth Assessment Report of the Intergovernmental Panel on Climate Change, Cambridge University Press, Cambridge, UK and New York, USA, ISBN: 978 0521 70596-7, 2007.

Kalnay, E., Kanamitsu, M., Kistler, R., Collins, W., Deaven, D., Gandin, L., Iredell, M., Saha, S., Whitea, G., Woolen, J., Zhu, Y., Chelliah, M., Ebisuzaki, W., Higgins, W., Janowiak, J., Mo, K., Ropelewski, C., Wang, J., Leetmaa, A., Reynolds, R., Jenne, R., and Joseph, D.: The NCEP/NCAR 40-year Reanalysis Project, B. Am. Meteorol. Soc., 77, 437–471, 1996.

Lambert, S.: The effect of enhanced greenhouse warming on winter cyclone frequencies and strengths, J. Climate, 8, 1447–1452, 1995.

Lau, N.-C.: Variability of the observed midlatitude storm tracks in relation to low-frequency changes in the circulation pattern, J. Atmos. Sci., 45, 2718–2743, 1988.

Lionello, P. and Giorgi, F.: Winter precipitation and cyclones in the Mediterranean region: future climate scenarios in a regional simulation, Adv. Geosci., 12, 153–158, doi:10.5194/adgeo-12-153-2007, 2007.

Lionello, P., Dalan, F., and Elvini, E.: Cyclones in the Mediterranean region: The present and the doubled CO_2 climate scenarios, Clim. Res., 22, 147–159, 2002.

Lionello, P., Malanotte-Rizzoli, P., and Boscolo, R. (Eds.): Mediterranean Climate Variability, Elsevier, 2006.

Lionello, P., Boldrin, U., and Giorgi, F.: Future changes in cyclone climatology over Europe as inferred from a regional climate simulation, Clim. Dynam., 30, 657–671, 2008.

Meehl, G., Stocker, T., Collins, W., Friedlingstein, P., Gaye, A., Gregory, J., Kitoh, A., Knutti, R., Murphy, J., Noda, A., Raper, S., Watterson, I., Weaver, A., and Zhao, Z.-C.: Global Climate Projections, in: Climate Change 2007: The Physical Science Basis. Contribution of Working Group I to the Fourth Assessment Report of the Intergovernmental Panel on Climate Change, edited by: Solomon, S., Qin, D., Manning, M., Chen, Z., Marquis, M., Averyt, K. B., Tignor, M., and Miller, H. L., Cambridge University Press, Cambridge, UK and New York, USA, Chap. 10, 747–846, 2007.

Petterssen, S.: Weather Analysis and Forecasting, Vol. I, Motion and Motion Systems, 2nd Edn., McGraw-Hill Book Company, 1956.

Pinto, J., Spangehl, T., Ulbrich, U., and Speth, P.: Assessment of winter cyclone activity in a transient ECHAM4-OPYC3 GHG experiment, Meteorol. Z., 15, 279–291, 2006.

Raible, C., Della-Marta, P., Schwierz, C., Wernli, H., and Blender, R.: Northern Hemisphere Extratropical Cyclones: A Compaision of Detection and Tracking Methods and Different Reanalyses, Mon. Weather Rev., 136, 880–897, 2008.

Randall, D., Wood, R., Bony, S., Colman, R., Fichefet, T., Fyfe, J., Kattsov, V., Pitman, A., Shukla, J., Srinivasan, J., Stouffer, R., Sumi, A., and Taylor, K.: Climate Models and Their Evaluation, in: Climate Change 2007: The Physical Science Basis. Contribution of Working Group I to the Fourth Assessment Report of the Intergovernmental Panel on Climate Change, edited by: Solomon, S., Qin, D., Manning, M., Chen, Z., Marquis, M., Averyt, K. B., Tignor, M., and Miller, H. L., Cambridge University Press, Cambridge, UK and New York, USA, Chap. 8, 589–662, 2007.

Rossby, C.: Relation Between Variations in the Intensity of the Zonal Circulation of the Atmosphäre and the Displacements of the Semi-Permanent Centers in Action, J. Mar. Res., 2, 38–55, 1939.

Schubert, M., Perlwitz, J., Blender, R., Fraedrich, K., and Lunkeit, F.: North Atlantic cyclones in CO_2 induced warm climate simulations: Frequency, intensity, and tracks., Clim. Dynam., 14, 827–837, 1998.

Sinclair, M.: Objective Identification of Cyclones and Their Circulation Intensity, and Climatology, Weather Forecast., 12, 595–612, 1997.

Trenberth, K., Jones, P., Ambenje, P., Bojariu, R., Easterling, D., Tank, A. K., Parker, D., Rahimzadeh, F., Renwick, J., Rusticucci, M., Soden, B., and Zhai, P.: Observations: Surface and Atmospheric Climate Change. In: Climate Change 2007: The Physical Science Basis. Contribution of Working Group I to the Fourth Assessment Report of the Intergovernmental Panel on Climate Change, edited by: Solomon, S., Qin, D., Manning, M., Chen, Z., Marquis, M., Averyt, K. B., Tignor, M., and Miller, H. L., Cambridge University Press, Cambridge, UK and New York, USA, Chap. 3, 235–336, 2007.

Ulbrich, U., Fink, A., Klawa, M., and Pinto, J.: Three extreme storms over Europe in December 1999, Weather, 56, 70–80, 2001.

Uppala, S., Kållberg, P., Simmons, A., Andrae, U., da Costa Bechtold, V., Fiorino, M., J., J. G., Haseler, Hernandez, A., Kelly, G., Li, X., Onogi, K., Saarinen, S., Sokka, N., Allan, R., Andersson, E., Arpe, K., Balmaseda, M., Beljaars, A., van de Berg, L., Bidlot, J., Bormann, N., Caires, S., Chevallier, F., Dethof, A.,

Dragosavac, M., Fisher, M., Fuentes, M., Hagemann, S., Hólm, E., Isaksen, B. H. L., Janssen, P., Jenne, R., McNally, A., Mahfouf, J.-F., Morcrette, J.-J., Rayner, N., Saunders, R., Simon, P., Sterl, A., Trenberth, K., Untch, A., Vasiljevic, D., Viterbo, P., and Woollen, J.: The ERA-40 re-analysis, Q. J. Roy. Meteor. Soc., 131, 2961–3012, 2005.

van Bebber, W.: Die Zugstrassen der barometrischen Minima, Meteorol. Z., 8, 361–366, 1891.

Wernli, H. and Schweiz, C.: Surface Cyclones in the ERA-40 Dataset (1958-2001). Part I: Novel Identification Method and Global Climatology, J. Atmos. Sci., 63, 2486–2507, 2006.

Permissions

 Every chapter published in this book has been scrutinized by our experts. Their significance has been extensively debated. The topics covered herein carry significant findings which will fuel the growth of the discipline. They may even be implemented as practical applications or may be referred to as a beginning point for another development.

The contributors of this book come from diverse backgrounds, making this book a truly international effort. This book will bring forth new frontiers with its revolutionizing research information and detailed analysis of the nascent developments around the world.

We would like to thank all the contributing authors for lending their expertise to make the book truly unique. They have played a crucial role in the development of this book. Without their invaluable contributions this book wouldn't have been possible. They have made vital efforts to compile up to date information on the varied aspects of this subject to make this book a valuable addition to the collection of many professionals and students.

This book was conceptualized with the vision of imparting up-to-date information and advanced data in this field. To ensure the same, a matchless editorial board was set up. Every individual on the board went through rigorous rounds of assessment to prove their worth. After which they invested a large part of their time researching and compiling the most relevant data for our readers.

The editorial board has been involved in producing this book since its inception. They have spent rigorous hours researching and exploring the diverse topics which have resulted in the successful publishing of this book. They have passed on their knowledge of decades through this book. To expedite this challenging task, the publisher supported the team at every step. A small team of assistant editors was also appointed to further simplify the editing procedure and attain best results for the readers.

Apart from the editorial board, the designing team has also invested a significant amount of their time in understanding the subject and creating the most relevant covers. They scrutinized every image to scout for the most suitable representation of the subject and create an appropriate cover for the book.

The publishing team has been an ardent support to the editorial, designing and production team. Their endless efforts to recruit the best for this project, has resulted in the accomplishment of this book. They are a veteran in the field of academics and their pool of knowledge is as vast as their experience in printing. Their expertise and guidance has proved useful at every step. Their uncompromising quality standards have made this book an exceptional effort. Their encouragement from time to time has been an inspiration for everyone.

The publisher and the editorial board hope that this book will prove to be a valuable piece of knowledge for researchers, students, practitioners and scholars across the globe.

List of Contributors

R. Floors
Wind Energy Division, Risø National Laboratory for Sustainable Energy, Technical University of Denmark, P.O. Box 49, 4000 Roskilde, Denmark

E. Batchvarova
Wind Energy Division, Risø National Laboratory for Sustainable Energy, Technical University of Denmark, P.O. Box 49, 4000 Roskilde, Denmark
National Institute of Meteorology and Hydrology, Sofia 1784, Bulgaria

S.-E. Gryning
Wind Energy Division, Risø National Laboratory for Sustainable Energy, Technical University of Denmark, P.O. Box 49, 4000 Roskilde, Denmark

A. N. Hahmann
Wind Energy Division, Risø National Laboratory for Sustainable Energy, Technical University of Denmark, P.O. Box 49, 4000 Roskilde, Denmark

A. Peña
Wind Energy Division, Risø National Laboratory for Sustainable Energy, Technical University of Denmark, P.O. Box 49, 4000 Roskilde, Denmark

T. Mikkelsen
Wind Energy Division, Risø National Laboratory for Sustainable Energy, Technical University of Denmark, P.O. Box 49, 4000 Roskilde, Denmark

S. Federico
Institute of Atmospheric Sciences and Climate, Italian National Research Council (ISAC-CNR), Section of Lamezia Terme, Italy
CRATI Scrl, Rende, Italy

L. Pasqualoni
Engineering Department, University of Rome "Tor Vergata", Italy

A. M. Sempreviva
Institute of Atmospheric Sciences and Climate, Italian National Research Council (ISAC-CNR), Section of Lamezia Terme, Italy
Wind Energy Department, Risoe-Danish Technical University, Roskilde, Denmark

L. De Leo
CRATI Scrl, Rende, Italy

E. Avolio
CRATI Scrl, Rende, Italy

C. R. Calidonna
Institute of Atmospheric Sciences and Climate, Italian National Research Council (ISAC-CNR), Section of Lamezia Terme, Italy

C. Bellecci
CRATI Scrl, Rende, Italy
Engineering Department, University of Rome "Tor Vergata", Italy

P. Domonkos
Centre for Climate Change (C3), Geography Dept., University Rovira i Virgili, Campus Terres de l'Ebre, C/ Betània 5 Tortosa, 43500, Spain

R. Poza
Centre for Climate Change (C3), Geography Dept., University Rovira i Virgili, Campus Terres de l'Ebre, C/ Betània 5 Tortosa, 43500, Spain

D. Efthymiadis
Centre for Climate Change (C3), Geography Dept., University Rovira i Virgili, Campus Terres de l'Ebre, C/ Betània 5 Tortosa, 43500, Spain

K. Radics
Geoinformation Service, Hungarian Defence Forces, Szilagyi E. fasor 7–9, 1024 Budapest, Hungary
Department of Meteorology, Eötvös Loránd University, Pazmany s. 1/A, 1117 Budapest, Hungary

J. Bartholy
Department of Meteorology, Eötvös Loránd University, Pazmany s. 1/A, 1117 Budapest, Hungary

Cs. N. Péliné
Geoinformation Service, Hungarian Defence Forces, Szilagyi E. fasor 7–9, 1024 Budapest, Hungary
Department of Meteorology, Eötvös Loránd University, Pazmany s. 1/A, 1117 Budapest, Hungary

C. W. Hutton
GeoData Institute, University of Southampton, UK

S. Kienberger
Centre for Geoinformatics, University of Salzburg, Salzburg, Austria

F. Amoako Johnson
School of Social Sciences, University of Southampton, UK

A. Allan
Centre for Water Law, Policy and Science, Dundee, UK

V. Giannini
Fondazione Eni Enrico Mattei, Venice, Italy

R. Allen
ICIMOD, Katmadu, Nepal

C. Schneider
RWTH Aachen Univ., Department of Geography, Faculty of Material Sciences and Geosciences, Germany

C. Balzer
RWTH Aachen Univ., Institute of Hygiene and Environmental Medicine, Faculty of Medicine, Germany

M. Buttstädt
RWTH Aachen Univ., Department of Geography, Faculty of Material Sciences and Geosciences, Germany

K. Eßer
RWTH Aachen Univ., Department of History, Faculty of Social Sciences and Humanities, Germany

S. Ginski
RWTH Aachen Univ., Institute of Planning Theory and Urban Development, Faculty of Architecture, Germany

J. Hahmann
RWTH Aachen Univ., Institute for Sociology, Faculty of Social Sciences and Humanities, Germany

G. Ketzler
RWTH Aachen Univ., Department of Geography, Faculty of Material Sciences and Geosciences, Germany

M. Klemme
Univ. Münster, Governance Research and Applied Geography, Department of Geosciences, Germany

A. Kröpelin
RWTH Aachen Univ., Construction Business and Building Services, Faculty of Civil Engineering, Germany

H. Merbitz
RWTH Aachen Univ., Department of Geography, Faculty of Material Sciences and Geosciences, Germany

S. Michael
RWTH Aachen Univ., Institute of Hygiene and Environmental Medicine, Faculty of Medicine, Germany

T. Sachsen
RWTH Aachen Univ., Department of Geography, Faculty of Material Sciences and Geosciences, Germany

A. Siuda
RWTH Aachen Univ., Department of Geography, Faculty of Material Sciences and Geosciences, Germany

M. Weishoff-Houben
RWTH Aachen Univ., Institute of Hygiene and Environmental Medicine, Faculty of Medicine, Germany

M. F. Brunk
RWTH Aachen Univ., Construction Business and Building Services, Faculty of Civil Engineering, Germany

W. Dott
RWTH Aachen Univ., Institute of Hygiene and Environmental Medicine, Faculty of Medicine, Germany

H. Hofmeister
RWTH Aachen Univ., Institute for Sociology, Faculty of Social Sciences and Humanities, Germany

C. Pfaffenbach
RWTH Aachen Univ., Department of Geography, Faculty of Material Sciences and Geosciences, Germany

C. Roll
RWTH Aachen Univ., Department of History, Faculty of Social Sciences and Humanities, Germany

K. Selle
RWTH Aachen Univ., Institute of Planning Theory and Urban Development, Faculty of Architecture, Germany

S. Alessandrini
RSE, Ricerca Sistema Energetico, Milano, Italy

P. Pinson
ECMWF, Reading, UK

R. Hagedorn
ECMWF, Reading, UK

G. Decimi
RSE, Ricerca Sistema Energetico, Milano, Italy

S. Sperati
RSE, Ricerca Sistema Energetico, Milano, Italy

R. Hamdi
Royal Meteorological Institute, 3 Avenue Circulaire, 1180 Brussels, Belgium

H. Van de Vyver
Royal Meteorological Institute, 3 Avenue Circulaire, 1180 Brussels, Belgium

M. Prasch
Department of Geography, Ludwig-Maximilians-Universität Munich, Germany

T. Marke
Department of Geography, Ludwig-Maximilians-Universität Munich, Germany
Institute of Geography and Regional Science, University of Graz, Austria

U. Strasser
Department of Geography, Ludwig-Maximilians-Universität Munich, Germany
Institute of Geography and Regional Science, University of Graz, Austria

W. Mauser
Department of Geography, Ludwig-Maximilians-Universität Munich, Germany

Z. Spasova
National Center of Public Health and Analyses, Sofia, Bulgaria

A. Baklanov
Danish Meteorological Institute (DMI), Lyngbyvej 100, 2100, Copenhagen, Denmark

M. Lawrence
Max Planck Institute for Chemistry (MPIC), Germany

S. Pandis
Foundation for Research and Technology (FORTH), Hellas, University of Patras, Greece

A. Mahura
Danish Meteorological Institute (DMI), Lyngbyvej 100, 2100, Copenhagen, Denmark

S. Finardi
ARIANET Consulting, Italy

N. Moussiopoulos
Aristotle University Thessaloniki (AUTH), Greece

M. Beekmann
Centre National de Recherche Scientifique (CNRS), France
Laboratory Inter-universitaire des Systèmes Atmosphériques (LISA), France

P. Laj
Centre National de Recherche Scientifique (CNRS), France

L. Gomes
Centre National de Recherche Scientifique (CNRS), France

J.-L. Jaffrezo
Centre National de Recherche Scientifique (CNRS), France

A. Borbon
Laboratory Inter-universitaire des Systèmes Atmosphériques (LISA), France

I. Coll
Laboratory Inter-universitaire des Systèmes Atmosphériques (LISA), France

V. Gros
Laboratoire des Sciences du Climat et de l'Environnement (LSCE), France

J. Sciare
Laboratoire des Sciences du Climat et de l'Environnement (LSCE), France

J. Kukkonen
Finnish Meteorological Institute (FMI), Finland

S. Galmarini
Joint Research Center (JRC), Ispra, Italy

F. Giorgi
International Centre for Theoretical Physics (ICTP), Italy

S. Grimmond
King's College London (KCL), UK

I. Esau
Nansen Environmental and Remote Sensing Center (NERSC), Norway

A. Stohl
Norwegian Institute for Air Research (NILU), Norway

B. Denby
Norwegian Institute for Air Research (NILU), Norway

T. Wagner
Max Planck Institute for Chemistry (MPIC), Germany

T. Butler
Max Planck Institute for Chemistry (MPIC), Germany

U. Baltensperger
Paul Scherrer Institute (PSI), Switzerland

P. Builtjes
TNO-Built Environment and Geosciences, The Netherlands

D. van den Hout
TNO-Built Environment and Geosciences, The Netherlands

H. D. van der Gon
TNO-Built Environment and Geosciences, The Netherlands

B. Collins
UK MetOffice (MetO), UK

H. Schluenzen
University of Hamburg (UHam), Hamburg, Germany

M. Kulmala
University of Helsinki (UHel), Finland

S. Zilitinkevich
University of Helsinki (UHel), Finland

R. Sokhi
University of Hertfordshire - Centre for Atmospheric & Instrumentation Research (UH-CAIR), UK

R. Friedrich
University of Stuttgart (UStutt), Germany

J. Theloke
University of Stuttgart (UStutt), Germany

U. Kummer
University of Stuttgart (UStutt), Germany

L. Jalkinen
World Meteorological Organization (WMO), Switzerland

T. Halenka
Charles University Prague (CUNI), Czech Republic

A. Wiedensholer
Institute of Tropospheric Research (IfT), Germany

J. Pyle
Centre for Atmospheric Science, University of Cambridge (UCam), UK

W. B. Rossow
Centre for Atmospheric Science, University of Cambridge (UCam), UK

B. Brecht
Deutscher Wetterdienst, Offenbach, Germany

H. Frank
Deutscher Wetterdienst, Offenbach, Germany

F. Kaspar
Deutscher Wetterdienst, Frankfurter Str. 135, 63067 Offenbach, Germany

G. Müller-Westermeier
Deutscher Wetterdienst, Frankfurter Str. 135, 63067 Offenbach, Germany

E. Penda
Deutscher Wetterdienst, Frankfurter Str. 135, 63067 Offenbach, Germany

H. Mächel
Deutscher Wetterdienst, Frankfurter Str. 135, 63067 Offenbach, Germany

K. Zimmermann
Deutscher Wetterdienst, Frankfurter Str. 135, 63067 Offenbach, Germany

A. Kaiser-Weiss
Deutscher Wetterdienst, Frankfurter Str. 135, 63067 Offenbach, Germany

T. Deutschländer
Deutscher Wetterdienst, Frankfurter Str. 135, 63067 Offenbach, Germany

G. V. Iungo
Wind Engineering and Renewable Energy Laboratory (WIRE), ´ Ecole Polytechnique Fédérale de Lausanne (EPFL), Switzerland

F. Porté-Agel
Wind Engineering and Renewable Energy Laboratory (WIRE), ´ Ecole Polytechnique Fédérale de Lausanne (EPFL), Switzerland

A. Boilley
MINES ParisTech, O.I.E. - Centre Observation, Impacts, Energie, CS 10207 rue Claude Daunesse, 06904 Sophia Antipolis Cedex, France

T. Ranchin
MINES ParisTech, O.I.E. - Centre Observation, Impacts, Energie, CS 10207 rue Claude Daunesse, 06904 Sophia Antipolis Cedex, France

A. Ghennioui
Institut de Recherche en Energie Solaire et Energies Nouvelles, Rabat, Morocco

J. T. Matsangouras
Laboratory of Climatology and Atmospheric Environment, Faculty of Geology and Geoenvironment, University of Athens, Panepistimiopolis, 157 84 Athens, Greece

P. T. Nastos
Laboratory of Climatology and Atmospheric Environment, Faculty of Geology and Geoenvironment, University of Athens, Panepistimiopolis, 157 84 Athens, Greece

J. N. Nissen
Risø National Laboratory for Sustainable Energy, Technical University of Denmark, Frederiksborgvej 399, P.O. Box 49, 4000 Roskilde, Denmark

S.-E. Gryning
Risø National Laboratory for Sustainable Energy, Technical University of Denmark, Frederiksborgvej 399, P.O. Box 49, 4000 Roskilde, Denmark

M. M. Miglietta
Institute of Atmospheric Sciences and Climate, Italian National Research Council (ISAC-CNR), Section of Padova, Italy

S. Zecchetto
Institute of Atmospheric Sciences and Climate, Italian National Research Council (ISAC-CNR), Section of Padova, Italy

F. De Biasio
Institute of Atmospheric Sciences and Climate, Italian National Research Council (ISAC-CNR), Section of Padova, Italy

K. V. Varotsos
Institute for Environmental Research and Sustainable Development, National Observatory of Athens, Greece
Division of Environmental Physics and Meteorology, National and Kapodistrian University of Athens, Greece

C. Giannakopoulos
Institute for Environmental Research and Sustainable Development, National Observatory of Athens, Greece

M. Tombrou
Division of Environmental Physics and Meteorology, National and Kapodistrian University of Athens, Greece

P. B. Rutkevich
Space Research Institute (IKI), RAS, Moscow, Russia

P. P. Rutkevych
Institute of High Performance Computing, A*STAR, Singapore

P. Domonkos
Centre for Climate Change, University Rovira i Virgili, Campus Terres de l'Ebre, Tortosa, Spain

D. Efthymiadis
Centre for Climate Change, University Rovira i Virgili, Campus Terres de l'Ebre, Tortosa, Spain

J. G. Pedersen
Department of Wind Energy, Risø Campus, Technical University of Denmark, Roskilde, Denmark

M. Kelly
Department of Wind Energy, Risø Campus, Technical University of Denmark, Roskilde, Denmark

S.-E. Gryning
Department of Wind Energy, Risø Campus, Technical University of Denmark, Roskilde, Denmark

R. Floors
Department of Wind Energy, Risø Campus, Technical University of Denmark, Roskilde, Denmark

E. Batchvarova
Department of Wind Energy, Risø Campus, Technical University of Denmark, Roskilde, Denmark
National Institute of Meteorology and Hydrology, Sofia, Bulgaria

A. Peña
Department of Wind Energy, Risø Campus, Technical University of Denmark, Roskilde, Denmark

I. T. Matsangouras
Laboratory of Climatology and Atmospheric Environment, Faculty of Geology and Geoenvironment, National and Kapodistrian University of Athens, Greece

P. T. Nastos
Laboratory of Climatology and Atmospheric Environment, Faculty of Geology and Geoenvironment, National and Kapodistrian University of Athens, Greece

I. Pytharoulis
Department of Meteorology and Climatology, School of Geology, Aristotle University of Thessaloniki, Greece

H. O. Hygen
Norwegian Meteorological Institute, Postboks 43 Blindern, 0313 Oslo, Norway

C. F. Øyen
SINTEF Building and Infrastructure, Postboks 124 Blindern, 0314 Oslo, Norway

A. J. Almås
Department of Civil and Transport Engineering, Norwegian University of Science and Technology (NTNU), 7491 Trondheim, Norway

P. E. Bett
Met Office Hadley Centre, Exeter, UK

H. E. Thornton
Met Office Hadley Centre, Exeter, UK

R. T. Clark
Met Office Hadley Centre, Exeter, UK

A. Oliver
Laboratori de C`alcul Num`eric (LaC`aN), Departament de Matem`atica Aplicada III, Universitat Polit`ecnica de Catalunya - BarcelonaTech, Jordi Girona 1-3, 08034 Barcelona, Spain

G. Montero
University of Las Palmas de Gran Canaria, University Institute for Intelligent Systems and Numerical Applications in Engineering (SIANI), Spain

R. Montenegro
University of Las Palmas de Gran Canaria, University Institute for Intelligent Systems and Numerical Applications in Engineering (SIANI), Spain

E. Rodríguez
University of Las Palmas de Gran Canaria, University Institute for Intelligent Systems and Numerical Applications in Engineering (SIANI), Spain

J. M. Escobar
University of Las Palmas de Gran Canaria, University Institute for Intelligent Systems and Numerical Applications in Engineering (SIANI), Spain

A. Perez-Foguet
Laboratori de Càlcul Num`eric (LaCàN), Departament de Matemàtica Aplicada III, Universitat Politècnica de Catalunya - BarcelonaTech, Jordi Girona 1-3, 08034 Barcelona, Spain

E. Kolozsi-Komjáthy
Department of Meteorology, Eötvös Lorànd University, Budapest, Hungary

R. Mészáros
Department of Meteorology, Eötvös Lorànd University, Budapest, Hungary

I. Lagzi
Department of Meteorology, Eötvös Lorànd University, Budapest, Hungary

P. Cerralbo
Maritime Engineering Laboratory, Polytechnic University of Catalonia (LIM/UPC), c./Escar, 6, 08039 Barcelona, Spain

M. Grifoll
Maritime Engineering Laboratory, Polytechnic University of Catalonia (LIM/UPC), c./Escar, 6, 08039 Barcelona, Spain
International Centre of Coastal Resources Research, Polytechnic University of Catalonia (CIIRC/UPC), c./ Jordi Girona, 1-3, 08034 Barcelona, Spain

J. Moré
Servei Meteorològic de Catalunya (SMC), c./Berlín, 38-46, 08029 Barcelona, Spain

M. Bravo
Servei Meteorològic de Catalunya (SMC), c./Berlín, 38-46, 08029 Barcelona, Spain

A. Sairouní Afif
Servei Meteorològic de Catalunya (SMC), c./Berlín, 38-46, 08029 Barcelona, Spain

M. Espino
Maritime Engineering Laboratory, Polytechnic University of Catalonia (LIM/UPC), c./Escar, 6, 08039 Barcelona, Spain
International Centre of Coastal Resources Research, Polytechnic University of Catalonia (CIIRC/UPC), c./ Jordi Girona, 1-3, 08034 Barcelona, Spain

F. De Biasio
Institute of Atmospheric Sciences and Climate, Italian National Research Council, Corso Stati Uniti 4, 35127 Padova, Italy

M. M. Miglietta
Institute of Atmospheric Sciences and Climate, Italian National Research Council, Corso Stati Uniti 4, 35127 Padova, Italy

S. Zecchetto
Institute of Atmospheric Sciences and Climate, Italian National Research Council, Corso Stati Uniti 4, 35127 Padova, Italy

A. della Valle
Institute of Atmospheric Sciences and Climate, Italian National Research Council, Corso Stati Uniti 4, 35127 Padova, Italy

D. Hill
Dept. of Electronic and Electrical Engineering, University of Strathclyde, Glasgow, UK

K. R. W. Bell
Dept. of Electronic and Electrical Engineering, University of Strathclyde, Glasgow, UK

D. McMillan
Dept. of Electronic and Electrical Engineering, University of Strathclyde, Glasgow, UK

D. Infield
Centre for Doctoral Training in Wind Energy Systems, University of Strathclyde, Glasgow, UK

S. Alessandrini
RSE, Ricerca Sistema Energetico, Milano, Italy
National Centre for Atmospheric Research, Boulder, CO, USA

F. Davò
RSE, Ricerca Sistema Energetico, Milano, Italy

S. Sperati
RSE, Ricerca Sistema Energetico, Milano, Italy

M. Benini
RSE, Ricerca Sistema Energetico, Milano, Italy

L. Delle Monache
National Centre for Atmospheric Research, Boulder, CO, USA

H. Nakayama
Japan Atomic Energy Agency, Ibaraki, Japan

H. Nagai
Japan Atomic Energy Agency, Ibaraki, Japan

F. Ruget
INRA, UMR 1114 EMMAH, 84914 Avignon cedex 9, France

J.-C. Moreau
Institut de l' ´ Elevage, Auzeville 31326 Castanet-Tolosan et 149 Rue de Bercy 75011, Paris, France

M. Ferrand
Institut de l' ´ Elevage, Auzeville 31326 Castanet-Tolosan et 149 Rue de Bercy 75011, Paris, France

S. Poisson
Institut de l' ´ Elevage, Auzeville 31326 Castanet-Tolosan et 149 Rue de Bercy 75011, Paris, France

P. Gate
Arvalis Institut du végétal, La Minière 78210 Guyancourt/31450 Baziège/91170 Boigneville, France

B. Lacroix
Arvalis Institut du végétal, La Minière 78210 Guyancourt/31450 Baziège/91170 Boigneville, France

J. Lorgeou
Arvalis Institut du végétal, La Minière 78210 Guyancourt/31450 Baziège/91170 Boigneville, France

E. Cloppet
Météo-France, 42 Av G Coriolis 31000 Toulouse, France

F. Souverain
Météo-France, 42 Av G Coriolis 31000 Toulouse, France

F. Kreienkamp
Climate & Environment Consulting Potsdam GmbH, David-Gilly-Str. 1, 14469 Potsdam, Germany

A. Spekat
Climate & Environment Consulting Potsdam GmbH, David-Gilly-Str. 1, 14469 Potsdam, Germany

W. Enke
Climate & Environment Consulting Potsdam GmbH, David-Gilly-Str. 1, 14469 Potsdam, Germany